Lecture Notes in Physics

Springer
Berlin
Heidelberg
New York
Hong Kong
London
Milan
Paris
Tokyo

Physics and Astronomy

ONLINE LIBRARY

springeronline.com

The Editorial Policy for Edited Volumes

The series *Lecture Notes in Physics* (LNP), founded in 1969, reports new developments in physics research and teaching - quickly, informally but with a high degree of quality. Manuscripts to be considered for publication are topical volumes consisting of a limited number of contributions, carefully edited and closely related to each other. Each contribution should contain at least partly original and previously unpublished material, be written in a clear, pedagogical style and aimed at a broader readership, especially graduate students and nonspecialist researchers wishing to familiarize themselves with the topic concerned. For this reason, traditional proceedings cannot be considered for this series though volumes to appear in this series are often based on material presented at conferences, workshops and schools.

Acceptance

A project can only be accepted tentatively for publication, by both the editorial board and the publisher, following thorough examination of the material submitted. The book proposal sent to the publisher should consist at least of a preliminary table of contents outlining the structure of the book together with abstracts of all contributions to be included. Final acceptance is issued by the series editor in charge, in consultation with the publisher, only after receiving the complete manuscript. Final acceptance, possibly requiring minor corrections, usually follows the tentative acceptance unless the final manuscript differs significantly from expectations (project outline). In particular, the series editors are entitled to reject individual contributions if they do not meet the high quality standards of this series. The final manuscript must be ready to print, and should include both an informative introduction and a sufficiently detailed subject index.

Contractual Aspects

Publication in LNP is free of charge. There is no formal contract, no royalties are paid, and no bulk orders are required, although special discounts are offered in this case. The volume editors receive jointly 30 free copies for their personal use and are entitled, as are the contributing authors, to purchase Springer books at a reduced rate. The publisher secures the copyright for each volume. As a rule, no reprints of individual contributions can be supplied.

Manuscript Submission

The manuscript in its final and approved version must be submitted in ready to print form. The corresponding electronic source files are also required for the production process, in particular the online version. Technical assistance in compiling the final manuscript can be provided by the publisher's production editor(s), especially with regard to the publisher's own LaTeX macro package which has been specially designed for this series.

LNP Homepage (springerlink.com)

On the LNP homepage you will find:
−The LNP online archive. It contains the full texts (PDF) of all volumes published since 2000. Abstracts, table of contents and prefaces are accessible free of charge to everyone. Information about the availability of printed volumes can be obtained.
−The subscription information. The online archive is free of charge to all subscribers of the printed volumes.
−The editorial contacts, with respect to both scientific and technical matters.
−The author's / editor's instructions.

R. Haberlandt D. Michel
A. Pöppl R. Stannarius (Eds.)

Molecules in Interaction with Surfaces and Interfaces

 Springer

Editors

Reinhold Haberlandt, Dieter Michel,
Andreas Pöppl and Ralf Stannarius
Universität Leipzig
Linnéstrasse 5
04103 Leipzig, Germany

R. Haberlandt, D. Michel, A. Pöppl, R. Stannarius (eds.), *Molecules in Interaction with Surfaces and Interfaces*, Lect. Notes Phys. **634** (Springer-Verlag Berlin Heidelberg 2004), DOI 10.1007/b13926

Cataloging-in-Publication Data:
Molecules in interaction with surfaces and interfaces / R. Haberlandt...[et al.] (eds.)
p.cm. – (Lecture notes in physics, ISSN 0075-8450; 643) Includes bibliographical references and index. ISBN 3-540-20539-X (acid-free paper) 1. Surface chemistry. 2. Molecular dynamics 3. Molecules–Surfaces. 4. Adsorption. 5. Catalysts. I. Haberlandt, R. (Reinhold), 1936 - II. Series. QD508.M65 2004 541'.33–dc22

Bibliographic information published by Die Deutsche Bibliothek Die Deutsche Bibliothek lists this publication in the Deutsche Nationalbibliografie; detailed bibliographic data is available in the Internet at <http://dnb.ddb.de>

ISSN 0075-8450
ISBN 3-540-20539-X Springer-Verlag Berlin Heidelberg New York

Springer-Verlag is a part of Springer Science+Business Media

springeronline.com

© Springer-Verlag Berlin Heidelberg 2004
Printed in Germany

Typesetting: Camera-ready by the authors/editor
Data conversion: PTP-Berlin Protago-TeX-Production GmbH
Cover design: *design & production*, Heidelberg

Printed on acid-free paper
54/3141/ts - 5 4 3 2 1 0

Preface

Research on the interaction of molecules with surfaces and interfaces plays an important role in various scientific disciplines. This field is of great importance not only for a better understanding of physical and chemical processes in connection with adsorption and catalysis on solid interfaces, but also for the study of the complex properties of molecules on fluid interfaces, as well as for the elucidation of relationships between structure and functionality in macromolecular biological systems. On the basis of long-standing research experience at Leipzig in the field of molecular physics and in studies of liquid and porous solid interfaces, the collaborative research center *Sonderforschungsbereich* (SFB) 294 of the Deutsche Forschungsgemeinschaft, entitled *Moleküle in Wechselwirkung mit Grenzflächen* ("Molecules in Interaction with Interfaces"), was founded at the University of Leipzig in 1994. Since this field is still rapidly growing, it is the aim of these Lecture Notes to present not only the most important results of our research activities but also to include and analyze in a comprehensive way the overall progress in this field, and to point out important methodical achievements.

Within the research field presented in these lecture notes, the *Grenzflächen* (interfaces) exhibit various levels of complexity, and the interactions of molecules with interfaces and the properties of the interfaces themselves have been studied by applying a great variety of modern physical and chemical methods.

Porous solids (such as zeolitic molecular sieves, mesoporous materials, layered materials, and porous glasses), characterized by their large internal surfaces, are comparatively well defined. Thus they lend themselves to use as model systems for the elucidation of fundamental relationships between the structure of the embedded molecules and molecular systems, their conformation, the reorientational and translational mobility, transport processes, and their catalytic properties. Moreover, the internal structure of zeolites may also be modified in a definite manner, for instance by inserting different metal atoms. Hence, we have an interface system where the embedded molecules may be influenced in their electronic density or in their spatial structure without essentially changing the solid surfaces themselves. As is well known, this situation explains also the great importance of porous solids as catalysts and adsorbents with shape-selective properties.

The behavior of fluid interfaces is generally much more complex because in the case of fluid interfaces (such as ordered molecular monolayers, internal and external surfaces of polymer systems, liquid-crystalline thin films, gel systems, and mem-

branes) we arrive at a more complex situation: their properties may change dynamically in response to interactions with molecules in the adjacent bulk phases. Besides the fundamental questions mentioned, the interaction of the associated molecules of molecular systems with the surfaces gives rise to subtle changes of the interface itself, which may lead to new relationships between the structure and properties of the systems considered. This area of research has the potential for important novel results for both, the materials and the life sciences, since such diverse systems as polymer and ferroelectric liquid-crystal phases and biomembrane models are under study. Moreover, complex macromolecular systems are being investigated here, such as cartilage, where the studies are especially concerned with the functionality of the system as a whole, which is determined by the cooperative action of the various compartments. In this sense, the studies are also related to medical questions, such as functional distortions and aging processes. Thus, the variety of systems is a prerequisite for a better understanding of the properties of more complex systems. In this sense, the interdisciplinary character of the research plays an important role, not only with respect to the close cooperation between physicists, biophysicists, and chemists but also including the theoretical work on the basis of molecular dynamic investigations.

One of the aims of studying the interaction between molecules and solid surfaces consists in the possibility of studies on a microscopic level. A large variety of nanoporous materials host channel networks with pore diameters on the order of typical molecular dimensions. Molecular reorientation, diffusion, and reaction under this type of confinement exhibit a number of peculiarities. Because nanoscopic channels may serve as routes of particle propagation in many more systems, with the ion channels in biological membranes and reptation paths in macromolecular systems as two well-known other examples, the interest in these peculiarities is not confined to porous materials.

From the methodical point of view, the work presented in these Lecture Notes strongly benefits also from the great experience achieved in the application of the methods of magnetic resonance and relaxation (NMR and EPR) – initiated in Leipzig by Professors Artur Lösche and Harry Pfeifer about fifty years ago – to complex molecular systems and to solids, and from efforts in specific developments in recent years with respect to modern methods of two-dimensional NMR spectroscopy in solids and systems with restricted internal mobility, and in the development and application of pulsed magnetic field gradient (PFG) NMR techniques and in the use of various pulsed EPR methods. Thus, the development and improvement of modern methods of magnetic resonance also play an essential role in this book.

Besides the application of these magnetic resonance methods, various other physical and chemical methods are also used in the context of this work. Hence, since a broad variety of methods have been developed and applied in the framework of studies of structure, dynamics, transport processes, and reactivity at interfaces with different topologies and properties, it is the aim of this publication also to include a comprehensive description of the methodological background of these interface studies.

The main research activities which are intended to be represented in this publication are characterized by distinct yet interconnected pillars, namely

- the theory of the interactions of guest molecules with surfaces;
- translations and rotational diffusion and dynamics of molecules adsorbed in zeolitic systems;
- the reactivity of the internal surfaces and study of catalytic processes;
- thin films, membranes, and biopolymers in interaction with interfaces;
- supramolecular organization and biological compartmentation.

In total, we intend to show that the great variety of the systems included will contribute to a better understanding of the complex phenomena and may lead to a mutual stimulation of investigations with a strong interdisciplinary character. We hope that the broad scope in the selection of topics will help nonexpert readers to become familiar with the complex field of molecular interactions with interfaces. The organization of the presentation is as follows:

The *first chapter* "Modeling and Simulation of Structure, Thermodynamics, and Transport of Fluids in Molecular Confinements" (S. Fritzsche, R. Haberlandt, H.L. Vörtler) deals with the molecular modeling and simulation of fluids confined to restricted geometries – such as micropores, porous media, or membranes – and comprises both the equilibrium structural and thermodynamic properties and the transport behavior of the enclosed particles. Recent simulation studies of diffusion processes in zeolites are reported and discussed in some detail. The aim of this chapter is twofold: first, to give a review of recent molecular simulation techniques and the underlying statistical-mechanical concepts, and secondly, to demonstrate the possibilities and limitations of the simulation methods discussed, showing recent results for systems and properties selected from the fields of research of the authors. The statistical mechanics and molecular simulation of inhomogeneous fluids provides basic information about the structure, thermodynamics, and phase behavior of molecules interacting with interphases. Particularly, the molecular modeling of associating (aqueous) phases confined to molecularly rough (hydrophobic and hydrophilic) interface layers is crucial for a theoretical understanding of hydration phenomena at biointerfaces (e.g. biomembranes). The intermolecular forces are modeled between the fluid molecules and the interface molecules in a uniform way, extending novel hierarchical potentials of aqueous bulk fluids (primitive models of association) to interfacial systems. On this basis, molecular models of hydrated interface layers with molecular roughness are simulated. The structural organization and the thermodynamics of the water–interface system are studied.

The *second chapter*, "Diffusion in Channels and Channel Networks" (P. Bräuer, S. Fritzsche, J. Kärger, G. Schütz, S. Vasenkov) may be considered partially as an extension of the first chapter, specifying in detail the theoretical tools for treating molecular propagation and reaction under confinement by channels. With respect to the prospects for experimental observation, it is also closely related to the *third chapter* "Structure-Mobility Relations of Molecular Diffusion in Interface Systems" (J. Kärger, C.M. Papadakis, F. Stallmach), where the potential of PFG (pulsed field gradient) NMR as a very sensitive technique for experimentally tracing these peculiarities is described. Beginning with the presentation of some peculiarities of molecular propagation in the individual channels – in particular under the so-called single-file condition (where the individual diffusing species are unable to exchange their positions) – some special features of transport and reaction in mutually intersecting channel arrays are treated, viz. the structure-correlated anisotropy of diffusion and the transport-induced reactivity enhancement by the so-called "molecular traffic control". In addition to the conventional way of modeling such a situation by dynamic Monte Carlo simulations, inital attempts at an analytic treatment are included.

Both types of theoretical treatment, which are applied to modeling the internal dynamics of single-file systems, are complemented by molecular dynamics (MD) simulations, following the quite general introduction to this method in the first chapter. Key experiments leading to our present knowledge of the real pore structure of nanoporous materials are described, until recently, these materials have been assumed to represent ideal channel host systems, in accordance with their textbook character.

An important question in this context concerns the investigation of structural properties of the inner surfaces of zeolitic molecular sieves. In the *fourth chapter*, "^{17}O NMR Studies of Zeolites" (D. Freude and T. Loeser), multiple-quantum magic-angle spinning (MQ MAS) and double rotation (DOR) NMR techniques were applied to structural studies of oxygen-17 enriched zeolites A, LSX, and sodalite. Although oxygen is the most abundant element in the earth's crust and an important local probe for the characterization of the real structure of internal solid surfaces, only a relatively small number of applications are known. The demand for characterization of inorganic materials and some recently developed experimental techniques are now causing a growing interest in high-resolution solid-state NMR spectroscopy of quadrupole nuclei with half-integer spin, and it is shown here that an additional valuable tool is obtained for the structural probing of the oxygen framework of inorganic materials and for study of the basic properties of porous catalysts as well.

In this context, the spectroscopic characterization of the interaction of adsorbate molecules with surface sites is a major topic in the study of microporous materials. Such studies provide valuable information about the adsorptive and catalytic properties of the surface sites on a microscopic scale. Among the various kinds of adsorption centers on the inner surface of zeolites, acid sites have attracted special interest because they give rise to the unique acid properties of these crystalline solids, making these materials attractive for specially tailored heterogeneous catalytic applications. For catalytic applications of microporous materials, the determination of the structure and concentration of the acid sites, as well as of their acidity, is of the utmost importance. Consequently, in the *fifth chapter*, entitled "Paramagnetic Adsorption Complexes in Zeolites as Studied by Advanced Electron Paramagnetic Resonance Techniques" (A. Pöppl, M. Gutjahr, and T. Rudolf), it is shown that some recently developed techniques of pulsed EPR spectroscopy are valuable tools for the characterization and structural elucidation of electron pair acceptor centers in zeolites, the so-called true Lewis acid sites, constituting, after the Brønsted acid sites, the second major group of acid surface sites in zeolites.

The interaction of adsorbed molecules with adsorption centers in the internal surfaces of porous solids not only may lead to changes in the reorientational and translational mobility of the molecular species but may also influence the molecular conformation. The *sixth chapter*, "Study of Conformation and Dynamics of Molecules Adsorbed in Zeolites by ^{1}H NMR" (D. Michel, W. Böhlmann, J. Roland and S. Mulla-Osman), is concerned with a combined or alternative application of conventional high-resolution NMR methods and of high-resolution (HR) solid-state NMR techniques, including magic-angle sample spinning (MAS), cross-polarization (CP), high-power decoupling and appropriate multi-pulse sequences for two- or higher- dimensional NMR and multiquantum spectroscopy. Examples will be given of simple olefins in interaction with inner zeolite surfaces. The conclusions about the correlation times of the internal reorientational and translational dynamics are in complete aggrement with the conclusions derived from diffusion coefficients measured by means of PFG NMR as discussed in the *second chapter*. Since the methodical approach of HR MAS NMR for heterogeneous systems presented here is also valuable for the investigation of lyotropic crystalline phases using HR MAS NMR (in chapter 12) and for NMR studies of cartilage (in chapter 13), it seems to be appropriate to elucidate the methodical background of these measurements in some more detail. "Molecular Dynamics of Liquids in Confinement" (F. Kremer and R. Stannarius), is studied in the *seventh chapter* by means of broadband dielectric spectroscopy. With its extraordinary dynamic range (in frequency and intensity), the studies enable one to unravel the subtle interplay between surface and confinement effects for glass-forming liquids. Confining geometries were realized in various ways. Ethylene glycol molecules in zeolites of type sodalite, silicalite-I, and zeolite beta and in an aluminophosphate of type $AlPO_4$-5 show a pronounced confinement effect. Propylene, butylene, and pentyleneglycols in nanoporous glasses show interactions with the hydrophilic inner surfaces. In untreated hydrophilic sol–gel glasses, the quasi-van der Waals liquid salol shows a dynamics characterized by an exchange dynamics between a bulk-like phase and an interfacial phase in the vicinity of the pore wall. Confinement effects are also studied in the *eighth chapter*, "Liquid Crystals in Confining Geometries" (R. Stannarius and F. Kremer). In general, the restricted-volume effects that are observed in isotropic liquids confined in porous matrices (seventh chapter) can be found in confined mesogenic materials as well. This contribution will focus on the description of a few selected systems, from a recollection of surface-induced orientation in ordered and disordered systems, via experiments that study the induction or suppression of mesogenic order, to the investigation of dynamic processes in confined liquid crystals. The experiments described in this chapter mainly involve spectroscopic (bulk) methods, polarizing microscopy, and electro-optic measurements.

In the next chapters, thin ordered molecular structures are investigated. The *ninth chapter* deals with "Surfaces and Interfaces of Free-Standing Smectic Films" (H. Schüring and R. Stannarius). Ordered molecular structures with lateral extensions of up to several square centimeters and with uniform thickness can be formed by only a few (even two) molecular layers. Thin free-standing smectic films, with their robust and stable layer structure, their exceptionally large surface-to-volume ratio, and their macroscopically ordered molecular arrangement, allow one to measure the surface tension of these anisotropic fluids with a variety of methods. In particular, Langmuir's principle can be tested, and contributions to the surface tension can be attributed to individual parts of the molecules. Another aspect is the investigation of the surface tension in the vicinity of liquid–liquid phase transitions. Deviations from the normal temperature dependence, connected with entropic contributions to the surface tension, can be observed. The chapter presents an overview of various methods used for surface tension measurements in liquid crystals, reports on results on the temperature dependence of surface tensions, and deals with interface tensions between different fluid phases derived from the study of isotropic inclusions in thin smectic films. "Pattern Formation in Langmuir Monolayers Due to Long-Range Electrostatic Interactions" (T.M. Fischer and M. Lösche) is the subject of the *tenth chapter*. Langmuir monolayers are monomolecular layers of insoluble amphiphiles at the air/water interface. Molecular self-organization causes these amphiphiles to have their hydrophilic headgroup immersed in the water and their hydrophobic tail dangling into the air. As such, Langmuir monolayers are interesting because, on the one hand, they enable us to study the peculiarities of quasi-two-dimensional (2D) systems. On the other hand, Langmuir monolayers represent half of a lipid bilayer, which in turn is prevalent in biology and forms the local environment where, for example signal cascades and signal transduction reactions occur. The authors report on recent progress in the understanding of the unusual and rather unexpected behavior of a quasi-2D system by reviewing recent experimental results obtained from optical microscopy on equilibrium phase shapes, nonequilibrium phenomena (such as relaxation of the shape after distortions caused by Laser tweezers or local impulse heating), and rheological properties of the system. Long-range electrostatic interactions in Langmuir monolayers cause the development of mesoscopic patterns. Shape transitions and topological transformations triggered by change of the area per molecule, electrostatic contrast, or line tensions between coexisting phases affect the morphology of the monolayer critically: a wealth of nontrivial dynamic reorganization events are observed upon manipulation of these quantities. It is thus clear that Young's equation – established over two centuries ago for 3D systems – does not hold in a straightforward extrapolation to 2D arrays of molecules at interfaces. The theoretical analysis of the underlying molecular interactions leads to a comprehension of the observed phenomena and describes microscopic properties of the system in quantitative terms. The *eleventh chapter* is closely related to the tenth chapter and deals with the "Characterization of Floating Surface Layers of Lipids and Lipopolymers by Surface-Sensitive Scattering" (P. Krüger et al.). Surface-sensitive scattering techniques are employed for the investigation of planar lipid membranes – floating monolayers on aqueous surfaces – to correlate structural, functional, and dynamic aspects of biomembrane models. This chapter surveys recent work on the submolecular structure of floating phospholipid monolayers, where the advent of third-generation synchrotron X-ray sources has driven the development of realistic, submolecular-scale quasi-chemical models, as well as work on more complex systems. The latter includes cation binding to anionic lipid surfaces; conformational changes of lipopolymers undergoing phase transitions; the conformational organization of phosphatidylinositol and phosphatidylinositides, as examples of physiologically important lipids; and the adsorption of peptides (neuropeptide Y, NPY) and solvents (dimethylsulfoxide, DMSO) onto phospholipid surface layers. The contribution "Studying Lyotropic Crystalline Phases Using High-Resolution MAS NMR Spectroscopy" (A. Pampel and F. Volke), the *twelfth chapter*, focuses on an experimental approach that is based on techniques that were primarily developed for investigations of the liquid-crystalline phases in combination with MAS. These techniques reveal properties that are related to the dynamic, liquid character of liquid-crystalline phase, which may easily be overlooked when these systems are studied with methods intended to observe real solid systems. The reader will be introduced in these methods. Their general applicability and their limits are discussed. Their use is demonstrated with some examples, covering biophysical studies as well as practical applications. The major focus is on problems that are related to molecules interacting with the lipid–water interfaces. The methods discussed, range form two-dimensional NOE spectroscopy for structure determination, via polarization transfer, to the latest development, the combination of MAS with pulsed field gradient NMR spectroscopy to study diffusional properties. Finally, in the *thirteenth chapter*, entitled "NMR Studies of Cartilage – Dynamics, Diffusion, Degradation" (D. Huster, J. Schiller, L. Naji, J. Kaufmann, and K. Arnold), it is shown that the molecular dynamics of biological macromolecules and their interactions with water play a decisive role in the viscoelastic properties of biological tissues. For example, articular cartilage consists of a variety of biopolymers with varying degrees of molecular mobility. Various NMR methods have been used to characterize the physical properties of cartilage on a molecular level. The studies have revealed very heterogeneous molecular dynamics of cartilage, which can serve as a basis for the development of artificial cartilage by tissue engineering methods.

The authors of the various contributions to these Lecture Notes are greatly indebted to their numerous collaboration partners, whether named in the variuos chapters here, or unnamed. This work would not have been possible without the financial support of the Deutsche Forschungsgemeinschaft (DFG) within the framework of the *Sonderforschungsbereich* 294. We express our sincere gratitude not only to the DFG but also to the University of Leipzig and to the Saxon State Ministry for Science and the Fine Arts for their continuous help and great mutual understanding. Many contributions were stimulated by the close collaboration between different institutes of the university and with numerous colleagues and friends from abroad who cannot be mentioned here. Thank you all very much for the excellent cooperation. In particular, the majority of the scientific work presented in this volume was connected with the preparation of Ph.D. theses and hence with the scientific careers of very talented young researchers. This gives us also the opportunity to express our gratitude to them and to wish them great success in their future scientific activities. In this context, we are glad to congratulate Dr. Christian Rödenbeck for having been awarded the 1999 FEZA Prize for Ph.D. Work in Zeolites or Related Materials by the Federation of European Zeolite Associations. We are greatly indebted to Mrs. Katrin Kunze for her great efforts in the technical organization of this research project. Above all, we would like to emphasize the excellent work and continuous support by the first spokesman of this SFB, Prof. Dr. Gotthard Klose.

Leipzig,
January 2004

Reinhold Haberlandt
Dieter Michel
Andreas Pöppl
Ralf Stannarius

Contents

Paramagnetic Adsorption Complexes in Zeolites as Studied by Advanced Electron Paramagnetic Resonance Techniques

Study of Conformation and Dynamics of Molecules Adsorbed in Zeolites by ^{1}H NMR

Molecular Dynamics of Liquids in Confinement

Liquid Crystals in Confining Geometries

Surfaces and Interfaces of Free-Standing Smectic Films

Pattern Formation in Langmuir Monolayers Due to Long-Range Electrostatic Interactions

Characterization of Floating Surface Layers of Lipids and Lipopolymers by Surface-Sensitive Scattering

Studying Lyotropic Crystalline Phases Using High-Resolution MAS NMR Spectroscopy

NMR Studies of Cartilage – Dynamics, Diffusion, Degradation

List of Contributors

Klaus Arnold
Universität Leipzig
Institut für Medizinische Physik
und Biophysik
Liebigstr. 27
04103 Leipzig, Germany
arnold@medizin.uni-leipzig.de

Peter Bräuer
Universität Leipzig
Institut für Experimentalphysik I
Linnéstr. 5
04103 Leipzig, Germany
brauer@chemie.uni-leipzig.de

Thomas M. Fischer
Florida State University
Department of Chemistry
and Biochemistry,
Tallahassee, FL 32306–4390, USA
tfischer@chem.fsu.edu

Dieter Freude
Universität Leipzig
Institut für Experimentalphysik I
Linnéstr. 5
04103 Leipzig, Germany
freude@uni-leipzig.de

Reinhold Haberlandt
Universität Leipzig
Institut für Theoretische Physik
Vor dem Hospitaltore 1
04103 Leipzig, Germany
Reinhold.Haberlandt@
physik.uni-leipzig.de

Siegfried Fritzsche
Universität Leipzig
Institut für Theoretische Physik
Vor dem Hospitaltore 1
04103 Leipzig, Germany
Siegfried.Fritzsche@
uni-leipzig.de

Jörg Kärger
Universität Leipzig
Institut für Experimentalphysik I
Linnéstr. 5
04103 Leipzig, Germany
Kaerger@physik.uni-leipzig.de

Friedrich Kremer
Universität Leipzig
Institut für Experimentalphysik I
Linnéstr. 5
04103 Leipzig, Germany
kremer@physik.uni-leipzig.de

Mathias Lösche
Universität Leipzig
Institut für Experimentalphysik I
Linnéstr. 5
04103 Leipzig, Germany
loesche@physik.uni-leipzig.de

Dieter Michel
Universität Leipzig
Institut für Experimentalphysik II
Linnéstr. 5
04103 Leipzig, Germany
michel@physik.uni-leipzig.de

André Pampel
Universität Leipzig
Institut für Experimentalphysik II
Linnéstr. 5
04103 Leipzig, Germany
anpa@physik.uni-leipzig.de

Andreas Pöppl
Universität Leipzig
Institut für Experimentalphysik II
Linnéstr. 5
04103 Leipzig, Germany
poeppl@physik.uni-leipzig.de

Heidrun Schüring
Universität Leipzig
Institut für Experimentalphysik I
Linnéstr. 5
04103 Leipzig, Germany
pge91dsf@
studserv.uni-leipzig.de

Ralf Stannarius
Otto-von-Guericke-Universität
Magdeburg
Institut für Experimentalphysik
Universitätsplatz 2
39106 Magdeburg, Germany
ralf.stannarius@
physik.uni-magdeburg.de

Modeling and Simulation of Structure, Thermodynamics, and Transport of Fluids in Molecular Confinements

Siegfried Fritzsche, Reinhold Haberlandt, and Horst Ludger Vörtler

Institut für Theoretische Physik, Universität Leipzig
Siegfried.Fritzsche@uni-leipzig.de,
Reinhold.Haberlandt@physik.uni-leipzig.de,
Horst.Voertler@physik.uni-leipzig.de

Abstract. The theoretical understanding of the properties of molecular confined fluids is crucial for many applications in science and technology, which reach from molecules enclosed in porous media (zeolites) to aqueous phases at biological active interfaces (biomembranes). We discuss these complicated many-particle systems – which usually are to complex for an analytical statistical-mechanical treatment– by means of molecular simulation methods. We focus on advanced Monte Carlo techniques to study structure and thermodynamics of inhomogeneous fluids and on recent molecular dynamics methods to describe transport phenomena in microporous materials.

The state of the art in the field is demonstrated by reviewing selected results of our recent computer simulations. We present both Monte Carlo studies of equilibrium properties of geometrically restricted fluids (such as spatial distribution functions, thermodynamic pressures and phase equilibria) and molecular dynamics studies of dynamical processes in nanoporous media, particularly diffusion of guest molecules in zeolites. The diffusion mechanism is analyzed in detail by computer simulations and theoretical analytical treatment as well.

1 Introduction

This chapter deals with the molecular modeling and simulation of fluids confined to molecular containments, such as micropores, porous media, or membranes, and comprises both the equilibrium structural and thermodynamic properties and the transport behavior of the enclosed particles.

The aim of the chapter is twofold: first, to give a review of recent molecular simulation techniques and the underlying statistical-mechanical concepts and second, to demonstrate the possibilities and limitations of the simulation methods discussed here, showing recent results for some systems and properties selected from the fields of research of the authors.

The chapter starts with a short review of the basic ideas and concepts of statistical physics relevant to the treatment of interfacial problems.

The next section deals with structural and thermodynamic properties of confined fluids. Starting with a short discussion of specific Monte Carlo methods, we present recent simulation results for the molecular structure and the phase behavior of several classes of confined systems, ranging from simple to associating fluids.

S. Fritzsche, R. Haberlandt, H.L. Vörtler, Modeling and Simulation of Structure, Thermodynamics, and Transport of Fluids in Molecular Confinements, Lect. Notes Phys. **634**, 1–88 (2004)
http://www.springerlink.com/

The last section focuses on the transport of particles through porous media. After an introduction to molecular dynamics simulation techniques, some results of recent simulation studies of diffusion processes in zeolites are reported and discussed in some detail.

2 Statistical Physics

An aim of statistical physics is to understand the macroscopic behavior of an N-particle system (N in the order of 10^{23}), governed by the laws of probability and classical or quantum mechanics. . Thus we start by sketching some features of probability theory [1–6] and classical mechanics [7, 8].

2.1 Some Notions of Probability Theory

Probability Distributions

The *probability p* of a *discrete* event i is given by the ratio of the number of *favorable* cases n_i to the total number of *possible* cases n:

$$p(i) = \frac{n_i}{n}, \qquad n_i \le n, \qquad \sum_i n_i = n. \qquad (1)$$

This is a definition of the probability *a priori*. The condition $\sum_i n_i = n$ of normalization is valid without any loss of generality because *one* of the possible events *must* occur. One says that an event that is *certain* has the probability 1, and if an event cannot be the outcome – in other words it is *impossible* – it has a probability 0. The statement about the value of the probability becomes more precise the larger the class of the individual events is. This yields another definition of the probability *a posteriori* (after the event, from experience):

$$\tilde{p}(i) = \lim_{n\to\infty} \frac{n_i}{n}, \qquad n_i \le n, \qquad \sum_i n_i = n. \qquad (2)$$

In generalization of (1) we define

$$p(i) = \frac{\sum_{n_i} g_r}{\sum_n g_r}, \qquad \text{where } g_r : \text{is the statistical weight of the state } r. \qquad (3)$$

Let x be a *continuous* property, then we ask for the probability p_x to find x. One cannot define p_x in the same manner as before, because the number of *possible* states is infinite (and thus p_x will vanish by use of the previous definition).

We define the probability, to find the event between x and $x + \Delta x$ to be

$$p_{x,x+\Delta x} = p(x)\,\Delta x. \qquad (4)$$

$p(x)$ is defined like a density. Thus we call it the *probability density*

$$p(x) = \lim_{\Delta x \to 0} \frac{p_{x,x+\Delta x}}{\Delta x}. \qquad (5)$$

Average Values and Fluctuations

The *average*, or *mean value*, or *expectation value* (as in quantum mechanics), of a property x in the distribution p_x or $p(x)$ is defined by

$$\langle x \rangle = \sum x p_x, \qquad \sum p_x = 1 \tag{6}$$

or

$$\langle x \rangle = \int x p(x)\,\mathrm{d}x, \qquad \int p(x)\,\mathrm{d}x = 1, \tag{7}$$

respectively.

Summation or integration, respectively, must be performed over all values of x in the whole range of validity of the distribution p_x or $p(x)$.

The *deviations from the averages* are called *fluctuations*, $\langle f(x) - \langle f(x) \rangle \rangle$. The mean values of these fluctuations are of interest. From (6), (7) it follows

$$\langle f(x) - \langle f(x) \rangle \rangle = 0.$$

Therefore, this quantity is of no use. To describe the desired *mean* fluctuation the so-called *mean square fluctuation* (variance) is of importance. It is equal to

$$\langle (f(x) - \langle f(x) \rangle)^2 \rangle = \sum (f(x) - \langle f(x) \rangle)^2 p_x \tag{8}$$

$$\langle (f(x) - \langle f(x) \rangle)^2 \rangle = \int (f(x) - \langle f(x) \rangle)^2 \,\mathrm{d}x \tag{9}$$

for discrete or continuous systems, respectively. More important are the *relative mean square fluctuations* of $f(x)$, which are defined by division of (8), (9) by $\langle f \rangle^2$:

$$\frac{\langle (f(x) - \langle f(x) \rangle)^2 \rangle}{\langle f(x) \rangle^2} = \frac{\langle (f(x))^2 \rangle - \langle f(x) \rangle^2}{\langle f(x) \rangle^2}. \tag{10}$$

For macroscopic systems these quantities will vanish as a rule, because they are inversely proportional to the square of the particle number. Therefore, the mean values are representative of those systems.

2.2 Some Notions of Classical Mechanics

Classical Mechanics serves as the basis of classical statistical physics, therefore we will present the most important features of classical mechanics [7, 8].

Equations of Motion

The generalized coordinates q_i and the generalized velocities $\dot{q}_i$ will be used as the $2f$ quantities to derive the equations of motion:

$$q_1, q_2, q_3, \ldots, q_f, \quad \text{for short } q; \quad \dot{q}_1, \dot{q}_2, \dot{q}_3, \ldots, \dot{q}_f, \quad \text{for short } \dot{q}; \quad (11)$$

here the $\dot{q}_i$ are related to the generalized momenta p_i:

$$p_i = \frac{\partial L}{\partial \dot{q}_i}, \quad i = 1, 2, \ldots, f. \quad (12)$$

$L = L(q_i, \dot{q}_i, t)$ is the *Lagrange* function, related by a Legendre transformation to the *Hamilton* function

$$H(q_i, p_i, t) = \sum_i p_i \dot{q}_i - L(q_i, \dot{q}_i, t). \quad (13)$$

The mechanical motion is governed by *equations of motion*, for instance

$$\frac{\mathrm{d}}{\mathrm{d}t} \frac{\partial L}{\partial \dot{q}_i} - \frac{\partial L}{\partial q_i} = 0 \qquad \text{(Lagrange)}, \quad (14)$$

$$\dot{q}_i = \frac{\partial H}{\partial p_i}, \quad \dot{p}_i = -\frac{\partial H}{\partial q_i} \qquad \text{(Hamilton)}, \quad (15)$$

$$m_i \ddot{\mathbf{r}}_i = \mathbf{F}_i, \quad \text{with} \quad F_i = -\frac{\partial U}{\partial r_i} \qquad \text{(Newton)}, \quad (16)$$

where m_i are the particle masses, F_i are the force components, $U(r_i)$ is the potential energy.

The last equation is Newton's second law for the position vector $\mathbf{r}_i$ of the ith particle and serves as an important starting point for molecular dynamics.

Hamilton Functions for Systems with N Particles

To describe the behavior of a given system one needs its *Hamilton function* (Hamiltonian), consisting of three parts:

1. The kinetic energy:

$$E = \frac{1}{2} \sum_i m_i \dot{q}_i^2 = \frac{1}{2} \sum_i \frac{p_i^2}{m_i}. \quad (17)$$

2. The potential energy of the interaction of the particles. As a rule the potential energy can be written for pairwise interaction of two particles – here assumed to depend only on the distance $r_{ij} = |\mathbf{r}_i - \mathbf{r}_j|$ between these particles $i, j (i \neq j)$ – as

$$U = \frac{1}{2} \sum_i^N \sum_j^N u_2(q_i, q_j) = \frac{1}{2} \sum_i^N \sum_j^N u_2(r_{ij}). \quad (18)$$

In reality, this is an approximation (for a generalization to long-ranged forces or nonspherical potentials see e. g. Sects. 3 and 4.3). For instance the often used Lennard–Jones potential (see Sect. 4) can be written

$$U = 4\epsilon \left[\left(\frac{\sigma}{r} \right)^{12} - \left(\frac{\sigma}{r} \right)^{6} \right], \tag{19}$$

where ϵ is U_{min} and σ is defined by $U(\sigma) = 0$ [9].

3. The potential energy caused by the walls:

$$U_{\mathbf{W}} = \sum_i u_1(q_i). \tag{20}$$

2.3 Some Notions and Relations of Statistical Thermodynamics

In this section the *radial distribution function* $g(r)$ ((30), (31)) and the *correlation function* $K(r, t)$ (61), which are two fundamental quantities of statistical thermodynamics and nonequilibrium theory, are introduced and correlated with the observables and system properties that can be evaluated ((30), (31), (64)).

Phase Space

We shall discuss two kinds of hyperspaces which are used for visualizing the mechanical state or the dynamics, respectively, of a complex systems.

The first is called the *molecule phase space*, or *μ-space* for short (μ stands for "molecule") with the molecules as the elements of the statistics. This method of statistical physics for ideal systems using the μ-space is called the method of *Boltzmann*.

The second kind – introduced by Gibbs for interacting particles – is the *gas phase space* (due to Ehrenfest), or *Γ-space* for short (Γ stands for "gas").

In this case the *whole system* is considered as a *single supermolecule* with f generalized coordinates and f generalized momenta. That means that this supermolecule (the whole system) has f degrees of freedom and its mechanical state therefore can be represented by *one* point in a $2f$-dimensional system of the f generalized coordinates and the f generalized momenta.

As in the μ-space the mechanical state of the system described by the q_i, p_i ($i = 1, 2, \ldots, f$) is called the *phase* of that system. The corresponding point in the Γ-space is called the *phase point*. In the Γ-space, $2f$-dimensional position vectors and velocity vectors

$$\mathbf{r} = \{q_1, q_2, \ldots q_f, p_1, p_2, \ldots p_f\} \tag{21}$$

$$\mathbf{v} = \{\dot{q}_1, \dot{q}_2, \ldots \dot{q}_f, \dot{p}_1, \dot{p}_2, \ldots \dot{p}_f\} \tag{22}$$

can be defined.

Following Gibbs, a large number of physically equivalent systems – the *Gibbs ensemble* – are considered. Let us imagine a *multitude* of such *physically equivalent systems*. This *imaginary multitude of physically equivalent systems* is called a *Gibbs virtual ensemble* of physically equivalent systems, differing in their mechanical states only [1, 6]. This ensemble can be visualized in the gas phase space by a *cloud* of phase points distributed with a particle density $\rho(q, p, t)$ (defined by (4), (24)). Its motion will characterize the mechanical state of the system.

Gibbs Ensembles

There are types of systems in different physical situations that must be treated by taking different starting points in computer simulations [6, 8–10]:

- Isolated systems – *systems without either energy exchange or particle exchange.*
- Closed systems – *systems with energy exchange, but without particle exchange.*
- Open systems – *systems with energy exchange and particle exchange.*

It is possible to create *Gibbs ensembles* for all of these different physical situations.

The analytical form of the phase density (N-particle distribution function) $\rho(q, p, t)$ (see (24)) will be determined for several physical ensembles and serves to determine macroscopic quantities by averaging the corresponding (7), (41).

Initially, microcanonical ensembles serve as the starting point for the molecular dynamical (MD) calculations (section 4), and canonical ensembles serve as the starting point for the Monte Carlo(MC) procedures (section 3.1). Today both MD and MC calculations can use several ensembles [8, 10–13]. Now, let us summarize the most important ensembles;

1. *Microcanonical ensemble.* This consists of a multitude of isolated systems with a given energy E, volume V and particle number N. The *characteristic function*, which describes the physical situation of this ensemble, is the entropy $S(E, V, N)$. The analytical form of the *phase density* is $\rho \sim \delta(H - E)$.
2. *Canonical ensemble.* This consists of a multitude of closed systems in contact with a heat bath to preserve the given temperature T, and with a given volume V and particle number N. The *characteristic function* of this ensemble is the free energy $F(T, V, N)$. The analytical form of the *phase density* is given by $\rho(q, p) \sim \exp\{-\beta H\}$ with $\beta = 1/kT$ (k is Boltzmann's constant, sometimes written as k_B to avoid confusion).
3. *Grand canonical ensemble.* This consists of a multitude of open systems in contact with a heat bath to preserve the given temperature T, and in contact with a particle reservoir to preserve the given chemical potential μ and the given volume V. The *characteristic function* - of this ensemble is the Massieu function $J(T, V, \mu)$. The analytical form of the *probability density* is given by $\rho(q, p) \sim \exp\{-\beta(H - \mu N)\}$.

For macroscopic systems – which means in the case of the thermodynamic limit $N \to \infty$ and N/V = const – all ensembles give equivalent results, differing from each other only by fluctuations, that are not important [1, 14]. Examples of the application of other ensembles for computer simulations will be given in Sects. 3.1 and 4.

In order to calculate the desired average values (see e.g. (7)) in the next sections, distribution functions of the kind of (4) will be defined and linked to different thermodynamic quantities.

Distribution Functions

N-Particle Distribution Functions. Considering an ensemble of ν systems with f degrees of freedom each

$$\rho_\nu^{(N)}(q, p, t) \, dq \, dp \tag{23}$$

is the number of systems in the range $q \ldots q+dq, p \ldots p+dp$ of Γ-space. If we introduce the (normalized) *phase space density* or the *(N-particle) distribution function,*

$$\rho^{(N)}(q, p, t) = \frac{1}{\nu} \rho_\nu^{(N)}(q, p, t), \tag{24}$$

$\rho^{(N)}(q, p, t) \, d \, q d p$ is the probability to find the system (consisting of N particles) in $q \ldots q+dq, p \ldots p+dp$ of the Γ-space. Because it is *certain* that the system will be found somewhere in the *whole* Γ-space, one finds the condition for the normalization to be

$$\int \rho^N(q, p, t) \, dq dp = 1 \,. \tag{25}$$

The integration must be performed over the whole accessible range of the Γ-space.

Reduced Distribution Functions. More often it is sufficient to use *reduced distribution functions* for finding k particles $(k \leq N)$ only, defined by a *k-particle distribution function* [9]

$$\rho^{(k)}(q, p, t) = \int \rho^{(N)}(q, p, t) \, dq^{(N-k)} dp^{(N-k)}. \tag{26}$$

Particularly important are the *two-particle distribution function* $(k = 2)$ and the *one-particle distribution function* $(k = 1)$ [9, 15].

$$\rho^{(2)}(q, p, t) = \int \rho^{(N)}(q, p, t) \, dq^{(N-2)} dp^{(N-2)}, \tag{27}$$

$$\rho^{(1)}(q, p, t) = \int \rho^{(N)}(q, p, t) \, dq^{(N-1)} dp^{(N-1)}. \tag{28}$$

The use of $\rho^{(2)}(q, p, t)$ is sufficient for practically all dilute systems. The *one-particle distribution function* plays an important role in gas kinetic theory.

Integration over the momentum space gives

$$n^{(2)}(q) = \int \rho^{(2)}(q, p, t)\,\mathrm{d}p, \qquad n^{(1)}(q) = \int \rho^{(1)}(q, p, t)\,\mathrm{d}p. \qquad (29)$$

Here $n^{(2)}(q_i, q_j)\mathrm{d}q_i\mathrm{d}q_j$ is the probability to find a particle i in $q_i \ldots q_i + \mathrm{d}q_i$ and another particle j in $q_j \ldots q_j + \mathrm{d}q_j$. $n^{(1)}(q_i)\mathrm{d}q_i$ is the probability to find a single particle i in $q_i \ldots q_i + \mathrm{d}q_i$.

Starting with (24), (27), one can derive the *pair distribution function* and the *radial distribution function* in the following manner [9, 15]. For $N \gg 1$, one can introduce a distribution function $g(q_i, q_{ji})$ by

$$n^{(2)}(q_i, q_j) = n^{(1)}(q_i)n^{(1)}(q_j)g(q_i, q_{ji}). \qquad (30)$$

$g(q_i, q_{ji})$ tends to one for $q_{ji} = |\mathbf{q}_i - \mathbf{q}_j| \to \infty$. The deviation from one is a measure of the correlation of the pairs of molecules (*the pair distribution function*).

The total potential energy U_N is assumed to be pairwise additive:

$$U_N(q_i, q_j) = \sum_{i<j} u(q_{ij}).$$

In isotropic systems, $u(q_{ij})$ and $g(q_i, q_{ji})$ are functions of the distance $q_{ji} = r$ only. Thus

$$g(q_i, q_{ji}) \Rightarrow g(r) \qquad \text{with} \qquad q_{ji} = r. \qquad (31)$$

The *radial distribution function* $g(r)$ is very important for evaluating structural and thermodynamic data and for comparing these data with experimental data.

Integration over the whole momentum space gives the probability to find the system in the range $q \ldots q + \mathrm{d}q$ of the configuration space; this means the *density number n* of the gas

$$n(q) = \int \rho(q, p, t)\,\mathrm{d}p. \qquad (32)$$

A very important and useful quantity of statistical thermodynamics, the *partition function $Q(T, V, N)$* (see e.g. [1, 6]), will be defined here by integration of the nonnormalized timeindependent phase density or N-particle density (24) over the whole phase space:

$$Q(T, V, N) = \int\!\!\int \rho_\nu(q, p)\,\mathrm{d}q\,\mathrm{d}p. \qquad (33)$$

Equivalent terms for the partition function are *Zustandssumme* or *sum over states*.

Liouville Equation

The N-particle density (the phase density) $\rho(q, p, t)$ (24) and the velocity vector $\mathbf{v}$ (22) derived from the configuration vector $\mathbf{r}$ obey the theorem of Liouville.

Using Hamilton's canonical equations (15) one can derive the following (equivalent) equations known as different forms of the *Liouville theorem*:

– The ensemble streams like an incompressible liquid:

$$\operatorname{div}\mathbf{v} = \nabla \cdot \mathbf{v} = 0 \quad \text{where} \quad \nabla = \sum_{i=1}^{2F} \mathbf{e}_i \frac{\partial}{\partial r_i}. \tag{34}$$

– The continuity equation is

$$\frac{\partial \rho_\nu(q,p,t)}{\partial t} + \mathbf{v} \cdot \nabla \rho_\nu = 0. \tag{35}$$

– For the local change of the density with respect to time, we obtain the *Liouville equation*

$$\frac{\partial \rho(q,p,t)}{\partial t} = -[\rho, H] = -i\mathsf{L}\rho, \tag{36}$$

where the Liouville operator (*Liouvillean*), is

$$\mathsf{L}\rho = -i[H, \rho] \tag{37}$$

and the Poisson bracket for $A(q,p)$, $B(q,p)$, is

$$[A, B] = \sum_{i=1}^{i=3N} \left(\frac{\partial A}{\partial q_i} \frac{\partial B}{\partial p_i} - \frac{\partial B}{\partial q_i} \frac{\partial A}{\partial p_i} \right). \tag{38}$$

The theorem of Liouville is of fundamental importance in the investigation of many-particle systems in equilibrium, and in the form of (36) it is useful in nonequilibrium systems too.

Ensemble, Time, and Measurement Averages

To consider the behavior of many-particle systems, one can use the following average values (see e.g. [6]):

1. *Time and measurement averages*,

$$\langle A \rangle_\tau = \frac{1}{\tau} \sum_{n=1}^{n=l} A(n\,\Delta t). \tag{39}$$

If we let $\tau \to \infty$, we call the resulting value,

$$\langle A \rangle_t = \lim_{\tau \to \infty} \frac{1}{\tau} \int_0^\tau A(q(t), p(t))\,\mathrm{d}t, \tag{40}$$

the *time average*. Assuming appropriate values for the time steps and the duration of the measurement in (39), both mean values – the measurement and time averages – are equivalent. In order to use time averages, we must solve the important task of finding *physically correct* values for τ and Δt (see Sects. 3.1, and 4).

2. *Ensemble averages.* Another possible way to describe manyparticle systems is to use *ensembles.* Let us construct an *ensemble* (see Sect. 2.3), distributed with the probability density (or probability distribution) $\rho(q(t), p(t))$ (24) with respect to our original system. Following the example of (7), we define the *ensemble average* as

$$\langle A \rangle_{\Gamma} = \int \int A(q(t), p(t))\rho(q(t), p(t)) \, \mathrm{d}q \, \mathrm{d}p. \tag{41}$$

Following the *ergodic hypothesis* [1, 8, 16, 17], which states that every accessible phase point in Γ-space will be occupied by the system at some time, one can assume that for physical systems in general, time and ensemble averages are equal. Hence one can use either one of them as the starting points for a computer simulation.

We shall use in Sects. 3 and 4 the average that is suitable for the situation discussed. In general, (39) is the starting point for molecular dynamics and (41) is the starting point for Monte-Carlo procedures in the calculation of mean values for various ensembles.

Let us give a useful example of some relations between mean values.

Virial Theorem. Let us average over the quantities

$$E_{\mathrm{kin}} = \frac{1}{2} \sum_{i=1}^{n} p_i \frac{\partial E}{\partial p_i}, \qquad [V] = \frac{1}{2} \sum_{j=1}^{n} q_j \frac{\partial E}{\partial q_j}. \tag{42}$$

The result is the *virial theorem.* The mean value of the kinetic energy E_{kin} of a system is equal to its virial $[V]$ (an application is given in Sect. 4.3):

$$\overline{E}_{\mathrm{kin}} = \overline{[V]}. \tag{43}$$

Thermodynamic Functions and Relations

We give here a summary of the important thermodynamic quantities only [6, 18, 19]:

Using a *Legendre transformations* where the product of the substituted variables (say TS) is subtracted one obtains one thermodynamic function (say $U(S, V, N)$) from another one (say $F(T, V, N)$). The most important functions are defined below, where they are related to the partition function $Q(T, V, N)$ (33) by use of the well-known thermodynamic relations [1, 6]:

$$U(S, V, N_i) = kT^2 \left(\frac{\partial \ln Q}{\partial T} \right) \qquad \text{internal energy,} \tag{44}$$

$$F(T, V, N_i) = U - TS = -kT \ln Q(T, V, N_i) \quad \text{free energy (Helmholtz),} \tag{45}$$

$$H(S, p, N_i) = U + pV = kT \left\{ T \left(\frac{\partial \ln Q}{\partial T} \right) + V \left(\frac{\partial \ln Q}{\partial V} \right) \right\} \quad \text{enthalpy,} \tag{46}$$

$$G(T, p, N_i) = H - TS = kT \left\{ -\ln Q + V \left(\frac{\partial \ln Q}{\partial V} \right) \right\} \text{free enthalpy (Gibbs).} \tag{47}$$

Here, we define the *chemical potential*

$$\mu = \left(\frac{\partial F}{\partial N}\right)_{T,V}.$$ (48)

Using (45) for one component $\mu_1 = \mu,\quad \mu_i = 0\ (i \neq 1)$,

$$dF = -S\,dT - p\,dV + \mu\,dN.$$ (49)

After some calculation we obtain

$$\frac{\mu}{kT} = \ln(n\Lambda^3) + \left(\frac{pV}{NkT} - 1\right) + \int_{n_0}^{n} \left(\frac{pV}{NkT} - 1\right)\frac{dn}{n}.$$ (50)

Λ is the de Broglie wavelength, defined by

$$\Lambda = \left(\frac{h^2}{2\pi mkT}\right)^{1/2}.$$ (51)

Equation (50) is very often used to determine the chemical potential μ in computer simulations (see Sect. 3.1).

Thermodynamic Quantities Using the Radial Distribution Function

The *radial distribution function* $g(r)$ defined by (30), (31) will be used to evaluate structural and thermodynamic data and to compare the results with experimental data as well. As examples, some relations are given for the internal energy U, the pressure p, the chemical potential μ and the *static structure factor* $S(k)$, which particularly important for comparing theoretical data with experiments (see, e.g. Sects. 3 and 4). These relations are obtained by starting from the the partition function (33).

Internal Energy U. The *internal energy U* can be written using $g(r)$ as

$$U = \frac{3}{2}NkT + \frac{1}{2}Nn\int_0^{\infty} u(r)g(r,n,T)4\pi r^2\,dr.$$ (52)

Pressure p. Analogously one can determine the pressure p:

$$\frac{p}{kT} = \left(\frac{\partial \ln Q}{\partial V}\right)_{T,N} = \frac{N}{V} - \frac{1}{6kT}\int\int r_{12}n(r_1)n(r_2)\frac{du(r_{12})}{dr_{12}}g(r_1,r_2)\,dr_1dr_2.$$ (53)

Thus we have the *pressure equation*:

$$p = kTn - \frac{n^2}{6}\int_0^{\infty} ru'(r)g(r,n,T)4\pi r^2\,dr.$$ (54)

One should note that $g(r)$ depends not only on T but also on the density n.

Chemical Potential μ. Corresponding to the definition of the chemical potential μ (48) the particle number of the system must be changed. Ensembles with $N - 1$, N, and $N + 1$ particles will be considered, and one test particle will be added or subtracted in each case. The starting points are the (33), (45), (48). Thus

$$\frac{\mu}{kT} = \ln(n\Lambda^3) + \frac{n}{kT} \int\limits_0^1 \int\limits_0^\infty u(r)g(r,\xi)4\pi r^2 \, dr \, d\xi. \tag{55}$$

There exist other possible ways to determine the chemical potential – e.g. using the grand canonical ensemble [20], the Widom method [21] (see Sect. 3.1), or the equation of state (see (49), (50)) as the starting point.

The determination of the radial distribution function $g(r)$ by use of computer simulations will be described in Sects. 3.1, and 4. For some possible ways to determine it analytically, the reader is reffered to the literature (e.g. [20]). It should be mentioned that the *static structure factor*

$$S(k) = 1 + n \int \{g(r) - 1\} \exp(i\mathbf{k} \cdot \mathbf{r}) \, d\mathbf{r} = 1 + nh(k) \tag{56}$$

plays an important role in comparing the theory with experimental data (neutron and X-ray scattering data; (see, e.g. , [22,23]). In this equation n is the density and $h(k)$ is the Fourier transform of $\{g(r) - 1\}$. The limit for $k \to 0$ gives

$$\lim_{k \to 0} S(k) = 1 + n \int \{g(r) - 1\} \, d\mathbf{r}. \tag{57}$$

2.4 Fundamentals of the Statistical Theory of Irreversible Processes

Almost all processes in nature are *irreversible*. They are defined by the condition that the *direction of time* cannot be changed without an additional force [24].

Irreversible processes are driven by *generalized forces* X and characterized by *transport coefficients* L [20, 22, 25]. It is most important to determine transport coefficients – e.g. using the well-known linear response theory (Kubo et al. [26]) – via correlation functions [27–29]. They can be calculated by the use of *equilibrium* ensembles. The correlation functions are as powerful in nonequilibrium as partition functions or radial distribution functions are in equilibrium (see e.g. (30), (31)).

The transport coefficients L_{ik} are defined by linear relations between the flux densities J_i and their corresponding generalized driving forces X_k:

$$J = L \cdot X + \text{(higher order terms)}, \qquad J_i = \sum_k L_{ik} X_{ik}. \tag{58}$$

In the case of *diffusion*, the *particle transport* is caused by concentration gradients:

$$J_z = -D \cdot \frac{\partial n_1}{\partial z} + \text{(higher-order terms)}. \tag{59}$$

The *ensemble average* in Γ-space,

$$\langle A \rangle_\Gamma = \int \int A(q(t), p(t)) \rho(q(t), p(t)) \, \mathrm{d}q \, \mathrm{d}p, \tag{60}$$

is calculated using the *equilibrium* phase space density (24). The *time correlation functions*

$$K_{AB}(t) = \langle A(t) B(0) \rangle_\Gamma = \int \cdots \int A(q, p, t) B(q, p, 0) \rho(q, p) \, \mathrm{d}q \, \mathrm{d}p \tag{61}$$

for the phase space functions $A\{q(t), p(t)\} = A(t)$, $B\{q(t), p(t)\} = B(t)$ are defined by (61). For $A = B$, $K_{AA}(t)$ is called the *autocorrelation function,*: and for $A = B = \mathbf{v}$ (velocity), it is called the *velocity autocorrelation function* $K_{vv}(t)$. The *spectral density* $f(\omega)$, which is important for comparisons with experimental data, is a Fourier transform of the velocity autocorrelation function:

$$f(\omega) = \int\limits_0^\infty \frac{\langle \mathbf{v}(0) \cdot \mathbf{v}(t) \rangle}{\langle \mathbf{v}(0)^2 \rangle} \cos(\omega t) \, \mathrm{d}t. \tag{62}$$

The *linear response theory* [26] can be outlined as follows:

- The linear response theory assumes *near-equilibrium* situations so that linear relations are sufficient (e.g. to describe relaxation back to equilibrium).
- This linear response is determined by fluctuations from *equilibrium*.

Transport coefficients are Fourier transforms of correlation functions:

$$\sigma(\omega) = \int\limits_0^\infty \exp(-i\omega t) \langle \dot{A}(t) \dot{B}(0) \rangle_\Gamma \, \mathrm{d}t. \tag{63}$$

In the limit of long waves ($\omega \to 0$, $k \to 0$), which is sufficient for transport coefficients,

$$\sigma = \int\limits_0^\infty \langle \dot{A}(t) \dot{B}(0) \rangle_\Gamma \, \mathrm{d}t \qquad L_{ij} = \int\limits_0^\infty \langle \dot{J}_i(t) \dot{J}_j(0) \rangle_\Gamma \, \mathrm{d}t. \tag{64}$$

In the long-time limit ($t \to \infty$) (64) can be replaced by *Einstein relations*,

$$2t\sigma = \langle (A(t) - A(0))^2 \rangle. \tag{65}$$

Further quantities used in comparing theoretical and experimental data are the *dynamic structure factor* ((66) [20, 23, 30, 31]) and the *scattering function* $F(\mathbf{k}, t)$ (67)

$$S(\mathbf{k}, \omega) = \frac{1}{2\pi} \int\limits_{-\infty}^\infty F(\mathbf{k}, t) \exp i\omega t \, \mathrm{d}t, \tag{66}$$

$$F(\mathbf{k}, t) = \frac{1}{N} \langle \rho(\mathbf{k}, t) \rho(-\mathbf{k}, 0) \rangle . \tag{67}$$

$F(\mathbf{k}, t)$ is related to the *van Hove correlation function* $G(\mathbf{k}, t)$ by

$$F(\mathbf{k}, t) = \int G(\mathbf{r}, t) \exp\left(-i\mathbf{k} \cdot \mathbf{r}_j\right) d\mathbf{r}. \tag{68}$$

and can be determined by ensemble averaging $\langle \cdots \rangle$ over the Fourier transform $\rho(\mathbf{k}, t)$ of the local one-particle density,

$$\rho(\mathbf{k}, t) = \sum_j \exp\left(i\mathbf{k} \cdot \mathbf{r}_j(t)\right) = \int \rho(\mathbf{r}, t) \exp\left(i\mathbf{k} \cdot \mathbf{r}_j\right) d\mathbf{r}. \tag{69}$$

In the limit $\omega \to 0$, $k \to 0$ one has the usual static structure factor (56).

As an example of the general theory of irreversible processes (see Table 1), some relations for the (self-) diffusion coefficient are given here:

$$D = \frac{1}{3} \int_0^\infty \langle \mathbf{v}(t) \cdot \mathbf{v}(0) \rangle \, dt; \qquad 2tD = \frac{1}{3} \langle (\mathbf{r}(t) - \mathbf{r}(0))^2 \rangle. \tag{70}$$

Table 1. Correlation functions – theoretical and experimental quantities

Observable	Correlation function	Equation
Spectral density $f(\omega)$	$\langle \mathbf{v}(0) \cdot \mathbf{v}(t) \rangle$	(62)
Structure factor $S(\mathbf{k}, \omega)$	$\langle \rho(\mathbf{k}, t) \rho(-\mathbf{k}, 0) \rangle$	(66)
Transport coefficient L_{ij}	$\langle \dot{J}_i(t) \dot{J}_j(0) \rangle_\Gamma$	(70)

3 Structure and Thermodynamics of Confined Fluids

3.1 Monte Carlo Techniques

In this section we shall briefly discuss important Monte Carlo simulation techniques. Particularly, we deal with recent sampling techniques relevant to the efficient simulation of equilibrium properties of confined fluids, where we focus on the contributions of the authors to recent methodical developments. Basic information about molecular dynamics simulations of transport properties of fluids in nanoporous media are given in Sect. 4. For more details we refer to a recent review on the subject [32].

Ensemble Averages and Importance Sampling

MC simulation methods in the classical sense are based on equilibrium statistical mechanics, which calculates ensemble averages of physical (thermodynamic) quantities of many body systems by means of stochastic methods. The average of an observable O takes, in general, the form of an integral (see (60)),

$$\langle O \rangle = \int \int O(\mathbf{r}^N, \mathbf{p}^N) f_0^{(N)}(\mathbf{r}^N \mathbf{p}^N), \mathrm{d}\mathbf{r}^N \mathrm{d}\mathbf{p}^N, \tag{71}$$

over the phase density $f_0^{(N)}(\mathbf{r}^N, \mathbf{p}^N)$ of the statistical ensemble under consideration.

Ensemble averages are high-dimensional integrals depending in general on all coordinates $\mathbf{r}^N$ and momenta $\mathbf{p}^N$ of the N particles of the system. The phase densities vary over many orders of magnitude, such that often only limited ranges of the phase space provide contributions to the ensemble averages. For example, for a microcanonical ensemble the phase density is a δ-function of the configurational energy.

To simulate ensemble averages, special stochastic integration methods have been introduced using importance sampling, based on distributions of specially weighted random numbers which preferentially sample those parts of the phase space which provide the largest contributions to the phase integrals.

Usually, the Hamiltonian H of a classical many-particle system is given by

$$H = E_{\mathrm{kin}}(\mathbf{p}^N) + U_N(\mathbf{r}^N), \tag{72}$$

where the kinetic energy E_{kin} depends only on the momenta $\mathbf{p}^N$ and the potential energy depends only on the coordinates $\mathbf{r}^N$. In this case ensemble averages may be written as a sum,

$$\langle O \rangle = O^{\mathrm{id}} + \langle O^{\mathrm{ex}} \rangle,$$

where the term O^{id} describes the behavior of an ideal gas, which is given analytically by integration over the momenta of the particles. The calculation of the ensemble average reduces to the estimation of the average over the excess part $\langle O^{\mathrm{ex}} \rangle$, which is called configurational average. Although configurational averages depend on the coordinates of the particles only, the typical properties of phase integrals (high dimensionality, and strongly varying integrand) apply to these averages too. Therefore the above-mentioned special importance sampling techniques have to be used for the simulation of configurational averages.

In the following we represent the basic ideas of the importance sampling (the Metropolis algorithm [33]) for the canonical ensemble of N particles in a given volume V at a temperature T. The extension to more general conditions is straightforward.

In a general notation (Binder [12, 34]), we characterize every particle i of the system ($i = 1, \ldots, N$) by a set of dynamical variables $\{\alpha_i\}$ (spatial vectors $\mathbf{r}_i$, orientations Ω_i, spin vectors $\mathbf{S}_i$, etc.).

The set $X = \{\{\alpha_1\}, \{\alpha_2\}, \ldots, \{\alpha_N\}\}$ is called configuration X (point X in the configurational space of the system).

If we denote the configurational part of the Hamilton function (usually the potential energy) by $\mathcal{H}_N(X)$, the configurational average of a quantity $O(X)$ is given by

$$\langle O \rangle = \frac{\int O(X) \exp[-\beta \mathcal{H}_N(X)] \, dX}{\int \exp[-\beta \mathcal{H}_N(X)] \, dX}. \tag{73}$$

The integrals may be approximated by sums,

$$\langle O \rangle \approx \bar{O} = \frac{\sum\limits_{j=1}^{M} A(X_j) P^{-1}(X_j) \exp[-\beta \mathcal{H}_N(X_j)]}{\sum\limits_{j=1}^{M} P^{-1}(X_j) \exp[-\beta \mathcal{H}_N(X_j)]}, \tag{74}$$

over M randomly chosen phase points X_j corresponding to a given probability distribution $P(X)$ (importance sampling). To implement the importance sampling one generates – according to Metropolis [12, 33, 34], a chain of configurations using

$$P(X_j) = P_{\text{eq}}(X_j) \propto \exp[-\beta \mathcal{H}_N(X_j)] \tag{75}$$

as the probability distribution for the random choice of the phase points. It follows, that

$$\bar{O} = \frac{1}{M} \sum_{j=1}^{M} O(X_j) \quad \text{(arithmetic average)}. \tag{76}$$

Since P_{eq} is not explicitly known, a random walk of points $\{X_j\}$ through the phase space (Markov chain) is constructed, consisting of points $\{X_j\}$ which has the property $P(X_j) \to P_{\text{eq}}(X_j)$ for $M \to \infty$. A sufficient condition for this behavior is that – in accordance with the principle of microreversibility – the transition probability $W(X_j \to X_{j'})$ for $X_j \to X_{j'}$ obeys the following criterion (Metropolis algorithm [33])

$$W(X_j \to X_{j'}) = \begin{cases} \exp[-\beta \delta \mathcal{H}] & \text{for} \quad \delta \mathcal{H}_N > 0 \\ 1 & \text{otherwise} \end{cases}. \tag{77}$$

A detailed discussion of the Metropolis algorithm may be found in the monographs [12, 34], for example.

We proceed to an isobaric–isothermal ensemble where the number of particles N, the pressure P and the temperature T are kept fixed. The configurational average of a quantity O in such an NPT–ensemble is formally of the same structure as the canonical average if an additional variable V and a modified weight factor are introduced. If we considering V formally as a dynamical variable of the system by setting

$$X = \{\{\alpha_1\}, \{\alpha_2\}, \dots, \{\alpha_N\}, V\},$$

the relations of the canonical average hold quite analogously for the isobaric–isothermal ensemble. In particular, we obtain

$$W(X_j \rightarrow X_{j'}) = \begin{cases} \exp\{-\beta[\delta V P + \delta\mathcal{H}_N]\} & \text{for} \quad \delta\mathcal{H}_N > 0 \\ 1 & \text{otherwise} \end{cases} \quad . \tag{78}$$

In generating the chain of configurations (i.e. performing transitions from microstate X to X'), we have to consider a random change of the volume of the system as a basic MC move, where the Metropolis acceptance check (78) has to be applied (see [35]).

To study coexisting fluid phases (phase equilibria) in the bulk and phase coexistence (adsorption, capillary condensation, etc.) in confined fluids, we consider more general open systems, statistical–mechanically described by the grand canonical ensemble (see Sect. 2.3), where we simulate the average number of particles $\langle N_k \rangle$ in the k coexisting phases while the volume V, the temperature T, and the chemical potentials μ_i are kept fixed. In this case the set of dynamical variables X comprises besides the coordinates, the particle numbers N_k in the k phases involved.

As a basic MC move we have to consider in the grand canonical simulation a random variation of the number of particles N. Since N can take only discrete (integer) values, at least two changes have to be performed:

1. Insertion of a particle at a randomly chosen position (i.e. $X_N \rightarrow X_{N+1}$), accepted with the probability

$$p(N \rightarrow N + 1) \propto \exp(-\beta[\mathcal{H}_N - \mathcal{H}_{N+1}]). \tag{79}$$

2. Removal of a (randomly chosen) particle (i.e. $X_N \rightarrow X_{N-1}$), accepted with the probability

$$p(N \rightarrow N - 1) \propto \exp(\beta[\mathcal{H}_N - \mathcal{H}_{N-1}]). \tag{80}$$

The generation of the chain of configurations is schematically shown in Fig. 1. The transition from one configuration to the next one takes place by a small random change of one dynamical variable (one degree of freedom) of a randomly chosen particle of the system.

In the canonical ensemble for spherically symmetrical fluids, this means a random displacement of the center of a particle; for molecular fluids, additionally a random change of the orientation of a molecule has to be performed. For the implementation of the orientation move, several algorithms are available, depending on the geometry of the molecules [10, 36, 37].

In the NPT ensemble the chain of configurations comprises, besides the (canonical) displacement moves, random changes of the volume, which have to be accepted according to (78).

In the μVT ensemble, we have to generate a Markov chain consisting of particle displacements and random insertions (79) and removals (80) of particles.

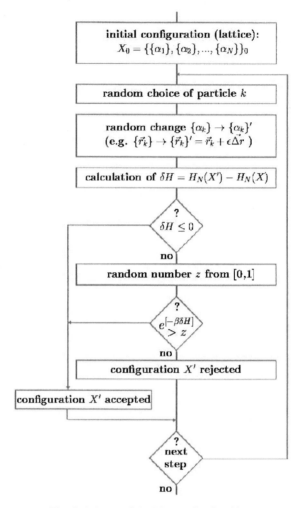

Fig. 1. Scheme of the Metropolis algorithm

The estimation of configurational (ensemble) averages requires that the generated chain of configurations (microstates) represents thermal equilibrium. The Metropolis algorithm ensures convergence to equilibrium for sufficient long Markov chains (rigorously only if the number of configurations $M \to \infty$). If the simulation is started with a special initial configuration (e.g. regular lattice) an equilibration run has to be performed, where a sufficiently large number of configurations has to be generated before the measurement of quantities is started. The necessary length of the chains for both the equilibration and the averaging of observables cannot be specified a priori but depends on the interaction potentials and the state conditions.

The structure of a fluid is described by spatial distribution functions, where the one-particle densities and the pair distribution functions are of special interest, since from a knowledge of these functions the thermodynamic properties of the fluid can be estimated. While in bulk fluids the one-particle distributions $\rho^{(1)}$ are usually constant within the entire fluid volume and given by $\rho^{(1)} \equiv \rho = N/V$, in confined fluids the one-particle distributions $\rho^{(1)}(\mathbf{r})$ represent the most important structural quantities.

Technically, distribution functions are calculated by means of histograms. In particular, the one-particle distributions of fluids confined to micropores may be estimated by dividing the pore volume into suitable volume bins, and by counting the number of particles observed in these bins during the course of the simulation. By registering these numbers in suitable histograms approximations of the (average) spatial distributions of the particles may be obtained.

The setting up of histograms is one of the basic techniques used to analyze MC configurations.

Simulation of Fluid Phase Equilibria

Partition Function Derivatives by Virtual Parameter Variation. Some key quantities for the description of phase equilibria such as the free energy and the chemical potential, cannot be represented directly by an average over a statistical ensemble. Therefore the simulation of these quantities is not possible by means of the usual Metropolis algorithm.

To directly calculate the relevant derivatives of the configurational partition function (i.e. of the free energy) Vörtler and Smith [38] introduced simulation techniques using virtual parameter variations, where, as a special case, Widom's test particle method [21] for estimating the chemical potential is recovered.

Under canonical conditions, the internal energy U, the pressure P, and the chemical potential μ are defined as partial derivatives of the Helmholtz free energy A (sometimes, as in (45), denoted by F) with respect to the temperature T, the negative volume $-V$, and the number of particles N, respectively, and related to the configurational partition function $\ln Q_N$. The following relations hold (compare (44-46))

$$d\beta A^{\mathrm{ex}} = U^{\mathrm{ex}}\, d\beta - \beta P^{\mathrm{ex}}\, dV + \beta \mu^{\mathrm{ex}}\, dN \tag{81}$$

$$\beta A^{\mathrm{ex}} = -\ln Q_N, \tag{82}$$

where the superscript ^{ex} denotes an excess quantity above the ideal-gas value. It can be seen that, formally, the derivatives of $-\ln Q_N$ with respect to parameters x selected from the set $\{\beta, V, N\}$ are the corresponding coefficients of the differentials in (81). Consider a system being simulated in the N, V, T ensemble as the *reference system*, denoted by the subscript 0; it can be shown that the first derivative of A^{ex} with respect to the parameter x is given by

$$\left(\frac{\partial \beta A^{\mathrm{ex}}}{\partial x}\right) = -\lim_{\Delta x \to 0} \frac{\ln \langle \exp(-\beta\, \Delta\mathcal{U})\rangle_0}{\Delta x}, \tag{83}$$

where $\langle \ldots \rangle_0$ denotes an average over configurations of the reference system, and $\Delta\mathcal{U}$ is the change of the potential energy if the parameter x is changed from the reference system value x_0 by Δx. If x and $\partial \beta A / \partial x$ form a conjugate intensive–extensive variable pair, one may perform computer simulations at fixed values of either quantity by means of a Legendre transformation (see (45)). One example of such a variable pair for the canonical ensemble variables $\{N, V, \beta\}$ with a thermodynamic potential βA is $\{N, \beta \mu\}$. Thus, $\beta \mu$ may be measured in an (N, V, β) canonical ensemble simulation using the test particle method (which we could refer to as virtual particle variation method), which is a special case of (83). Correspondingly, grand canonical ensemble simulations correspond to fixed $\beta \mu \equiv \partial(\beta A)/\partial N$. Similarly, P may be measured in a canonical ensemble simulation using (83) (the virtual volume variation method), and (N, P, T) simulations correspond to fixed $\beta P \equiv -\partial \beta A / \partial V$. A third conjugate pair of variables is $\{\beta U, \beta\}$. As in the case considered by Eppenga and Frenkel [39], (83) may provide a convenient method to measure $\beta U \equiv \partial \beta A / \partial \beta$ in a computer simulation for systems of complex geometry. Finally, simulations at fixed βU may be implemented in an analogous manner. In all these cases, the simulations at specified values of $\beta \mu$, P, or U may be performed in a canonical ensemble using fluctuating values of V, N, or β, respectively.

For confined fluids the application of the virtual volume variation technique opens up a new way to estimate thermodynamic pressures directly e.g. the spreading pressure of fluids confined to micropores (see Sect. 3.3).

Direct Simulation of Phase Equilibria in Bulk and Confined Fluids. The simulation methods presented so far, such as grand canonical MC or virtual particle insertion (the Widom method), permit the study of phase equilibria only indirectly via the estimation of the chemical potentials.

Recently Panagiotopoulos has proposed a computer experiment – the Gibbs ensemble approach – which directly simulates the average densities of coexisting fluid phases not only for bulk systems [40] but also for confined fluids in equilibrium with a bulk region [41] and avoids the expensive (time-consuming) calculations of the chemical potentials.

The basic idea of the Gibbs ensemble method consists in the representation of a macroscopic system (fluid) with two coexisting phases (vapor I and liquid II) by means of two microscopic regions within the corresponding phases which are far from the physical contact between the phases, these regions are modeled by two separate simulation boxes for the gas and the liquid phase, respectively.

Obeying the conditions of coexistence of the two phases

$$T_{\mathrm{I}} = T_{\mathrm{II}}, \qquad p_{\mathrm{I}} = p_{\mathrm{II}}, \qquad \mu_{\mathrm{I}} = \mu_{\mathrm{II}}, \tag{84}$$

the computer experiment is performed by the generation of a chain of configurations. This chain consists of the basic moves (i) particle displacements within the corresponding boxes, (ii) volume fluctuations and (iii) particle transfer between the boxes, and converges to the properties of the coexisting phases in equilibrium.

The investigation of phase equilibria of bulk fluids by means of the Gibbs ensemble technique has become in recent years a standard simulation technique for the study of such equilibria on a molecular level [42].

Recent modifications of the Gibbs method permit the study of molecular fluids with highly elongated molecules (e.g. Kihara fluids [43]) and associating fluids.

By way of contrast, the simulation method used for the study of fluids in confined regions in equilibrium with a corresponding bulk fluid has typically been the grand canonical ensemble technique, which allows the study of the properties of a confined fluid in equilibrium with a bulk fluid of specified chemical potential and temperature. The grand canonical method is typically only useful in the study of single-phase systems in each region. As an extension of the original Gibbs technique for confined fluids [41], Smith and Vörtler [44] recently developed a more general version to simulate inhomogeneous phase equilibria involving two coexisting fluid phases either in the bulk or in a confined region. This method allows the study of adsorption equilibria between bulk fluids and fluids confined in pores and of capillary condensation phenomena (phase equilibria between confined fluid phases).

We consider a fluid distributed over a bulk region (I) and a pore (II) with preset temperature T, total volume $V = V_I + V_{II}$, and total number of particles $N = N_I + N_{II}$, i.e. the bulk and pore region together form a canonical ensemble. The thermodynamic equilibrium between both regions requires:

$$T_I = T_{II}, \qquad \mu_I = \mu_{II}. \tag{85}$$

As discussed in detail by Panagiotopoulos [41], pressure is not a relevant variable for the description of bulk–pore equilibrium. This means that both the total volume of the two bulk boxes and the volume of the pore box must be kept fixed during the course of the simulation. Two simulation boxes represent the bulk (I) and the pore (II) in regions "far from the physical contact area" between both regions. If there are coexisting phases within a region these are described by two boxes representing coexisting vapor and liquid phases, either in the bulk (I^v, I^l) or in the pore (II^v, II^l).

In close analogy to the bulk Gibbs ensemble, we have to consider a chain of configurations consisting of three types of moves:

1. Random displacements of the particles (independently in their boxes) accepted with probability

$$p_d = \min \left\{ 1, \exp(-\beta \, \Delta U) \right\}, \tag{86}$$

   where ΔU is the change of the configurational energy.
2. transfer of a randomly chosen particle from region II to a random location in region I, accepted with probability

$$p_{tf} = \min \left\{ 1, \exp(-\beta(\Delta U_I + \Delta U_{II} - \ln \frac{(N_I + 1)V_{II}}{N_{II}V_I} \right\}. \tag{87}$$

3. Moves in sub-boxes, if vapor and liquid phases coexist either in (I) or in (II):
   a) Increase of the volume V_l by a randomly chosen ΔV (and a decrease of the volume V_v by ΔV), accepted with probability

$$p_v = \min \left\{ 1, \exp(-\beta(\Delta U_l + \Delta U_v)) + N_l \ln \frac{V_l + \Delta V}{V_l} + N_v \ln \frac{V_v - \Delta V}{V_v} \right\}. \tag{88}$$

b) Particle transfer between subboxes of one region, accepted with probability analogous to that of move 2.

This random walk converges to the equilibrium distribution of the particles. In Sect. 3.3 results [44] of three-box simulations of the phase equilibria of square-well fluids in planar slits will be discussed. A generalization of the Gibbs method to systems with chemically reacting components uses the reaction ensemble recently introduced by Smith and Triska [45] which studies simultaneously phase equilibria and chemical equilibria.

The extension of the reaction ensemble to chemical reactions in confined fluids is straightforward and is at present under consideration [46, 47].

Efficiency Problems

MC methods for the simulation of chemical potentials, for grand canonical simulations, and for phase and reaction equilibria require as basic moves the (virtual or real) insertion of test particles or the transfer (exchange) of molecules between simulation boxes.

In general, the efficiency of these simulation techniques depends essentially on the probability of finding a cavity in the system which is able to accommodate an additional particle. This probability decreases rapidly with increasing number density of the fluid phase and may for usual liquid densities (e.g. at the triple point) become so small that the resulting acceptance rate for the particle insertion/transfer steps becomes so small that the convergence of the corresponding Markov chain breaks down in the high–density range. Similarly, in confined fluids, the low particle insertion probability in complex interface systems considerably hinders the direct simulation of phase equilibria.

A number of approaches have been proposed to overcome these efficiency problems. These techniques comprise modifications of the particle insertion move, the introduction of extended ensembles and the use of modified importance sampling methods with non-Boltzmann probability distributions which completely avoid the insertion of particles. In the following we shall discuss some of these methods in more detail. Although these methods show an improved efficiency for a number of important fluid models a generally valid, universal methodology to overcome the basic problem of the low insertion rate in high dense and complex fluids does not exist so far.

Gradual Particle Insertion and Extended Ensembles. To enhance the efficiency recently a gradual–particle–insertion technique has been proposed recently [48–50].

This method performs the transition of a particle between two canonical ensembles with N and $N + 1$ particles in an efficient way, via a set of k unphysical subensembles, containing N normal particles plus one partially coupled (scaled) particle. The excess chemical potential (over the ideal gas) is given by

$$\beta\mu^{\text{ex}} = \ln\left[w_{k+1}\frac{\text{Prob}[N]}{\text{Prob}[N + 1]}\right], \tag{89}$$

where Prob[N] is the probability of observing a state with N fully coupled particles (where the scaled particle is of size 0, not interacting with the fluid molecules). Prob[$N + 1$] is the probability of observing a state with $N + 1$ fully coupled particles (where the scaled particle is the size of a normal particle, with full interactions with the N original molecules). w_{k+1} represents the statistical weight of the $(N + 1)$–particle ensemble.

The most important physical difference between Widom's test particle method and the gradual–insertion technique lies in the fact that in the original Widom method the insertion of the test particle is performed virtually for the sole purpose of the estimation of the average Boltzmann factor, while in the gradual-insertion method, the test particle (scaled particle) is a real (partly) coupled $(N + 1)$th fluid particle. Therefore the implementation of the gradual particle insertion is straightforward in comparison with other simulation methods which require particle insertion or transfer, such as grand canonical [48, 51] and Gibbs simulations (see below).

Very similar to the gradual–particle–insertion method is the scaled-particle MC method of Labik and Smith [52], which has been applied to the simulation of the chemical potential of dense fluids confined to slit-like pores [53].

The implementation of gradual particle insertion in the Gibbs ensemble [43, 54] leads us to the notion of extended ensembles which are created by the addition of (unphysical) states (ensembles) to an existing statistical ensemble. The extended Gibbs ensemble combines the standard Gibbs ensemble with the gradual–particle–insertion method discussed above. As described above, the gradual–particle–insertion estimation of chemical potentials may be considered as a simulation of a set of canonical subensemble which differ by the extent of coupling (the "size") of a scaled test particle.

In close analogy, we introduce here a set of modified Gibbs ensembles where each of them contains a scaled (fluctuating) particle and the subensembles differ by the extent of coupling of the fluctuating particle to the other fluid molecules.

A Markov chain is generated which permits transitions between the modified Gibbs ensembles. The goal is the construction of a path through the set of subensembles (with a partially coupled scaled particle) which increases the probability of particle transfer from one simulation box to the other in comparison with the standard Gibbs ensemble.

In order to construct such a chain of configurations instead of the usual particle transfer step, the following transitions are considered:

1. Change of the extent of coupling of the fluctuating particle remaining in the same box (this corresponds, in the gradual particle insertion, to a transition in the chain of subensembles).
2. Insertion of the fluctuating particle in the other simulation box (possibly by a simultaneous change of the extent of coupling).

For the set of Gibbs ensembles with fluctuating particles described (the extended Gibbs ensemble), the partition function can be written quite analogously to the standard Gibbs ensemble. From the form of this partition function, the acceptance probabilities for the above–mentioned transitions for the changes of coupling of the scaled particle and the insertion probabilities of the scaled particles can be derived [54,55].

For several classes of fluids, such as simple spherical systems (mixtures of square-well fluids [54]), fluids with non-spherical particles (Kihara fluids with highly elongated molecules [55]), and associating fluids [56] gradual insertion significantly increases the efficiency of particle insertion (transfer) moves.

Advanced Sampling Methods. Closely related to the extended Gibbs ensemble is the combination of the Gibbs simulation with Valeau's concept [57] of MC sampling with thermodynamic scaling [58]. This method permits the simultaneous estimation of the coexistence densities for a number of similar systems (i.e. systems with similar intermolecular potentials) at a number of different temperatures during a single simulation run.

The basic idea of the method is the use of MC sampling with non-Boltzmann distributions (similar to the umbrella sampling technique [59,60]), which generates a chain of configurations representing simultaneously the microstates of a number of similar systems. The configuration averages of interesting quantities (particularly the densities of coexisting phases) for a number of neighboring state points (or a set of similar pair potentials) can then be estimated by an analysis of the uniform Markov chains generated.

Another advanced MC method developed for complex fluids where the Metropolis algorithm is inefficient is the so-called biased sampling technique [17]. The basic idea of this method is to generalize the importance sampling assuming that the trial configurations are generated by a probability which depends on the configurational Hamilton function $\mathcal{H}_N$ through an arbitrary function $f(\mathcal{H}_N)$. In this case the detailed balance condition – which ensures the convergence of the Markov chain – takes a more general form, and the corresponding acceptance criterion for the biased transition is given by

$$
W(X_j \to X_{j'}) = \begin{cases} \frac{f(\mathcal{H}_N(X_{j'}))}{f(\mathcal{H}_N(X_j))} \exp[-\beta\delta\mathcal{H}] & \text{for} \quad \delta\mathcal{H}_N > 0 \\ 1 & \text{otherwise} \end{cases} \tag{90}
$$

as a direct generalization of (77). Hence an arbitrary biasing function can be introduced in the importance sampling scheme in order to enhance the acceptance rate of the corresponding MC moves.

Examples are orientation bias MC [61], where the acceptance of an orientation change of a particle – which may be very small in complex molecules – is increased by a biased orientation move, and configurational-bias MC [62–65], where a biased segment-by-segment insertion step for macromolecules is used. This is of special importance for chain molecules, particularly in confined systems, where the adsorption of chain molecules in zeolites [46,66,67] has been studied.

Despite the efforts to improve the efficiency of the particle insertion moves, there are state conditions where the conventional methods of simulation of phase equilibria fail. This refers on the one hand to the properties near the critical point and on the other hand to the range of low (strongly subcritical) temperatures where liquid phases of high density may appear. To deal with these problems, a number of advanced sampling techniques have been proposed in the last years. Although these methods are not yet beeing used on a large scale so far we mention these approaches because of their methodical importance for future developments.

For an investigation of the critical and the weakly subcritical two-phase regions, finite-size scaling techniques [68] have been combined with so-called histogram reweighting MC methods [69–71].

The starting point is the introduction of a coupled probability density $p_L(\rho, U)$ to find a configuration with both a given density ρ and a configurational energy U. To determine $p_L(\rho, U)$, a series of grand canonical simulations is performed where always, after a distinct number of configurations (ρ, U), histograms are recorded. The wanted probability density can be represented by a weighted average of the distributions $p_L^{(i)}(\rho, U)$ $(i = 1, ..., R)$ of the R simulation runs performed, where the weight of a special contribution essentially depends on the frequency of occurrence K_m in run m. From $p_L(\rho, U)$, the corresponding distributions of the state conditions of interest can be estimated by means of specific histogram reweighting methods. The position of the critical point may be roughly estimated by an analysis of the density histograms. While above the critical temperature a density distribution around a single equilibrium value is found, we observe in the subcritical region a bimodal distribution, where the maxima represent the coexistence densities of the liquid and the gas phase. [72, 73].

The critical parameters may be accurately determined by finite–size scaling techniques [74, 75] which were originally developed for Ising-like lattice models using the concept of field mixing [68, 76]. The essence of this method consists of the study (simulation) of a series of systems of different sizes (number of particles), where, from the properties of the series of finite systems with increasing numbers of particles, the properties of the infinite system can be concluded by the use of theoretically founded general scaling relations.

For example, the following relation holds between the critical temperature $T_c^*(L)$ of a system with a finite box length L and the critical temperature $T_c^*(\infty)$ of the infinite system:

$$T_c^*(\infty) - T_c^*(L) \propto L^{-(\theta+1)/\nu}, \tag{91}$$

where θ and ν are system parameters. For simple (two- and three-dimensional) systems, this relation could be proved explicitly [72, 73, 77]. By extrapolation to the infinite system the critical temperature can be calculated with high precision.

The described methods described for the estimation of the coexistence densities are restricted to state conditions near the critical point only, where the barrier of the free energy between the coexisting fluid phases becomes not too large.

A way to deal with subcritical states of high density is provided by the so-called multicanonical simulations [78]. To avoid the problems resulting from the high barrier in the free energy, the usual Boltzmann distribution is replaced by an a priori weighted distribution which is flat enough to permit transitions between the coexisting phases. Explicitly, an effective Hamiltonian is introduced which defines a suitable pseudo-Boltzmann distribution $P'(\rho)$. Choosing the effective Hamiltonian in such a way that $P'(\rho) \approx$ const holds for all ρ, the wanted Boltzmann weighted density distribution $P(\rho)$ can be estimated efficiently by the use of an iterative procedure. One starts with a state point near T_c where the coexistence densities can be determined by a conventional grand canonical simulation (using recent histogram-reweighting methods, if necessary). From this starting point, one gets successively to lower and lower subcritical temperatures, where an increasing improvement of the efficiency against the conventional μVT simulation is reached [79].

With respect to confined fluids advanced sampling methods are of growing importance. Particularly, properties related to the change of dimensionality in thin adsorbed fluid films (the transition from 3d to 2d systems) are studied by finite size–scaling and similar techniques. The critical properties of thin fluid films are also under consideration using the equivalence of lattice gas models with the Ising ferromagnet [80].

3.2 Hierarchical Modeling of Fluid–Interface Systems

General Aspects

A basic problem of the statistical mechanics of fluids lies in the fact that in the case of the intermolecular interactions in real (molecular or associating) fluids, realistic potentials are too complex for simple application in statistical-mechanical theories. In order to construct tractable molecular models of fluids which capture the basic structural and thermodynamic features, we consider general properties of the intermolecular potentials. Starting with simple fluids it is well–known that the dominant part of the intermolecular forces is given by the short-ranged hard-core repulsion between the particles, which determines the molecular structure of simple (Lennard-Jones-like) fluids, whereas the longer-ranged attractive part of the potential may be treated as a perturbation.

Recent studies of different types of realistic intermolecular potentials [56, 81–83], comparing simulations using the full potential with simulations using the same potential functions truncated at some distance, give clear evidence that the structure (and to some extent even the thermodynamics) of more complex (molecular and associating) fluids is determined primarily by the short-ranged parts of the interaction potential, quite similarly to the case of simple fluids.

These findings suggest a systematic way to represent the full intermolecular potential of a real fluid in a perturbed form, using a hierarchy of approximations with respect to the range of the intermolecular forces, starting with a short-ranged reference term which captures the basic structural features of the real fluid under consideration.

This notion may be understood as an extended form of the thermodynamic perturbation theory, characterized by

– splitting of the full pair potential into a short-range part $u_{SR}(1, 2)$ and a long-range part $u_{LR}(1, 2)$,

– mapping of the short-range part u_{SR} onto a tractable short-range reference potential u_{Ref} and solving (approximately) the statistical-mechanical problem of the short-range reference fluid,
– treating the long-range part $u_{LR}(1, 2)$ as a perturbation.

In the case of atomic and weakly polar molecular fluids, this procedure recovers the usual thermodynamic perturbation theory, where hard sphere (hard body) reference potentials are used [30].

More complicated is the local molecular structure in associative (aqueous) fluids, where, besides the hard excluded–volume repulsions, highly localized attractive interactions are present which form (H-bonded) molecular clusters and networks.

Several short-ranged molecular models implement H-bonding by means of strongly orientation–dependent attractive forces of short–range between (off-center) interaction sites of different kinds [84, 85]. Particularly, the so-called primitive models (PM's) of association due to Nezbeda and coworkers [86] are able to describe the basic structural features of water and similar associating fluids on the basis of purely repulsive potentials.

A direct link to realistic intermolecular potentials is provided by the most sophisticated so-called extended primitive models (EPM's), which may be considered as short-ranged versions of the most common type of realistic potentials used in computer simulations, the transferable potential models (OPLS) of Jorgensen [87].

Therefore the EPM models may serve as short-ranged reference potentials comprising explicitly the following contributions:

– hard–sphere (body) repulsion (excluded volume of the molecules);
– two types of interaction sites (called e-sites and H-sites) arranged off-center, e.g. at the surface of the hard core, to describe association;
– short-range strong directional (square-well) attractions between unlike sites (H-bonding);
– short-range directional (hard sphere) repulsions between like sites.

Particularly, for water an entire family of EPM's has been introduced [86, 88, 89], essentially as truncated (short-ranged) approximations of realistic water potentials (ST2, TIP, etc.). By extensive computer simulations [81, 82] it could be shown that the local fluid structure of water (represented by site-site correlation functions) is determined nearly exclusively by the short-range part of the corresponding full realistic point charge potentials.

Summarizing the recent research on short-ranged molecular models of fluids (particularly on primitive models of association) based on hierarchical potential approximations these models have to be considered as a sound physical basis for a statistical-mechanical theory of several classes of real bulk fluids reaching from simple atomic to associating fluids.

Particularly, the basically correct description of the fluid structure of homogeneous aqueous phases by means of the PM's has given us enough reasons to apply this approach to fluid phases in inhomogeneous (interfacial) situations. In the next paragraph we discuss the extension of short-ranged hierarchical potential models to molecularly confined fluids in some detail.

Short-Ranged and Primitive Models of Confined Fluids

Before we discuss the different kinds of intermolecular forces present in confined fluids, we have to deal with the modeling of basic types of molecular confinements itself. In order to understand on a molecular level the structure and thermodynamics of fluid phases under restricting geometrical conditions, idealized models of interfaces and micropores are of special importance. Such models are on the one hand simple enough to permit a statistical-mechanical treatment and on the other hand able to model the typical properties of a confinement i.e. the geometrically determined features in the behavior of the confined fluid.

The most common models used to study the behavior of confined fluids are slit-like and cylindrical micropores, where the width of the slit or the diameter of the pore, respectively, ranges from a few to the order of a hundred nanometers.

Most of the theoretical studies discussed in this paper deal with fluids enclosed in slit-like pores (or confined to planar interface layers). Therefore we present these models in some detail.

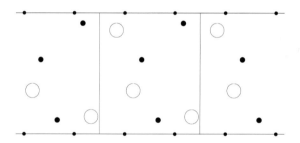

Fig. 2. Fluid in a slit-like pore with periodic continuation

Figure 2 shows a scheme of a fluid confined to a slit-like pore. A Cartesian coordinate system is introduced, where a slit pore is modeled by two parallel interfaces (planar walls) in the (x, y) plane, which have a distance L in the z-direction. In this infinitely extended slit, a fluid is enclosed whose intermolecular forces are described by a potential U_{ff}. The interactions between the walls and the confined fluid particles are given by a potential function U_{fw}, and possible forces between the atoms at the surface (interface) are represented by the function U_{ww}. The total potential energy of the fluid–pore system is given by the sum

$$U_{\mathrm{p}} = U_{\mathrm{ff}} + U_{\mathrm{fw}} + U_{\mathrm{ww}}. \tag{92}$$

If we restrict our consideration of the intermolecular forces to site–site models with N interaction sites (atoms) per fluid molecule – as is done in most simulation studies of confined fluids – the potential energy U_{ff} can be written in general as

$$U_{\text{ff}} = \sum_{i<j} \sum_{\alpha i,\beta j} u(r_{\alpha i,\beta j}), \tag{93}$$

where $u(r_{\alpha i,\beta j})$ is the site–site potential between site α of particle i and site β of particle j, which depends only on the (scalar) distance of the sites involved.

The interaction of a fluid particle with the wall u_{fw} in the simplest case depends only on the z-coordinates of the centers of the fluid particles inside the pore, i.e. $u_{\text{fw}} \to u_{\text{w}}(z)$. That means we consider smooth walls, and in the case of (hard-core) molecules of diameter σ confined to slits with hard walls $u_{\text{w}}(z)$ takes the form

$$u_{\text{w}}(z) = \begin{cases} \infty & \text{for} \quad |z| > (L - \sigma)/2 \\ |u_{\text{w}}^{\text{in}}(z)| < \infty & \text{for} \quad |z| < (L - \sigma)/2, \end{cases} \tag{94}$$

where the walls of the pore are assumed to be at $z = \pm L/2$ and. $u_{\text{w}}^{\text{in}}(z)$ is the fluid–wall potential in the interior of the slit.

Molecularly rough interfaces can be modeled by discrete molecules additionally distributed over the smooth walls which interact via a pair potential u_{sf} with the fluid molecules. The total fluid–wall interaction is then given by

$$U_{\text{fw}} = \sum_{i=1}^{N} u_{w}(z_i) + \sum_{i=1}^{N} \sum_{j=1}^{K} u_{\text{sf}}(r_{i,j}). \tag{95}$$

In (95) i denotes one of the N fluid molecules and j one of the K interface molecules.

If the interface molecules are assumed to be movable along the surfaces of the pore – where they interact pairwise via a pair potential u_{ww} – the influence of the nature and the amount of the enclosed fluid on the organization of the structure of the interface can be investigated.

In order to illustrate the modeling of the molecular roughness of interfaces (micropores) as they are observed at so-called "fluid" interfaces [90] (for example at the amphiphilic interfaces of phospholipid bilayers), we consider the following model with freely movable surface molecules.

A given number of interaction sites (spheres) may be placed at the walls of a slit-like pore, which are assumed to be freely movable along the walls. This results in a distribution of surface molecules which permanently changes depending on both the interaction between the interface molecules themselves and the interaction with the molecules of the confined fluid. The fixing of the wall molecules near the interfaces is reached by an attractive force in the wall region (of width $\Delta L + \sigma$) which is modeled in the simplest case by a potential well of suitable depth and width. Variants of the model consider diverse types of pair potentials u_{fw} and u_{ww}, which may be hard-core-like or of Lennard–Jones type. If associating forces are involved, the modeling of "hydrophobic" and "hydrophilic" interfaces is possible in a simple way, exploiting the notion of hierarchical potentials.

Figure 3 represents the basic model of molecularly rough interfaces discussed. The mobility of the wall molecules permits the molecular modeling of the structural self-organization of the interface during its interplay with the confined fluid, which

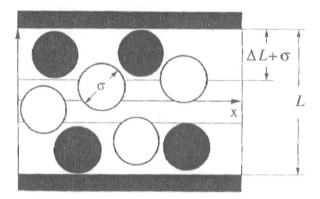

Fig. 3. Basic model of molecularly rough interface

is crucial for the modeling of flexible ("fluid") interfaces appearing in biologically relevant interface structures, e.g. biomembranes.

Penetrable interfaces are obtained if the (hard) wall is replaced by a finite potential barrier, i.e. if in (95) for all z we set $U_w(z) \leq C$ with $C < \infty$, and for the discrete surface-forming particles a suitable pair potential is applied [91].

As a simulation cell for a fluid in a slit-like pore of wide L, a prism is used with a square base and an edge length l_{xy} in the (x, y)–plane and a height of L in the z-direction. This simulation box, which is filled with the fluid under consideration is periodically repeated in two dimensions (in the x- and y-directions), whereby an infinitely extended system is emulated.

Owing to the periodic continuation for fluids confined in micropores of simple geometry (e.g. slit-like, cylindrical, or cubic pores), the transition to infinite extended systems is possible. Therefore a thermodynamic limit ($N \rightarrow \infty$) can be defined which ensures the validity of the laws of statistical mechanics.

In contrast to the above systems, fluids in closed cavities (e.g. in spherical micropores) have to be considered as real finite systems, where the relations of the thermodynamics of small systems have to be applied [92].

In general, MC simulations of fluids confined to the molecular interface structures discussed above, allow the study of equilibrium properties of the fluids (density distribution, order of the molecules, pressure at the wall, etc.) in terms of the structure (roughness) of the interface, including feed–back effects from the fluid properties to the interface structure. Some illustrative examples of simulation studies of confined fluids in molecularly rough slit-like pores will be discussed in the next section.

Concerning the representation of microporous media we have to distinguish between systems with a regular lattice-like arrangement of cavities and channels, such as zeolites and materials with a random distribution of micropores. While in the first case the structure can be represented by regular geometrical models, in the second case the modelling of these random media is a serious problem.

Since extensions of lattice-like models which describe random pore arrangements by disturbed regular matrix models are not able to reproduce adequately the disordered nature of these materials, a computer-based stochastic geometrical modeling of fluids enclosed in microporous media has been developed (Madden and Glandt [93]). The approach considers a binary mixture where one component represent the constituents (hard particles) forming the microporous matrix and the other component consists of the fluid molecules. By simulation of this mixture, random arrangements of the constituent fluid systems are generated. By stopping the simulation and keeping the constituents fixed, the properties of the fluid confined in these random arrangements of constituents representing a disordered porous medium can be studied.

In order to study (simulate) the basic types of confined fluids, we have to specify explicitly the intermolecular potentials involved (cf. (92)). We start with the fluid–fluid potential $u_{\mathrm{ff}}(r)$, where we exploit the recent knowledge on a hierarchical modeling of intermolecular forces and deal with basic short-range molecular models.

To represent simple fluids, we use hard-sphere and square-well pair potentials, which in general are given by

$$u_{\mathrm{ff}}(r) = \begin{cases} \infty & \text{for} \quad r \leq \sigma \\ -\epsilon & \text{for} \quad \sigma < r \leq (\lambda_{\mathrm{sw}} - 1)\sigma \\ 0 & \text{for} \quad r > \sigma, \end{cases} \tag{96}$$

where σ is the diameter of the particles, ϵ is the depth, and $(\lambda_{\mathrm{sw}} - 1)\sigma$ is the width of the (attractive) well. The special case of hard spheres is recovered by setting $\lambda_{\mathrm{sw}} = 1$.

We describe water-like fluids by means of the above-mentioned short-ranged Primitive Models using both the original PM of water [94] and the extended 5-site version EPM5 [86, 88]. The original PM represents the water molecules by a hard sphere (diameter σ_{w}) with four tetrahedrally arranged interaction sites (2 H- and 2 e-sites) on the surface of the sphere. Hydrogen bonding is modeled by short-ranged square-well attractions between e- and H-sites.

The pair potential of the original PM of water is given by

$$u_{\mathrm{PM}}(1,2) = u_{\mathrm{hs}}(1,2) + \sum_{i,j}^{1,2} [u_{\mathrm{sw}}(r_{\mathrm{H}i,ej}^{1,2}) + u_{\mathrm{sw}}(r_{ei,\mathrm{H}j}^{1,2})]. \tag{97}$$

The hard-sphere potential u_{hs} describes the excluded volume of the water molecules, and the coreless square-well potential

$$u_{\mathrm{sw}}(r) = \begin{cases} -\epsilon & \text{for} \quad \sigma_{\mathrm{w}} < r \leq \lambda\sigma_{\mathrm{w}} \\ 0 & \text{for} \quad r > \sigma_{\mathrm{w}} \end{cases} \tag{98}$$

describes hydrogen bonding.

The EPM5 model is obtained if, additionally, hard–sphere repulsions u_{hsl} (diameter σ_{r}) between like sites are introduced [88].

The pair potential takes now the form

$$u_{\mathrm{EPM}}(1,2) = u_{\mathrm{PM}}(1,2) + \sum_{i,j}^{1,2} [u_{\mathrm{hsl}}^{(e)}(r_{ei,ej}^{1,2}) + u_{\mathrm{hsl}}^{(\mathrm{H})}(r_{\mathrm{H}i,\mathrm{H}j}^{1,2})]. \tag{99}$$

By switching off the associating forces ($\epsilon = 0$), a fluid of so-called pseudo-hard bodies [95] is obtained. The model parameters used in our studies are $\lambda = 0.4\sigma_w$, and $\sigma_r = 0.8\sigma_w$, and $\sigma_w = 1$.

The range of the influence of interfaces on the structure of bulk fluids (in particular, inhomogeneities in the number densities) depends essentially on the specific interactions of the class of fluids under consideration. For dense fluids, it typically amounts to several times the average distance between the centers of the particles.

To simulate such interface or surface effects on a molecular level, one usually considers fluids enclosed in basic cells of simple geometry with dimensions in the order of the expected range of these effects [96].

In the simple case of slit-like micropores with smooth hard walls the interfaces interact only with the hard cores of the enclosed molecules via a potential $u_w(z)$, given in (94), where, according to the hierarchical approximation, a short-ranged potential $u_w^{in}(z)$ is applied to the interior of the slit. The properties of several classes of fluids confined to hard slits are discussed in Sect. 3.3.

Slit-like pores with a slit width L may be considered as simple highly idealized models of membrane pores. Real (bio)membrane pores do not show such an ideal geometry. For instance, the hydrophilic headgroups of the membrane-forming amphiphiles form fluid and molecularly rough surfaces which are self-organized [97–99]. Therefore, following the concepts of [100], we have introduced two new types of slit-like pores, describing the molecular roughness of membrane pores by layers of discrete mobile "surface particles" located at the walls according to the basic model of molecularly rough interfaces shown in Fig. 3.

First, we construct a hydrophobic pore model (model A), filling hard spheres in a slit-like pore with hard walls. As in [100], these surface spheres are allowed to move with some restrictions between the two hard walls to model a system able to self-organize. If an aqueous fluid is confined to this pore it behaves hydrophobically, because there are no specific association forces between the confined fluid molecules and the surface spheres and the (inert) surface spheres hinder the water from forming extended molecular clusters and network structures (the hydrophobic effect [101]).

Similarly, as in real membrane pores [102], two types of interactions are present:

1. The pair interaction between the surface spheres:

$$u_{ss}^{(A)}(r_{ss}) = \begin{cases} \infty & \text{for} \quad r_{ss} < \sigma_s \\ 0 & \text{for} \quad r_{ss} \geq \sigma_s \end{cases}, \tag{100}$$

where r_{ss} denotes the distance between a pair of surface sphere centers and σ_s is the hard-core diameter of the spheres.

2. The surface sphere – wall interaction:

$$u_{sw}(z) = \begin{cases} 0 & \text{for} \quad |z| \leq ((L-\sigma)/2 - \zeta_w) \\ -\varepsilon_w & \text{for} \quad |z| > ((L-\sigma)2 - \zeta_w) \\ \infty & \text{for} \quad |z| > (L-\sigma_s)/2 \end{cases}, \tag{101}$$

where z describes the z-component of the position vector $\mathbf{r}$ of the center of the particle, and the walls are extended in the (xy)–plane at $z = \pm L/2$. To model realistically the distribution of the surface spheres, we introduce an additional (square-well) attraction between the surface spheres and the hard walls. The parameters ζ_w and ε_w describe the range and the strength of this potential, respectively.

To ensure that the pore roughness has a molecular length scale of about one fluid diameter σ – as suggested by experiments on real membrane pores [103, 104] – we have set $\sigma_s = \sigma$ in all our studies.

When the pore is filled with "primitive" water, the interaction of the water molecules with the walls $u_{fw}(z)$ is assumed to be hard-sphere-like. Additionally a hard-sphere repulsion between the surface molecules and the water particles is present,

$$u_{fs}^{(A)}(r_{fs}) = \begin{cases} \infty & \text{for} \quad r_{fs} < (\sigma + \sigma_s)/2 \\ 0 & \text{for} \quad r_{fs} \geq (\sigma + \sigma_s)/2, \end{cases} \tag{102}$$

where r_{fs} denotes the distance between the centers of a fluid and a surface sphere. This excluded-volume interaction is responsible for the surface reorganization in the presence of a fluid which is studied below.

Our second model (model B) represents a "hydrophilic" pore. Consequently, the surfaces spheres of model A were spiked with local interaction sites, comparable to the modeling of the association sites in PM–water. But to mimic a "hydrophilic" behavior of the surface spheres (to define specific molecular axis) they were spiked with only three sites, two e- and one H-sites, as in case of "primitive methanol" [94]. Therefore, the surface sphere – surface sphere interactions of this model are described by those of model A plus the sum over all terms of the pair potential

$$u_{ss}^{(B)}(r_{ss}) = \begin{cases} -\varepsilon_s & \text{for} \quad r_{ss} < \zeta_s \\ 0 & \text{for} \quad r_{ss} \geq \zeta_s, \end{cases} \quad \text{with} \quad r_{ss} = | \mathbf{r}_{(H)i,n} - \mathbf{r}_{(e)k,m} |, \tag{103}$$

where the vectors $\mathbf{r}_{(H)i,n}, \mathbf{r}_{(e)k,m}$ represent the position vectors of the n–th H-site of the i–th and the m–th e-site of the k–th surface molecule, respectively. ζ_s denotes the interaction range and ε_s the strength of the square-well potential localized at the sphere surface.

The surface sphere–wall interaction is assumed to be the same as for model A given by (101).

The fluid–surface sphere interactions of this model are completely described by those of model A plus the sum over all terms of the pair potential

$$u_{fs}^{(B)}(r_{fs}) = \begin{cases} -\varepsilon_{fs} & \text{for} \quad r_{fs} < \zeta_{fs} \\ 0 & \text{for} \quad r_{fs} \geq \zeta_{fs} \end{cases} \quad \text{with} \quad r_{fs} = | \mathbf{r}_{(H)i,n} - \mathbf{r}_{(e)k,m} |, \tag{104}$$

where the position vectors $\mathbf{r}_{(H)i,n}$ and $\mathbf{r}_{(e)k,m}$ belong always to pairs consisting of one surface and one fluid molecule. ζ_{fs} and ε_{fs} denote the range and strength of the potential.

The parameters ζ_i and ε_i of all localized square-well potentials ($i = s$, f and sf) are set to be equal: $\zeta_i = 0.1\sigma$ and $\varepsilon_i/k_BT = 8$. Therefore, the strength of all modeled hydrogen bonds of the system (fluid–fluid, surface–surface and fluid–surface) are assumed to be equal.

3.3 Recent Results

Cavity Distribution Functions

Cavity distribution functions are basic structural quantities for both bulk and confined fluid systems, which may be measured relatively easily by computer simulation and which provide an efficient route to calculating the corresponding excess chemical potentials and background correlation functions.

The cavity concept was originally introduced by Reiss [105] and Speedy [106] and is sometimes called the geometric approach to fluid theory. We consider a hard–sphere (HS) fluid in a hard (slit) pore [107].

An HS cavity is an HS not interacting with the walls but with all other HS's. Outside a wall it is equivalent to an HS particle. The simplest cavity distribution function is the singlet function

$$n(\mathbf{1}) \equiv n(z) = \exp[-\beta\mu^e(z)], \tag{105}$$

where $n(z)$ is the probability to place a cavity at a distance z from a wall. $\mu^e(z)$ is the corresponding excess chemical potential.

An m-cavity is formed by a fixed set of $m > 1$ (overlapping) HS cavities, which may be denoted as a fused hard sphere (FHS) cavity. The m-cavity distribution function $n(1, 2, \dots, m)$ is the insertion probability of the FHS cavity, related to its excess chemical potential at infinite dilution, $\beta\mu^e(\mathbf{1}, \mathbf{2}, \dots, \mathbf{m})$, via

$$n(\mathbf{1}, \mathbf{2}, \dots, \mathbf{m}) = \exp[-\beta\mu^e(\mathbf{1}, \mathbf{2}, \dots, \mathbf{m})]. \tag{106}$$

For $m = 1$, $\beta\mu^e(z)$ may be calculated by Widom's test-particle insertion method. More efficient is a recently introduced scaled-particle MC method due to Labik and Smith (cf. Sect. 3.1), exploiting test insertions of spheres of smaller sizes. For the more inefficient case $m > 1$ we use work-related ideas similar to scaled-particle MC to calculate $\beta\mu^e(\mathbf{1}, \dots, \mathbf{m})$ (Smith and Vörtler [107]). The basic relationship exploited is

$$\beta\mu^e(\mathbf{1}, \dots, \mathbf{m} + \mathbf{1}) = \beta\mu^e(\mathbf{1}, \dots, \mathbf{m}) + \beta w(\mathbf{m}, \mathbf{m} + \mathbf{1}), \tag{107}$$

where $w(\mathbf{m}, \mathbf{m} + \mathbf{1})$ is the work required to "grow" an m-particle cavity into an $(m + 1)$-particle cavity, related to the conditional cavity distribution function $n^*(\mathbf{1}, \dots, \mathbf{m} + \mathbf{1})$, via

$$n^*(\mathbf{1}, \dots, \mathbf{m} + \mathbf{1}) \equiv \frac{n(\mathbf{1}, \dots, \mathbf{m} + \mathbf{1})}{n(\mathbf{1}, \dots, \mathbf{m})} = \exp[-\beta w(\mathbf{m}, \mathbf{m} + \mathbf{1})]. \tag{108}$$

Here $n^*(1, \dots, m+1)$ is the conditional probability to add a sphere to an m-cavity to form an $(m+1)$-cavity. n^* is easily measured by simulation. We focus in this study on pair cavity functions.

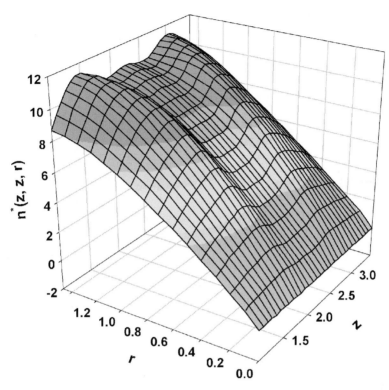

Fig. 4. "In-plane" cavity pair distribution function $n^*(z, z, r)$ (slit width $L = 3.5\sigma$; confined fluid in equilibrium with a bulk fluid of density $\rho_b\sigma^3 \approx 0.8$)

We have introduced novel computer simulation techniques to calculate efficiently cavity distribution functions of HS's in a slit with hard walls. We generated a chain of configurations by canonical Metropolis MC and measured the wanted structural quantities by means of virtual particle/cavity insertions, which do not affect the course of the simulations.

To measure singlet cavity functions $n(z)$ virtual insertion of cavities was performed at random positions in the slit.

To estimate cavity pair functions $n(1, 2)$, the corresponding conditional cavity distribution function

$$n^*(1, 2) \equiv n^*(z_1, z_2, r)$$

was measured by virtual insertion of a cavity at z_2 at a distance r from the center of a given HS center at z_1, where z_i are the distances of the sphere/cavity centers from a (specified) slit wall. More technical details are given in our original paper [107].

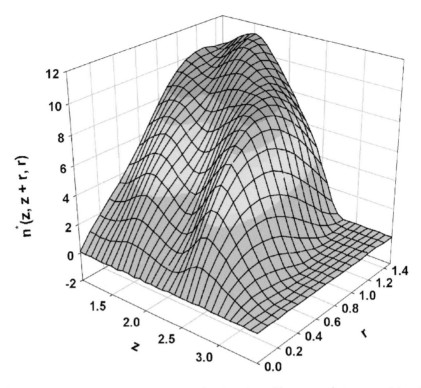

Fig. 5. "Perpendicular" cavity pair distribution function $n^*(z, z + r, r)$ (same model as in Fig. 4)

Since cavity pair functions are complicated functions of three variables, we restricted our calculation to the following important special cases:

- The conditional probability $n^*(z, z, r)$ of inserting an HS cavity at a distance r from a given HS located in a plane at z parallel to the walls. We call this the "in-plane" conditional cavity pair distribution function; the dimensionless excess chemical potential of the corresponding 2-particle cavity is $\beta\mu^e(z, z, r)$.
- The conditional probability $n^*(z, z + r, r)$ of inserting an HS cavity at distance r from an HS located at z, where the line of sphere centers is perpendicular to the walls. We call this the "perpendicular" conditional cavity pair distribution function; the dimensionless excess chemical potential of the corresponding 2-particle cavity is $\beta\mu^e(z, z + r, r)$.
- The conditional probability $n^*(z, z + \cos(\theta), 1)$ of inserting an HS cavity in contact with a sphere located at z, where the line of sphere centers forms an angle θ with the line perpendicular to the wall through the sphere at z. We call this the "rolling-sphere" conditional cavity pair distribution function; the dimensionless excess chemical potential of the corresponding 2-particle cavity is $\beta\mu^e(z, z + \cos(\theta), r)$.

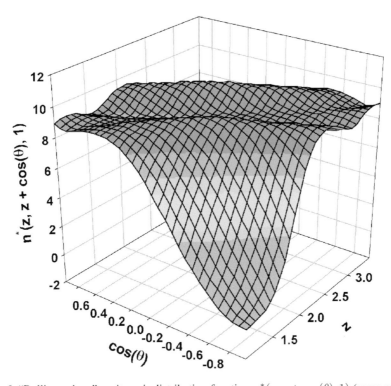

Fig. 6. "Rolling-sphere" cavity pair distribution function $n^*(z, z + \cos(\theta), 1)$ (same model as in Fig. 4)

Typical results for these functions for HS's confined to a hard slit are given in Figs. 4, and 5, and 6, respectively. The results are presented and discussed in our original paper in detail [107]. To the best of our knowledge this is the first calculation of such functions, which are of interest as reference data for the development of statistical-mechanical theories and provide basic information for a modeling of the solubility of polyatomic molecules in simple molecularly confined solvents. We have checked the validity of Kirkwood's superposition approximation for the cavity functions obtained using the BGY integral equation hierarchy. The one- and two-body distribution functions are related by the first member of the BGY integral equation hierarchy. In case of HS's in a hard slit for a cavity moving outwards from the wall, it is

$$\frac{d \ln[n(z)]}{dz} = 2\pi \exp(\beta \mu_t) \int_{\cos(\theta_{\min})}^{\cos(\theta_{\max})} n^*(z, z + x, 1)\, x\, dx, \qquad (109)$$

where $n^*(z, z + x, 1)$ is the "rolling-sphere" conditional cavity pair function. (109) is a starting point for approximate closures for n^*.

Figure 7 shows the derivative of the logarithm of the single cavity function vs. z. The solid curve represents the left side of (109), calculated by numerical differ-

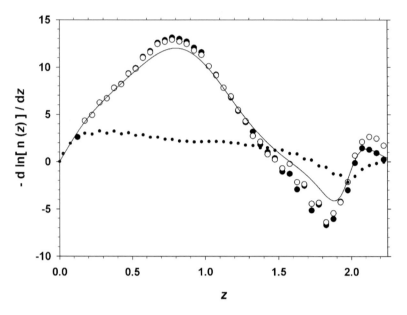

Fig. 7. Derivative of the logarithm of the single cavity function

entiation of the corresponding simulation data. The points represent the right side of (109), calculated by numerical integration of the simulation data; the open and filled circles are data from opposite sides of the slit. The dashed curve is Kirkwood's superposition approximation for the right side of (109).

The approximation is found to be exact in certain limiting geometrical situations, but in general it is quantitatively poor.

Since hard spheres in a slit represent a basic reference model for confined fluids, the cavity distribution functions studied provide new basic information for a statistical-mechanical study of correlation functions, chemical potentials, solubilities, and related properties.

Aqueous Phases at Planar Interfaces and in Thin Films

This section deals with the simulation of the structural properties of associating (water-like) fluids in narrow molecularly rough planar slits and in (monomolecular) planar films. The molecular modeling of the rough interfaces starts from the basic model introduced in Sect. 3.2, and the water-like fluids are represented by primitive models of association (primitive water) which were also introduced in Sect. 3.2. While in the case of aqueous systems at molecularly rough interfaces the original primitive water model of Kolafa and Nezbeda was used, the more recent studies of thin water films are based on the recently developed extended primitive models.

The motivation of the selection of results is to present basic information on the structure of hydrated fluids in terms of molecular models which describe the relevant physical features in a simple uniform way by short-ranged (hierarchical) inter-molecular potentials for the fluid–fluid, fluid–surface, and surface–surface molecular interactions. In order to meet fundamental properties of experimental hydration phenomena two variants of molecularly rough slit pores – introduced in Sect. 3.2 – have been studied:

- model A: a "hydrophobic" model, where the interactions between the interface molecules themselves and between the fluid and the interface particles are hard-core forces only; and
- model B: a "hydrophilic" model, which permits – besides the excluded-volume interactions mentioned – associating forces (hydrogen bonds) between both the surface particles themselves and between fluid and interface particles.

Figure 8 shows the one-particle densities $\rho_i(z)$ for slit widths of $L = 2.31\sigma$, 3.02σ, and 3.82σ for both models. The index i denotes surface molecules (full lines) and fluid particles (dotted lines), where the right plots always mean "hydrophobic" and the left ones mean "hydrophilic" pore models.

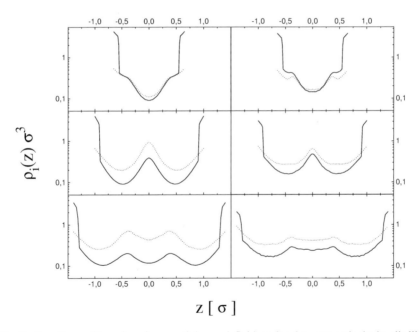

Fig. 8. Density profiles of surface particles and fluid molecules, respectively in slit-like pores of different widths (surface particles, full lines; fluid molecules, dotted lines; left plots,"hydrophobic" pores; right plots, "hydrophilic" pores)

The figure shows significant differences between the density profiles of both models. While the fluids in the "hydrophobic" pores have properties of $\rho(z)$ which are comparable with those of hard-sphere fluids, in "hydrophilic" pores the hard-core-like oscillations are distinctly weakened, which is caused by the presence of additional associating surface forces in this model.

As mentioned in Sect. 3.2, models with movable surface particles are able to organize the surface roughness by themselves in its interplay with the confined fluid. Despite the differences between the two models mentioned above, two basic effects of self-organization are observed.

First, a reorganization of the surfaces with increasing width of the slit is observed, reflecting the decrease of the correlation between the walls, and second, in the centers of the wider slits a nearly homogeneous mixture of surface and fluid particles is observed. Both effects of self-organization are observed experimentally in real molecular interface layers (e.g. biomembranes [103, 104, 108]).

A deeper insight into the molecular structure of the fluid–pore system permits the study of the distribution of the sizes of molecular clusters $w(N)$ connected by hydrogen bonds. For both "hydrophobic" and "hydrophilic" clusters some results are presented in Fig. 9. Owing to the construction of the pore model for "hydrophobic" pores there can exist only clusters consisting of fluid molecules, i. e. water clusters (filled circles). In the case of "hydrophilic" pores there are several types of clusters present: besides pure water clusters (circles), there are also clusters consisting of surface molecules only (squares) and, finally, mixed fluid-surface clusters.

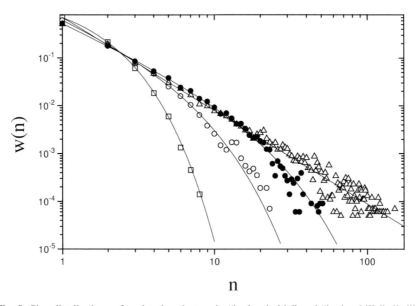

Fig. 9. Size distributions of molecular clusters in "hydrophobic" and "hydrophilic" slit-like pores (explanations of the symbols in the text)

The analysis of this picture shows that for both types of pores, large clusters (networks) exist which span the entire pore. In the case of "hydrophilic" pores, however, the big clusters are composed of both surface and fluid particles, while the pure water clusters are even smaller, as in the "hydrophobic" system.

Summarizing the simulation results, the molecular structure of the fluid–pore system can be classified in terms of the slit width as follows:

- In very narrow slits, the structure of both pore surfaces is highly correlated. The confined fluid is not able to form its own structure, but it appears in the form of single molecules only, which affect with increasing concentration the formation of the interface structures.
- When the width of the layer reaches a value of about $L = 3\sigma$, then in the center of the slit a bulk-like water structure can be formed, which shows – similarly to smooth pores [109] but strongly affected by the surface molecules – a transition from small molecular clusters to network structure.
- In the case of wider slits, the formation of separate wall layers separated by an extended bulk fluid in the inner part of the pore is observed. The wall region and the central part are nearly decoupled and may be considered as separate phases to some extent. Under some thermodynamic conditions there may appear solid-like layers at the walls which are in equilibrium with a fluid state in the center of the pore [96]. Analogous situations are found experimentally in water–lipid systems.

Another important class of confined (geometrically restricted) fluid systems is fluid monolayers. We report recent unpublished simulation results of structural properties of water-like fluids at the level of primitive models in the important limiting case of two-dimensional monolayers. We use in our study recently introduced extended primitive models (EPM5) which include additional hard-sphere repulsion between like e- and H-sites.

Conventional Metropolis Monte Carlo sampling [110, 111] in the canonical ensemble has been used in order to estimate the site–site pair distribution functions and other structural quantities, such as coordination numbers and the average saturation of hydrogen bonds of both three- and two-dimensional fluids.

While in the case of three dimensions the usual cubic simulation cell $V_{3\mathrm{d}} = L^3$ with periodic boundary conditions is used, the two-dimensional fluid is modeled as a system of real three-dimensional particles (diameter σ), where the centers of the molecules are located in the (xy)–plane and the motion of the centers of the particles is assumed to be restricted to this plane. Therefore the simulation cell used is a square $A_{2\mathrm{d}} = L^2$ in that plane and periodic boundary conditions are applied in the x- and the y-direction.

In three dimensions the site–site distribution functions $g_{s_1,s_2}(r)$ were calculated from the simulations as usual, by generating independent configurations and collecting histograms of the numbers of all pairs of respective sites s_1 and s_2. In bin i of the histogram, the number of pairs $N_{s_1,s_2}(r_i)$ which lie in a distance range $(r_i \dots r_i + \Delta r)$ was stored. From this pair number the number density of the sites in a spherical shell of radius r_i (thickness Δr) is estimated and divided by the mean number density in the system. This density ratio represents, by definition, the site–site distribution function $g_{s_1,s_2}(r_i)$.

To estimate adequate site–site distribution functions in the case of two-dimensional arrangements of EPM water molecules, where the centers of the particles may lie in the (x,y) plane, we have to distinguish between the center–center functions $g_{00}(r)$ and functions which involve off-center (e, H) sites. In the first case we have a regular two-dimensional distribution function which may be calculated in the standard way, quite analogously to three dimensions. If off-center sites are concerned, we have to take into account that these are real three-dimensional functions defined within the layer of thickness σ and that there is a density profile in the z-direction $w_x(z)$ (where $x = H, e$) for the distribution of the off-center sites. In case of the EPM5 model the density profiles of both e- and H-sites are equal. Results for these density profiles are shown in Fig. 10.

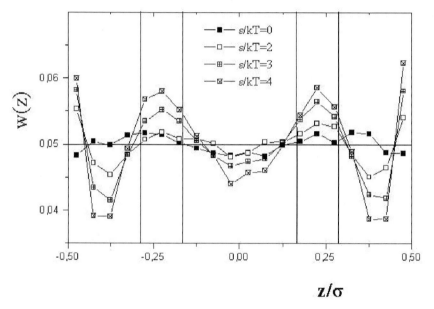

Fig. 10. Density profiles of the off-center e- and H-sites in two-dimensional EPM5 water layers

These features suggest that we should sample the off-center functions in a special way: we start with particle 1 and site s_1, which may have a position z_1 and collect the distances to the sites s_2 of all other particles in a histogram, and calculate the number densities of the sites with respect to those volumes of the spherical shells around s_1 which are accessible for the respective sites, i.e. localized within the layer. These volumes depend, for small distances r, on the position of the specified site z_1. Now we take site s_2 of particle 1 with position z_2 and measure the number densities of sites s_1 around the specified site s_2 in the same way as above.

We have to repeat this procedure for all particles. To get the desired (averaged) site–site distributions, we have to take an arithmetic average over all of the single results and to divide this mean value by the number of configurations analyzed.

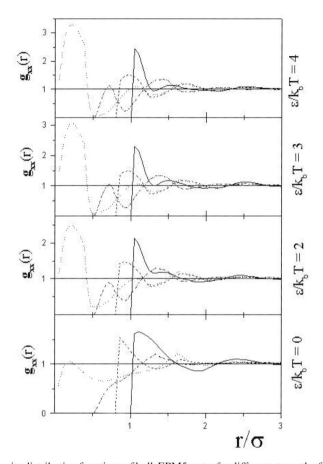

Fig. 11. Site–site distribution functions of bulk EPM5 water for different strength of association

Figure 11 shows our simulation results for the site–site distribution functions of EPM5 water in three dimensions for different ratios $\epsilon/kT = 2, 3, 4$. Taking into account that EPM5 water is known to reproduce the structural properties of real water at room temperature best at ϵ/kT =5–6 [88], we study in this paper a high temperature range. But even in this temperature range, the systems are known to exhibit strong association. As a consequence it was found that most of the water molecules of these systems belong to large clusters spanning the entire system.

For comparison purposes, we included in our study the high-temperature limit of EPM5, i.e. the model without hydrogen bonding ($\epsilon/kT = 0$). The results for this pseudo-hard-body (PHB) fluid [95] are shown in Fig. 11 as well.

For pseudo-hard bodies, as well as for EPM5 water with $\epsilon/kT = 3$ the results are found to be in full agreement with those of the original papers.

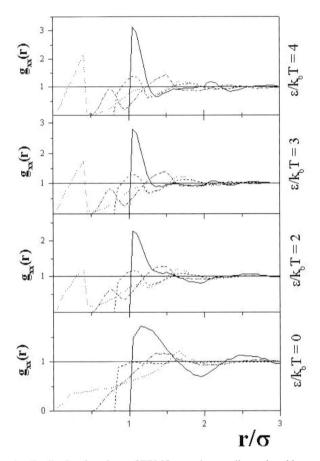

Fig. 12. Site–site distribution functions of EPM5 water in two-dimensional layers for different strength of association

Figure 12 presents the results for site–site distribution functions for two-dimensional layers of EPM5 molecules for values of $\epsilon/kT = 0, 2, 3, 4$, which correspond directly to the above-discussed three-dimensional site–site distribution functions. It is seen that even the PHB fluid ($\epsilon/kT = 0$) shows differences in most of the site–site distribution functions in comparison with three dimensions. While in this case g_{00} is comparable, g_{HH} and g_{eH} differ significantly. Both exhibit no association peaks. This behavior indicates a decreased contribution of the repulsive interaction of the embedded sites in two dimensions, which is responsible for these peaks in three dimensions. This is responsible for a preferred molecular orientation within the two-dimensional layer, which is in good agreement with the observed density profile for the off-center sites, showing a maximum around $\pm\sigma/(2 * \sqrt{3})$ caused by such a preferred orientation.

A direct consequence of the orientation would be a decrease of the effective packing fraction of the system. As shown in [88], a decrease of the effective packing for EPM5 in 3d causes small changes in g_{00}. Particularly, as a result one should find

1. the first minimum at slightly larger distances and

2. the first minimum to be more pronounced.

Both effects are found in our simulations comparing the results for two dimensions and three dimensions at the same number density. This supports the assumption of a decreased repulsion, i.e. an inefficient repulsive interaction, and this results in a molecular orientation in two-dimensional layers.

In case of the full (associating) EPM5 in two dimensions the situation is similar. Qualitatively, g_{00} shows an analogous association behavior to that in 3d, if the ratio ϵ/kT is increased. The system is found to exhibit a first minimum of g_{00} at 1.3 – 1.4 σ, and this minimum gets more pronounced with decreasing temperature. But this minimum is located (for corresponding ratios ϵ/kT)) at slightly larger distances than in three dimensions and is found to be deeper than there. Hence, as for the PHB system, a decreased effective packing fraction of the molecules can be concluded. Furthermore, for the finite temperatures applied, one finds further changes of the distribution functions compared with three dimensions:

1. The second peak of g_{00} at 1.5 – 1.7 σ – typical of bulk water – is seen for intermediate values of $\epsilon/kT = 2 - 3$ only.

2. For values of ϵ/kT above 3 the second peak of g_{OO} vanishes under a rising peak at 2.1 – 2.2 σ, which is not present in three dimensions. The second shell is thus finished after this new peak at about 2.3 – 2.4 σ. The distance found for the second shell corresponds to that of the second nearest neighbors in a hexagonal lattice, which is $\sqrt{10}/2$ times the bond length. Correspondingly, the coordination number of the molecules N_c in 2d is found to decrease from about 6 in the case of PHB to 3 for associating molecules, in agreement with a hexagonal arrangement. We have found this kind of molecular arrangement recently for the original PM water in narrow slit-like pores, too. But the arrangement of EPM5 in two dimensions is not equal to the graphite-like structure obtained for PM water in wider slit-like pores. In contrast to that structure with relaxed bonds, it has "stretched bonds", which try to expand in three dimensions.

Summarizing our structural results we conclude that as a counterpart of the basically tetrahedral structure of EPM5 water in three dimensions, a comparable hexagonal arrangement of the molecules in two-dimensional layers was observed. Because of the geometrical restrictions, the repulsion of the off-center sites is significantly decreased in two dimensions, which causes a preferred orientation of the molecules in thin films which does not exist in three dimensions.

The importance of thesimulation study presented here is twofold. On the one hand the results present pseudo-experimental reference data which provide a sound basis for the development of statistical-mechanical theories of associating fluids in two dimensions, and on the other hand the two-dimensional water layers represent an important limiting case for water confined to interfaces that are interesting for both microporous materials and hydrated biointerfaces (model membranes).

In general, the simulation studies demonstrate the ability of relatively simple molecular-based models to represent the basic structural features of complex molecular-fluid interface situations and thin fluid films, providing fundamental pseudo-experimental data for the statistical-mechanical description of the thermo-dynamics and phase behavior of confined fluids in real molecular interface systems which exceed the possibilities of conventional heuristic lattice-like models [112–114].

Thermodynamics and Phase Equilibria

Concerning thermodynamic quantities, we focus on our recent results for chemical potential and phase equilibria. Our studies on the pressure problem in inhomogeneous fluids were reported in several original papers [100, 115, 116] and in a recent review [32], and will be mentioned only briefly. The basis of our treatment is recent methodical developments in MC sampling methods, such as gradual particle insertion, virtual parameter variations and extensions of the Gibbs ensemble, which have already been discussed in the methodical part on MC techniques.

Regarding the pressure definition, it is well–known that in confined fluids a (uniform) mechanical pressure does not exist in general, but has to be replaced by a tensor depending on the position and direction in space [117]. Restricting our discussion to the case of slit-like pores, we have to consider two important problems, the pressure at the wall and the spreading pressure, which is the averaged mechanical pressure parallel to the walls and has the meaning of a thermodynamic pressure in phase equilibria comprising bulk and confined fluids.

The estimation of the pressure at the wall of the slit is straightforward for fluids consisting of hard-core molecules, via the contact theorem, where in the case of smooth interfaces the value of the particle density at the wall (the contact density) is proportional to the pressure at the wall [117, 118], which may be extended to molecularly rough (hard) interfaces, where the pressure has to be calculated by the analysis of the distribution of the (different) local slit widths, averaging over the corresponding contact values. Some of our results of simulations of the pressure for several classes of confined fluids at the walls of slit pores are given in [100].

The most interesting feature of these results is that for hard-core fluids, the value of the pressure oscillates versus the slit width rather than it monotonoically decays as one might expect. Although the exact form of this function depends on the roughness of the interfaces, the general behavior is not changed on going from smooth to molecularly rough walls.

Lennard–Jones fluids in narrow slits show a very similar behavior of the pressure at the walls versus the width of the slit [119]. Even in more complex systems such as alkanes in thin planar slits (films), the force exerted by the fluid molecules on the walls versus the wall distance behaves very similarly [67].

In general, the problem of the behavior of the pressure in terms of the distance of the walls is an actual problem in experimental studies of planar interface structures e.g. biologically active phospholipid bilayers [120].

More important in the context of phase equilibria in confined systems is the thermodynamic pressure, since the equality of the thermodynamic pressure in coexisting confined fluid phases is one of the conditions of phase equilibrium (e.g. of capillary condensation). In the simple case of a slit-like pore, the thermodynamic pressure reduces to the spreading pressure, which is the averaged mechanical pressure parallel to the walls. The direct calculation (simulation) of this quantity was done for the first time by Vörtler and Smith using the virtual parameter variation method (virtual volume change) introduced in Sect. 3.1. The virtual volume variation method calculates the excess of the spreading pressure of the slit fluid Π^{ex} by means of

$$\beta\Pi^{\text{ex}} \equiv -\left(\frac{\partial\beta A^{\text{ex}}}{\partial V}\right) = \lim_{\Delta V \to 0} \frac{\ln\langle\exp(-\beta\Delta\mathcal{U})\rangle_0}{\Delta V} \tag{110}$$

in agreement with the general expression (83) given in Sect. 3.1. An indirect route to the spreading pressure is the thermodynamic (Gibbs–Duhem) integration of the chemical potential μ with respect to the density ρ,

$$\beta\Pi = \beta\Pi^{\text{id}} + \beta\Pi^{\text{ex}}$$
$$= \rho + \rho\beta\mu^{\text{ex}} - \int_0^\rho \beta\mu^{\text{ex}}\,\mathrm{d}\rho. \tag{111}$$

In general, the chemical potential μ has to be considered as a central quantity for the description of confined fluids. Particularly, fluid phase equilibria are governed by the equality of μ in the coexisting phases. Therefore a large number of simulation studies of chemical potentials using both the grand canonical ensemble and Widom's test particle method have been performed for confined fluid models [121].

According to our general concept, we have simulated the chemical potential of short-range hard-core fluids (by means of Widom's test particle method) and primitive-associating fluid models (by means of gradual particle insertion). For several reasons, we study the chemical potential as a function of density at a number of temperatures. On the one hand we use the data to calculate the spreading pressure by means of Gibbs–Duhem integration, and on the other hand we estimate the densities of coexisting fluid phases in the subcritical temperature range by means of a Maxwell construction in terms of μ and ρ. Figure 13 shows $\beta\mu$ for the square-well slit fluid as a function of density in a hard planar slit at several temperatures. The van der Waals loops indicate subcritical temperatures.

Results for the spreading pressure of a square-well fluid confined to a hard slit – calculated directly by virtual volume variation (VVV) and indirectly by Gibbs–Duhem integration of the simulated $\mu(\rho)$-isotherms – are given in Fig. 14. The agreement of Π obtained by the Gibbs–Duhem integration with the Π obtained by direct simulation is excellent, even in the two-phase region. Thesimulation data presented [38] are the first direct quantitative estimates of the spreading pressure of a

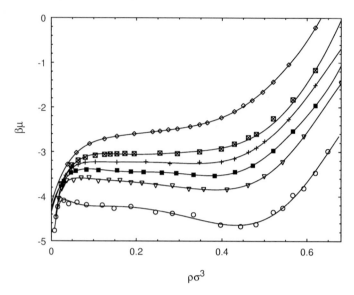

Fig. 13. Total chemical potential of the square-well fluid with $\lambda = 1.5\sigma$ in a hard planar slit of width $L = 4\sigma$ as a function of density at several reduced temperatures $T^* = kT/\epsilon$, calculated using the TPI method. The curves are smoothed numerical fits to the computer simulation data. Open circles denote results at $T^* = 0.8$, inverted triangles results at $T^* = 0.9$, filled squares results at $T^* = 0.95$, crosses results at $T^* = 1.0$, crosses in squares results at $T^* = 1.05$, and diamonds results at $T^* = 1.2$

confined fluid model so far. The special importance of these results consists in the interpretation of the spreading pressure as the thermodynamic pressure governing the corresponding fluid phase equilibria in the confinement (compare the next section). The meaning of Π as a thermodynamic pressure is explicitly confirmed by the numerical consistency of the directly simulated spreading pressure with the pressure obtained by a thermodynamic integration of the Gibbs–Duhem equation.

In general, phase equilibria of fluids confined to micropores can be divided into two main groups: (i) shifted bulk fluid phase equilibria, where the position of the critical point of the usual bulk vapor–liquid transition has been changed by the geometrical restrictions of the confinement (capillary condensation), and (ii) adsorbed fluid phases near the interface caused by specific interaction forces between the wall and the fluid molecules (e.g. layering and wetting). While capillary condensation phenomena are typical for narrow pores, adsorbed surfaces phases are present in wide pores and at single interfaces too. In narrow pores (with adsorbing walls), both types of phase transitions are generally superimposed. In order to demonstrate the abilities of recent simulation techniques we shall focus in the following on capillary condensation equilibria in narrow slit pores with smooth (hard) walls, where adsorbed wall phases play a minor role only. For the presentation of simulation results on adsorption phenomena at the walls, which are usually obtained by standard

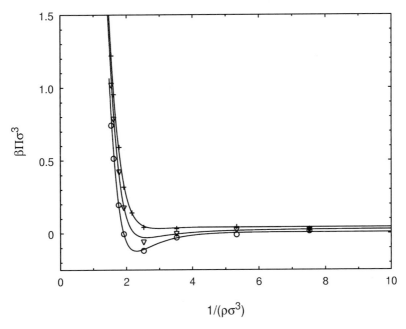

Fig. 14. Spreading pressure of a square-well fluid with $\lambda_{sw} = 1.5$ in a hard planar slit of width $L = 4\sigma$ as a function of reduced volume at the temperatures of Fig. 13. The symbols are the VVV results, and the curves are obtained from the Gibbs–Duhem equation (111), using the smoothed chemical potential data of Fig. 13.

MC methods (grand canonical ensemble and Widom insertion), we refer to recent comprehensive reviews [121–123].

   We shall demonstrate in some detail the abilities of recent MC simulation techniques in the more general case of condensation phenomena in pore–bulk systems involving more than two coexisting phases. We focus on both the direct simulation of fluid phase equilibria by means of the Gibbs ensemble, including recent extensions to multi–box approaches (cf. Sect. 3.1), and an indirect route analyzing μ-versus-ρ isotherms. As emphasized in the previous section, we are interested here, in the first place in capillary condensation effects due to the shift of the bulk fluid phase equilibrium caused by the restricting geometry of the confinement (micropore). For that reason – to exclude surface phases – pores with smooth, hard walls are under consideration. As in other parts of this paper, we discuss simulation studies of square-well fluids in slit-like pores. To demonstrate the extension of the Gibbs simulation technique from the usual (two-phase) vapor-liquid equilibrium in the bulk to a (three–phase) bulk–pore equilibrium, a (slightly) subcritical square-well bulk fluid has been coupled to a slit filled with the same fluid. By performing a three-box Gibbs simulation (according to Sect. 3.1) the equilibrium densities of the three phases involved were estimated. The bulk phase equilibria properties are well known from extensive computer simulations of Vega et al. [124]. Using the same temperatures studied by Vega, the bulk coexistence properties have been recovered in the slit-bulk simulation and in the slit, additionally, a dilute vapor phase in equilibrium with the bulk was found. The corresponding ρT diagram of the inhomogeneous three phase equilibrium in the temperature range from $T^* = 1.05$ to $T^* = 1.20$ ($T^* = k_B T/\epsilon$) is shown in Fig. 15.

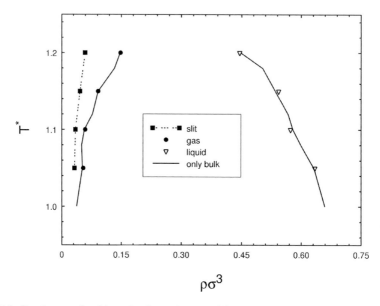

Fig. 15. Coexistence densities of a three-phase equilibrium between a free bulk fluid and a fluid in a slit-like pore in the subcritical temperature range

Recently McGrother and Gubbins [125] extended the constant-pressure Gibbs MC technique to inhomogeneous fluids. They studied the adsorption of a simple model of water in narrow slit-like pores, presetting the bulk pressure and permitting volume fluctuations in the bulk, whereas the slit volume remained constant. Despite the simplicity of the model used, the estimated adsorption isotherms are in qualitative agreement with experimental results.

As discussed above in general we have to expect in micropores of suitable size the appearance of gas–liquid equilibria (capillary condensation) which may be understood as a shift of the bulk fluid phase equilibrium caused by the confinement. In micropores with hard walls, usually a decrease of the bulk critical temperature is found [122]. The simulation of capillary condensation has been performed by Gibbs ensemble MC too [41, 67].

A very general method – recently proposed by Vörtler and Smith [116] – to estimate the densities of the coexisting (fluid) phases in both homogeneous bulk fluids and confined fluids exploits (simulated) isotherms of the chemical potential μ/kT versus the packing fraction $\eta = N\pi\sigma_w^3/6V$ in the (expected) subcritical range, where these isotherms show van der Waals-like loops, quite analogous to the isotherms of the pressure p versus volume V. The coexistence densities were calculated by a numerical integration of the simulated $\mu/kT(\eta)$ isotherms, which is equivalent to an equal-area Maxwell construction in terms of μ and N. To demonstrate the estimation of equilibrium densities from chemical potential isotherms, we present a subcritical $\mu/kT(\eta)$ isotherm of EPM5 water which was simulated by very long simulation runs (up to 1 million cycles), using gradual particle insertion, ensuring an accuracy of the results for μ of about one percent [126].

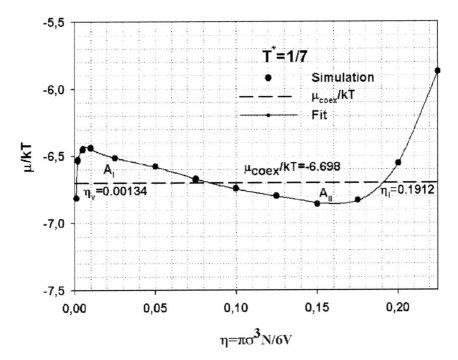

Fig. 16. Maxwell construction for vapor–liquid coexistence in the $\mu/kT-\eta$ diagram

In Fig. 16, we show these results together with an accurate polynomial fit of the simulated isotherm. The fitted curve was used to perform the Maxwell integration to locate the densities of the two coexisting (fluid) phases. Graphically, we illustrate the Maxwell integration in the $\mu/kT-\eta$ diagram, where the two indicated areas, A_{I} and A_{II}, are equal. In the figure we have indicated both the coexistence densities and the corresponding value of the common equilibrium chemical potential. The numbers given were obtained numerically from the Maxwell integration. This method was applied to several bulk and pore equilibria. We show the results for the capillary condensation of the square-well fluid in a hard slit discussed above.

The resulting coexistence densities of the slit vapor and liquid phase are presented in Fig. 17 together with direct Gibbs simulation results. Both sets of equilibrium densities are in good mutual agreement.

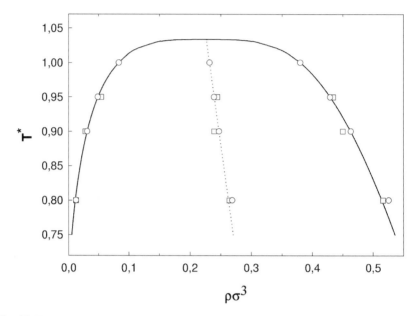

Fig. 17. Temperature–density vapor–liquid coexistence curve (solid line) for the square-well fluid with $\lambda = 1.5$ in a hard planar slit of width $L = 4\sigma$. The rectilinear diameter line given by $(\rho_\ell + \rho_v)/2$ is also shown (dotted line). The curves are the result of a numerical fit to (112); the circles are the phase equilibrium data from the analysis of the μ vs. ρ simulation data and the squares are the Gibbs Ensemble simulation results

Finally, the slit critical point properties may be calculated from the phase equilibrium data in the usual way [124], by a nonlinear regression procedure to fit the coexistence densities to the expression

$$\rho_{\ell/v} = \rho_c + C_2(1 - T/T_c) \pm 0.5B_0(1 - T/T_c)^\beta, \; T < T_c, \qquad (112)$$

where ℓ and v denote the liquid and vapor phases, respectively, T_c and ρ_c are the critical temperature and density, respectively, and $\{C_2, B_0, \beta\}$ are constants. Subsequently one estimates P_c by fitting the saturation pressures to the Antoine equation,

$$\ln P^* = A + B/T^*, \qquad (113)$$

where $P^* = P\sigma^3/\epsilon$, and using the calculated value of T_c^*.

The critical point properties of the square-well slit fluid estimated in this way from the phase equilibrium densities obtained by the analysis of the smoothed $\beta\mu^{ex}$–results are also included in Fig. 17. The critical temperature, pressure, and density are all lower than in the case of the bulk square-well fluid (cf. [38, 124]), as expected. Phase equilibria in aqueous bulk systems and water films represented by extended primitive water models along these lines are at present under consideration [126].

4 Transport of Molecules in Nanopores

4.1 Introduction

Zeolites are porous solids that contain a regular network of pores or channels. They are used in many technological applications, e.g. as molecular sieves, catalysts, and adsorbents in the petroleum and chemical industries and in environment protection, for ion exchange in washing agents, and in many other fields. The framework of zeolites consists of silicon and aluminum atoms that are called T-atoms and oxygen bridges between them. To balance the electrical charge, additional exchangeable cations appear in some types of zeolites.

Typical members of this family of materials are the zeolites of structure type Linde A (LTA). The geometry of these zeolites is demonstrated in Fig. 18 which shows a structure scheme of a fragment of such a zeolite. It must be thought that it is continued periodically in all directions.

The knowledge of diffusion processes in zeolites is of crucial importance for many technological applications [23, 32, 127, 128] because it controls the timescale and therefore the efficiency of such processes. The most popular applications are use as ion exchangers in phosphate–free washing agents and use as catalysts in gasoline production. Because of some very special features the migration of particles inside

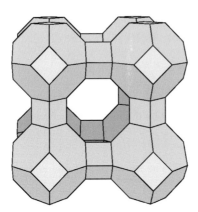

Fig. 18. Structure scheme of LTA zeolites. The vertices mark the positions of the T-atoms, and the lines symbolize the oxygen bridges between them. The edges of the cube shown are formed by sodalite units consisting of 4- and 6-rings. The sodalite units surround the large α-cage of ≈ 1 nm diameter. The 8-rings which are often called windows have a free diameter of ≈ 0.41 nm. They build the connection between the cages and can be passed by small hydrocarbons such as ethane

the zeolite is of great scientific interest too. Therefore, diffusion of guest molecules in zeolites has been the subject of many experimental and theoretical investigations. Reviews can be found in [23, 32, 128–132]. Molecular dynamics computer simulations (MD) [8, 10] that give insight into the dynamics of kinetic processes in zeolites will be discussed on a molecular level in the following sections. MD is a well-established tool for such investigations that, within the limits of the model, allows one, to to obtain more detailed information than from real experiments. Furthermore, MD allows variations in the system parameters which are not possible in real experiments. So, interrelations and dependencies can be examined. One goal of our investigations is to understand the experimentally observed behavior of diffusing guest molecules in zeolites [23] by MD simulations and to derive more general knowledge from the results. Moreover, we treated the problems also by theoretical derivations, where possible.

4.2 Molecular Dynamics Techniques (MD)

What is MD?

Starting in 1957 with the classical papers of Alder and Wainwright [133–135] the method of molecular dynamical computer simulations [8, 10, 17] became a very powerful tool in the examination of equilibrium and non-equilibrium properties of many particle systems.

The method consists of the numerical calculation of the classical trajectories of several hundreds or thousands of particles at discrete times separated by a time increment of 10^{-16} - 10^{-14} s. Until now, the largest number of particles involved in an MD simulation has been 5 180 116 000 according to our knowledge. However, the trajectory had a length of only a few time steps [136].

Here we can give only a very short overview of the technical fundamentals of this method. Besides the textbooks [8, 10, 17], a more detailed description of the method can also be found in [32].

The Calculation of Trajectories

The trajectories are calculated by the integration of Newton's equations (see (16))

$$m_i \ddot{\mathbf{r}}_i(t) = \mathbf{F}_i(\mathbf{r}_1, ..., \mathbf{r}_N) \tag{114}$$

using numerical algorithms. The most popular one is at present the velocity Verlet algorithm [137]

$$\mathbf{r}_i(t+h) = \mathbf{r}_i(t) + h\mathbf{v}_i(t) + \frac{h^2}{2m_i}\mathbf{F}_i(t) \tag{115}$$

$$\mathbf{v}_i(t+h) = \mathbf{v}_i(t) + \frac{h}{2m_i}[\mathbf{F}_i(t) + \mathbf{F}_i(t+h)]. \tag{116}$$

Here $\mathbf{r}_i$, $\mathbf{v}_i$ and $\mathbf{F}_i$ mean the sites, velocities and forces, respectively. The main advantages of this simulation algorithm are its simplicity and an excellent stability of the average total energy. The total energy fluctuates in all simulations in the micro canonical ensemble because of numerical artefacts. In some algorithms the average even shows a permanent drift.

Very sophisticated techniques (reviewed in [138]) introduce multiple time steps for separate treatment of rapidly changing and slowly changing forces.

Boundary Conditions

As in MD runs usually only a few hundreds or thousands of particles are included one must take care of artefacts connected with the small size of the system. In a regular cubic lattice of $8\times8\times8 = 512$ particles, 338 of them form the surface of the cube. This is about 66 per cent. Thus, surface effects would dominate all properties of this tiny troplet of matter if special techniques would not prevent this. A trick that is sometimes also used in analytical calculations is the invention of periodic boundary conditions .

Using this method, identical replicas of the original MD box including all of the particles are thought to be arranged around this original one (Fig. 19). The sites of any one of the original particles and its imaginary images in the other boxes differ by integer multiples of the edge length L of the cubic MD box. The velocities of all "images" of one original particle are the same. Other box shapes are possible but, will not be discussed here.

Each particle in the central box interacts with the other particles in the central box, but, also with the "images" in the surrounding boxes. Therefore, no surface effects appear. However, it has to be checked whether or not results could be influenced by the artificial periodicity which has been introduced by this trick.

If one of the particles leaves the central box, one of the "images" will enter the central box on the opposite side. Hence, the number of particles in the central box is constant. Of course, only the sites and velocities of the particles in the central box must be stored. Additionally, the number and directions of border crossings must

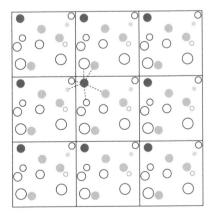

Fig. 19. An example of periodic boundary conditions in two dimensions

be registered if the total distance over which a particle has moved during a longer time is asked for example. This particle shift is e.g. used to calculate the diffusion cofficient (see Sect. 4.3).

Evaluation of the Runs

The trajectories obtained from MD are analyzed by statistical mechanical methods to obtain equilibrium properties but also to calculate time correlation functions, transport coefficients, and other properties of nonequilibrium systems.

One example of an equilibrium quantity is the temperature which is usually evaluated from the kinetic energy per degree of freedom. Another example is the pressure, which can be evaluated by the virial theorem (see (42)). Because of the wall forces, the pressure is an ambiguous quantity for particles in cavities. Instead, the chemical potential μ is used, because it can be evaluated by insertion of imaginary particles. μ is obtained from the change in the potential energy caused by the insertion (Widom [21]).

Unlike Metropolis Monte Carlo simulations (see Sect. 3.1) in MD simulations a sequence of representative configurations is not produced but, instead, the evolution in time of the simulated system is obtained. Therefore, MD simulations not only are suited to examining equilibrium properties but also can be used to calculate time correlation functions, transport coefficients, and other properties of nonequilibrium systems. In the present paper the diffusion coefficient plays a central role. The evaluation of this quantity from moments of the particle shifts is discussed in more detail in Sect. 4.3.

Dynamical Monte Carlo (DMC)

Recently, hybrid methods called dynamical Monte Carlo simulation have become more and more important for processes on timescales that are beyond the possibilities of MD.

In such simulations the real continuous trajectory is replaced by a network of discrete sites and the real movements are replaced by sequences of jumps.

DMC methods, far from replacing MD, are often derived from MD simulations and often used in connection with them [139]. The jump models treated in the present article are examples of some steps in this direction.

4.3 Simulation of Diffusion in Zeolites

The Propagator – A Tool to Describe the Dynamics of Diffusion

Definition of the Propagator. The propagator is defined as the conditional probability density to find a particle at time t at place $r_0 + r$ if it has been at time $t = 0$ at r_0. So the propagator is a function of the four dimensions of space and time. For a complete visualization a five-dimensional coordinate system would be required.

For a pure random walk the propagator is a Gaussian distribution

$$P(\mathbf{r}, t) = (4\pi Dt)^{-3/2} \exp\left\{-\frac{\mathbf{r}^2}{4Dt}\right\}. \tag{117}$$

Figure 20 shows P for different times in a plane that cuts a cavity center and 4 windows in the cation-free LTA zeolite (xy plane). For the diffusion of guest molecules in zeolites the propagator is a more complex function.

Powers of the Particle Shift – The Moments of the Propagator. As the propagator is a probability distribution, the expectation values are

$$\langle f(\mathbf{r}) \rangle = \int P(\mathbf{r}, t) f(\mathbf{r}) \, d\mathbf{r}. \tag{118}$$

Particularly, $f(\mathbf{r})$ can be any power of the particle shift. It has been explained how the diffusion coefficient can be evaluated in the case of isotropic diffusion, from each one of the powers of this shift if the diffusion equation is applicable (see e.g. [8,32,140]). Moreover, the agreement of the values of the diffusion coefficient D from different moments shows whether or not the diffusion equation is applicable. In [141] these considerations are extended to the more general case of anisotropic diffusion. The propagator for the x-component and similarly for the y- and z-direction is

$$P_x(x, t) = \frac{1}{\sqrt{4\pi D_x t}} \exp\left\{\frac{-x^2}{4D_x t}\right\}, \qquad \int_{-\infty}^{\infty} dx \, P_x(x, t) = 1. \tag{119}$$

If $D_x = D_y = D_z = D$, the spherically symmetric propagator can be defined as

$$P(\mathbf{r}, t) = P_x(x, t) P_y(y, t) P_z(z, t) = (4\pi Dt)^{-3/2} \exp\left\{\frac{-r^2}{4Dt}\right\}. \tag{120}$$

The moments of displacement yield in this case [140]

$$<|\,\mathbf{r}\,|> = 4\sqrt{\frac{Dt}{\pi}}, \tag{121}$$

$$\langle (\mathbf{r})^2 \rangle = 6Dt, \tag{122}$$

$$\langle |\,(\mathbf{r})^3\,| \rangle = \frac{32(Dt)^{3/2}}{\sqrt{\pi}}, \tag{123}$$

$$\langle (\mathbf{r})^4 \rangle = 60(Dt)^2. \tag{124}$$

D can be obtained from each one of these. The first and second moments yield:

$$D = \frac{\pi}{16} \frac{d}{dt} \langle |\,\mathbf{r}\,| \rangle^2, \tag{125}$$

$$D = \frac{1}{6} \frac{d}{dt} \langle \mathbf{r}^2 \rangle. \tag{126}$$

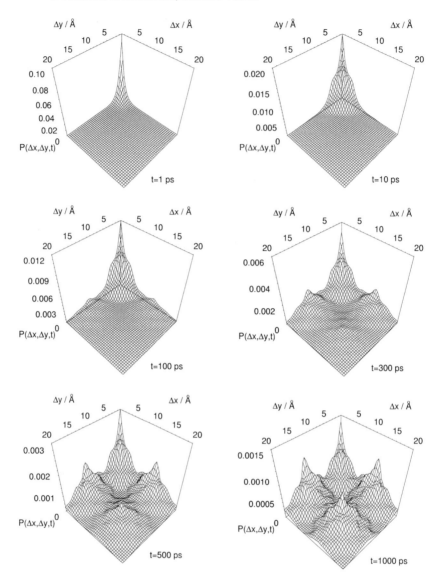

Fig. 20. Time development of P for ethane in the cation-free LTA zeolite

In the case of anisotropy, one can assume again that the three components of the probability distribution (propagator) are uncorrelated and hence

$$P(\mathbf{r},t) = P_x(x,t)P_y(y,t)P_z(z,t) = \frac{(4\pi t)^{-3/2}}{\sqrt{D_x D_y D_z}} \exp\left\{ -\frac{x^2}{4D_x t} - \frac{y^2}{4D_y t} - \frac{z^2}{4D_z t} \right\}.$$

$$(127)$$

As $r^2 = x^2 + y^2 + z^2$ is a sum of three contributions, with each summand depending only on one variable, it is easy to show that

$$\int\limits_{-\infty}^{\infty} \int\limits_{-\infty}^{\infty} \int\limits_{-\infty}^{\infty} P(\mathbf{r}, t)\, r^2\, dx\, dy\, dz = 2 D_x t + 2 D_y t + 2 D_z t. \tag{128}$$

In analogy to the spherically symmetrical case, a general D can be defined by

$$\int\limits_{-\infty}^{\infty} \int\limits_{-\infty}^{\infty} \int\limits_{-\infty}^{\infty} P(\mathbf{r}, t)\, r^2\, dx\, dy\, dz = 6\, D\, t. \tag{129}$$

Owing to (128), D is connected to D_x, D_y and D_z by

$$D = \frac{D_x + D_y + D_z}{3}. \tag{130}$$

Such a simple procedure is not possible for the other moments, as they do not consist of a sum of contributions depending only on one coordinate. The moments can be evaluated for each component. The first four moments yield:

$$D_x = \frac{d}{dt}\, \frac{\pi \langle |\, x - x_0\, | \rangle^2}{4}, \tag{131}$$

$$D_x = \frac{d}{dt}\, \frac{\langle (x - x_0)^2 \rangle}{2}, \tag{132}$$

$$D_x = \frac{d}{dt}\, \frac{(\pi \langle |\, (x - x_0)\, |^3 \rangle^2)^{1/3}}{4}, \tag{133}$$

$$D_x = \frac{d}{dt} \sqrt{\frac{\langle (x - x_0)^4 \rangle}{12}}. \tag{134}$$

The van Hove Correlation Function. A more general quantity than the propagator is the van Hove correlation function

$$G(\mathbf{r}, t) = \frac{1}{N} \left\langle \sum_{j=1}^{N} \sum_{\ell=1}^{N} \delta[\mathbf{r} + \mathbf{r}_\ell(0) - \mathbf{r}_j(t)] \right\rangle. \tag{135}$$

$G(\mathbf{r}, t)$ can be split into the *self* part G_s and the *distinct* part G_d. G_s correlates positions of the *same* particle ($j = \ell$) at different times, while G_d correlates positions of *different particles* ($j \neq \ell$). G_s is essentially the discrete version of the definition of the propagator. It counts how often a shift $\mathbf{r}$ appears after a time t. This becomes even more clear in the continuos version:

$$G(\mathbf{r}, t) = G_s(\mathbf{r}, t) + G_d(\mathbf{r}, t), \tag{136}$$

$$G_s(\mathbf{r}, t) = \frac{1}{N} \left\langle \sum_{j=1}^{N} \delta[\mathbf{r} + \mathbf{r}_j(0) - \mathbf{r}_j(t)] \right\rangle, \tag{137}$$

$$G_d(\mathbf{r}, t) = \frac{1}{N} \left\langle \sum_{j=1}^{N} \sum_{\ell=1(\neq j)}^{N} \delta[\mathbf{r} + \mathbf{r}_\ell(0) - \mathbf{r}_j(t)] \right\rangle, \tag{138}$$

$$G(\mathbf{r}, t) = \frac{1}{N} \int \langle \rho(\mathbf{r}', 0) \rho(\mathbf{r}' + \mathbf{r}, t) \rangle \, d^3\mathbf{r}'. \tag{139}$$

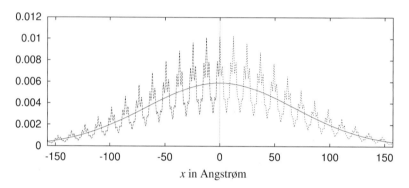

Fig. 21. Comparison of the self-part of the van Hove correlation function evaluated in an MD run at 300 K with 3 guest molecules per cavity at $t = 4095$ ps with the approximated version according to (143). The right-hand side shows the true G_s and the left-hand side the approximation. The smooth solid line is the hydrodynamic limit

In the hydrodynamic limit (which means, in the case of LTA zeolites, for times larger than the residence time of a guest molecule in a cavity and for distances larger than the cavity diameter), we have

$$G_s^{\text{hyd}}(r, t) = \frac{1}{(4\pi D_s t)^{3/2}} e^{-r^2/4D_s t}. \tag{140}$$

In [142–145], a factorization of G_s could be proposed. Because of the cubic symmetry it is sufficient to examine one dimension, e.g. the x-direction.

In the following $P(x_0)$ is the probability density to find a particle at x_0. The quantity $p(x_0 + x, t | x_0)$ is the transition probability density that the particle moves from x_0 to $x_0 + x$ during the time interval t. Using these quantities the self-part of the van Hove correlation function is

$$G_s(x, t) = \int_0^a p(x_0 + x, t | x_0) P(x_0) \, dx_0. \tag{141}$$

In [142–145], an approximation could be derived that reads

$$p(x_1, t|x_0) \approx P(x_1)\, G_{\mathrm{s}}^{\mathrm{hyd}}(x_1 - x_0, t). \tag{142}$$

This yields

$$G_{\mathrm{s}}(x, t) \approx G_{\mathrm{s}}^{\mathrm{hyd}}(x, t) I(x) \tag{143}$$

where

$$I(x) = a \int_0^a P(x_0 + x) P(x_0)\, \mathrm{d}x_0. \tag{144}$$

a is the lattice constant in the x direction.

This means $G_{\mathrm{s}}(x, t)$ can be approximated by its Gaussian hydrodynamic limit, modulated with the space autocorrelation function of the density (see Fig. 21).

Influence of the Exchangeable Cations on the Diffusion of Neutral Molecules

For a long time, it has been assumed that the diffusion of neutral molecules in zeolites is not influenced by exchangeable cations unless the molecules have unsaturated bonds [146]. However, the dynamics even of small neutral molecules which possess only saturated bindings is strongly influenced by the presence of exchangeable cations [147, 148]. For these investigations, we have chosen the NaCaA zeolite with 4 Na and 4 Ca so that the windows are free from cations. The strong effect can clearly be seen in Fig. 22 and has also been confirmed experimentally by [149]. In comparison with the cation-free LTA zeolites , the self-diffusivity decreases by up to two orders of magnitude.

It should be noted that the computational effort is much larger in this case than in our previous simulations for the cation-free analogue zeolite, since much longer trajectories (up to 5–10 ns) are necessary to evaluate such small D's. Additionally,

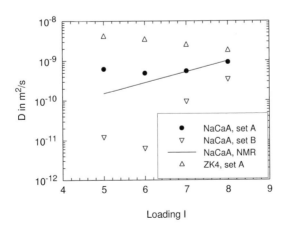

Fig. 22. Comparison of the self-diffusion coefficient D of methane in NaCaA and in cation-free LTA zeolites with D-values from NMR experiments

the calculation of the forces resulting from the polarization energy is very time-consuming although we were able to replace the full Ewald sum for our system by a corrected r space part of this sum (see e.g. [32]).

Figure 22 demonstrates MD results for two different sets of parameters and compares these data with experiments. The agreement with experimental results from NMR measurements [150] is satisfactory in the case of set A.

The Influence of Interaction Parameters and Lattice and Molecule Vibrations on Diffusion

Methane in the Cation-Free LTA Zeolite. It could be shown that the self-diffusion coefficient D of methane in the cation-free LTA zeolite depends strongly on the Lennard–Jones parameter σ used to describe the interaction of the methane with the oxygen atoms in the zeolite lattice [151]. If this parameter was changed within the range of the values proposed in the literature, not only did D change by nearly one order of magnitude but also the density dependence of D was reversed. This could be explained by the interplay of two effects that have opposite influences on diffusion [151]. These effects have a different density dependence. The first effect is that particles reflected at the window entrance will sometimes move again toward the window after collisions with other guest molecules. This enhances the diffusion with increasing concentration of guest molecules. In contrast, the diffusion is diminished if a particle that has passed through the window to an adjacent cavity is pushed back by other guest molecules [151].

For the diffusion of methane in the cation-free LTA zeolite an influence of the lattice flexibility has been found in some papers [152,153]. This finding was corrected

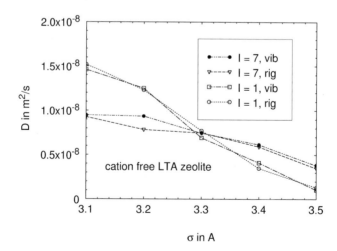

Fig. 23. The diffusion coefficient D as a function of the Lennard–Jones σ parameter of the CH_4–O interaction for loadings of $I = 1$ and $I = 7$ guest molecules per cavity, from runs with a rigid and a vibrating lattice at 300 K.

in [154, 155]. It turned out that there is nearly no influence of the lattice vibrations on the diffusion of methane in the cation-free LTA zeolite.

Figure 23 illustrates both the strong influence of the interaction parameter σ on the self-diffusion (including the change in its density dependence) and the fact that D is not influenced by lattice vibrations. This is true even in the case of large σ, which means small window diameters, where a different finding could be expected. More details can be found in [155].

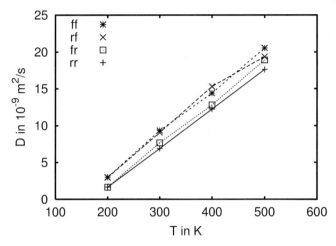

Fig. 24. The self-diffusion coefficient D of methane in silicalite-1 at 300 K in 10^{-9} m^2/s for $I = 1$ averaged over all directions. ff means lattice and molecule flexible, "rf" means molecule rigid and lattice flexible, "fr" means molecule flexible and lattice rigid, and "rr" means both lattice and molecule rigid

Methane in Silicalite-1. In silicalite-1, the Lennard–Jones parameter σ is less important because of the size and shape of the channels. However, in contrast to the cation-free LTA zeolite , the self-diffusion coefficient is influenced by lattice vibrations. Earlier findings in the literature could be confirmed in this case. Additionally, the influence of the internal molecular vibrations has been investigated. Owing to the anisotropy of the channel system, the self-diffusion was investigated for the different directions separately [141]. However, as an example, D as averaged over all directions is shown in Fig. 24. It turns out that the lattice vibrations have some influence on D at low concentrations of guest molecules while at higher loadings there is no influence. This can be due to the strong mutual thermalization of guest molecules at high concentrations [141]. The influence of the internal vibrations of the methane molecules turns out to be negligible.

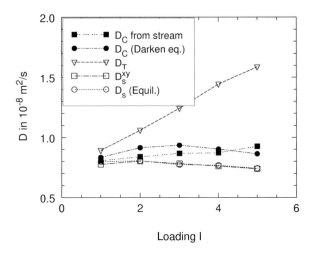

Fig. 25. The self-diffusion coefficient D_s, the corrected diffusion coefficient D_c and the transport diffusion coefficient D_T evaluated in various ways

Self-diffusion and Transport Diffusion

Besides the self-diffusion coefficient D_s there are other kinds of diffusion coefficients. The velocity of the stream caused by an external force field **F** is

$$\langle \mathbf{v} \rangle = B\,\mathbf{F}. \tag{145}$$

B is the mobility. The Kubo theory (see e.g. Sect. 2.4, (70), [26]) yields

$$B = \frac{1}{3Nk_{\mathrm{B}}T} \int_0^\infty \sum_{i=1}^{N} \sum_{j=1}^{N} \langle \mathbf{v}_i(0) \cdot \mathbf{v}_j(\xi) \rangle \, \mathrm{d}\xi. \tag{146}$$

The self-diffusion coefficient D_s in terms of an autocorrelation function is

$$D_{\mathrm{s}} = \frac{1}{3N} \int_0^\infty \sum_{i=1}^{N} \langle \mathbf{v}_i(0) \cdot \mathbf{v}_i(\xi) \rangle \, \mathrm{d}\xi. \tag{147}$$

The so-called corrected diffusion coefficient D_c is defined as

$$D_{\mathrm{c}} = B\,k_{\mathrm{B}}T = \frac{1}{3N} \int_0^\infty \sum_{i=1}^{N} \sum_{j=1}^{N} \langle \mathbf{v}_i(0) \cdot \mathbf{v}_j(\xi) \rangle \, \mathrm{d}\xi. \tag{148}$$

Therefore

$$D_{\mathrm{c}} = D_{\mathrm{s}} + \frac{1}{3N} \int_0^\infty \sum_{i=1}^{N} \sum_{j \neq i} \langle \mathbf{v}_i(0) \cdot \mathbf{v}_j(\xi) \rangle \, \mathrm{d}\xi. \tag{149}$$

For the stream caused by a gradient $\nabla\mu$ of the chemical potential the Kubo theory yields

$$\langle \mathbf{v} \rangle = B\,\nabla\mu. \tag{150}$$

With Fick's law which defines the transport diffusivity D_T one has

$$\mathbf{j} = -D_\mathrm{T}\nabla n \qquad \mathbf{j} = n\langle \mathbf{v} \rangle. \tag{151}$$

For the gradient of the chemical potential the following is valid analogously:

$$\mathbf{j} = n\langle \mathbf{v} \rangle = nB\,\nabla\mu = -D_\mathrm{T}\nabla n. \tag{152}$$

With

$$\nabla\mu = \frac{\mathrm{d}\mu}{\mathrm{d}n}\nabla n \tag{153}$$

and $D_\mathrm{c} = B\,k_\mathrm{B}T$ it follows finally that

$$D_\mathrm{T} = D_\mathrm{c}\frac{n}{k_\mathrm{B}T}\frac{\mathrm{d}\mu}{\mathrm{d}n}. \tag{154}$$

The transport diffusion coefficient has been investigated to our knowledge for the first time by a simulated stationary density gradient in [156]. This work has been continued in [142, 143]. Results are displayed in Fig. 25. As an alternative, D_T was calculated from the relaxation of a periodic inhomogenity in the particle density [157]. Meanwhile, these phenomena have been discussed in many papers, e.g. [158–171].

Self-diffusion of a Mixture in Silicalite-1

In [172] the diffusion of a mixture of methane and xenon in the zeolite silicalite-1 was studied by combining MD simulations and pulsed field gradient (PFG) NMR measurements in a close cooperation of experimental and theoretical physicists.

The results show that the self-diffusion coefficient of methane is strongly affected, if, at a constant total number of guest molecules, the concentration of xenon is changed. If the concentration of methane is changed at constant total number of guest molecules then the self-diffusion coefficient of xenon is only slightly influenced.

The reason for the dominance of xenon is the larger local heat of adsorption of xenon and the larger mass of xenon compared with methane, in combination with the channel size and topology in silicalite-1. Both of these effects contribute nearly equally [172]. This could be shown making use of the outstanding possibility of MD simulations to change single system parameters leaving the other ones unchanged [172]. Simulated and experimental data are in good agreement with each other (Fig. 26).

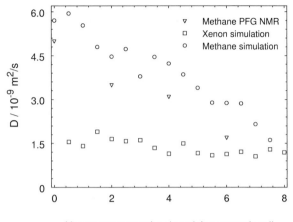

Fig. 26. A comparison of experimental and simulated data for the mean diffusion coefficient of methane and xenon in different compositions at $T = 293$ K (experiments), respectively $T = 300$ K (simulations) for a total loading of eight particles per unit cell

An Entropic Diffusion Barrier

Ethane in the Cation-Free LTA Zeolite. MD simulations of ethane in the cation-free LTA zeolite show an unusual dependence of the self-diffusion coefficient D_s upon the temperature. As diffusion is the movement of particles caused by their thermal motion, D should be expected to increase with increasing temperature. However, for the case treated in this section, there was a region of density and temperature where D decreased with increasing temperature. Figure 27 shows a logarithmic plot of D

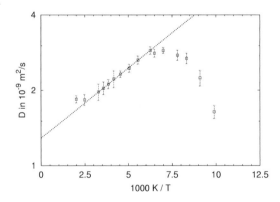

Fig. 27. Arrhenius diagram for D at a loading of one ethane molecule per cage. Between $T = 150$ K and 400 K the Arrhenius law can be fitted with $E_A = -1.1$ kJ/mol

as a function of the inverse temperature $1/T$. The straight line would correspond to an Arrhenius law

$$D(T) = D_0 \exp\left\{-\frac{E_A}{k_B T}\right\}.$$

E_A is often called the activation energy and k_B is Boltzmann's constant. If such a law is valid and E_A is positive (as is usually expected) then D increases with increasing temperature. In Fig. 27 this is clearly not the case [173].

A first hint concerning the nature of this effect is given by the distribution of the angle Φ between the window axis and the ethane molecule. The window has the shape of a tube that can be passed by an ethane molecule only with orientations nearly parallel to the axis of this tube. Figure 28 shows that at higher temperatures there is an increasing trend to orientations perpendicular to the window axis at the entrance to the window.

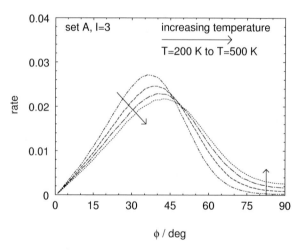

Fig. 28. Histogram of the distribution of angles between the ethane and the symmetry axis of the window at several temperatures in the transition region from the cavity to the window

This behavior can be understood by taking into account that for a particle close to the cavity wall, the orientation of lowest potential energy is along the wall. A molecule approaching the window with such orientation will be led into the window with an orientation nearly parallel to the window axis. At higher temperatures states of higher potential energy connected with other orientations become more probable.

In the following this effect will be treated analytically using the local entropy.

The Reversible Work Theorem – Local Entropy. If $\overline{K}_x$ is the average force component in the x-direction on particle 1 with x-coordinate value x_1 then

$$\overline{K}_x = \left\langle \left(-\frac{\partial U(\mathbf{r}_1, \ldots, \mathbf{r}_N)}{\partial x_1}\right)\right\rangle_{x_1} = k_B T \frac{d}{dx_1} \ln p(x_1),$$

where $p(x_1)$ is the probability density to find particle 1 at this site [174]. The potential energy $U(\mathbf{r}_1, \ldots, \mathbf{r}_N)$ includes all interactions (walls, and other particles), i. e. also many-body effects.

The free-energy difference between states where the particle is at x_1 or x_2, is the reversible work for this change [174]

$$\Delta F = \int_{x_1}^{x_2} \overline{K}_x \, \mathrm{d}x = k_\mathrm{B} T [\ln p(x_2) - \ln p(x_1)].$$

Thus, the local free energy defined for the single particle within the many body system is

$$F(x) = k_\mathrm{B} T \ln p(x) + \mathrm{const.} \tag{155}$$

In this picture the particle moves on the Helmholtz free-energy surface [175], which coincides only in the low density limit with the usual potential-energy surface.

By applying the local-equilibrium approximation, the equilibrium thermodynamics can be used to analyze this result. The free energy $F(\mathbf{r})$ (see (45)) of a particle at site $\mathbf{r}$ is

$$F(\mathbf{r}) = U(\mathbf{r}) - TS(\mathbf{r}), \tag{156}$$

where U is the internal energy and S is the entropy.

$$TS(\mathbf{r}) = U(\mathbf{r}) - F(\mathbf{r}), \tag{157}$$

$$F(\mathbf{r}) = k_\mathrm{B} T \ln p(\mathbf{r}), \tag{158}$$

$$U(\mathbf{r}_i) = \left\langle U_\mathrm{W}(\mathbf{r}_i) + \frac{1}{2} \sum_{j \neq i} u_{ij}(r_{ij}) + \frac{m}{2} v_i^2 \right\rangle. \tag{159}$$

$U_\mathrm{W}(\mathbf{r}_i)$ is the external potential (including, for example, the wall potential) acting on a particle at $\mathbf{r}_i$.

The Free-Energy Landscape. The one-particle density can easily be evaluated in the MD run. The result is shown in Fig. 29. Then the local Helmholtz free energy can be calculated using (155). Also, the local internal energy can be evaluated in the MD run and the local entropy can be obtained using (157). The three quantities defined above, namely the local Helmholtz free energy, the local entropy and the local internal energy are plotted in Fig. 30. The values are given along a line that connects two cavity centers. The center of the window is situated in the middle of the picture.

It turns out that the local free energy shows a maximum at the window center that increases with increasing temperature. This maximum corresponds to a threshold for the diffusing particle and thus reduces the diffusion. A comparison with the other two curves given in Fig. 30 shows that this threshold is not caused by an energetic effect. Instead, it is of purely entropic origin.

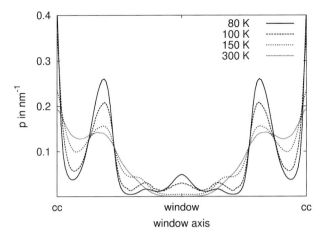

Fig. 29. The one-particle density distribution at different temperatures along the symmetry axis

A Jump Model for This Diffusion Process. Figure 29 shows, at low temperatures local maxima of the one-particle density in front of the windows. This means that the probability to find the center of mass of the diffusing ethane molecules at such places is higher than at other sites. Therefore, one can try to reduce the diffusion process to jumps between such places.

Jump models, as a tool to describe diffusion processes are becoming more and more popular [32]. This is mainly due to the following reasons

- Extension of the timescale, the number of particles, and the system size is possible
- More importantly in our case, the description is reduced to essentials. This makes it easier to understand interdependences and mechanisms
- Random models usually allow an easier theoretical treatment including compact analytical results.

A very simple random model for diffusion can be derived from the Einstein relation (see (65)) for the displacement which is identical to the expression for the second moment of the displacement in Sect. 4.3. The Einstein relation that is valid for zeolite diffusion only for large times t reads

$$\langle \mathbf{r}^2(t) \rangle = 6Dt. \tag{160}$$

Let us now introduce the quantity τ for the average time of residence of a particle in a cavity. The number of moves of the particle from one cavity to another one per unit time will be called k_w.

Now the diffusion is modeled by a discrete process in which each particle always moves to another cavity after a time period τ. The time needed for the move is neglected. Let $t = \nu\tau$, with integer ν. As $\langle \mathbf{r}^2(t) \rangle$ grows in proportion to t, a constant of proportionality L can be introduced, i.e. $\langle \mathbf{r}^2(t) \rangle = \nu L^2$ with, up to now, unknown L.

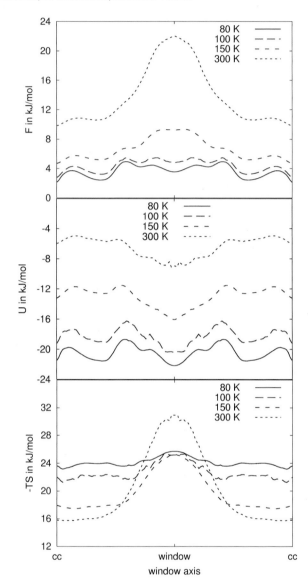

Fig. 30. Free energy, internal energy, and entropy of a molecule along an axis from the center of one cage to another cage for different temperatures. The centers of the cages are labeled by "cc"

If (160) could be used for $\nu = 1$, then L could be identified with the cage-to-cage distance and

$$D = \frac{1}{6\tau}L^2 = \frac{1}{6}k_{\mathrm{w}}L^2. \tag{161}$$

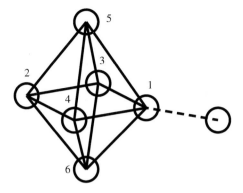

Fig. 31. Site connectivity for the jump model of ethane in LTA zeolite

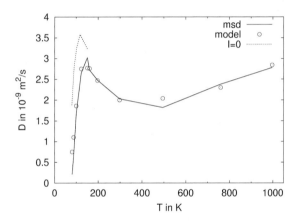

Fig. 32. Comparison of self-diffusion coefficients calculated from mean square displacements (msd) via the Einstein equation and from the jump model. $I = 0$ shows the self-diffusion coefficients at infinite dilution.

However, (161) is only an approximate formula because (160) is not valid for short times. Instead, correlations between the single jumps must be taken into account. The resulting mean square displacement after n jumps of a particle is

$$\langle r^2(n) \rangle = \left\langle \left(\sum_{i=1}^{n} \mathbf{l}_i \right)^2 \right\rangle$$

$$= \sum_{i=1}^{n} \langle l_i^2 \rangle + \sum_{i=1}^{n-1} \langle \mathbf{l}_i \cdot \mathbf{l}_{i+1} \rangle + \dots \qquad (162)$$

where $\mathbf{l}_i$ is the ith shift of the particle. If the jumps are not correlated, then with

$$l^2 = \frac{1}{n} \sum_{i=1}^{n} \langle l_i^2 \rangle \quad \text{it follows that} \quad \langle r^2(n) \rangle = nl^2. \qquad (163)$$

However, in general, they are correlated. For example, the geometry of the channels or cavities in zeolites allows only moves in a few possible directions. Because of the regular geometry of the zeolites, it is in many cases possible to define all moves as shifts from one cavity to an adjacent one and to assume that the length of each shift is the same. This length will be called l. Let $\Theta_{i,i+1}$ be the angle between the ith and the $(i+1)$th move of the particle. Then it can be shown [23] that, with $l_i \cdot l_j = l_j \cdot l_i = l^2 \cos \Theta_{i,j}$, (162) can be replaced by

$$\langle r^2(n) \rangle \approx nl^2 \left(1 + 2 \sum_{j=1}^{\infty} \langle \cos \Theta_{i,i+j} \rangle \right). \tag{164}$$

The average $\langle \cos \Theta_{i,i+j} \rangle$ does not, of course, depend upon i. The upper limit can be ∞ because there are no correlations for large $|j - i|$. If Θ is always the same and we have high symmetry then it follows from the cos law that on average $\langle \cos \Theta_{i,i+j} \rangle = \langle \cos \Theta \rangle^j$. Then the sum in (164) is a geometrical series and (164) yields

$$\langle r^2(n) \rangle = nl^2 \left(1 + \frac{2 \langle \cos \Theta \rangle}{1 - \langle \cos \Theta \rangle} \right), \tag{165}$$

$$\langle r^2(n) \rangle = nl^2 \frac{1 + \langle \cos \Theta \rangle}{1 - \langle \cos \Theta \rangle}. \tag{166}$$

Comparison with (163) shows that a correlation factor

$$f = \frac{1 + \langle \cos \Theta \rangle}{1 - \langle \cos \Theta \rangle} \tag{167}$$

has to be introduced. The diffusion coefficient becomes

$$D = \frac{\langle r^2(n) \rangle}{6n\tau} = \frac{nl^2 f}{6n\tau} = \frac{1}{6} L^2 k_w f. \tag{168}$$

To find an expression for f in terms of properties of the jump frequencies one has to take into account the geometry of the zeolite under consideration. Figure 31 shows the six adsorption sites that can be occupied by a molecule that entered a cavity along the dashed line. After arriving at site 1, the molecule has the possibility to move back through the window or to move to another adsorption site (2–6) within the cavity. In [176] a method has been derived to analyze such jump processes in terms of unbroken sequences of window passages or of jumps within the cavity. In [177] it could be shown that, in a very good approximation, the whole information about geometrical correlations is included in the average length $\langle n_w \rangle$ of window jump sequences. Detailed analysis shows that the corresponding f is [177]

$$f = \frac{1}{2\langle n_w \rangle - 1}. \tag{169}$$

Then the self-diffusion coefficient becomes:

$$D = \frac{L^2}{6} \frac{k_{\mathrm{w}}}{2\langle n_{\mathrm{w}} \rangle - 1}.$$
(170)

Figure 32 shows a comparison of the self-diffusion coefficient D that was evaluated directly in the MD run by means of the Einstein relation with the values given by (170). The agreement turns out to be very good. This means that the random model works successfully and leads to the simple formula (170) for the self-diffusion coefficient.

Diffusion Memory of Methane in Silicalite-1

Analytical Treatment. Figure 33 shows the topology of the channels in silicalite. Owing to the anisotropy of the channel systems in some zeolites (e.g. silicalite-1) one has to examine a diffusion tensor instead of the diffusion coefficient. The geometrical structure of the channel system in silicalite-1 zeolite is visualized in Fig. 33 (a more realistic shape of the channels can be seen in Fig. 35) showing three unit cells, including straight channels in the y-direction and zigzag channels in the x- and z-directions. See also Fig. 9 in Chap. 2. A consequence of this structure is that, for example, long-range movements of a particle in the z-direction are only possible as sequences of moves in both the straight and zigzag channels. Therefore, the components of the diffusion tensor are correlated [178]. In earlier papers, the memory effects in the sequence of random jumps have been neglected [179], or have been taken into account only partially by introduction of a two-step model of

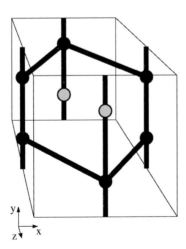

Fig. 33. Topology of channels in the unit cell of zeolites of type ZSM-5, e.g. silicalite. The black dots show the intersections between straight channels and zigzag channels in the unit cell and the grey dots show intersections of the straight channels with zigzag channels that are outside the unit cell

diffusion [180–182]. If such effects are neglected, the principal values of the diffusion tensor, i.e. the diffusivities in x-, y- and z-direction are determined by the topology and they are related to each other by simple reciprocal addition [178, 179, 183, 184]

$$\frac{l_z^2}{D_z} = \frac{l_x^2}{D_x} + \frac{l_y^2}{D_y} \tag{171}$$

with l_x, l_y and l_z denoting the unit cell extensions in x-, y- and z-directions.

In numerous experimental [127, 185] and simulation studies [139, 172, 176, 180–182, 186], (171) has in principle been confirmed, although some deviations have also been observed. In [139] these deviations were quantified by introducing a memory parameter

$$\beta = \frac{l_z^2/D_z}{l_x^2/D_x + l_y^2/D_y}. \tag{172}$$

β is obviously equal to 1 if (171) is valid. In cooperation with Prof. Kärger, also university of Leipzig, deviations of β from unity have been examined both in an analytical approach and in simulations (see also Chap.).

In this treatment, the trajectory of a given molecule has been replaced by a series of jumps from intersection to intersection. As in [176], conditional probabilities have been introduced in order to examine the memory effects. $p_{y,y}$, $p_{y,-y}$, $p_{y,x} = p_{y,-x}$ (and $p_{x,x}$, $p_{x,-x}$, $p_{x,y} = p_{x,-y}$) denote the probabilities that a displacement from intersection to intersection along a straight channel (zigzag channel) is followed by a displacement in the same direction, in the opposite direction, and along the other type of channels, respectively. These definitions are illustrated in Fig. 34. The probabilities that the next move of the diffusant is in the x- or y-direction (independent of the previous move) are named p_x and p_y. Let n_x and n_y be the number of moves of a molecule in the x- and y-direction during the time interval t. For the limiting

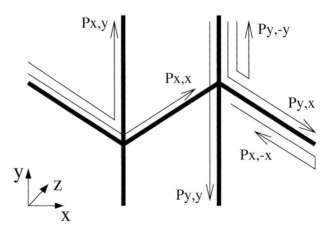

Fig. 34. Illustration of the transition probabilities

case of particles without memory (i.e. the case considered in [178, 179, 183, 184]) we have

$$p_{y,y} = p_{y,-y} = p_{x,y} = p_y/2 \tag{173}$$
$$p_{x,x} = p_{x,-x} = p_{y,x} = p_x/2, \tag{174}$$

By summing infinite series containing the conditional probabilities of passages through the straight and the zigzag channels, it is possible to derive formulas for the components of the diffusion tensor. Detailed analysis [187–189] leads to the result that the components of the diffusion tensor can be expressed in terms of these quantities by

$$D_x = \frac{a^2}{2t/n_x} \frac{1 + (p_{x,x} - p_{x,-x})}{1 - (p_{x,x} - p_{x,-x})}, \tag{175}$$

$$D_y = \frac{b^2}{2t/n_y} \frac{1 + (p_{y,y} - p_{y,-y})}{1 - (p_{y,y} - p_{y,-y})}, \tag{176}$$

$$D_z = \frac{c^2}{2t/n_x} \frac{p_{x,y}}{1 - p_{y,x} - p_{x,y}}. \tag{177}$$

Inserting these relations into the expression for the correlation factor (172) yields

$$\beta = \frac{p_1 \, C}{p_1 \, A + p_2 \, B} \tag{178}$$

The abbreviations

$$A = \frac{1 - (p_{x,x} - p_{x,-x})}{1 + (p_{x,x} - p_{x,-x})}, \tag{179}$$

$$B = \frac{1 - (p_{y,y} - p_{y,-y})}{1 + (p_{y,y} - p_{y,-y})}, \tag{180}$$

$$C = \frac{1 - p_{y,x} - p_{x,y}}{p_{x,y}} \tag{181}$$

have been used.

Test by MD Simulations. The relevance of the analytical expressions obtained has been investigated in extensive MD simulations, which are described in detail in [188]. The simulations have been carried out for a rigid lattice of pure silicalite-1 with methane as a spherically shaped guest molecule at a loading of one molecule per channel intersection (corresponding to four molecules per unit cell) and a temperature of 300 K. The interaction parameters for the methane/lattice interaction have been taken from the spherical model potential derived in [141]. In the simulation procedure we have essentially followed our previous studies of the same host–guest system [141, 172, 190].

In the present study we have considered runs with an unperturbed evaluation part (after thermalization) of 5×10^6 simulation steps. The temperature was adjusted in the

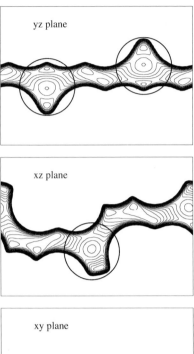

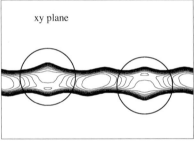

Fig. 35. Definition of the intersection regions. The circles mark those regions that are treated as intersection regions. The lowest energy values connected with small minima outside of the intersections are about minus 18 kJ/mol while the maximum in the center of the intersection is at about minus 9 kJ/mol. The energy difference between adjacent lines in the picture amounts to 1 kJ/mol. The outermost isopotential lines correspond to minus 5 kJ/mol

thermalization part of the run using a procedure described in [141, 191] which enables runs in the micro canonical ensemble with a predefined value of the temperature. As the time step was 5 fs, the length of the trajectory examined corresponded to a total time of 25 ns.

A declaration of the limits of the channel intersection regions is not free from arbitrariness. One possible choice is illustrated by Fig. 35. It shows the isopotential lines for the center of a single methane molecule in three planes: a cut through the straight channel in the yz-plane at $x = 0$, a cut through the zigzag channel in the xz-plane at $y = 0$, and a cut through the straight channel in the xy-plane at $z = 0$. The circles correspond to cuts through spherical regions of radius 3 Å, which can be

Table 2. Results of MD simulations for different radii r of intersection regions. The D values are in 10^{-8} m^2/s. The notations "no-mem", "mem" and MD refer to data analysis without memory effects, with memory effects, and the MD data

r in Å	2.0	2.5	3.0	5.0
$p_{x,x}$	0.19345	0.16695	0.13477	0.04369
$p_{x,-x}$	0.32031	0.43876	0.56070	0.87158
$p_{x,y}$	0.24078	0.19379	0.14724	0.04161
$p_{y,y}$	0.37762	0.33034	0.26652	0.08525
$p_{y,-y}$	0.34844	0.43512	0.54964	0.86070
$p_{y,x}$	0.13891	0.11877	0.09365	0.02751
p_y	0.63348	0.61635	0.60295	0.59967
p_x	0.36547	0.37775	0.38350	0.39653
n_x	3 634	4 768	6 661	30 353
n_y	6 443	8 007	11 023	47 611
D_x (no mem)	0.921	1.21	1.68	7.64
D_x (mem)	0.714	0.691	0.677	0.720
D_x (MD)	0.718	0.718	0.718	0.718
D_y (no mem)	1.61	1.99	2.74	11.81
D_y (mem)	1.70	1.62	1.53	1.49
D_y (MD)	1.68	1.68	1.68	1.68
D_z (no mem)	0.263	0.338	0.469	2.09
D_z (mem)	0.160	0.152	0.146	0.153
D_z (MD)	0.17	0.17	0.17	0.17
β (mem, (178))	1.40	1.42	1.42	1.41
β (MD, (172))	1.33	1.33	1.33	1.33

interpreted as intersection regions. The simulations have been carried out for different radii of such spheres. Table 2 provides a summary of the simulation results obtained. It particularly includes numerical values for all probabilities introduced in this study. To illustrate the data scattering, the values of both $p_{x,y}$ and $p_{x,-y}$ are presented, which are found to differ by several percent though they should coincide. The D values obtained from MD have been calculated from four moments of the displacement as described in [140] for the isotropic case, and in [141] for the anisotropic case. In addition to these diffusivities from the MD simulations, Table 2 also contains the diffusivity data which would result from use of (175), (176) with the indicated probabilities (case "with memory" marked by "(mem)" in the table) and with their simplifications by (173) and (174) (case "without memory" marked by "(nomem)" in the Table).

The analytical expressions derived yield values that are much closer to MD results than are those from all earlier analytical treatments of the problem. Comparison of the derived dependencies with the results of the MD simulations yields satisfactory agreement. Nevertheless, the agreement can probably be improved by further examinations. The interpretation of the results also remains a task for more detailed studies.

Butene in Silicalite-1

MD Simulations. The structure and connectivity of the channels of the zeolite silicalite-1 are visualized in Fig. 33 and explained in the previous section.

The diffusion of 1-butene in silicalite-1 has been examined by a jump model and MD simulations. In the simulations, the zeolite was modeled with a flexible lattice according to the model of Demontis et al. [192, 193]. Therein, harmonic potentials for the T-O bonds and the pairs of O-atoms bound to the same T-atom are assumed. The simulation box contained $2 \times 2 \times 4$ unit cells of silicalite-1.

1-butene was modeled in the united-atom approximation with four interaction sites. The molecules are flexible, with a dihedral potential

$$V_D(\phi) = V_0 + V_1 \cos \phi + V_2 \cos^2 \phi + V_3 \cos^3 \phi, \tag{182}$$

where $V_0 = 2.265$ kJ/mol, $V_0 = 7.755$ kJ/mol, $V_2 = 1.83$ kJ/mol and $V_3 = 9.38$ kJ/mol according to [194]. Furthermore, bending potentials

$$V_B(\Theta) = \frac{k_\Theta}{2}(\Theta - \Theta_0)^2 \tag{183}$$

were used, where we chose the force constant $k_\Theta = 520$ kJ/mol, used for propane in [46]. The equilibrium angles are $\Theta_0 = 123.7°$ for the angle in CH_2=CH-CH_2 and $\Theta_0 = 112.8°$ for the angle in CH-CH_2-CH_3. A Morse potential

$$V_{C-C}(r) = D \left[1 - e^{-\beta(r-r_0)} \right]^2 \tag{184}$$

was used for the C-C bonds. According to the geometry of 1-butene, we set $r_0 = 0.138$ nm for the C=C double bond, $r_0 = 0.152$ nm for the middle C-C bond and $r_0 = 0.154$ nm for the outer single C-C bond. The other parameters are $D = 83.9$ kJ/mol and $\beta = 18.4$ nm^{-1}, as used for ethane in [195]. For the guest–host interaction, the potential parameters listed in Table 3 were chosen.

Table 3. Lennard–Jones potential parameters for the guest–guest and guest–host interaction, "bfc" denotes butene force center

interaction	σ / nm	ϵ / kJ/mol
bfc–bfc	0.378	0.866
bfc–O	0.317	1.180
bfc–Si	0.212	0.683

Figure 36 shows positions of the center of mass of the butene molecules registered during the simulation. 1-butene is most likely to be found in the channels with a probability of $P_x \approx 0.5$ for the x-channel and $P_y \approx 0.4$ for the y-channel at 300 K. It turns out that not the intersections but, the central regions of the channels are favorable places for the center of mass of the butene molecules. Therefore, a jump model has been developed, based on these regions as adsorption sites.

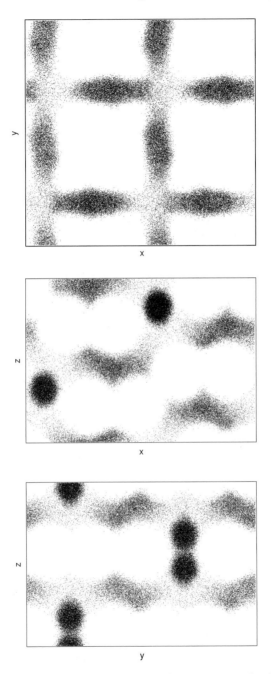

Fig. 36. Positions registered for the center of mass of the 1-butene molecules in the unit cell of silicalite-1 at $T = 300$ K in different projections

Analytical Derivations. As the particles spend relatively long times at the individual adsorption sites, memory effects are negligible and the correlations between the components of the diffusion tensor are only of geometrical nature. Therefore, it is not necessary to sum infinite jump sequences as in cases where the nodes in the model are the intersections, e.g. in [187–189]. If N_{xx} is the number of direct jumps from one x-channel segment to another one, N_{xy} the number of jumps from an x-channel segment to a y-channel segment, and so on, it can be shown [196, 197] that

$$D_x = \frac{1}{2t} \left(\frac{a}{2}\right)^2 \left(N_{xx} + \frac{1}{4}(N_{xy} + N_{yx})\right), \tag{185}$$

$$D_y = \frac{1}{2t} \left(\frac{b}{2}\right)^2 \left(N_{yy} + \frac{1}{4}(N_{xy} + N_{yx})\right), \tag{186}$$

$$D_z = \frac{1}{2t} \left(\frac{c}{2}\right)^2 \frac{1}{4}(N_{xy} + N_{yx}), \tag{187}$$

$$\beta = \left(\frac{(N_{xy} + N_{yx})/4}{N_{xx} + (N_{xy} + N_{yx})/4} + \frac{(N_{xy} + N_{yx})/4}{N_{yy} + (N_{xy} + N_{yx})/4}\right)^{-1}, \tag{188}$$

i.e., if

$$P_{x,xy} = \frac{(N_{xy} + N_{yx})/4}{N_{xx} + (N_{xy} + N_{yx})/4} \approx \frac{N_{xy}/2}{N_{xx} + N_{xy}/2} \tag{189}$$

is defined as the relative part of xy-jumps in the self-diffusion (and analogously $P_{y,xy}$), then

$$\beta = 1/(P_{x,xy} + P_{y,xy}). \tag{190}$$

Comparison and Conclusions. The analytical derivations have been compared with MD simulations. The results are displayed in Table 4.

Table 4. Self-diffusion coefficients (in $10^{-9}\,\mathrm{m}^2/\mathrm{s}$) from the evaluation of jumps between sites (N) and from the mean square displacement, and also the memory parameter β

T in K	$D_x^{(N)}$	$D_y^{(N)}$	$D_z^{(N)}$	$\beta^{(N)}$	$P_{x,xy}^{(N)}$	$P_{y,xy}^{(N)}$	D_x	D_y	D_z	β
300	1.0	2.3	0.23	1.34	0.53	0.22	0.8	2.2	0.20	1.36
400	2.9	5.8	0.68	1.28	0.52	0.26	2.7	6.0	0.65	1.29
500	5.5	10.4	1.3	1.23	0.53	0.28	5.4	10.1	1.3	1.21
600	7.9	14.6	1.9	1.19	0.55	0.29	8.0	16.0	2.0	1.20

The coincidence between the respective values is very good in most cases, $D_x^{(N)}$ is slightly overestimated at 300 K. This indicates that there might be correlations which are not considered in the model. Nevertheless, the comparison shows that the simple model without correlations except the geometrical ones works surprisingly well.

The memory parameter β is greater than unity in all cases. The interpretation of this fact is facilitated by the probabilities $P_{x,xy}$ and $P_{y,xy}$. $P_{x,xy} > 0.5$ means that there is a slight preference for xy-jumps when the molecule leaves the x-channel, but $P_{y,xy} < 0.5$ indicates that the step forward from one to the neighboring y-channel segment is significantly preferred. An analysis of the molecular orientations showed that the prefered orientation in the channel intersections is along the y-axis. This can explain the behavior described above.

Water in Chabazite

Zeolites, except purely siliceous kinds, usually contain water, and considerable effort is necessary to remove the water, e.g. by heating of the zeolite.

Water molecules play a very important role in cation–containing zeolites as, for example the mobility of the exchangeable cations depends strongly upon the water content [23]. As water is important for nearly all technological applications of zeolites, it is of great interest to examine the behavior of water molecules inside the zeolites. Particularly, the diffusion of water molecules in chabazite has been investigated. This system has been chosen because experimental data are accessible [184].

The simulation of this diffusional process [198] turned out to be at the limit of the computational capabilities, so we had to increase the temperature up to $T = 600$ K to get mean square displacements which were large enough to evaluate diffusion co-efficients. The multiple–time–step algorithm RESPA [199] has been used to enhance the performance of the computation. At $T = 600$ K the system shows a quite un-common dependence on the loading (see Fig. 37). For the almost dehydrated zeolite with only one quarter of the full loading, there is a very slow diffusion. Then the

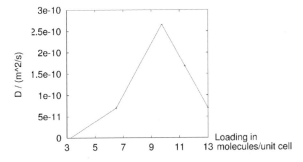

Fig. 37. Mean diffusion coefficient versus loading at $T = 600$ K.

diffusion coefficient increases with increasing loading, up to a maximum value for 75% of the full loading. Then it decreases with further increase of the loading.

This anomalous behavior can be explained with knowledge about the adsorption places. At approximately half of the maximum loading, almost all preferred places in the hydration shells of the cations are filled up. Therefore, at higher loadings there are some water molecules which are only loosely bound and relatively free to move. With further increase in the loading, the fraction of mobile molecules increases, leading to more diffusional motion, but with more molecules, the number of potential collision partners increases as well, which limits the increase of D and dominates for the highest loadings.

Quantum Chemical – Potential Calculations and MD Simulations

Simulation methods such as MD or MC need as input the interaction potentials of the particles involved. As these potentials cannot be measured directly, there are in principle only two possible ways to obtain them. The first possibility consists of fitting potential parameters to experiments, and the second way is to find their values by quantum mechanical calculations. Of course, various modifications of these two major routes exist. In a cooperation of theoretical and experimental groups at Leipzig university with scientists at Chulalongkorn University, Bangkok, quantum calculations of potentials have been carried out and the potentials then have been used in MD simulations to investigate the static and dynamic properties of water in silicalite-1 [200–204]. One question of interest was wether the water molecules form clusters in the channels of silicalite-1. In the cases investigated in [200–204] no clusters could be found. The diffusion coefficients obtained from the simulations agree satisfactorily with those from experiments [204].

Acknowledgements

The authors wish to thank the colleagues of the SFB 294 for a fruitful long-term collaboration and numerous helpful discussions and comments, particularly we would like to mention J. Kärger, G. Klose, M. Lösche, and D. Michel.

Further we acknowledge cooperation with S. Auerbach (Amherst), A. Bell (Berkeley), Ph. Bopp (Bordeaux), A.S. Cukrowski (Warsaw), P. Demontis (Sassari), H. Fischer (Wien), S. Hannongbua (Bangkok), K. Heinzinger (Mainz), H. Krienke (Regensburg), H. Jobic (Villeurbanne), I. Nezbeda (Prague), W.R. Smith (Guelph), R. Snurr (NWU), G.B. Suffritti (Sassari), D. Theodorou (Athens), M. Wolfsburg (Irvine).

Last but not leat we are indebted to our scientific co-workers in the projects M. Gaub, S. Jost, A. Schüring, G. Hofmann, J. Galle, M. Kettler, C. Bussai (Bangkok).

References

1. A. Münster: *Statistical Thermodynamics, Vol. 1* (Springer, Berlin, Heidelberg, 1969)
2. J. Doob: *Stochastic Processes* (New York, 1953)
3. A. Ramakrisnan: "Probability and Stochastic Processes", in *Encyclopedia of Physics, Volume III/2, Principles of Thermodynamics and Statistics*, ed. by S. Flügge (Springer, Berlin, Göttingen, Heidelberg, 1959), pp. 524–651
4. F. Schlögl: *Probability and Heat* (Vieweg, Braunschweig, Wiesbaden, 1989)
5. L.D. Landau, E.M. Lifshitz: *Course of Theoretical Physics, Vol. 5: Statistical Physics* (Pergamon, London-Paris, 1958)
6. R. Becker: *Theory of Heat* (Springer, Berlin, Heidelberg, 1967)
7. L.D. Landau, E.M. Lifshitz: *Course of Theoretical Physics, Vol. 1: Classical Mechanics* (Pergamon, London-Paris, 1958)
8. R. Haberlandt, S. Fritzsche, G. Peinel, K. Heinzinger: *Molekulardynamik - Grundlagen und Anwendungen, mit einem Kapitel über Monte-Carlo-Simulationen von H. - L. Vörtler* (Vieweg–Verlag, Wiesbaden, 1995)
9. J. Hirschfelder, C. Curtiss, R. Bird: *Molecular Theorie of Gases and Liquids* (John Wiley and Sons, New York, 1954)
10. M.P. Allen, D. Tildesley: *Computer Simulation of Liquids* (Clarendon Press, Oxford, 1989)
11. F.F. Abraham: Adv. Phys. **35**, 1 (1985)
12. K. Binder: *Monte Carlo Methods in Statistical Physics* (Springer Verlag, Berlin, Heidelberg, New York, Tokyo, 1986)
13. K. Binder, D. Heermann: *Monte Carlo Simulations in Statistical Physics* (Springer Verlag, Berlin, Heidelberg, New York, Tokyo, 1988)
14. M.P. Allen, D.J. Tildesley (Eds.): *Computer Simulation in Chemical Physics* (Kluwer Academic Publishers, Dortrecht, Boston, London, 1993) (1993), proceedings of the NATO Advanced Study Institute on New Perspectives in Computer Simulation in Chemical Physics, Alghero, Sardinia, Italy, September 14-24, 1992
15. F. Vesely: *Computerexperimente an Flüssigkeitsmodellen* (Physik–Verlag, Weinheim, 1978)
16. R. Kubo: *Statistical Mechanics* (North-Holland, Amsterdam, 1967)
17. D. Frenkel, B. Smit: *Understanding Molecular Simulation* (Academic Press, San Diego, London, Boston, New York, Sidney, Tokyo, Toronto, 1996)
18. M.W. Zemansky: *Heat and Thermodynamics* (McGraw-Hill, New York, Toronto, London, 1957)
19. K. Denbigh: *The Principles of Chemical Equilibrium* (University Press, Cambridge, 1955)
20. D. McQuarrie: *Statistical Mechanics* (Harper & Row, New York, Evanston, San Francisco, London, 1976)
21. B. Widom: J. Chem. Phys. **39**, 2808 (1963)
22. J.P. Boon, S. Yip: *Molecular Hydrodynamics* (McGraw-Hill, New York, 1980)
23. J. Kärger, D.M. Ruthven: *Diffusion in Zeolites and Other Microporous Solids* (Wiley, New York, 1992)
24. S. de Groot, P. Mazur: *Non-Equilibrium Thermodynamics* (North-Holland, Amsterdam, 1963)
25. H.J. Kreuzer: *Nonequilibrium Thermodynamics and its Statistical Foundations* (Clarendon Press, Oxford, 1981)
26. R. Kubo, M. Toda, N. Hashitsume: *Statistical Physics II Nonequilibrium Statistical Mechanics* (Springer Verlag, Berlin, Heidelberg, New York, London, Tokyo, 1991)
27. G.D. Harp, B.J. Berne: Phys. Rev. A **2**, 975 (1970)

28. R. Zwanzig: J. Chem. Phys. **33**, 1338 (1960)
29. R. Zwanzig: Annu. Rev. Phys. Chem. **12**, 67 (1961)
30. J. Hansen, I. McDonald: *Theory of Simple Liquids* (Academic Press, London, Orlando, New York, San Diego, Austin, Boston, Tokyo, Sidney, Toronto, 1986)
31. R. Haberlandt: "Statistical Theory and Molecular Dynamics of Free and Adsorbed Fluids-Diffusion in Zeolites", in *Diffusion in Condensed Matter*, ed. by J. Kärger, P. Heitjans, R. Haberlandt (Vieweg, Wiesbaden, 1998), pp. 363 – 382
32. R. Haberlandt, S. Fritzsche, H.L. Vörtler: "Simulation of Microporous Systems: Confined Fluids in Equilibrium and Diffusion in Zeolites", in *Handbook of Surfaces and Interfaces of Materials*, ed. by H.S. Nalwa, Vol. 5 (Academic Press, San Diego, London, Boston, New York, Sidney, Tokyo, Toronto, 2001), pp. 358–444
33. N. Metropolis, A.W. Rosenbluth, M.N. Rosenbluth, A.H. Teller, E. Teller: J. Chem. Phys. **21**, 1087 (1953)
34. K. Binder: *Applications of the Monte Carlo Methods in Statistical Physics* (Springer, Berlin, Heidelberg, New York, Tokyo, 1987)
35. I.R. McDonald: Mol. Phys **23**, 41 (1972)
36. H.L. Vörtler, J. Kolafa, I. Nezbeda: Mol. Phys **68**, 547 (1989)
37. I. Nezbeda, H.L. Vörtler: Mol. Phys **57**, 909 (1985)
38. H.L. Vörtler, W.R. Smith: J. Chem. Phs. **112**, 5168 (2000)
39. R. Eppenga, D. Frenkel: Molec. Phys. **52**, 1303 (1984)
40. A.Z. Panagiotopoulos: Mol. Phys **61**, 813 (1987)
41. A.Z. Panagiotopoulos: Molec. Phys. **62**, 701 (1987)
42. A. Panagiotopoulos: Molecular Simulations **9**, 1 (1992)
43. M. Kettler, H.L.V. Strnad, I. Nezbeda, M. Strnad: Fluid Phase Equilibria **181**, 83 (2001)
44. W.R. Smith, H.L. Vörtler: Chem. Phys. Letters **249**, 470 (1996)
45. W.R. Smith, B. Triska: J. Chem. Phys. **100**, 3019 (1994)
46. B. Smit: J. Chem. Phys. **99**, 5597 (1995)
47. W.R. Smith, I.Nezbeda, M. Strnad, B. Triska, S. Labik, A. Malijevsky: J. Chem. Phys. **109**, 1052 (1998)
48. I. Nezbeda, J. Kolafa: Molec. Simulation **5**, 391 (1991)
49. J. Kolafa, H.L. Vörtler, K. Aim, I. Nezbeda: Molec. Simulation **11**, 5 (1993)
50. H.L. Vörtler: Verhandl. der Dt. Phys. Ges. **17a**, 965 (1993)
51. H.L. Vörtler, M. Heuchel: "Grand canonical monte carlo simulations with gradually particle insertion: Mixtures of hard spheres and linear symmetric triatomics", Paper #4, 4th Liblice Conf. Statist. Mech. Liquids, June 1994) (1994)
52. S. Labik, W.R. Smith: Molec. Simulation **12**, 23 (1994)
53. S. Labik, W.R. Smith: Molec. Phys. **88**, 1411 (1996)
54. M. Strnad, I. Nezbeda: Molec. Simulation **22**, 193 (1999)
55. M. Kettler:(1998),"Monte-Carlo-Simulation molekularer Fluide: Struktur und Phasenverhalten im Bulk und in Poren", Ph.D. Thesis, University Leipzig,
56. M. Kettler, I. Nezbeda, A.A. Chialvo, P.T. Cummings: J. Phys. Chem. B **106**, 1537 (2002)
57. J.P. Valleau: J. Chem. Phys. **99**, 4718 (1993)
58. K. Kiyohara, T. Spyriouni, K.E. Gubbins, A.Z. Panagiotopoulos: Mol. Phys. **89**, 965 (1996)
59. G.M. Torrie, J.P. Valleau: J. Chem. Phys **66**, 1402 (1977)
60. G.M. Torrie, J.P. Valleau: J. Comp. Phys **23**, 187 (1977)
61. R.F. Cracknell, D. Nicholson, N.G. Parsonage, H. Evans: Mol. Phys. **71**, 931 (1990)
62. J. Harris, S.A. Rice: J. Chem Phys. **88**, 1298 (1988)
63. J.I. Siepman, D. Frenkel: Mol. Phys. **75**, 59 (1992)
64. D. Frenkel, G.C.A.M. Mooij, B. Smit: J. Phys. Condens. Matter **4**, 3053 (1992)

65. J.J. de Pablo, M. Laso, U.W. Suter: J. Chem. Phys. **96**, 2395 (1992)
66. B. Smit, T.L.M. Maesen: Nature **374**, 42 (1995)
67. M. Dijkstra: J. Chem. Phys. **107**, 3277 (1997)
68. N.B. Wilding, A.D. Bruce: J. Phys. Condens. Matter **4**, 3087 (1992)
69. A.M. Ferrenberg, R.H. Swendsen: Phys. Rev. Lett. **61**, 2635 (1988)
70. A.M. Ferrenberg, R.H. Swendsen: Phys. Rev. Lett. **63**, 1195 (1989)
71. R. Swendsen: Physica A **53**, 53 (1993)
72. N.B. Wilding: Phys. Rev. E **52**, 602 (1995)
73. J.J. Potoff, A.Z. Panagiotopoulos: J. Chem. Phys. **109**, 10914 (1998)
74. Privman: *Finite Size scaling and numerical simulation of statistical systems* (World Scientific, Singapore, 1990)
75. K. Binder: *Computational methods in Field theory (Eds. H. Gausterer, C.B. Lang)* (Springer, Berlin, 1990)
76. R.B. Griffiths, J.C. Wheeler: Phys. Rev. A **2**, 1047 (1970)
77. A.D. Bruce, N.B. Wilding: Phys. Rev. Lett. **68**, 193 (1992)
78. B.N. B. A. Berg: Phys. Rev. Lett. **68**, 9 (1992)
79. W. Janke: *Computer simulations in condensed matter physics VII (Eds. D.P. Landau, K.K. Mon, H.B. Schüttler)* (Springer, Heidelberg, Berlin, 1994)
80. O. Dillmann, W. Janke, M. Müller, K. Binder: J. Chem. Phys **114**, 5853 (2001)
81. I. Nezbeda, J. Kolafa: Molec.. Phys. **97**, 2205 (1999)
82. J. Kolafa, I. Nezbeda: Molec.. Phys. **98**, 1505 (2000)
83. J. Kolafa, M. Lísal, I. Nezbeda: Molec.. Phys. **99**, 1751 (2001)
84. W. Bol: Mol. Phys. **45**, 602 (1982)
85. W.R. Smith, I. Nezbeda: J. Chem. Phys. **81**, 602 (1984)
86. I. Nezbeda: J. molec. Liq. **73-74**, 317 (1997)
87. W.M. Jorgensen: J. Am. Soc. **10**, 335 (1981)
88. I. Nezbeda, J. Slovak: Mol. Phys **90**, 353 (1997)
89. J. Slovak, I. Nezbeda: Mol. Phys **91**, 1125 (1997)
90. W. E: *Hydration of Macromolecules* (McMillen Press, New York, 1993)
91. S. Murad, P. Ravi, J.G. Powles: J. Chem. Phys. **98**, 9771 (1993)
92. J.S. Rowlinson: J. Chem. Soc. Faraday Trans. 2 **82**, 801 (1986)
93. W.G. Madden, E.D. Glandt: J. Stat.Phys. **51**, 537 (1988)
94. J. Kolafa, I. Nezbeda: Mol. Phys. **61**, 161 (1987)
95. I. Nezbeda: Mol. Phys. **91**, 1125 (1997)
96. M. Schoen: *Computer simulations of condensed phases in complex geometries* (Springer, Berlin, 1993)
97. J.N. Israelachvili, S. Marcelja, R.G. Horn: Quart. Rev. Biophys. **13**, 121 (1980)
98. M. Bloom, E. Evans, O. Mouritsen: Quart. Rev. Biophys. **24**, 293 (1991)
99. G.C.E.W. ed.): *Hydration of Macromolecules* (McMillen Press, New York, 1993)
100. J. Galle, H.L. Vörtler, K.P. Schneider: Surface Sci. **387**, 78 (1997)
101. M. Predota, I. Nezbeda: Mol. Phys. **96**, 353 (1999)
102. T.J. McIntosh, A.D. Magid, S.A. Simon: Biochemistry **26**, 7325 (1987)
103. J.N. Israelachvili, H. Wennerström: Langmuir **6**, 873 (1990)
104. J.N. Israelachvili, H. Wennerström: J. Phys. Chem. **96**, 520 (1992)
105. H. Reiss: Adv. Chem. Phys. **9**, 1 (1965)
106. R.J. Speedy: J. Chem. Soc. Faraday II **76**, 693 (1980)
107. W. Smith, H. Vörtler: Molec. Phys. **101**, 805 (2003)
108. J.M. Pope, D. Dubro, J.W. Doane, P.W. Westerman: J. Am. Chem. Soc. **108**, 5426 (1986)
109. J. Galle, H.L. Vörtler: Surface Science **481**, 39 (2001)
110. K. Binder: *The Monte Carlo Methods in Condensed Matter Physics* (Springer, Berlin, Heidelberg, 1995)

111. H.L. Vörtler: *Abriss der Monte-Carlo-Methode (in R. Haberlandt, S. Fritzsche, G. Peinel, K. Heinzinger Molekulardynamik - Grundlagen und Anwendungen)* (Vieweg, Wiesbaden, 1995)
112. J. Galle, F. Volke: Biophys. Chem. **54**, 109 (1995)
113. Z. Zhang, M.J. Zuckermann, O.G.M.E.R. Brasseur): *Molecular Description of Biological Membrane Components by Computer Aided Conformational Analysis, Vol. 1* (CRC Press, 1988)
114. Z. Zhang, M.J. Zuckermann, O.G. Mouritsen: Mol. Phys. **80**, 1195 (1993)
115. H.L. Vörtler, M. Kettler: Chem. Phys. Letters **266**, 368 (1997)
116. H. Vörtler, W. Smith: J. Chem. Phys. **112**, 5168 (2000)
117. D. Nicholson, N.G. Parsonage: *Computer simulation and the statistical mechanics of adsorption* (Academic Press, London, 1982)
118. I. Nezbeda, M.R. Reddy, W.R. Smith: Mol. Phys **71**, 915 (1990)
119. R. Kjellander, S. Sarman: J. Chem. Soc. Faraday Trans. **87**, 1869 (1991)
120. S.H. Chen, R.Rajagopalan: *Micellar Solutions and Microemulations* (Springer, New York, 1990)
121. L.D. Gelb, K.E. Gubbins, R. Radhankrishnan, M. Sliwinska-Bartowiak: Rep. Prog. Phys **62**, 1573 (1999)
122. R. Evans: J. Phys. Cond. Matter **2**, 8989 (1990)
123. H. Dominguez, M.P. Allen, R. Evans: Molec. Phys. **96**, 209 (1999)
124. L. Vega, E. de Miguel, L. Rull, A.M. G. Jackson: J. Chem. Phys. **96**, 2296 (1992)
125. S.C. McGrother, K.E. Gubbins: Mol. Phys. **97**, 955 (1999)
126. H. Vörtler, M. Kettler: Chem. Phys. Letters **377**, 557 (2003)
127. J. Caro, M. Noack, K. Kölsch, R. Schäfer: Micropor. Mesopor. Mater. **38**, 3 (2000)
128. S. Bates, R. van Santen: Adv. Catal. **42**, 1 (1998)
129. N. Chen, J. T.F. Degnan, C. Smith: *Molecular Transport und Reaction in Zeolites* (VCH, New York, 1994)
130. D.N. Theodorou, R. Snurr, A.T. Bell: "Molecular Dynamics and Diffusion in Microporous Materials", in *Comprehensive Supramolecular Chemistry, Edt. G. Alberti and T. Bein*Vol. 7 (Pergamon, Oxford, 1996), pp. 507–548
131. P. Demontis, G.B. Suffritti: Chemical Reviews **97**, 2845 (1997)
132. F. Keil, R. Krishna, M.O. Coppens: Chem. Engin. Journal **16**, 71 (2000)
133. B. Alder, T. Wainwright: J. Chem. Phys. **27**, 1208 (1957)
134. B.J. Alder, T. Wainwright: J. Chem. Phys. **31**, 459 (1959)
135. B.J. Alder, T. Wainwright: J. Chem. Phys. **33**, 1439 (1960)
136. ZAM Jülich, Report No. **82** (February 2000)
137. H. Andersen: J. Comput. Phys. **52**, 24 (1983)
138. M.E. Tuckerman, G.J. Martyna: J. Phys. Chem. B **104**, 159 (2000)
139. E.J. Maginn, A.T. Bell, D.N. Theodorou: J. Phys. Chem **100**, 7155 (1996)
140. S. Fritzsche, R. Haberlandt, J. Kärger, H. Pfeifer, K. Heinzinger: Chem. Phys. Lett. **198**, 283 (1992)
141. S. Fritzsche, M. Wolfsberg, R. Haberlandt: Chem. Phys. **289**, 321 (2003)
142. M. Gaub: (1998), "Molekulardynamische Untersuchungen zur Diffusion von Methan in Zeolithen", Ph.D. Thesis, University Leipzig
143. M. Gaub, S. Fritzsche, R. Haberlandt, D.N. Theodorou: "Transport and Self-Diffusion Coefficients in Zeolites–An MD Study", in *Proceedings of the 12th International Zeolite Conference, Baltimore 1998*, ed. by M.M.J. Treacy, B.K. Marcus, M.E. Bisher, J.B. Higgins (Materials Research Society, Warrendale, 1999), pp. 371–378
144. M. Gaub, S. Fritzsche, R. Haberlandt, D.N. Theodorou: J. Phys. Chem. B **103**, 4721 (1999)
145. R. Haberlandt, J. Kärger: Chemical Engineering Journal **74**, 15 (1999)

146. General Discussion during the Faraday Symposium 26 on *Molecular Transport in Confined Regions and Membranes*, J. Chem. Soc. Faraday Trans. **87**(1991)1997–2010

147. S. Fritzsche, R. Haberlandt, J. Kärger, H. Pfeifer, M. Waldherr-Teschner: "An MD study of Methane diffusion in zeolites of structure type LTA", in *Zeolites and Related Microporous Materials: State of the Art, Studies in Surface Science and Catalysis, Vol. 84, 1994, p. 2139 - 2146*, ed. by J. Weitkamp, H.G. Karge, H. Pfeifer, W. Hoelderich (Elsevier, Amsterdam, 1994), pp. 2107–2113, proceedings of the the 10th International Zeolite Conference, Garmisch– Partenkirchen, 1994

148. S. Fritzsche, R. Haberlandt, J. Kärger, H. Pfeifer, K. Heinzinger, M. Wolfsberg: Chem. Phys. Lett. **242**, 361 (1995)

149. W. Heink, J. Kärger, S. Ernst, J. Weitkamp: Zeolites **14**, 320 (1994)

150. W. Heink, J. Kärger, H. Pfeifer, P. Salverda, K. Datema, A. Nowak: J. Chem. Soc. Faraday Trans. **88**, 515 (1992)

151. S. Fritzsche, R. Haberlandt, J. Kärger, H. Pfeifer, K. Heinzinger: Chem. Phys. **174**, 229 (1993)

152. P. Demontis, G.B. Suffritti: Chem. Phys. Lett. **223**, 355 (1994)

153. P. Demontis, G.B. Suffritti: "Molecular Dynamics Simulations of Diffusion in a Cubic Symmetry Zeolite", in *Zeolites and Related Microporous Materials: State of the Art*, ed. by J. Weitkamp, H.G. Karge, H. Pfeifer, W. Hoelderich (Elsevier, Amsterdam, 1994), pp. 2107–2113, proceedings of the the 10th International Zeolite Conference, Garmisch– Partenkirchen, 1994

154. S. Fritzsche, M. Wolfsberg, R. Haberlandt, P. Demontis, G.B. Suffritti, A. Tilocca: Chem. Phys. Lett. **296**, 253 (1998)

155. S. Fritzsche, M. Wolfsberg, R. Haberlandt: Chem. Phys. **253**, 283 (2000)

156. S. Fritzsche, R. Haberlandt, J. Kärger: Z. phys. Chem. **189**, 211 (1995)

157. E.J. Maginn, A.T. Bell, D.N. Theodorou: J. Phys. Chem **97**, 4173 (1993)

158. S. Xu, Z. Brandani, D. Ruthven: Microporous Materials **7**, 323 (1996)

159. P. Demontis, G.B. Suffritti: Studies in Surface Science and Catalysis **105**, 1843 (1997)

160. J. Kärger, D.M. Ruthven: "Self–Diffusion and Diffusive Transport in Zeolite Crystals", in *Progress in Zeolite and Microporous Materials*, ed. by Y.U. H. Chon S.-K. Ihm (Elsevier Science B.V., 1997), pp. 1843–1858, studies in Surface Science and Catalysis, **105**(1997)

161. R. Haberlandt: Thin Solid Films **330**, 34 (1998)

162. O.M. Coppens, A.T. Bell, A.K. Chakraborty: "Influence of Occupancy and Pore Network Topology on Tracer and Transport Diffusion in Zeolites", in *Scientific Computing in Chemical Engineering*, ed. by F. Keil, W. Mackens, H. Voss, J. Werther (Springer–Verlag, Berlin, 1999), pp. 200

163. D. Paschek, R. Krishna: Phys. Chem. Chem. Phys. **2**, 2389 (2000)

164. J.P. Hoogenboom, H.L. Tepper, N.F.A. van der Vegt, W.J. Briels: J. Chem. Phys. **113**, 6875 (2000)

165. D. Paschek, R. Krishna: PCCP **3**, 3185 (2001)

166. K. Malek, M.O. Coppens: Phys. Rev. Lett. **87**, 1255 051 (2001)

167. A.I. Skoulidas, D.S. Sholl: J. Phys. Chem. B **105**, 3151 (2001)

168. R. Krishna: Chem. Phys. Lett. **355**, 483 (2002)

169. A.I. Skoulidas, D.S. Sholl: J. Phys. Chem. B **106**, 5058 (2002)

170. T. Düren, F.J. Keil, N.A. Seaton: Mol. Phys. **100**, 3741 (2002)

171. T. Düren, S. Jakobtorweihen, F.J. Keil, N.A. Seaton: Phys. Chem. Chem. Phys. **5**, 369 (2003)

172. S. Jost, N.K. Bär, S. Fritzsche, R. Haberlandt, J. Kärger: J. Phys. Chem. B **102**, 6375 (1998)

173. A. Schüring, S.M. Auerbach, S. Fritzsche, R. Haberlandt: J. Chem. Phys. **116**, 10 890 (2002)
174. D. Chandler: *Introduction to Modern Statistical Mechanics* (Oxford University Press, New York, 1987)
175. D. Chandler: J. Chem. Phys. **68**, 2959 (1978)
176. F. Jousse, S.M. Auerbach, D.P. Vercauteren: J. Chem. Phys. **112**, 1531 (2000)
177. A. Schüring, S.M. Auerbach, S. Fritzsche, R. Haberlandt: submitted to J. Phys. Chem. B Capturing geometric correlations for ethane diffusion in cation-free LTA zeolite through the vacancy correlation factor
178. J. Kärger: J. Phys. Chem **95**, 5558 (1991)
179. D. Fenzke, J. Kärger: Z. Phys. D **25**, 345 (1993)
180. J. Kärger, P. Demontis, G.B. Suffritti, A. Tilocca: J. Chem. Phys. **110**, 1163 (1999)
181. P. Demontis, J. Kärger, G.B. Suffritti, A. Tilocca: Phys. Chem. Chem. Phys. **2**, 1455 (2000)
182. P. Demontis, G.B. Suffritti, A. Tilocca: J. Chem. Phys. **113**, 7588 (2000)
183. J. Kärger, H. Pfeifer: Zeolites **12**, 872 (1992)
184. N.K. Bär, J. Kärger, H. Pfeifer, H. Schäfer, W. Schmitz: Microporous Mesoporous Mater. **22**, 289 (1998)
185. J. Caro, M. Noack, J. Richter-Mendau, F. Marlow, D. Peterson, M. Griepentrog, J. Kornatowski: J. Phys. Chem. **97**, 13 685 (1993)
186. T.J.H. Vlugt, C. Dellago, B. Smit: J. Chem. Phys. **113**, 8791 (2000)
187. S. Fritzsche, J. Kärger: Europhys. Lett. **63**, 465 (2003)
188. S. Fritzsche, J. Kärger: J. Phys. Chem. B **107**, 3515 (2003)
189. S. Fritzsche, J. Kärger: Studies in Surface Science and Catalysis **142**, 1955 (2002)
190. S. Fritzsche, R. Haberlandt, S. Jost, A. Schüring: Molec. Sim. **25**, 27 (2000)
191. S. Fritzsche: *Untersuchung ausgewählter Nichtgleichgewichtsvorgänge in Vielteilchensystemen mittels statistischer Physik und Computersimulationen*, Habilitation Thesis, University of Leipzig, 1998
192. P. Demontis, G.B. Suffritti, S. Quartieri, E.S. Fois, A. Gamba: Zeolites **7**, 522 (1987)
193. P. Demontis, G.B. Suffritti, S. Quartieri, E.S. Fois, A. Gamba: J. Phys. Chem. **92**, 867 (1988)
194. W. Jorgensen, J. Madura, C. Swenson: J.Am.Chem.Soc. **106**, 6638 (1984)
195. P. Demontis, G.B. Suffritti, A. Tilocca: J. Chem. Phys. **105**, 5586 (1996)
196. A. Schüring: (2003), "Molekulardynamik-Simulationen und Sprungmodelle zur Diffusion in Zeolithen", Ph.D. thesis, University of Leipzig, 2003
197. A. Schüring, D. Michel, J. Roland, S. Fritzsche, R. Haberlandt, *On the coupling of rotational and translational diffusion in zeolite channels – a new means to measure intracrystalline self diffusion*, in preparation
198. S. Jost, S. Fritzsche, R. Haberlandt: Studies in Surface Science and Catalysis **142**, 1947 (2002)
199. M.E. Tuckerman, G.J. Martyna, B.J. Berne: J. Chem. Phys. **97**, 1990 (1992)
200. C. Bussai, S. Hannongbua, R. Haberlandt, J. Phys. Chem. B **105**, 3409 (2001)
201. C. Bussai, R. Haberlandt, S. Hannongbua, Jost, S. S. Fritzsche, Studies in Surface Science and Catalysis **106**, 15–P–28 (2001).
202. C. Bussai, S. Hannongbua, S. Fritzsche, R. Haberlandt, Chem. Phys. Lett. **354**, 310 (2002).
203. C. Bussai, S. Hannongbua, S. Fritzsche, R. Haberlandt, Studies in Surface Science and Catalysis **142**, 1979 (2002).
204. C. Bussai, S. Vasenkov, H. Liu, W. Böhlmann, S. Fritzsche, S. Hannongbua, R. Haberlandt, J. Kärger, Applied Catalysis A **232**

Diffusion in Channels and Channel Networks

Peter Bräuer[1], Siegfried Fritzsche[2], Jörg Kärger[1], Gunter Schütz[3],
and Sergey Vasenkov[1]

[1] Universität Leipzig, Institut für Experimentalphysik I,
    kaerger@physik.uni-leipzig.de, brauer@chemie.uni-leipzig.de,
    vasenkov@physik.uni-leipzig.de
[2] Institut für Theoretische Physik,
    Siegfried.Fritzsche@uni-leipzig.de
[3] Forschungszentrum Jülich, Institut für Festkörperforschung,
    G.Schuetz@fz-juelich.de

Abstract. A large variety of nanoporous materials host channel networks with pore diameters
on the order of typical molecular dimensions. Molecular diffusion and reaction in such systems
exhibits a number of peculiarities, which are exemplified in more detail in this chapter.

As a main feature of diffusion in channels with diameters small enough so that the individual diffusants cannot pass each other, their mean square displacement is found to increase with
the square root of time rather than with the time itself, as to be required for normal diffusion.
In channels of finite extension this peculiarity of "single-file" diffusion is soon masked by a
second mechanism, stipulated by the fast particle exchange at the boundary, which leads to
molecular displacements following the time dependence of normal diffusion. The interplay
of these two processes is discussed in Sect. 2. In mutually intersecting arrays of channels,
molecular diffusion in different directions may be correlated between each other. Section 3
in particular considers the question, up to which extent the "memory" of the diffusants may
affect this correlation. If the different arrays have different affinities to the constituents of
multicomponent systems, the rate of molecular reactions in such systems may be enhanced
in comparison with systems, where all parts of the pore system are equally accessible by the
involved components. The potentials and limits of reactivity enhancement by this effect of
"molecular transport control" are discussed in Sect. 4. In addition to the examples of experimental reference, which are appropriately included in each of the individual sections, Sect. 5 is
exclusively devoted to the question, how closely the textbook structure of nanoporous materials
with channel arrays is approached by reality.

1 Introduction

A large variety of nanoporous materials host channel networks with pore diameters
on the order of typical molecular dimensions. Molecular diffusion and reaction under this type of confinement exhibit a number of peculiarities. Because nanoscopic
channels may serve as routes for particle propagation in many more systems, with
the ion channels in biological membranes [1] and reptation paths in macromolecular
systems [2] as two famous examples, the interest in these peculiarities is by far not
confined to porous materials.

In some sense, the present chapter may be considered as an extension of the
initial chapter of this book, specifying in detail the theoretical tools for treating

P. Bräuer, S. Fritzsche, J. Kärger, G. Schütz, S. Vasenkov, Diffusion in Channels and Channel Networks, Lect. Notes
Phys. **634**, 89–125 (2004)
http://www.springerlink.com/   ⓒ Springer-Verlag Berlin Heidelberg 2004

molecular propagation and reaction under confinement by channels. With respect to the prospects for experimental observation, it is also closely related to the Chap. 2 and the potentials of PFG (pulsed field gradient) NMR described there (see also Chaps. 12, 13 as a most sensitive technique for experimentally tracing these peculiarities.

Beginning with the presentation of peculiarities of molecular propagation in the individual channels – in particular under the so-called single-file condition (i.e. if the individual diffusants are unable to mutually exchange their positions) – in Sect. 2, the subsequent sections are devoted to special features of transport and reaction in mutually intersecting channel arrays, viz. the structure-correlated anisotropy of diffusion in Sect. 3 and the transport-induced reactivity enhancement by the so-called "molecular traffic control" in Sect. 4. Before speculating about future tendencies in the development of this rather new field of interface science in the concluding Sect. 6, Sect. 5 describes some of the key experiments leading to our present knowledge of the real pore structure of nanoporous materials, which according to their textbook structure [3] until recently have been assumed to represent ideal channel host systems.

2 Isolated Channels

2.1 The Correlated Movement of Single-File Diffusion

Let us assume that the diameters of the channels accommodating the diffusants are so small that the individual diffusants are unable to pass each other. This condition of "single-file diffusion" leads to a high degree of mutual interaction of the diffusants and hence to significant deviations from the pattern of "normal" diffusion as described likewise by the Fick's laws or by the Einstein relation [4, 5]

$$\langle x^2(t) \rangle = 2Dt, \tag{1}$$

with $\langle x^2(t) \rangle$ denoting the mean square displacement during the observation time t and D denoting the self-diffusivity. (1) may be deduced by calculating the mean square distance of molecular propagation on the basis of the appropriate solution of Fick's second law, viz. the probability distribution of labeled molecules in their un-labeled surroundings if initially all labeled molecules have been at one and the same position. The proportionality between the observation time and the mean square displacement appearing in (1) may be easily rationalized on the basis of the fundamental presupposition of normal diffusion, i.e. by assuming that it is possible to divide the total observation time into equal time intervals so that the probability distribution of molecular displacement is identical for each of these time intervals, and that - moreover - the displacement probability is independent of previous displacements. Thus, the random walk of each individual particle may be considered as a Marko-vian process, i.e. a process whose further evolution is exclusively determined by the given state and not by the past. If we denote the molecular displacement in the ith interval of time by x_i, the square of the resulting molecular displacement after n time intervals of duration τ may thus be noted as

$$\left\langle \left(\sum_{i=1}^{n} x_i \right)^2 \right\rangle = \left\langle \sum_{i+1}^{n} x_i^2 \right\rangle + \left\langle \sum_{i \neq j=1}^{n} x_i x_j \right\rangle = n\langle x_i^2 \rangle \propto n\,\tau \qquad (2)$$

because $\langle x_i^2 \rangle$ does not depend upon i. Under the conditions of anomalous diffusion, subsequent displacements are correlated so that, as an immediate consequence of (2), the mean square displacement cannot be expected to increase in proportion to time t. In the case of single-file diffusion this correlation leads to an increased probability for subsequent displacements to be directed opposite to each other. This tendency may be understood by rationalizing that, as a consequence of the single-file condition, molecular displacement in one direction will more likely lead to an enhanced concentration of particles "in front" of it rather than "behind" it. Therefore, subsequent movement in the backward direction is facilitated in comparison with its continuation in the "forward" direction. It is interesting to note that – in contrast to the well-known correlation effect in solid-state diffusion [6,7] – this anticorrelation is conserved over arbitrarily long intervals of time.

Rigorous analysis shows that under the conditions of single-file diffusion in the long-time limit, the mean square displacement increases with the square root of the observation time [6, 8–11] with a probability distribution given by a Gaussian function [12], in complete analogy to the case of normal diffusion. Moreover, the "anomalous" time dependence of the mean square displacement under single-file conditions could also be confirmed by molecular dynamics simulations implying – for the sake of simplicity – periodic model potentials [13] rather than the genuine potential landscape of zeolites as generally considered in Chap. 1 of this book. Diffusion studies with zeolitic host–guest systems, which were assumed to guarantee the conditions of single-file diffusion, have been carried out by PFG NMR [14–16], quasi-elastic neutron scattering (QENS) [17], the tracer zero-length-column (T-ZLC) [18,19] technique, and the frequency response [20] method. The results obtained in these studies are partially contradictory and far from comprehensive. This is most likely caused by both the diverging temporal and spatial ranges of observation of these techniques and the deviations of the systems under study from their ideal textbook structure. The most recent evidence from interference microscopy about this very aspect will be provided in Sect. 5.

Following a Gaussian probability distribution, single-file diffusion may also be treated analytically on the basis of the one-dimensional Fick's relations of normal diffusion by simply considering its dependence on the square root of the time rather than the time itself. One should have in mind, however, that this analogy does holds for an infinitely extended single-file system. As soon as this requirement of infinite extension has to be abandoned and boundary or internal matching conditions come into play, sticking to this analogy leads to completely erroneous conclusions [21,22].

2.2 Single-File Channels of Finite Extension

In any practical case of application, one is concerned with the behavior of finite rather than infinite single-file systems. The intricate consequences of single-file confinement for the propagation patterns of the particles involved, however, have so far prohibited the invention of rigorous analytical solutions for the evolution of the probability function of the particle distribution in a finite system in a closed form. The present section aims to illustrate some of the currently investigated possibilities to overcome this limitation.

A Network of Characterizing Functions

Stationary populations in exchange with their surroundings may be identified as existing in quite diverse contexts, ranging from adsorbate–adsorbent systems in equilibrated atmospheres up to living organisms and even social communities [23]. Depending on the given purpose, the properties of these systems may be characterized by quite different functions. As an example, Fig. 1 illustrates in a schematic way the meaning of some of these "characterizing" functions [24]. Because all of them depend – though in different ways – on the internal dynamics of the system, it is not unexpected that these functions may be mutually transformed into each other. Figure 2 shows some of the relevant relations, which form a type of "network" of these functions. Whereas certain parts of this network are in common use, e.g. in chemical engineering [25], the wealth of these interrelations was illustrated in [23] for the first time by their application to adsorbate–adsorbent systems. The benefit of this network of mutual interrelations is particularly noteworthy for such systems where the processes of interest are not accessible by analytical means. Owing to the network, it is sufficient to determine one of the characterizing functions, from which all the others result by an appropriate use of the network relations. In this way it is sufficient to determine the function which is more easily accessible, e.g. by dynamical Monte Carlo simulation, and there is, in particular, no need for separate simulation efforts to determine the other functions. This way of mutual transformation has proved to be particularly valuable in considering the different features of single-file systems of finite lengths, as described in the subsequent sections.

Tracer Exchange and Particle Conversion

The tracer exchange curve $\gamma(t)$ has been introduced in the previous section as the relative number of molecules which have not yet been in the system at time $t = 0$. It represents one of the characterizing functions and its "time constant" may be identified as the mean molecular exchange time. For its experimental observation, at time $t = 0$ one has to replace, for example, the so-far unlabeled molecules in the surrounding gas phase by labeled (but otherwise identical) ones. Then, the quantity $\gamma(t)$ results as the probability that at time t an arbitrarily selected molecule of the system is a labeled one.

   For an exact treatment of the single-file system it is necessary to introduce joint probabilities $\Theta(\sigma_1, \sigma_2, \sigma_3, \ldots, \sigma_n)$ for the occupation of any individual site, where the variables σ_i may assume the values 1 and 0 for site i occupied and vacant, respectively. Being represented in terms of $\Theta(\sigma_1, \sigma_2, \sigma_3, \ldots, \sigma_n)$, the system evolution may now be considered as a Markovian process: For predicting the future file configurations, it is completely sufficient to know the present state $\Theta(\sigma_1, \sigma_2, \sigma_3, \ldots, \sigma_n)$.

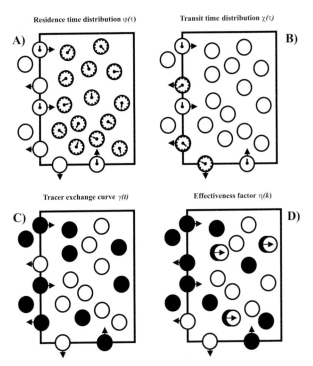

Fig. 1. Illustration of the definitions of the basic characteristic functions describing the stationary adsorbate–dsorbent system. (A) Each particle is equipped with a clock which is initialized at the moment of entrance of the particle. The residence time distribution $\varphi(\tau)$ is the probability density of the times obtained by reading all the clocks simultaneously at an arbitrary observation time. B) In the case of the transit time distribution $\chi(\tau)$, the clocks are, in contrast to A), read at the moment of exit of the respective particle. (C) Only labeled (black) particles have been entering the system during the past time interval of length t. The tracer exchange curve for the constant input function, $\gamma(t)$, gives the relative amount of these particles which have replaced the original unlabeled (white) ones. (D) Reactant particles (black) enter the system and have the chance there to be converted irreversibly into product particles (white). The effectiveness factor $\eta(k)$ gives, in dependence on the intrinsic rate k of this conversion, the relative amount of particles still unconverted, expressing the competition between the supply of black particles by the transport processes and their "consumption" by the reaction [23]

If particle conversion within the system is considered, it is useful to introduce, in analogy to the residence time distribution $\varphi(\tau)$ of Figs. 1 and 2, a conditional residence time distribution $\varphi(\sigma_1, \sigma_2, \ldots, \sigma_{i-1}, \sigma_{i+1}, \ldots, \sigma_n, \tau)$ [26]. It indicates the probability (density) that the molecule at site i has entered the system at a time t ago, if the set of variables $\sigma_1, \sigma_2, \ldots, \sigma_{i-1}, \sigma_{i+1}, \ldots, \sigma_n$ describes the occupation of the other sites. For simplicity, the mobility of the molecules is assumed to be unaffected by conversion. Consequently, there is no need to specify whether a site is occupied by a reactant or by a product molecule.

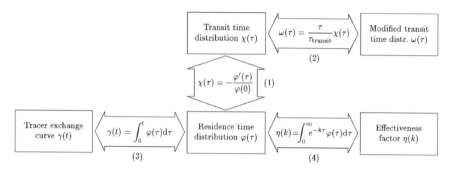

Fig. 2. The network of characterizing functions, interdependent through one-to-one relations (for definitions, see text) [24]

For determining the probability $\eta_i(\sigma_1, \sigma_2, \ldots, \sigma_{i-1}, \sigma_{i+1}, \ldots, \sigma_n)$ that a molecule residing at site i (with the given occupation set $\sigma_1, \sigma_2, \ldots, \sigma_{i-1}, \sigma_{i+1}, \ldots, \sigma_n$ of the other sites) is still a reactant molecule, adopting the relevant relation of Fig. 2, one obtains

$$\eta_i(\sigma_1, \sigma_2, \ldots, \sigma_{i-1}, \sigma_{i+1}, \ldots, \sigma_n)$$
$$= \int_0^\infty e^{-k\tau} \varphi_i(\sigma_1, \sigma_2, \ldots, \sigma_{i-1}, \sigma_{i+1}, \ldots, \sigma_n, \tau)\, d\tau. \tag{3}$$

Considering all possible configurations $\sigma_1, \sigma_2, \ldots, \sigma_{i-1}, \sigma_{i+1}, \ldots, \sigma_n$, in [27] the mean total number H of reactant molecules in single-file systems consisting of $n = 8$ sites (corresponding to 64 different configuration patterns $\sigma_1, \sigma_2, \ldots, \sigma_n$) has been calculated by solving the corresponding set of equations for the unknown variables $\eta_i(\sigma_1, \sigma_2, \ldots, \sigma_{i-1}, \sigma_{i+1}, \ldots, \sigma_n)$. Figure 3 displays this number as a function of the intrinsic reaction rate κ (in units of the mean time of jump attempts between adjacent sites) and the ratio υ of the adsorption and desorption rates on the marginal sites of the file. By applying the principle of detailed balance to the marginal sites, the adsorption/desorption ratio υ is easily found to be related to the site occupation probability Θ by the expression [28]

$$\Theta = \frac{\upsilon}{1+\upsilon}. \tag{4}$$

Not unexpectedly, the largest population number, 8, of reactant molecules (corresponding to the 8 sites of the file considered) is attained for the lowest reaction rates κ. With respect to the parameter υ (and thus – via (4) – to the site occupancy Θ), the total amount of reactant molecules passes a maximum.

The relevance of this finding appears more clearly in Fig. 4, showing the isoline representation of Fig. 3. In addition to Fig. 3, Fig. 4 shows two straight lines, representing two different cases of variation of υ (and hence of the occupation) and of κ with varying temperature due to Arrhenius' law. The quantity one is primarily

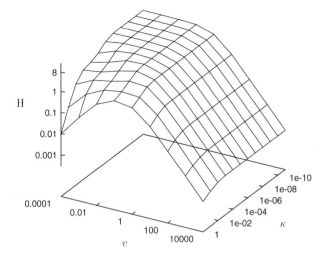

Fig. 3. Mean number H of reactant molecules within the channel in dependence on the intrinsic reaction rate κ (in units of the intracrystalline hop time $1/2\Gamma$) and the ratio v of desorption and adsorption rates. The behavior presented is valid for the case of rapid particle exchange between the marginal sites and the gas phase [27]

interested in when considering catalytic reactions is their yield, i.e. the number of product molecules generated per unit of time. This quantity may be easily determined by multiplying the total amount of reactant molecules H as provided by Fig. 3 and Fig. 4 by the intrinsic reactivity at the given temperature. Obviously, if the number of reactant molecules remains unaffected by temperature variation the "effective" activation energy of the reaction coincides with that of the intrinsic reactivity. The two straight lines in Fig. 4 represent two different cases for the temperature dependence of the total amount of reactant molecules. Let us first consider the situation illustrated by the broken line. With increasing temperature, i.e. with increasing values of κ and v, the total amount of reactant molecules H steadily decreases. As a consequence, with increasing temperature, the increase in the effective reactivity with increasing values of κ is partially compensated by the decrease in the total number of reactant molecules. This means that the effective activation energy is smaller than that of κ. This is the usual situation in heterogeneous catalysis under conditions of transport control, i.e. in a situation where the rate of molecular transportation from the reactive sites to the surrounding atmosphere is inferior to the conversion rate that would occur if the product molecules could be instantaneously replaced by new reactant molecules.

As a most remarkable finding, the dot–dash line in Fig. 4 indicates that in single-file systems the reverse situation is also possible, i.e. that the effective activation energy of catalytic reactions may exceed that of the intrinsic reaction. This unusual behavior is caused by the fact that with increasing temperature (and a corresponding decrease in the total loading) the severe restriction caused by single-file confinement

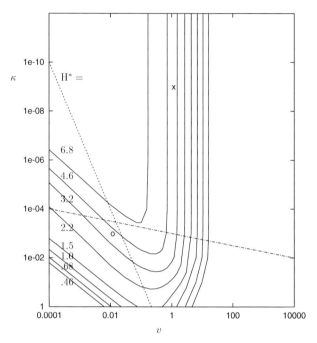

Fig. 4. The same information as in Fig. 3, presented as isolines of H over the plane of the parameters κ and v. The isolines refer to reactant particle numbers 6.8, 4.6, ... , down to 0.46. Since the parameters κ and v are temperature–dependent, a variation of the temperature corresponds to a displacement in the parameter plane along a straight line, two examples of which are indicated by the broken lines ($- - -$: $E_\kappa = 150$ kJ/mol, $\kappa = 10^{20}$, $E_v = 50$ kJ/mol, $v = 10^6$; $\cdot - \cdot -$: $E_\kappa = 37.5$ kJ/mol, $\kappa = 10^3$, $E_v = 150$ kJ/mol, $v = 10^{24}$) [27]

is reduced to such an extent that the number of reactant (i.e. unconverted) molecules may even increase. Indications of such a behavior have in fact been found in several experimental studies [29, 30].

Calculating Propagators via the Reflection Principle

The occupation state of a single-file system accommodating n particles may be represented by a single point in an n-dimensional space, with the coordinates $x_1, x_2, \ldots, x_n$ denoting the positions of all of these particles. The further evolution of this system may be considered as a random walk in this n-dimensional space. Because the single-file condition requires the maintenance of the once-established order

$$x_1 < x_2 < x_3 \cdots < x_n \tag{5}$$

this random walk has to be confined to a certain segment of this n-dimensional space. The boundaries of this segment are reflecting walls in the planes $x_i = x_{i+1}$

with $i = 1, 2, \ldots, n - 1$. In addition, the influence of the file boundaries, i.e. the existence of reflecting or absorbing boundaries or even the possibility of exchange with a surrounding reservoir, has to be considered. The single-particle propagator for any individual molecule in this file may be deduced from the solution of Fick's second law in this n-dimensional space with the relevant boundary conditions and with the initial condition given by the starting positions of the molecules in the file.

In [31], solutions have been derived for several special cases of initial particle distributions in finite and infinite single-file systems. As an example, Fig. 5 displays the distribution probability (the "propagator") of a particle starting in a single-file system between two absorbing boundaries at 1/5 of their distance. The comparison with the simulations illustrates the remarkable benefit of the proposed analytical procedure. Figure 6 displays the probability that a particle in a single-file system with absorbing boundaries has not yet been absorbed by them after a certain interval of time, as a function of the starting position. Moreover, the single-file results are compared with those for a system undergoing normal diffusion. If, it starts in the channel center, the particle subject to single-file confinement is with much less probability absorbed than the free particle, whereas the contrary situation is true for particles starting near the boundaries. These calculations are found to be much less time-consuming than dynamic Monte Carlo simulations (cf., e.g., [32–34]), whose results are also displayed in Fig. 6.

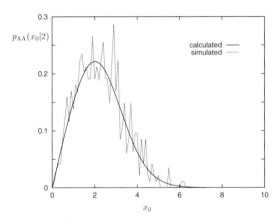

Fig. 5. Probability density $p_{AA}(x_0|2)$ of finding a tagged particle at position x_0 within a single file of finite length (10 units) at the end of which particles are irreversibly desorbed into the surrounding space. The tagged particle is assumed to have started its diffusion at position 2, at a time in the past such that a freely diffusing particle would have moved over a mean square distance of $(2.23 \text{ unit lengths})^2$. The channel is initially occupied by one particle per unit length on average. The two curves compare the analytically calculated and the Monte Carlo simulated propagator [31]

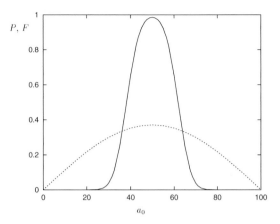

Fig. 6. For the finite single file of Fig. 5 (the length here is 100 units), the probability that a given particle is still inside the channel in dependence on its initial position a_0, after a time during which a freely diffusing particle would have moved over a mean square distance of $(50 \text{ units})^2$. The probabilities in the single file case (solid line) are compared with the case of normal diffusion (noninteracting particles, broken line) [31]

Tracer Exchange

From the analytical solution of the one-dimensional tracer exchange curve $\gamma(t)$ for normal diffusion, the intracrystalline molecular mean lifetime (defined as the first moment $\int_0^\infty [1 - \gamma(t)]\, dt$ of the tracer exchange curve) is easily found to obey the relation [35–37]

$$\tau_{\text{intra}} = \frac{L^2}{12D} \tag{6}$$

with L denoting the extension of the crystal in the direction of diffusional exchange. The proportionality $\tau_{\text{intra}} \propto L^2/D$ resulting from (6) may be most easily rationalized by considering that L and D are the only relevant quantities for the exchange process and that there is no other way of combining them into a quantity with the dimension of time than that provided by this proportionality. Adapting the same way of reasoning to single-file diffusion would lead to $\tau_{\text{intra,s.f.}} \propto L^4/F$, where the quantity F [38] is the equivalent of the self-diffusivity in the Einstein relation in the single-file case, serving as a factor of proportionality between the mean square displacement and the square root of the observation time. As mentioned already at the end of Sect. 2.1, this type of reasoning would only be correct for an infinitely extended system.

As soon as the finite size of the system comes into play, in addition to the movement within the whole file, molecular displacements may also be caused by the statistically occuring processes of molecular adsorption and desorption at the marginal sites, which lead to a shift of the file as a whole. Because subsequent events of adsorption and desorption may occur independently from each other, the total process of displacement is again Markovian, just as in the case of ordinary diffusion. Molecular displacements according to this mechanism may therefore be described by a diffusivity, which results to be [37, 39–42]

$$D_{\text{eff}} = D\frac{1-\Theta}{\Theta N} \tag{7}$$

with N denoting the total number of sites and D denoting the diffusivity of an isolated, single particle in the file. Thus, in addition to the concentration dependence given by the factor $(1-\Theta)/\Theta$, the effective diffusivity is reduced in comparison with the single-particle diffusivity D by the site number N. Since the site number N is proportional to the file length L, combination with (6) leads to the important result that the intracrystalline mean lifetime in single-file systems scales with the third power of the crystal size. One must have in mind, however, that in contrast to genuine diffusion, (7) cannot adequately describe the tracer exchange behavior of single-file systems from the very beginning.

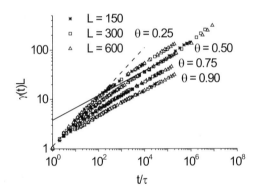

Fig. 7. The normalized tracer exchange curves in single-file systems obtained by dynamic Monte Carlo (DMC) simulations for different L and Θ (points). The dashed and solid lines show the best fit lines for $\Theta = 0.5$ with the slope of 1/2 and 1/4 expected for the mechanism, of normal and single-file diffusion, respectively, in the limit of short times [42]

This is demonstrated in particular by Fig. 7, showing the initial stages of tracer exchange for the total site numbers $N = 150, 300$ and 600 and the site occupancies $\Theta = 0.25, 0.50, 0.75,$ and 0.90 [42]. For comparison, the slopes to be expected for diffusion-limited exchange (broken line) and for exchange limited by single-file confinement (straight line) are also indicated. For the very first steps of the particles leaving the single-file system, single-file confinement is obviously not yet relevant, so that the time dependence is that of a diffusion-limited process. This initial part is followed by dependences, which may be shown analytically to result

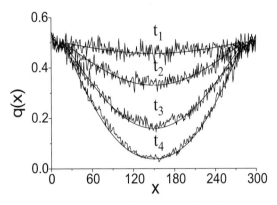

Fig. 8. Comparison of the concentration profiles of tagged particles obtained by DMC simulations for tracer exchange in single-file systems of length $L = 300$ (oscillating solid lines) with the concentration profiles for normal diffusion, with D_{sim} and N^* given in Table 1 (solid lines) at times: $t_1 = 0.93 \times 10^6 \, \tau$, $t_2 = 2.1 \times 10^6 \, \tau$, $t_3 = 3.7 \times 10^6 \, \tau$, and $t_4 = 7.6 \times 10^6 \, \tau$ (τ is the duration of the elementary diffusion step) [42]

from tracer exchange under single-file confinement [42]. As a strong argument that under these conditions the diffusion-like mechanism of displacement described above does not yet work, one has to consider the fact that for one and the same loading the product of the tracer exchange curve and the total number of sites is independent of the file length. Because this product yields nothing else but the total number of exchanged molecules, the total file length obviously has no influence yet on the observed behavior up to the given time. This is the situation encountered, for example, in exchange experiments with laser-polarized ^{129}Xe NMR spectroscopy in the nano-channels of organo-crystallites [43]. Only in the case of $\Theta = 0.50$, with $N = 300$ and 600, does Fig. 7 indicate the beginning of a transition to the diffusion-controlled regime of tracer exchange. It is demonstrated by Fig. 8 that, over by far the largest part of the file, the concentration profiles during the exchange process are astonishingly well represented by the corresponding solutions of the classic diffusion equation, i.e. of Fick's second law. In the approach used, a certain part at the two file margins was assumed to attain the equilibrium concentration instantaneously, so that the exchange had to be followed over N^* rather than over N sites. This approach corresponds to Fig. 6 which illustrates the extremely fast particle exchange between the marginal sites and the surroundings. Table 1 indicates the values of the ratios N^*/N and of the ratio $D_{\text{sim}}/D_{\text{eff}}$ between the diffusivity used in the analytical simulations and the effective file diffusivity, as given by (7), which have led to the best agreement with the simulation data. Not unexpectedly, the site number N^* relevant to the simulations approaches the total number. It is interesting to note that D_{sim}, though being of the order of D_{eff}, does not seem to approach this value.

Table 1. Relations between the results of the best fit D_{sim} and N^* and the center-of-mass diffusivity D_{eff} and site number N in the final time domain of tracer exchange (controlled by center-of-mass diffusion) [42]

L	$D_{\mathrm{sim}}/D_{\mathrm{eff}}$	N^*/N
150	1.51	0.83
300	1.49	0.87
600	1.80	0.92

3 Correlated Diffusion Anisotropy

As soon as the nanochannels considered exclusively in the previous section interpenetrate each other, they give rise to the formation of channel networks. In these new structural entities a number of novel features arise, whereas some of the problems with the treatment of molecular propagation in the individual channels – as so far considered – may disappear. This means in particular that the channel intersections allow molecular bypassing, so that when considering the total channel network one may abandon the requirement of a strict order of the diffusants. As a consequence, the molecular diffusivity may again become a meaningful quantity. The present section is going to explain why under certain conditions this diffusivity exhibits a remarkable property, which has been termed the phenomenon of correlated diffusion anisotropy [44].

3.1 The Origin of Correlated Diffusion Anisotropy

The nature of correlated diffusion anisotropy is best illustrated by the schematic representation of the channel network of the MFI-type zeolites (Fig. 9). This class of zeolites includes such important representatives as ZSM-5 and silicalite-1 [3]. MFI-type zeolites obviously contain two mutually intersecting sets of parallel channels, viz. the so-called straight channels in the crystallographic y-direction and the zigzag channels in the x-direction. Though there is no corresponding third channel system, molecular propagation is clearly also possible in the z-direction. Obviously, effective molecular displacements in the z-direction have to imply subsequent shifts along segments of the straight and zigzag channels. Correlating the relevant probabilities for displacements from intersection to intersection, one may deduce the following relation between the principal values of the diffusion tensor [44–46]:

$$\frac{c^2}{D_z} = \frac{a^2}{D_x} + \frac{b^2}{D_y} \,, \tag{8}$$

with a, b and c denoting the unit cell extension in the x-, y-, and z-direction as indicated in Fig. 9. As a condition for deriving (8), a molecule has to make its way from intersection to intersection independently of its past, i.e. the particle "memory" has to be shorter than the mean traveling time from intersection to intersection.

Though (8) has proved to be a reasonable starting equation for analyzing both MD simulations [47–50] and experimental studies [51, 52] of anisotropic diffusion in MFI, the possibility of deviations from this simple correlation rule could not be excluded. Therefore, in [53] a so-called memory parameter

$$\beta = \frac{c^2/D_z}{a^2/D_x + b^2/D_y} \tag{9}$$

has been introduced. Obviously, β is equal to 1 if (8) is correct. $\beta > 1$ indicates a value of D_z reduced in comparison with (8). Since propagation in the z-direction has been identified to occur as a consequence of molecular exchange between the two channel types, the case $\beta > 1$ obviously refers to a less frequent than random exchange between the two channel types at the points of intersection. Correspondingly, $\beta < 1$ refers to the opposite case, i.e. to a preferential change between the two channel types.

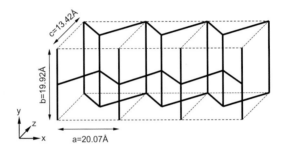

Fig. 9. Topology of the channels in silicalite-1

3.2 Memory Effects in Correlated Diffusion Anisotropy

Memory effects in correlated diffusion anisotropy leading to deviations from the no-memory case $\beta = 1$ may be best represented by a set of conditional probabilities, as illustrated by Fig. 10. Instead of the two parameters p_1 and p_2 introduced in [44] to describe the probabilities that a molecule after passing a channel intersection will enter a straight or a zigzag channel, conditional probabilities are now introduced [54,55]. They refer to the probabilities that a molecule entering a channel intersection along, for example, an element of a zigzag channel, will continue its trajectory backward $(p_{x,-x})$ or forward $(p_{x,x})$ along the same channel or will switch to the other channel $(p_{x,y})$, and to the analogous probabilities in the reverse case if the indices x and y are mutually exchanged.

In [56] these quantities are shown to lead to a simple possibility of correlating deviations of the memory parameter β from its no-memory value $\beta = 1$. In the first-order approximation one obtains

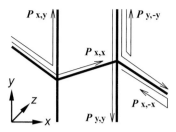

Fig. 10. Illustration of the transition probabilities in silicalite-1

$$\beta - 1 = 2(p_y \Delta\pi_x + p_x \Delta\pi_y + \Delta\pi_{x,y}) \tag{10}$$

with the notation

$$\Delta\pi_x = p_{x,x} - p_{x,-x} , \tag{11}$$
$$\Delta\pi_y = p_{y,y} - p_{y,-y} , \tag{12}$$
$$\Delta\pi_{x,y} = p_{x,x} + p_{x,-x} - 2p_{y,x} . \tag{13}$$

Obviously, $\Delta\pi_x$ and $\Delta\pi_y$ represent the probabilities that a molecule which has got to a certain channel intersection "prefers" to proceed to the subsequent intersection rather than to the preceding one. The quantity $\Delta\pi_{x,y}$ compares the probabilities that on passing a channel intersection a molecule remains in the same channel type or moves to the other one. In fact, each of the cases just considered occurring with the probabilities $\Delta\pi_x$, $\Delta\pi_y$ and $\Delta\pi_{x,y}$ have been intuitively identified above as leading to an enhancement of the memory parameter β. Equation (10) is in complete agreement with the expected behavior. It has been demonstrated by detailed MD simulations using methane as a probe molecule [55], see also Chap. 1 of this book) that the estimates on the basis of (10) are in excellent agreement with the rigorous calculation via (9) on the basis of the calculated principal elements of the diffusion tensor.

4 Molecular Traffic Control

4.1 The Phenomenon of Molecular Traffic Control

Mutually intersecting arrays of single-file systems can serve as a model system for the simulation of reactivity enhancement by confinement, i.e. by the phenomenon of the so-called "molecular traffic control" (MTC) [57,58]. The expression "MTC" has been chosen to describe chemical reactions in nanoporous materials where reactant and product molecules prefer different pathways for their diffusion. Important examples of materials which might exhibit such a behavior are zeolites of the MFI structure type such as ZSM-5 [3]. Such zeolites contain two different types of mutually intersecting channels as schematically represented by Fig. 9. Molecular dynamics simulations of gas adsorption of two components in silicalite-1 [59], in boggsite zeolite (BOG) [60] and in faujasite (FAU) [61] have in fact shown that the residence probabilities in different areas of the intracrystalline pore space may be notably different for the two fluid components. The concept of MTC implies that the difference in the residence probabilities of reactant and product molecules in the two-channel systems causes a reduced transport inhibition, which in the case of transport-controlled reactions [35, 62] leads to an enhanced output of product molecules. The MTC effect has remained a subject of controversial discussions over two decades in the literature [57, 58, 63–69] and has not found up to now a sound theoretical foundation.

4.2 The Model Applied

The model system is assumed to consist of a network of n_α and n_β mutually intersecting channels of types α and β with identical lengths. The number of sites between two neighboring channel intersections (X) and between a channel intersection and a marginal lattice point (m) is l.

For the lattice given as an example in Fig. 11, we have $n = n_\alpha = n_\beta = 5$ and $l = 5$. Each site of the lattice can be occupied by only one particle.

On the sites in channel intersections (X), molecules A can be converted to molecules B, assuming a unidirectional molecular reaction A⇒B with a rate of reaction

$$\frac{dN_A}{dt} = -k_c N_A , \tag{14}$$

where N_A is the number of the reactant molecules A, and $k_c = \kappa_c/\tau$ is the site reaction rate. τ denotes the length of an elementary time step of the simulation. Therefore, the mean time necessary for conversion of the reactant molecule A to the product molecule B is equal to τ/κ_c. κ_c represents the reaction probability at an intersection point.

The time unit τ of the simulation is set equal the mean value of the time interval between subsequent jump attempts of one and the same particle. These time intervals are identical for the A and B molecules and can be formally set equal to one. This quantity is denoted as the intrinsic reactivity.

We consider two kinds of lattices. In the so-called molecular traffic control system (MTC system) the molecules A only can occupy the α-channels and the molecules B only the β-channels. In the case of the so-called reference system (REF system), both channel types are equally accessible to both types of molecules.

The simulations were started with a random distribution of molecules A over the lattice, i.e. for the MTC system inside the α-channels and for the REF system in both the α- and the β-channels.

The stationary state is considered to be attained as soon as, with evolving simulation time, the average number of molecules B on the lattice remains constant.

The systems are surrounded by a gas stream, which contains only molecules of type A. Desorbing molecules B, which originate from molecules A, are spilled out

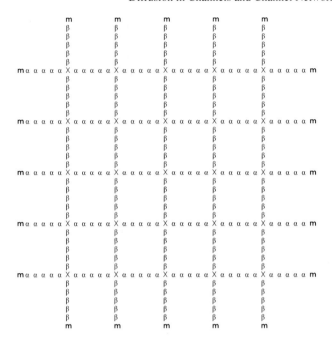

Fig. 11. Quadratic lattice consisting, as an example, of $n_\alpha = 5$ channels of type α and of $n_\beta = 5$ channels of type β with $l = 5$ sites between the intersections. The intersections of the channels are characterized by the symbol X and the marginal lattice sites by **m**

into the surrounding gas stream. In the simulation, this is achieved by removing all B molecules from marginal sites after each simulation step. Further, we assume that the marginal lattice sites have lost the memory of their occupation after any jump to or from them [28, 41], i.e., after each jump attempt all marginal lattice sites are vacated and subsequently again occupied with a probability $\langle \Theta_A(t = 0) \rangle$. In this way, their permanent equilibrium with the surrounding atmosphere is ensured.

All simulations were repeated 1000 times with different sequences of random numbers, i.e. all calculated values are averages over 1000 equal systems. Further details of the simulation may be found in [70].

4.3 Reactivity Enhancement by Molecular Traffic Control

We have calculated the output of B-molecules per lattice site and per elementary time step for the MTC system, ξ_B^{MTC}, and for the REF system, ξ_B^{REF}, for a lattice with $n = 3$ α- and β-channels in dependence on the number l of sites between two neighboring intersection points for a reaction probability $\kappa_c = 0.5$ and for an initial lattice occupation $\langle \Theta_A(t = 0) \rangle = 0.25$. Figure 12a shows the ratio of the outputs ξ_B^{MTC}/ξ_B^{REF} as a function of l. Obviously, for small values of l, the output in the REF system is greater than in the MTC system, while for $l > 6$ the ratio becomes larger than one, making the MTC system more favorable than the REF system. As an example, Fig. 12b shows the ratio between the effective reactivities of the MTC and REF systems for $l = 1$ as function of the total amount n of α- and β-channels. It passes a maximum for $n = 2$. Even at this point the ratio is less than one, indicating that in this case the efficiency of the MTC system is notably inferior to that of the REF system. This is to be expected on the basis of Fig. 12a, where for small values of l (and in particular for $l = 1$) this ratio has been found to be distinctly smaller than 1. In Fig. 12b it is exhibited that the further enhancement of n monotonically further impairs rather than improves the benefit of the MTC system in comparison with the reference system. In the following, the options under which the reactivity in the MTC system may exceed that of the REF system will be discussed.

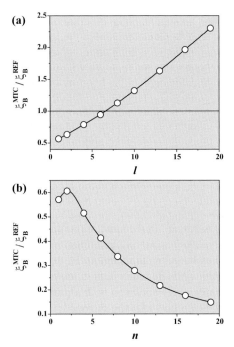

Fig. 12. Benefit ξ_B^{MTC}/ξ_B^{REF} of the MTC system in comparison with the REF system for $n = 3$ as a function of the number l of sites in the channel segments between two neighboring intersections (a) and for $l = 1$ in dependence on the number n of α- and β-channels (b). In both graphs the curves were calculated for a reaction probability $\kappa_c = 0.5$ and an initial lattice occupation $\langle \Theta_A(t = 0) \rangle = 0.25$ [70]

Figure 13 displays the distribution of the A-molecules (a and d), of the B-molecules (b and e) and of both molecules (c and f) in the α- and β-channels for an initial lattice occupation $\langle \Theta_A(t = 0) \rangle = 0.25$ in the case of the MTC system (a – c) and of the REF system (d – f) for a reaction probability $\kappa_c = 0.01$. The lattice

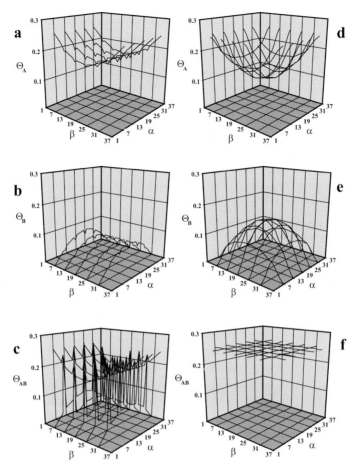

Fig. 13. Concentration profiles of the A-molecules (a and d), of the B-molecules (b and e), and of both molecules (c and f) in the α- and β-channels for a reaction probability $\kappa_c = 0.01$ and for an initial lattice occupation of $\langle \Theta_A(t=0) \rangle = 0.25$ in the case of the MTC system (a–c), and of the REF system (d – f) [70]

consists of $n = 5$ α- and β-channels and $l = 5$ sites between the channel intersections (cf. Fig. 11). As expected for transport-dependent processes, in both the MTC and the REF systems the concentration of the reactant molecules decreases toward the network center (a and d), whereas the concentration of the product molecules decreases toward the margins of the lattice (b and e). As a remarkable feature, in contrast to the behavior of the REF system (f), where the total concentration of the molecules A and B is constant all over the lattice, molecular traffic control (c) leads to a pronounced deviation from a constant occupancy over the network.

The concentration profiles shown in Fig. 13 allow a semiquantitative prediction of the interrelation of the effective reactivity under MTC and REF conditions. According to the Thiele concept [35, 62, 71] of catalytic reactions, the so-called effectiveness factor

$$\eta = \frac{\langle \Theta_A \rangle}{\langle \Theta_{AB} \rangle}, \tag{15}$$

i.e. the percentage of molecules which in a given instant of time may be subjected to a chemical reaction, may be represented as a function of a single quantity, which involves the intrinsic reactivity, the diffusity, and a geometrical factor. This quantity is generally referred to as the Thiele modulus Φ and represented as [71]

$$\Phi = \frac{V}{S} \sqrt{\frac{\langle k \rangle}{D}} \tag{16}$$

with V and S denoting, respectively, the volume and the surface of the catalyst grain. $\langle k \rangle$ stands for its average reactivity, and D denotes the diffusion coefficient. For a slab of catalyst, the following holds [71]:

$$\eta = \frac{\tanh \Phi}{\Phi} . \tag{17}$$

Quite generally, i.e. independent of the grain geometry, we have

$$\lim_{\Phi \to \infty} \eta = \frac{1}{\Phi} . \tag{18}$$

Intrinsic reaction and intercrystalline exchange are the two competing processes during conversion by catalytic particles. This fact becomes more obvious if the Thiele modulus is expressed in terms of the intracrystalline mean lifetime τ_{intra} rather than in terms of the diffusivity [38, 71]. In the case of a slab catalyst of half-thickness a, for example, one has $V/S = a$ and [35, 71] (see also (6))

$$D = \frac{a^2}{3\tau_{intra}} . \tag{19}$$

From (16) one finally obtains

$$\Phi = \sqrt{3 \langle k \rangle \tau_{intra}} . \tag{20}$$

Thus, it turns out that introducing the intracrystalline mean lifetime into (16) allows one to simultaneously take account of the particle size and the intrinsic diffusivity.

As an alternative to the analytical expression resulting from (19), i.e. independent of the geometry of the catalyst grain, the intracrystalline mean lifetime τ_{intra} can be calculated in the case of the REF system from the tracer exchange curve $\gamma(t)$. Then, τ_{intra} is given via the relation [35, 36]

$$\tau_{intra} = \int_0^\infty [1 - \gamma(t)] dt . \tag{21}$$

As an example, Fig. 14 displays the intracrystalline residence times determined thus for different numbers of files ($2n$) and of sites (l) between the file intersections. In addition, Fig. 14 also contains the values of the intracrystalline residence times calculated analytically on the basis of the corresponding solution of the diffusion equation with the diffusity

$$D = \frac{l_{cc}^2}{4\tau_{cc}} \qquad (22)$$

with l_{cc} and τ_{cc} denoting the distance between adjacent channel intersections and the mean time it takes a molecule to get from one channel intersection to an adjacent one [35, 36, 70]. The values of τ_{cc} have also been determined from the simulation runs.

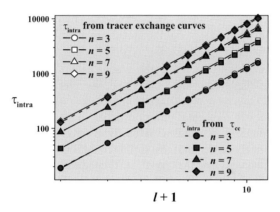

Fig. 14. Mean intracrystalline residence time τ_{intra} (in units of the time step of the simulation) of the molecules A in the REF system with $n = 3$ (circles), $n = 5$ (squares), $n = 7$ (triangles), and $n = 9$ (diamonds) and a total occupation of 0.25 as function of the file length $l_{cc}/\lambda = l+1$ between two intersections, calculated from the tracer exchange curves (open symbols) and from the mean passage times τ_{cc} between adjacent channel intersections (filled symbols). λ is the distance between adjacent sites [70]

As a most interesting feature of Fig. 14, the intracrystalline mean lifetime is found to increase with a power of 2.57, rather than only with the square of the distance between the intersections as would be the case for normal diffusion (cf. (6)). This anomalous power of 2.57 has to be attributed to the single-file conditions under which molecular propagation occurs in the REF system. To rationalize this enhanced power, let us recall that above ((6) with D replaced by (7)) the mean lifetime within a single-file system has been shown to scale with the third power of the file length, rather than with the second as in the case of normal diffusion. Though the mean travel time τ_{cc} from one intersection to an adjacent one cannot be expected to coincide with the mean lifetime within a file, it can surely be used as an estimate.

In view of the excellent agreement between the values of τ_{intra} determined from the simulated tracer exchange curves and via the diffusivity given by (22) (cf. Fig. 14), τ_{intra} is easily understood to follow the scaling of τ_{cc}, so that in the REF system the anomalous scaling of τ_{intra} with l_{cc} has to be anticipated.

From Fig. 13 it can be seen that, as a most significant feature, the overall concentration in the REF system is spatially constant (Fig. 13f, whereas the MTC system operates under the condition of permanent concentration gradients along both channel types, viz. along the α-channels into the catalyst and along the β-channels out of the catalyst (Fig. 13c. Therefore, molecular transport in the MTC system may proceed according to the laws of transport diffusion, where the single-file confinement is of no relevance [38,72]. Uptake of the reactant molecules into the catalyst and release of the product molecules into the surrounding atmosphere, therefore, occur under conditions of ordinary diffusion with a (transport) diffusivity coinciding with the single-particle self-diffusivity [35]. Thus, the exchange times tend to scale with the second power of l rather than with 2.57, as observed with the REF systems. For sufficiently large values of the Thiele modulus Φ, i.e. under diffusion control, with (18) and (20) the effectiveness factor of catalytic reaction turns out to decrease with increasing intracrystalline mean lifetime. Therefore, the relative gain in the effectiveness factor of the MTC system as displayed in Fig. 12 may be easily referred to the fact that increasing values of l lead to a more pronounced enhancement of τ_{intra} in the REF sytem.

The peculiarities of molecular transport under single-file conditions have thus been verified as a possible mechanism for reactivity enhancement by MTC. By doing so, we have deliberately chosen the most simple conditions, i.e. a network of only two arrays of parallel channels, and the "hard" MTC condition, i.e. the total exclusion of the accommodation of A molecules in the β-channels and, vice versa, of B molecules in the β-channels. It is demonstrated by Fig. 15, that partially abandoning the "hard" MTC condition may lead to effective reactivities which may still be notably higher than the values for the REF system.

In the present considerations, it has turned out that the MTC system becomes progressively beneficial over the REF system with increasing separation between the intersections. The benefit of the MTC system is thus purchased by stronger transport inhibition and a reduced density of active sites (which in the model have been allocated to the intersections). Therefore, in the light of the results obtained so far, the benefit of MTC seems to be of minor practical importance. With Fig. 15 it has been demonstrated, however, that even by partially abandoning the "hard" MTC condition effective reactivities may be obtained which are still notably higher than in the REF system. This finding is of particular relevance, since all practically feasible systems, including those considered in [59–61], are expected to possess features of "soft" MTC. Moreover, soft MTC systems would allow the existence of active sites in the channel segments as well, so that their density may be dramatically enhanced. Finally, so far all simulations have only exploited a two-dimensional channel arrangement. Thus, in addition to the options already mentioned, considering the phenomena of MTC in the three-dimensional space is most likely to open up a new dimension of research in the twofold sense of the word.

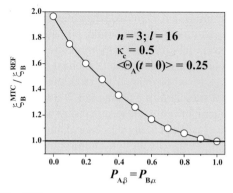

Fig. 15. Benefit $\xi_{\mathrm{B}}^{\mathrm{MTC}}/\xi_{\mathrm{B}}^{\mathrm{REF}}$ of the MTC system in comparison with the REF system for $n = 3, l = 16, \kappa_{\mathrm{c}} = 0.5$ and $\langle \Theta_{\mathrm{A}}(t = 0)\rangle = 0.25$ in dependence on the probabilities $P_{\mathrm{A},\beta}$ that an A molecule enters a β-channel and the probability $P_{\mathrm{B},\alpha}$ that a B molecule enters an α-channel (which, for the sake of simplicity, are assumed to be identical). The limiting cases $P_{\mathrm{A},\beta} = P_{\mathrm{B},\alpha} = 0$ and 1 represent the "hard" MTC system so far considered and the REF system, respectively [70]

4.4 Analytical Treatment

From a theoretical point of view, it is crucial to develop analytical tools to describe the MTC effect as only this ensures that all the underlying mechanisms are properly understood. Identification of the ingredients necessary for a theory of MTC will ultimately allow the derivation of optimization strategies and hence allow addressing the question of whether the MTC effect can be employed for practical purposes.

Since the MTC effect has been observed in a model system defined by the Markov process described above, a natural starting point for a theoretical description is an analytical treatment of the exact master equation for this process. However, owing to the intrinsic nonequilibrium nature of the steady state (which manifests itself in the flow of product particles out of the MTC array), one cannot write the stationary distribution of the process in terms of a standard Gibbs measure with some known energy function. This difficulty is compounded by the well-known fact (cf. Sect. 2 of this chapter) that in single-file systems diffusive mixing is inefficient and therefore correlations persist for long (and possibly infinite) times. Therefore mean-field predictions which neglect correlations frequently fail in one dimension. The subdiffusive motion of a tracer particle is just one of many manifestations of this phenomenon.

In order to overcome these conceptual difficulties, we note that the physical properties of a hopping process with exclusion in a single channel with open boundaries at both ends are actually well known. The master equation for the joint probabilities $\Theta(\sigma_1, \ldots, \sigma_n)$ is equivalent to a Schrödinger equation in imaginary time, with a quantum Hamiltonian H generating the time evolution of Θ. Using the exact equivalence of H with the quantum Hamiltonian of the isotropic Heisenberg ferromagnet, one may use the Bethe ansatz and the related dynamical matrix product ansatz to explicitly calculate the exact time evolution of joint probabilities; for a review see [73].

In the present context, only the stationary properties need to be considered [74, 75]. Open boundary conditions, with particle exchange with external reservoirs of constant densities $\rho_{1,2}$, are implemented by rates of injection and absorption of particles at the terminal sites 1, L of a channel of L sites. Because of the exclusion interaction, particles attempt to enter the terminal sites with rates $D\rho_i$ and leave with rates $D(1 - \rho_i)$ where D is the bulk hopping rate of an individual particle. With the latter notation, we are following here the convention of the relevant literature [73–75], in which – assuming unit step lengths – the single-particle diffusivity (cf. (1) and (2)) coincides with the site exchange rate. As in ordinary diffusion without interaction, this results in a stationary current

$$j = \frac{D \, \Delta\rho}{L + 1}, \tag{23}$$

where $\Delta\rho = \rho_1 - \rho_2$, divided by $L + 1$, is the macroscopic density gradient between the boundaries, referred to unit step legths. The density profile $\rho_k = \langle\sigma_k\rangle$ is linear,

$$\rho_k = \rho_1 - \Delta\rho \frac{k}{L + 1}, \tag{24}$$

and the correlations $C_{k,l} = \langle\sigma_k\sigma_l\rangle - \langle\sigma_k\rangle\langle\sigma_l\rangle$ of the particle occupancies of different sites are negative,

$$C_{k,l} = -\frac{(\Delta\rho)^2}{L} \frac{k}{L + 1} \left(1 - \frac{l}{L + 1}\right). \tag{25}$$

Notice that close to the boundaries (both k and l are of order 1 or of order L), the anticorrelations are only of order $1/L^2$. The current is of order $1/L$.

While the results in (23 – 25) are exact only for the exclusion process, i.e. for particles with only on-site hard-core interaction, one expects the gross finite-size scaling properties of this system to be valid also for more sophisticated short-range interactions. These features may be used to describe the full MTC system in terms of interconnected single-file channels. The idea is to use small boundary correlations in order to justify the replacement of the catalytic sites by special sites with fixed but unspecified densities $\rho_i^{A,B}$ of reactant and product particles. Equation (23) and (24), together with the exclusion rule and conservation of the total number of particles within the MTC array, then yield a set of equations which determine self-consistently the densities $\rho_i^{A,B}$. Thus the states of the channels between the catalytic sites and of those connected to the external reservoirs are completely determined and known exactly within the framework of this approximation. The number Q of degrees of freedom in the system is reduced from exponential in the system size, $Q = 2^{m_{\text{tot}}}$, to algebraic, $Q = 2n^2$, and one obtains all steady-state properties in terms of the external-reservoir densities ρ^A of reactants and ρ^B of product particles. In the simulations discussed in Sect. 4.3, $\rho^B = 0$ was used.

This approximation scheme, which combines exact results (for the individual one-dimensional channels) with a mean-field approximation (neglecting correlations between the two-dimensional array of catalytic intersections and the adjacent sites belonging to the channels), has not been fully carried out yet, but preliminary DMC simulations confirm the theoretical prediction in the diffusion-controlled regime of large conversion rate k_c [76] for sufficiently large L. For a single intersection one finds an extremal principle for the current, which is somewhat reminiscent of an extremal principle for the selection of the current in bulk-driven diffusive systems [77, 78].

To derive this scheme, it is instructive to slightly generalize the setting discussed above and assume that the two channels have different lengths L, L'. Reflection symmetry along the diagonal axis of the cross allows us to reduce the cross to a single chain with an α-segment of length L connected to a β-segment of length L'. We label the sites from 1 through $M = L + L' + 1$ and denote the catalytic site by $N = L + 1$.

Because of the particle conservation in the bulk of each channel, we have $j_A = j_B \equiv j$. Equation (24) then may be written

$$\rho_k^A = \rho^A - \frac{jk}{D} , \tag{26}$$

$$\rho_k^B = \rho^B - \frac{j(M + 1 - k)}{D} . \tag{27}$$

Here $\rho^{A,B}$ are the predescribed boundary densities, and j is the stationary current that the system selects and which has to be determined. Since the catalyst acts like a sink for particles A and a source for particles B, one has $j \geq 0$. Equations (26), (27) together with the exclusion constraint $0 \leq \rho_k \leq 1$ imply the rigorous bounds

$$0 \leq j \leq \frac{D\rho^A}{L} , \quad 0 \leq j \leq \frac{D(1 - \rho^B)}{L'} . \tag{28}$$

Examining the exact master equation of the process and assuming $L, L' \gg 1$ (and therefore neglecting correlations between the occupancy of the catalytic site N and its two neighboring sites) yields a quadratic equation $\rho_{N-1}^A (1 - \rho_N^B) = 0$ for the current with the two solutions

$$j = \begin{cases} D\rho^A / L \\ D(1 - \rho^B)/L' \end{cases} . \tag{29}$$

The question to be addressed therefore is which of these solutions the system selects. The answer follows from the inequalities (28). They can be satisfied only if j selects the smaller of the two values (29). Hence

$$j = \min \left\{ \frac{D\rho^A}{L}, \frac{D(1 - \rho^B)}{L'} \right\} , \tag{30}$$

which may alternatively be written

$$j = \min \left\{ j_{\max}^A, j_{\max}^B \right\} . \tag{31}$$

This has an intuitive physical meaning. In each channel the system tries to maximize its current, but the current actually selected is limited by the smaller of the two possible maximal currents.

To understand the origin of this selection principle, notice that the stationary probability of finding a particle A on a catalytic site is almost zero (i.e. at most of order $1/L$), which comes from the model assumption that a particle A is very likely to be converted into a particle B before returning to the α-channel. Now let us assume, first, that ρ^A, i.e. the reservoir density of particles A, is very small. For small ρ^A, the system tries to sustain the current $j = j_{\max}^A$, which is possible as long as the β-channel can support this current. This is the case if the density of particles B at the site N required to generate this current in the β-channel is less than 1. Notice that in this regime, the density of particles A both on site N and on the last site of the α-channel is nearly zero. Now we assume that ρ^A is increased adiabatically, i.e. so slowly that the system reaches stationarity before a further increase occurs. As long as $j_{\max}^A < j_{\max}^B$, essentially nothing changes on the last site of the α-channel. On the site N, the density ρ_N^B increases in order to sustain the enhanced stationary current. This density therefore is approximately given by the relation $(\rho_N^B - \rho^B)/L' = \rho^A/L$, which follows from (26), (27). If ρ^A becomes so large that $j_{\max}^A = j_{\max}^B$, one has $\rho_N^B = 1$. Then the current is limited by the capacity of the β-channel, and increasing ρ^A further does not increase the current. Instead, the last site L in the α-channel acquires a finite density of particles A.

5 Experimental Evidence About the Real Structure of Nanoporous Channel Systems

Among different types of nanoporous materials, zeolites are famous for their well-defined morphology. While the ideal structure of zeolites is routinely used to elucidate their adsorption and transport properties, it was only recently appreciated that these properties can be influenced to a great extent by building defects into the crystals. Particularly important in this respect are regular intergrowth effects, which may affect the properties of the majority of the crystals in macroscopic zeolite systems, such as zeolite membranes and zeolite beds.

Interference microscopy and IR microscopy have proved to be powerful tools to study the internal structure of zeolite crystals, as well as the influence of this structure on intracrystalline molecular transport.

5.1 Experimental Details of Interference Microscopy and IR Microscopy

The interference and IR microscopy techniques are the only techniques which have proved to be suitable for monitoring intracrystalline concentration profiles of guest molecules in zeolite crystals. A detailed description of these techniques may be found in [79–82]. Briefly, the interference microscopy method is based on the changes of the optical density of porous crystals following adsorption or desorption of guest molecules. Assuming proportionality between the local adsorbate concentration and the change of the local refractive index, the experimental data allow one to determine a value proportional to the integral of the local concentration in the direction of light propagation. The concentration integrals are recorded with a spatial resolution of $\sim 0.5 \times 0.5\ \mu m^2$. The concentration integrals can also be monitored in a somewhat more direct way by recording IR absorbance spectra of guest molecules residing in a crystal. This idea has found its implementation in the IR microscopy method. Despite a poor spatial resolution ($\sim 20 \times 20\ \mu m^2$), the IR microscopy technique presents an extremely useful tool to study intracrystalline concentration profiles, owing to its ability to distinguish between different adsorbates. This is achieved by relating adsorbates to their characteristic IR absorption bands.

For the measurements and the zeolite activation, the zeolite sample was introduced into a specially made optical or IR cell connected to the vacuum system. Prior to the measurements, the sample was activated by keeping it under high vacuum at elevated temperature (typically at around 200 °C) for over 12 h. The measurements of the concentration integrals were always performed with one selected zeolite crystal.

5.2 Concentration Profiles in AFI-Type Crystals

The ideal framework topology of the AFI-type zeolites can be described as a packing of identical oriented cylinders [3]. This textbook structure can hardly be found compatible with the equilibrium intracrystalline concentration profiles of methanol in CrAPO-5 crystals (Fig. 16). The profiles in Fig. 16 were recorded using the interference microscopy technique in equilibrium with methanol vapor in the gas phase surrounding the crystal [82]. The highly inhomogeneous profiles in Fig. 16 exhibit a reproducible regular pattern characteristic of regular intergrowth effects. The occurrence of such effects in the crystals investigated renders part of the channel system inaccessible to methanol molecules. On the basis of the recorded profiles, an internal structure of the CrAPO-5 crystals was proposed [82]. This structure is shown in Fig. 17. The semipyramid in Fig. 17 outlines schematically the crystal section accessible to methanol in the lower part of the crystal. Owing to the crystal symmetry with respect to reflection in the geometrical center of the crystal, the structure of the upper part of the crystal is not shown. The channel direction in the semipyramids coincides with the z-direction.

In order to confirm the results obtained by interference microscopy, the equilibrium intracrystalline concentration profiles of methanol in CrAPO-5 crystals were recorded under the same measurement conditions by IR microscopy [82]. Owing to the much lower spatial resolution of IR microscopy ($\sim 20 \times 20\ \mu m^2$), a direct comparison of two-dimensional concentration profiles obtained by these two techniques is not feasible. Instead, the one-dimensional profiles along the y- and z-directions were compared (Fig. 18). Figure 18 demonstrates good agreement between the results obtained by both techniques.

The results discussed above show clearly that the crystals investigated cannot, by far, be considered as ideal AFI-type crystals. Intergrowth effects similar to those

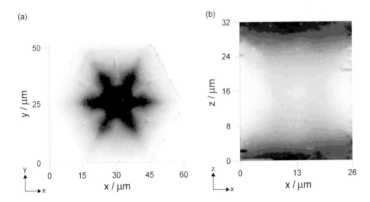

Fig. 16. Equilibrium intracrystalline concentration profile of methanol in CrAPO-5 crystal. The color intensity is proportional to the integrals of local concentration in the z-direction (a) and in the y-direction (b). Darker regions correspond to larger concentration integrals. x, y, and z are the crystallographic directions (the channel direction is z) [82]

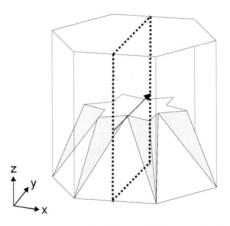

Fig. 17. The internal structure of a CrAPO-5 crystal (shown only for the lower part of the crystal). x, y, and z are the crystallographic directions. Some planes of the semipyramidal component have been colored in a shade of gray to improve the three-dimensional presentation of the shape of this component. Broken lines outline the observation plane of the interference microscopy measurements of the transient profiles shown in Fig. 19, with the arrow indicating the direction of observation [82, 86]

shown in Fig. 17 were frequently observed in other AFI-type crystals synthesized in solution [83–85]. For some of these crystals, it was shown that only the internal crystal components which geometrically resemble the semipyramid in Fig. 17 could be loaded with adsorbate [85].

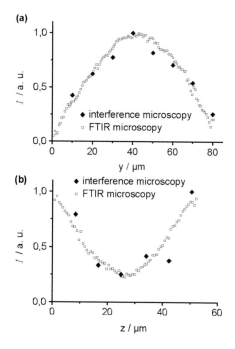

Fig. 18. The mean concentration integrals I recorded by FTIR and interference microscopy: (a) along the y-direction for x-values in Fig. 16 between 35 and 55 μm; (b) along the z-direction for x-values in Fig. 16 between 12 and 32 μm. x, y, and z are the crystallographic directions [82]

It is remarkable that interference microscopy, when combined with Monte Carlo simulations, can provide quantitative information about molecular transport even in the crystals considered so far with a complex, nonideal structure. As an example, Fig. 19 shows the results of the best fit of the transient concentration profiles of methanol recorded in CrAPO-5 crystals by the simulation results [86]. These profiles present the intracrystalline concentration of methanol integrated along the y-direction in the plane schematically shown by the broken lines in Fig. 17. The arrow in Fig. 17 shows the direction of observation (which coincides with the direction of integration). The concentration profiles in Fig. 19 were measured at different times after the start of the methanol adsorption. The profile recorded after the largest time $(4 \times 10^4 \text{ s})$ represents the equilibrium concentration profile. The Monte Carlo simulations were performed assuming (i) the internal crystal structure shown in Fig. 17 and (ii) the existence of an additional transport resistance on the crystal surface. The second assumption proved to be necessary in order to get a good agreement between the experimental results and the results of the simulations. The permeability of the transport barriers on the crystal surface and the intracrystalline diffusivity were assumed to be independent of molecular concentration. Figure 19 shows that at short

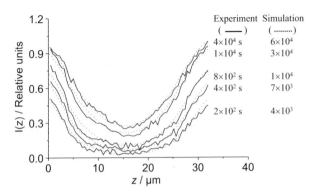

Fig. 19. Intracrystalline concentration of methanol integrated along the crystallographic y-direction in CrAPO-5 at different times after the start of the methanol adsorption. The concentration integrals were measured by the interference microscopy method (full lines) and were also obtained by the dynamic MC method (dotted lines). For the calculated profiles, the time unit is the time of one elementary diffusion step [86]

times there is a reasonably good agreement between the experimental profiles and their simulated counterparts when both the relative time intervals between the consecutive profiles and their shape are considered. However, with increasing time, the relative time intervals between the consecutive simulated profiles get significantly smaller than those between the corresponding measured profiles. These discrepancies signal that, for sufficiently large methanol concentrations, the permeability of the transport barriers and/or the intracrystalline diffusivity of methanol may depend on the methanol concentration. Comparison of the measured concentration profiles with those simulated by the dynamic Monte Carlo method allows us to obtain quantitative information at the limit of low methanol concentration about the intracrystalline diffusivity of methanol and the permeability of the transport barriers on the crystal surface [86].

5.3 Monitoring the Dynamics of Guest Distributions in MFI-Type Crystals

Regular intergrowth in MFI crystals is one of the commonly known examples of intergrowth effects in zeolites. The textbook channel structure of MFI-type zeolite consists of straight and zigzag channels (see the discussion in Sect. 3).

Large MFI crystals usually exhibit a morphologically perfect shape, which is common for single crystals. However, a careful optical examination of these crystals often reveals an hourglass structure (Fig. 20), indicating regular intergrowth effects [52, 81, 87–89]. The hourglass structure in MFI crystals is usually assigned to the division lines, or interfaces between several intergrowth components of the crystal. The existence of such interfaces opens up the possibility of fast molecular uptake through these interfaces, provided that they are directly accessible to the guest molecules from the surrounding gas phase. The role of the interfaces in molecular up-

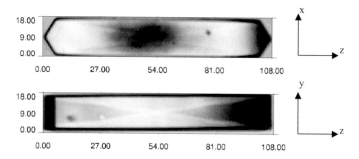

Fig. 20. Microscopic images of a typical silicalite-1 crystal in two different orientations. The hourglass structure was made visible by using the shearing mechanism of the microscope. The length scale in the x-, y-, and z-directions is shown in μm [81]

take was investigated using interference microscopy combined with dynamic Monte Carlo simulations.

There are two models of MFI crystal morphology currently discussed in the literature. According to one of the models [52, 87, 89], each crystal is composed of three components, i.e. of two identical pyramid units and of a central component (Fig. 21a). According to another [88], the central component is not a single section but, in its turn, consists of four components (Fig. 21b), two of which are identical to the pyramid units in Fig. 21a.

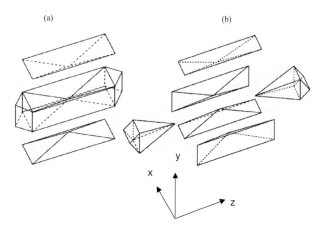

Fig. 21. Schematic representations of the internal structure of silicalite-1 crystals: (a) according to [52, 87, 89]; (b) according to [88]

Figure 22 shows an example of the measured concentration profiles in the rectangular part of a typical silicalite-1 crystal (z between 13 and 88 μm in Fig. 20)

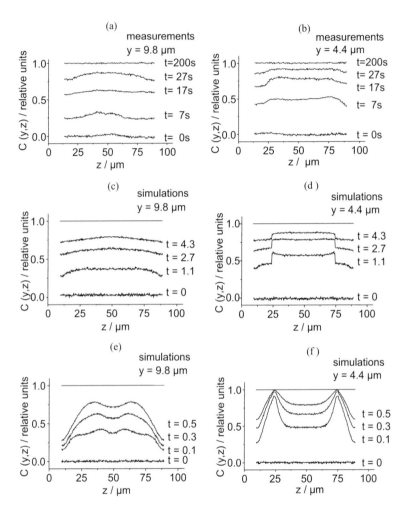

Fig. 22. Intracrystalline concentration profiles of isobutane in an MFI-type crystal along the z-direction during adsorption: (a), (b) profiles measured by interference microscopy; (c), (d) simulated profiles, assuming that the internal interfaces serve only as transport barriers; (e), (f) simulated profiles, assuming that adsorption/desorption may also occur through the internal interfaces. For the simulated profiles, the time unit is 10^3 elementary diffusion steps [81]

during isobutane adsorption, and the corresponding results of the Monte Carlo simulations [81]. The integrals of the isobutane concentration in Fig. 22a,b were measured in the middle part of the crystal ($y = 9.8$ μm) and near the crystal border ($y = 4.4$ μm), respectively. The profiles in the middle and the lower part of the figure show the corresponding results of the simulations. The simulations were performed either by assuming molecular uptake to proceed only through the external crystal surface, with the internal interfaces being transport barriers (Fig. 22c,d) or by assuming molecular uptake to occur both through the external crystal surface and through the internal interfaces (Fig. 22e,f). The ratio between the isobutane diffusivities along the straight and the zigzag channels was assumed to be equal to that calculated in [90] by transition path sampling. The diffusivity along the direction perpendicular to that of the straight and the zigzag channels was determined using the correlation rule of diffusion anisotropy in MFI (cf. [44] and Sect. 3.1).

Comparison between the measured concentration profiles and the corresponding simulation results allows us to rule out isobutane uptake through the internal interfaces in the crystals investigated. The results of this study show that fitting of the experimental concentration integrals by the results of dynamic MC simulations can provide qualitative and quantitative information about intracrystalline transport even in the most complex systems, such as the MFI-type zeolites.

6 Conclusion

At first glance, molecular propagation in channels and channel networks with diameters comparable to those of the diffusants might appear to represent a particularly easily manageable case of molecular dynamics in interfaces. Irrespective of the sole, seemingly simple condition of mutual particle exclusion in one-dimensional arrays (i.e. the "single-file" condition), the theoretical treatment of such a situation yields a wealth of surprising results and, moreover, a substantial number of still unsolved problems.

Starting from an introduction to the peculiarities of single-file diffusion, we have highlighted the occurrence of completely new features of molecular dynamics as soon as a multitude of single-file systems are composed into channel networks. Together with the occurrence of a new type of diffusion anisotropy, it is in particular the possibility of an increase in reactivity thus achieved which may open up novel routes for performance enhancement in heterogeneous catalysis. The situation ensuring such a possibility has been referred to as the regime of "molecular traffic control" (MTC). In addition to the conventional way of modeling such a situation by dynamic Monte Carlo simulations, the first attempts at an analytic treatment have been included as well. Both types of theoretical treatment, which have been applied as well to modeling the internal dynamics of single-file systems, are complemented by MD simulations following the quite general introduction to this method in Chap. 1.

The challenging issue of investigating systems of the type considered also includes the transformation of the principles detected and their confirmation by those experiments. Over the last few years, zeolites have generally been considered to be excellent candidates for the experimental verification and technological exploitation of the phenomena considered. However, the most recent investigations, primarily by interference and IR microscopy, have revealed dramatic deviations of these materials from the ideal textbook structure. This has been illustrated for both zeolites of type AFI, which turned out to deviate substantially from the expected structure of bundles of macaroni of atomic dimensions, and zeolites of type MFI, which are found to host a number of relevant structural items such as internal transport resistances, acting in addition to the diffusion resistance exerted by a network of intersecting channels. The present lack of ideal host systems for tracing the predicted phenomena of diffusion in channels and channel networks implies a twofold challenge. On the one hand it concerns the field of zeolite synthesis, from which real crystals with the ideal – i.e. textbook – structure are expected. On the other hand, further progress in the experimental techniques of diffusion measurements, as illustrated, for example in Chap. 2, 4, and 3, would likewise help to contribute to a more efficient comparison with theory by ensuring the performance of crucial experiments even with incomplete materials.

Acknowledgments

We are very grateful to all our students and colleagues at Leipzig University for numerous stimulating discussions during nine years of most agreeable collaboration in the SFB 294 and in the Graduiertenkolleg "Physikalische Chemie der Grenzflächen". We are particularly obliged to S. Auerbach (Amherst), K. Binder (Mainz), A. Bunde (Gießen), W. Dieterich (Konstanz), P. Heitjans (Hannover), H. Karge (Berlin) and R. Snurr for helpful suggestions and advice.

References

1. B. Sakmann: Science **256**, 503 (1992)
2. P.G. de Gennes: J. Chem. Phys. **55**, 572 (1971)
3. C. Baerlocher, W.M. Meier, D.H. Olson: *Atlas of Zeolite Structure Types* (Elsevier, London, 2001)
4. J. Kärger, P. Heitjans, R. Haberlandt (Eds.): *Diffusion in Condensed Matter* (Vieweg, Braunschweig, 1998)
5. P. Heitjans, J. Kärger (Eds.): *Diffusion in Condensed Matter: from the Elementary Step to Transport in Complex Systems* (Springer, Berlin, Heidelberg, 2004) in press
6. K.W. Kehr, K. Mussawisade, G.M. Schütz, T. Wichmann: in *Diffusion in Condensed Matter: from the Elementary Step to Transport in Complex Systems*, ed. by P. Heitjans, J. Kärger (Springer, Berlin, Heidelberg, 2004) in press
7. H. Mehrer, F. Wenwer: *Diffusion in metals* in *Diffusion in Condensed Matter*, ed. by J. Kärger, P. Heitjans, R. Haberlandt (Vieweg, Braunschweig, 1998), pp. 1–39
8. P.A. Fedders: Phys. Rev. B **17**, 40 (1978)
9. R. Arratia: Ann. Prob. **11**, 362 (1983)
10. K. Binder: *Application of the Monte Carlo Method in Statistical Physics* (Springer, Berlin, Heidelberg, 1987)
11. J. Kärger: Phys. Rev. A **45**, 4173 (1992)
12. J. Kärger: Phys. Rev. E **47**, 1427 (1993)
13. K. Hahn, J. Kärger: J. Phys. Chem. **100**, 316 (1996)
14. V. Gupta, S.S. Nivarthi, A.V. McCormick, H.T. Davis: Chem. Phys. Lett. **247**, 596 (1995)
15. K. Hahn, J. Kärger, V. Kukla: Phys. Rev. Lett. **76**, 2762 (1996)

16. V. Kukla, J. Kornatowski, D. Demuth, I. Girnus, H. Pfeifer, L.V.C. Rees, S. Schunk, K.K. Unger, J. Kärger: Science **272**, 702 (1996)
17. H. Jobic: "Diffusion studies using quasi-elastic neutron scattering", in *Recent Advances in Gas Separation by Microporous Ceramic Membranes*, ed. by N.K. Kanellopoulos (Elsevier, Amsterdam, 2000), pp. 109–137
18. S. Brandani, D.M. Ruthven, J. Kärger: Zeolites **15**, 494 (1995)
19. S. Brandani, M. Jama, D.M. Ruthven: Micropor. Mesopor. Mater. **6**, 283 (2000)
20. L.V.C. Rees, L. Song: "Frequency response methods for the characterization of microporous solids", in: *Recent Advances in Gas Separation by Microporous Ceramic Membranes*, ed. by N.K. Kanellopoulos (Elsevier, Amsterdam, 2000), pp. 139-186
21. R. Roque-Malherbe, V. Ivanov: Micropor. Mesopor. Mater. **47**, 25 (2001)
22. J. Kärger: Micropor. Mesopor. Mater. **56**, 321 (2002)
23. C. Rödenbeck, J. Kärger, K. Hahn: Ber. Bunsen-Ges. Phys. Chem. Chem. Phys. **102**, 929 (1998)
24. C. Rödenbeck, J. Kärger, H. Schmidt, T. Rother, M. Rödenbeck: Phys. Rev. E **60**, 2737 (1999)
25. G.F. Froment, K.B. Bischoff: *Chemical Reactor Analysis and Design* (Wiley, New York 1979)
26. C. Rödenbeck, J. Kärger, K. Hahn: Phys. Rev. **55**, 5697 (1997)
27. C. Rödenbeck, J. Kärger, K. Hahn: J. Catal. **176**, 513 (1998)
28. C. Rödenbeck, J. Kärger, K. Hahn: J. Catal. **157**, 656 (1995)
29. G.D. Lei, W.M.H. Sachtler: J. Catal. **140**, 601 (1993)
30. G.D. Lei, B.T. Carvill, W.M.H. Sachtler: Appl. Catal. A – General **142**, 347 (1996)
31. C. Rödenbeck, J. Kärger, K. Hahn: Phys. Rev. E **57**, 4382 (1998)
32. F.J. Keil, R. Krishna, M.O. Coppens: Rev. Chem. Eng. **16**, 71 (2000)
33. P.H. Nelson, A.B. Kaiser, D.M. Bibby: J. Catal. **127**, 101 (1991)
34. P.H. Nelson, D.M. Bibby, A.B. Kaiser: Zeolites **11**, 337 (1991)
35. J. Kärger, D.M. Ruthven: *Diffusion in Zeolites and Other Microporous Solids* (Wiley, New York, 1992)
36. R.M. Barrer: *Zeolites and Clay Minerals as Sorbents and Molecular Sieves* (Academic Press, London, 1978)
37. K. Hahn, J.Kärger: J. Phys. Chem. **102**, 5766 (1998)
38. J.Kärger, M. Petzold, H. Pfeifer, S. Ernst, J. Weitkamp: J. Catal. **136**, 283 (1992)
39. P.H. Nelson, S.M. Auerbach: J. Chem. Phys. **110**, 9235 (1999)
40. P.H. Nelson, S.M. Auerbach: Chem. Eng. J. **74**, 43 (1999)
41. C. Rödenbeck, J. Kärger: J. Chem. Phys. **110**, 3970 (1999)
42. S. Vasenkov, J. Kärger: Phys. Rev. E **66**, 052601 (2002)
43. T. Meersmann, J.W. Logan, R. Simonutti, S. Caldarelli, A. Comotti, P. Sozzani, L.G. Kaiser, A. Pines: J. Phys. Chem. **104**, 11665 (2000)
44. J. Kärger: J. Phys. Chem. **95**, 5558 (1991)
45. J. Kärger, H. Pfeifer: Zeolites **12**, 872 (1992)
46. D. Fenzke, J. Kärger: Z. Phys. D **25**, 345 (1993)
47. S. Jost, N.K. Bär, S. Fritzsche, R. Haberlandt, J. Kärger: J. Phys. Chem. B **102**, 6375 (1998)
48. R. Haberlandt, J. Kärger: Chem. Eng. J. **74**, 15 (1999)
49. D.N. Theodorou, R.Q. Snurr, A.T. Bell: "Molecular dynamics and diffusion in microporous materials", in *Comprehensive Supramolecular Chemistry*, ed. by G. Alberti, T. Bein (Pergamon, Oxford, 1996), pp. 507–548
50. R.L. June, A.T. Bell, D.N. Theodorou: J. Phys. Chem. **94**, 8232 (1990)

124   P. Bräuer et al.

51. U. Hong, J. Kärger, R. Kramer, H. Pfeifer, G. Seiffert, U. Müller, K.K. Unger, H.B. Lück, T. Ito: Zeolites **11**, 816 (1991)
52. J. Caro, M. Noack, J. Richter-Mendau, F. Marlow, D. Peterson, M. Griepenstrog, J.J. Kornatowski: J. Phys. Chem. **97**, 13685 (1993)
53. E.J. Maginn, A.T. Bell, D.N. Theodorou: J. Phys. Chem. **100**, 7155 (1996)
54. F. Jousse, S.M. Auerbach, D.P. Vercauteren: J. Chem. Phys. **112**, 1531 (2000)
55. S. Fritzsche, J. Kärger: J. Phys. Chem. B **107**, 3515 (2003)
56. S. Fritzsche, J. Kärger: Europhys. Lett. **63**, 465 (2003)
57. E.G. Derouane, Z. Gabelica: J. Catal. **65**, 486 (1980)
58. E.G. Derouane: Appl. Catal. A **115**, N2 (1994)
59. R.Q. Snurr, J. Kärger: J. Phys. Chem. B **101**, 6469 (1997)
60. L.A. Clark, G.T. Ye, R.Q. Snurr: Phys. Rev. Lett. **84**, 2893 (2000)
61. L.A. Clark, G.T. Ye, A. Gupta, L.L. Hall, R.Q. Snurr: J. Chem. Phys. **111**, 1209 (1999)
62. N.Y. Chen, T.F. Degnan, C.M. Smith: *Molecular Transport and Reaction in Zeolites* (VCH, New York, 1994)
63. C.G. Pope: J. Catal. **72**, 174 (1981)
64. E.G. Derouane: J. Catal. **72**, 177 (1981)
65. F.J. Keil: *Diffusion und chemische Reaktion in der Gas/Feststoff-Katalyse* (Springer, Berlin, Heidelberg, 1999), p. 200
66. J. Weitkamp, S. Ernst, L. Puppe: *Shape-selective catalysis in zeolites*, in *Catalysis and Zeolites*, ed. by J. Weitkamp, L. Puppe (Springer, Berlin, Heidelberg, 1999), p. 346
67. N. Neugebauer, P. Bräuer, J. Kärger: J. Catal. **194**, 1 (2000)
68. P. Bräuer, J. Kärger, N. Neugebauer: Europhys. Lett. **53**, 8 (2001)
69. J. Kärger, P. Bräuer, H. Pfeifer: Z. Phys. Chem. **214**, 1707 (2000)
70. P. Bräuer, A. Brzank, J. Kärger: J. Phys. Chem. B **107**, 1821 (2003)
71. R. Aris: *The Mathematical Theory of Diffusion and Reaction in Permeable Catalysts* (Clarendon Press, Oxford, 1975)
72. R. Kutner: Phys. Lett. **81A**, 239 (1981)
73. G.M. Schütz: "Exactly solvable models for many-body systems far from equilibrium", in *Phase Transitions and Critical Phenomena*, Vol. 19, ed. by C. Domb, J. Lebowitz (Academic Press, London, 2000)
74. H. Spohn: J. Phys. A **27**, 4275 (1983)
75. R.B. Stinchcombe, G.M. Schütz: Phys. Rev. Lett. **75**, 140 (1995)
76. A. Brzank, G.M. Schütz, P. Bräuer, J. Kärger: Phys. Rev. E, in press (2004)
77. J. Krug: Phys. Rev. Lett. **67**, 1882 (1991)
78. V. Popkov, G.M. Schütz: Europhys. Lett. **48**, 257 (1999)
79. U. Schemmert, J. Kärger, C. Krause, R.A. Rakoczy, J. Weitkamp: J. Europhys. Lett. **46**, 204 (1999)
80. U. Schemmert, J. Kärger, J. Weitkamp: Micropor. Mesopor. Mater. **32**, 101 (1999)
81. O. Geier, S. Vasenkov, E. Lehmann, J. Kärger, U. Schemmert, R.A. Rakoczy, J. Weitkamp: J. Phys. Chem. B **105**, 10217 (2001)
82. E. Lehmann, C. Chmelik, H. Scheidt, S. Vasenkov, B. Staudte, J. Kärger, F. Kremer, G. Zadrozna, J.J. Kornatowski: J. Am. Chem. Soc. **124**, 8690 (2002)
83. G.L. Klap, M. Wübbenhorst, J.C. Jansen, H. van Konnigsveld, H. van Bekkum, J. van Turnhout: J. Chem. Mater. **11**, 3497 (1999)
84. G.L. Klap, H. van Konnigsveld, H. Graafsma, A.M.M. Schreurs: Micropor. Mesopor. Mater. **38**, 403 (2000)
85. L. Girnus, K. Jancke, R. Vetter, J. Richter-Mendau, J. Caro: Zeolites **15**, 33 (1995)
86. E. Lehmann, S. Vasenkov, J. Kärger, G. Zadrozna, J.J. Kornatowski: J. Chem. Phys. **118**, 6129 (2003)

87. C. Weidenthaler, R.X. Fischer, R.D. Shannon, O. Medenbach: J. Phys. Chem. **8**, 12687 (1994)
88. E.R. Geus, J.C. Jansen, H. van Bekkum: Zeolites **14**, 82 (1994)
89. M. Kocirik, J.J. Kornatowski, V. Masarik, P. Novak, A. Zikanova, J. Maixner: Micropor. Mesopor. Mater. **23**, 295 (1998)
90. T.J.H. Vlugt, C. Dellago, B. Smit: J. Chem. Phys. **113**, 8791 (2000)

Structure–Mobility Relations of Molecular Diffusion in Interface Systems

Jörg Kärger, Christine M. Papadakis, and Frank Stallmach

Universität Leipzig, Institut für Experimentalphysik I,
Kaerger@physik.uni-leipzig.de,
Christine.Papadakis@physik.uni-leipzig.de,
Stallmac@physik.uni-leipzig.de

Abstract. The behaviour of molecules in contact with interfaces may significantly deviate from the patterns known from their behaviour in the bulk. This is in particular true with respect to molecular diffusion, i.e. to the random movement of the constituents of the interface systems. As a non-invasive technique and owing to the large spectrum of rather specific information accessible by this technique, pulsed field gradient (PFG) NMR has proved to be the method of choice for such investigations.

After a short introduction into the principles of PFG NMR and its particular strengths and limitations, various examples of the benefit of exploiting this technique in diffusion studies of interface systems are given. They include the investigation of nanoporous materials such as zeolites, where the application of PFG NMR has revolutionized the understanding of intracrystalline diffusion. As typical examples for structure-mobility relations, the propagator representation and the various (structure-dependent) patterns of concentration dependence of self-diffusion are presented. In the investigation of sediments (such as beds of sand grains), the influence of the interface between the fluid and solid phases is shown to affect both molecular propagation and nuclear magnetic relaxation, which is shown to eventually lead to a new route of attaining fractal geometries. Finally, the internal structurization of self-organized multicomponent polymer systems containing diblock copolymers is shown to give rise to peculiarities in the propagation patterns of the individual constituents. The conclusion summarizes the presented principles of structure-mobility correlations and refers to the numerous further examples in nature and technology, some of which – see e.g. the contributions by Fritzsche et al., Bräuer et al., Freude et al., Kremer et al., Michel et al., and Arnold et al. – may be found in further chapters of the book.

1 Introduction

Molecular diffusion, i.e. the irregular microscopic motion of matter, is among the fundamental phenomena in nature and technology [1, 2]. It occurs in essentially all types of matter and may be associated with timescales from femtoseconds up to ages. Besides its fundamental relevance, the continuously increasing interest in diffusion research is an immediate consequence of the limiting role diffusion processes may play in numerous technological applications, including semiconductor production [3, 4], metallurgy [5, 6], biotechnology [7], polymer chemistry [8], mass separation [9], and catalysis [9, 10]. It is not unexpected, therefore, that *Current Contents* (Physical, Chemical, and Earth Sciences), for example, reports over 8000 publications per year concerning this topic, and the number of publications continues to steadily increase.

J. Kärger, C.M. Papadakis, F. Stallmach, Structure–Mobility Relations of Molecular Diffusion in Interface Systems,
Lect. Notes Phys. **634**, 127–162 (2004)
http://www.springerlink.com/

Particularly fascinating patterns of diffusion may result if molecular transport occurs in systems dominated by the influence of interfaces. In this case it is often impossible to quantify the phenomenon of diffusion by a sole parameter. This is an immediate consequence of the fact that molecular propagation may most decisively be affected by the presence of the interfaces, leading to a temporal and spatial dependence of the diffusion phenomena under study. Following the numerous facets of the theoretical treatment, as introduced in Chap. 1, the present chapter deals with a most powerful technique of diffusion measurement, the pulsed field gradient (PFG) NMR method [1,2,9–13]. The wealth of information provided by this technique is particularly impressive in its application to heterogeneous systems with internal interfaces. This is due, in particular, to the ability of PFG NMR to operate noninvasively and to be sensitive over both space and time scales covering several orders of magnitude, in the range of micrometers and milliseconds [14,15].

After an introduction to the fundamentals of this technique in Sect. 2, which includes both its strengths and the limitations, the present chapter is devoted to a general overview of the potentials of this technique, covering its application to nanoporous host–guest systems (section 3), to the elucidation of the different types of solid–fluid interaction in sediments (section 4), and to the multitude of transport patterns as a consequence of the internal structuring of self-organized multicomponent systems containing diblock copolymers (section 5). The manifold correlations of both the technique and the communicated results with the general topic of the book will be highlighted in the concluding Sect. 6.

2 Fundamentals of Self-Diffusion Studies by PFG NMR

2.1 Theoretical Approach to Quantifying Diffusion Processes

Propagator, Mean Square Displacement, and Self-Diffusion Coefficient

Self-diffusion, i.e. the Brownian (thermal) motion of molecules, may be described by Fick's first and second laws (diffusion equation) if one considers the motion of labeled particles in an environment of unlabeled but otherwise completely identical particles. In complex interface systems, where the interfaces generally represent barriers for the diffusing particles, it is often very difficult or even impossible to solve these equations. Alternatively, self-diffusion processes may be quantified on the basis of the so-called propagator or (as it is generally referred to in neutron scattering and Mössbauer spectroscopy) van der Hove self-correlation function. If we know the propagator, the time dependence of the mean square displacement $\langle r^2(t) \rangle$ and the self-diffusion coefficient D, which are physical quantities describing the diffusion processes in a more compact form, can be derived. As we shall see later, the propagator averaged over all possible starting positions within a heterogeneous sample may be measured directly by PFG NMR, which enables us to determine experimental values for $\langle r^2(t) \rangle$ and D even in complex interface systems.

Generally, the propagator $P(r_2, r_1, t)$, which denotes the (conditional) probability density that a particle originally located at position coordinate r_1 diffuses (propagates) to a position in a volume element $\mathrm{d}\boldsymbol{r}$ at $\boldsymbol{r}_2$ during the time t, may be obtained by solving the diffusion equation [1,2] (compare also Chap. 1)

$$\frac{\partial P}{\partial t} = D \, \boldsymbol{\nabla}^2 P, \tag{1}$$

with the initial condition $P(\boldsymbol{r}_2, \boldsymbol{r}_1, t = 0) = \delta(\boldsymbol{r}_2 - \boldsymbol{r}_1)$ and the boundary condition

$$0 = D \, \boldsymbol{n} \, \boldsymbol{\nabla} \, P + \rho P|_{\text{surface}}. \tag{2}$$

In (2), $\boldsymbol{n}$ denotes the unit vector normal to the surface at the interface and ρ describes how the molecules interact with the interface, i.e. whether they are absorbed, penetrate, or bounce off. Depending on the actual situation, the physical meaning of ρ is the transmission (or escape) probability (section 3.4) or the surface relaxivity (section 4.1).

Knowing the propagator for each starting position $\boldsymbol{r}_1$ allows us to determine the averaged probability $P(\boldsymbol{r}, t)$ of particle displacement for an arbitrary initial distribution of diffusing particles $p_0(\boldsymbol{r}_1)$. Denoting the particle displacement by the vector $\boldsymbol{r} = \boldsymbol{r}_2 - \boldsymbol{r}_1$, $P(\boldsymbol{r}, t)$ is calculated by multiplying p_0 by the propagator and integrating over all possible starting values:

$$P(\boldsymbol{r}, t) = \int_V P(\boldsymbol{r}_1 + \boldsymbol{r}, \boldsymbol{r}_1, t) p_0(\boldsymbol{r}_1) \, \mathrm{d}\boldsymbol{r}_1. \tag{3}$$

$P(\boldsymbol{r}, t)$ is the propagator averaged over all starting positions. It is referred to as the mean propagator [16] and represents the probability density that an arbitrarily selected particle in the sample is shifted over a distance r during the time t. In heterogeneous samples, this probability function of molecular displacement is understood as a mean value over the sample.

The mean square displacement $\langle r^2(t) \rangle$ of the particles follows from the mean propagator by calculating its second moment:

$$\langle r^2(t) \rangle = \int_V P(\boldsymbol{r}, t) r^2 \, \mathrm{d}\boldsymbol{r}. \tag{4}$$

In the case of normal diffusion, the individual diffusion steps of a particle leading to the total displacement $r(t)$ are independent of each other. On the basis of the central limit theorem [17], the resulting probability distribution of displacements may be shown to be a Gaussian ([9], pp. 63–67). In agreement with this consideration, the solution of the diffusion equation (1) for an infinitely extended homogeneous medium and the subsequent calculation of the averaged propagator by (3) yields

$$P(\boldsymbol{r}, t) = \frac{1}{\sqrt{(4\pi Dt)^3}} \exp\left(-\frac{r(t)^2}{4Dt}\right). \tag{5}$$

By inserting (5) into (4), one obtains the well-known Einstein relation:

$$\langle r^2(t) \rangle = 6Dt. \tag{6}$$

Equation (6) may be regarded as an alternative definition of the self-diffusion coefficient D, which, however, is completely equivalent to the definition via Fick's laws used in (1).

Self-Diffusion in Interface Systems

Generally, for self-diffusion under the influence of interfaces, the mean propagator as defined via (1) and (3) may deviate from being a Gaussian, and the mean square displacement increases less than in proportion to the time t [1, 15]. Nevertheless, following the idea that Einstein's relation (6) defines the self-diffusion coefficient in the case of normal diffusion, the mean square displacement may be used to obtain an effective, time-dependent diffusivity $D_{\text{eff}}(t)$ [14, 15]

$$D_{\text{eff}}(t) \equiv \frac{\langle r^2(t) \rangle}{6t}, \tag{7}$$

for quantification of diffusion processes under the influence of interfaces. Owing to the restriction of the diffusion path at the interfaces, D_{eff} is expected to be smaller than the corresponding diffusivity in the interface-free system (D_0).

The actual time dependence of $D_{\text{eff}}(t)$ depends on the system under study. Its prediction requires the solution of the diffusion equation incorporating the interactions at the interfaces with the boundary conditions. However, the interrelation between the diffusion path ($l_{\text{Dif}}(t) = \sqrt{\langle r^2(t) \rangle}$) and a typical length scale R describing the structure of the interface system may define conditions under which the interpretation of measured mean square displacements and effective diffusivities is simplified to a large degree [14]. In the following, four such conditions are introduced, where – depending on the system – R can be identified as, for example, the pore radius, the ratio of the pore volume to the surface area ($V_{\text{p}}/S \propto R$), the radius of curvature of the interface, the unit cell dimensions, or the domain size.

a) *Unrestricted Diffusion*, $l_{\text{Dif}}(t) \ll R$. During very short diffusion times, the majority of the diffusing molecules do not feel the presence of the interfaces. Thus, they behave as in the bulk or neat fluid. The mean square displacement obeys the Einstein relation (6) with a time-independent self-diffusion coefficient coinciding with the value of D_0.

b) *Short-diffusion-time approximation*, $l_{\text{Dif}}(t) \lesssim R$. It can easily be rationalized that the relative amount of molecules restricted by the interface increases linearly with l_{Dif}, i.e., with the square root of the diffusion time. Consequently, the effective diffusivity is expected to decrease with $\sqrt{t}$. The first-order series expansion of the effective diffusivity yields [18–20]

$$\frac{D_{\text{eff}}(t)}{D_0} = 1 - \frac{4}{9\sqrt{\pi}} S_{\text{V}} \sqrt{D_0 t} + \dots, \tag{8}$$

which allows one to determine the surface-to-pore volume ratio (S_V) of the interface system hosting the diffusing molecules. In this chapter, this relation will be applied to rationalize time-dependent intracrystalline diffusion measurements in zeolite crystals (section 3.4) and to study pore–grain interfaces in natural sediments (sections 4.2 and 4.3).

c) *Long diffusion time, but no complete restriction,* $l_{Dif}(t) \gg R$. If the interfaces do not represent total diffusion barriers and/or the diffusing molecules form an interconnected fluid phase, diffusion times t may be realized where l_{Dif} exceeds the typical length scale of the interface system. The reduction of the effective diffusivity compared with D_0 results from the interaction of the diffusing molecules with the interfaces. This situation applies for most diffusion studies in nanoporous crystals (section 3.2) and in polymer blends (section 5) if one identifies R with the unit cell dimension and the domain size, respectively.

If the effective diffusivity is not very much smaller than D_0 and independent of the diffusion time, the ratio D_0/D_{eff} may be used to determine the tortuosity factor τ [21],

$$\frac{D_0}{D_{eff}} = \tau \geq 1 \,, \tag{9}$$

which represents a geometrical pore structure parameter measuring the square of the length of the pathway connecting two points in the interface system compared with the length of the connecting straight line.

d) *Long diffusion time and complete restriction,* $l_{Dif}(t \to \infty) = f(R)$. If the interfaces form isolated areas which trap the diffusing molecules, the mean square displacement at very long diffusion times is limited by the geometrical size of the "traps". For example, for spherical boundaries of radius R_s one obtains (see e.g. [9, 11, 14])

$$l^2{}_{Dif} = \langle r^2(t = \infty) \rangle = \frac{6}{5} R_s{}^2 \,. \tag{10}$$

Inserting (10) into the definition of the effective diffusivity yields

$$D_{eff}(t) = \frac{R_s{}^2}{5t} \,, \tag{11}$$

which shows a typical behavior in this diffusion regime: the effective diffusivity decreases in proportion to $1/t$. If such a behavior is observed experimentally, the characteristic size of the restricting geometry may be determined via (10) and (11). Similar relations exist also for other simple restricting geometries [9,11,14].

2.2 Principle of PFG NMR Diffusion Measurements

Typical pulse sequences used for PFG NMR diffusion studies are drawn in Fig. 1 [11, 22–24]. They consist of radio frequency (rf) pulses to excite the nuclear spin

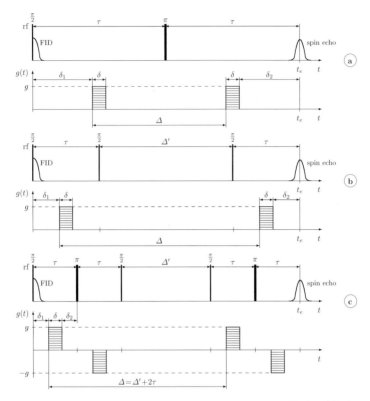

Fig. 1. NMR pulse sequences (rf and pulsed-gradient schemes) suitable for diffusion studies: primary (a) and stimulated (b) spin echo with unipolar pulsed field gradients; 13-interval sequence (c) with bipolar pulsed field gradients [22, 23]

system and to generate a spin echo NMR signal, and pulsed magnetic field gradients for encoding of translational motions of the molecules carrying the nuclear spins under investigation. The application of NMR to the study of molecular transport is based on the Larmor condition

$$\omega = \gamma B \qquad (12)$$

between the magnetic field ($\mathbf{B} = (0, 0, B)$) and the resonance frequency ($\omega = 2\pi f$) for transitions between the Zeeman levels of the nuclear spins in the magnetic field. The gyromagnetic ratio γ is a characteristic quantity of the nucleus under study (e.g. 2.67×10^8 $T^{-1}s^{-1}$ for ^1H). The Larmor frequency ω may be intuitively understood as the precessional frequency of the nuclear spins (and hence of the nuclear magnetization $\mathbf{M}$) about the direction of the magnetic field. For positive values of γ, this precessional motion is clockwise with respect to $\mathbf{B}$, i.e. $\omega \downarrow\uparrow \mathbf{B}$. If the magnetic field B consists of a homogeneous contribution B_0 superimposed by an inhomogeneous field $B_{\mathrm{add}} = gz$, where $g = \partial B / \partial z$ denotes the amplitude of the field gradient in the z-direction, the Larmor frequency becomes space-dependent:

$$\omega = \omega(z) = \gamma(B_0 + gz) = \omega_0 + \gamma gz \, . \tag{13}$$

For the PFG NMR diffusion studies, the field gradient is applied over two short time intervals (duration δ, separation Δ) either with opposite signs (i.e. the amplitudes $+g$ and $-g$) or with rf pulses of suitable duration (the π-pulse in Fig. 1a or the two $\pi/2$-pulses in Fig. 1b) in between. In both cases, the effect of the second field gradient pulse has to be subtracted from that of the first one. The phase shift φ of a nuclear spin which during the observation time Δ has been displaced over a distance $(z_2 - z_1)$ in the z-direction, in comparison with a spin, which has remained at the same position, is therefore

$$\varphi = \gamma g \delta(z_2 - z_1). \tag{14}$$

The quantity monitored in PFG NMR is the amplitude of the "spin echo" as generated, for example, in the NMR sequences shown in Fig. 1. The amplitude of this spin echo is proportional to the total magnetization, i.e. to the vector sum of the contributions of the individual spins. The application of field gradient pulses thus leads to a signal (spin-echo) attenuation Ψ (see e.g. [11, 22, 24]):

$$\Psi(g\delta, \Delta) = \frac{M(g\delta, \Delta)}{M_0(\Delta)} = \int \int p(z_1) P(z_2, z_1, \Delta) \cos\left(\gamma g \delta(z_2 - z_1)\right) \, \mathrm{d}z_2 \, \mathrm{d}z_1$$

$$= \int P(z, \Delta) \cos(\gamma g \delta z) \, \mathrm{d}z \, , \tag{15}$$

where $P(z_2, z_1, \Delta)$ and $P(z, \Delta)$ denote the one-dimensional propagator and mean propagator, respectively, as defined via (1) and (3) for three spatial dimensions. In this context, $P(z, \Delta)$ is the probability density that an arbitrarily selected particle in the sample is shifted over a distance z in the z-direction (which is the direction of the applied field gradient) during the time interval between the two field gradient pulses. Thus, Δ represents the diffusion time in the PFG NMR experiments and may easily be varied by the timing in the NMR pulse sequences (Fig. 1).

Equation (15) may be considered as the Fourier transform of the mean propagator. Thus, by Fourier inversion of (15), the mean propagator may be directly deduced from the spin echo attenuation observed experimentally [16]:

$$P(z, \Delta) = \frac{1}{2\pi} \int \Psi(\delta g, \Delta) \cos(\gamma \delta g z) \, \mathrm{d}(\gamma \delta g). \tag{16}$$

The pair of (15) and (16) represent the key advantage of the PFG NMR method for diffusion studies: the spin echo attenuation and the mean propagator are Fourier conjugates [16, 24], which connect data obtained in a straightforward experiment to theoretical descriptions of diffusion processes based on the diffusion (1).

A first, simplified analysis of the spin echo attenuation for an unknown propagator is obtained by series expansion of the $\cos(\gamma g \delta z)$ term in (15). For small pulsed-field-gradient intensities, $\cos(\gamma g \delta z)$ is approximated by $1 - \frac{1}{2}(\gamma g \delta z)^2$. Inserting this expression in the second line of (15) and using the relation $\int P(z, \Delta) \, \mathrm{d}z = 1$ yields

$$\Psi(g\delta, \Delta) = 1 - \frac{(\gamma g \delta)^2}{2} \int P(z, \Delta) z^2 \, dz$$

$$= 1 - \frac{(\gamma g \delta)^2}{2} \langle z^2(\Delta) \rangle, \tag{17}$$

where $\langle z^2(\Delta) \rangle$ denotes the mean square displacement in the z-direction (compare with (4)). Thus, for small pulsed-field-gradient intensities, the initial spin echo attenuation depends only on the mean square displacement in the z-direction.

If the heterogeneous system studied is isotropic, i.e. $\langle r^2(\Delta) \rangle = 3 \langle z^2(\Delta) \rangle$, time-dependent measurements of $P(z, \Delta)$ and, thus, of $\langle r^2(\Delta) \rangle$ and $D_{eff}(\Delta)$ by PFG NMR are possible. Equations (15)–(17) allow us to obtain a wealth of information on diffusion processes and fluid/interface interaction as described in Sect. 2.1, without knowing the solution of the diffusion equation.

If the heterogeneous system is anisotropic, leading to an anisotropy of diffusion over the length scales accessible by PFG NMR measurements, the individual contributions of the three orthogonal directions to the displacement in the z-direction have to be evaluated. An example of the spin echo attenuation for an axisymmetric diffusion tensor in nanoporous materials and the determination of its tensor elements is presented in Sect. 3.2. In Chap. 3 of this book, the corresponding relation for diffusion in randomly oriented sheets is given and applied to study drug diffusion in lamellar lipid/water phases.

The general considerations above are correct for the narrow-gradient pulse approximation, i.e. for the case that $\delta \ll \Delta$ in the pulse sequences drawn in Figs. 1a and 1b. Using PFG NMR in interface systems, where the interaction of the diffusing molecules often shortens the nuclear magnetic relaxation times (which limit Δ: see below) and where these systems often require long and intense pulsed field gradients, this condition cannot always be fulfilled. The exact relations for the spin echo attenuation for the three pulse sequences given in Fig. 1 are [11, 22, 24]

$$\Psi(g\delta, \Delta) = \exp \left\{ -(\gamma g \delta)^2 D_{eff} \left(\Delta - \frac{1}{3} \delta \right) \right\} \tag{18}$$

for sequences (a) and (b), and [11, 23]

$$\Psi(g\delta, \Delta) = \exp \left\{ -(\gamma g 2\delta)^2 D_{eff} \left(\Delta - \frac{1}{2} \tau - \frac{1}{6} \delta \right) \right\} \tag{19}$$

for sequence (c), where it has been assumed that the diffusion process described by the effective diffusivity D_{eff} is sufficiently well approximated by a Gaussian propagator. If the pulsed field gradients deviate from the rectangular shapes drawn in Fig. 1, the spin echo attenuations are controlled by the time integrals over the pulsed field gradients $\int g(t) \, dt$. We refer the reader to [11, 12, 24], which describe the approach used to derive the corresponding equations, and to Chap. 3 of this book, where pulsed field gradients of sinusodal shape are applied.

2.3 Ultrahigh-Intensity PFG NMR Experiments

In order to reliably measure small diffusivities in interface systems and to overcome experimental difficulties due to (i) internal magnetic field gradients in heterogeneous samples and (ii) short nuclear magnetic relaxation times due to fluid/interface interactions, ultrahigh-intensity alternating pulsed magnetic field gradients ($|g_{max}| >$ 10 T/m) with short rise and fall times are often necessary. The home-built NMR spectrometer *FEGRIS 400 NT* is designed for such NMR diffusion studies [25]. It was constructed using commercially available electronic components for the rf synthesizer (PTS 500, Programmed Test Sources, Inc., USA), for the rf power amplifier (RF200400-150P, R.F.P.A. S.A., France), and for the spectrometer console (MARAN ULTRA, Resonance Instruments, UK). The spectrometer is equipped with a wide-bore 9.4 T sc magnet (Bruker, Germany) and operates at a ^{1}H resonance frequency of 400 MHz. The RF probe design and temperature control of the samples are described in [26].

The combination of the gradient current power supply and the gradient coil determines the maximum pulsed gradient amplitude as well as the rise and fall times. To increase the output voltage, two TECHRON 8606 [27] gradient current power supplies are connected in the so-called push–pull configuration (PSPPC), where the gradient coil is wired in the middle of the power supplies [28]. Each TECHRON 8606 amplifier is rated for currents of up to ±100 A and voltages of up to ±150 V. The push–pull configuration does not change the output current but it doubles the maximum available voltage across the gradient coil to a value of ±300 V. This provides the advantage that the rise and fall times of the pulsed gradients are shorter than with a single amplifier.

The gradient coil driven by the PSPPC has an ohmic resistance of 1.4 Ω, an inductance of 220 µH and a current-to-gradient conversion ratio of 0.35 (T/m)/A. With the maximum available output current of the PSPPC of ±100 A, it yields a maximum gradient of ±35 T/m over a cylindrical sample volume of 1 cm length and 0.8 cm diameter. The gradient rise time from 0 to ±35 T/m is 120 µs. To maintain the maximum current after the initial rise period, a voltage of ±140 V is required across the coil, which is about a factor of two less than the maximum output voltage of the PSPPC. This ensures that the power amplifiers are still well protected from overloading when they supply their maximum output current. Moreover, this provides enough voltage gain for the fast switching times required for nearly rectangularly shaped pulsed field gradients [25, 29].

Using such ultrahigh-intensity pulsed field gradients requires much care in "matching" the time integrals of the pulsed field gradients. Any mismatch in this quantity would lead to an additional spin echo attenuation and, thus, to an overestimate of the self-diffusion coefficient [30–32]. By observing the spin echo in the time domain in the presence of a small read gradient, we developed an efficient method which automatically detects and subsequently corrects possible mismatches of the pulsed field gradients [25].

This newly developed method was first applied to measure the self-diffusion coefficient of a poly(ethylethylene)–poly(dimethylsiloxane) diblock copolymer melt

(PEE–PDMS: molar mass 10.4 kg/mol) in a temperature range between 285.5 K and 317.5 K. In this interval, the PEE–PDMS forms lamellae. However, the resulting structure-related anisotropy of diffusion is small [33], and close to single-exponential spin echo attenuations are expected. The results for the applied 13-interval sequence (see Fig. 1c, with $\delta = 1$ ms, $\Delta = 500$ ms, $\tau = 2.5$ ms, and $g_{max} = \pm 35$ T/m) are shown in Fig. 2. All spin echo attenuations follow very well an exponential decay (19) with increasing gradient intensity. Especially, no additional signal intensity loss due to a pulsed-gradient mismatch is observed. Thus, there are no indications of artifacts in the diffusion measurements with ultrahigh-intensity pulsed field gradients.

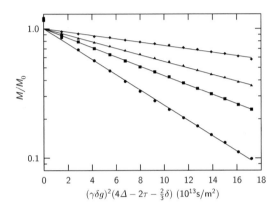

Fig. 2. Spin echo attenuation M/M_0 versus gradient intensity for PEE–PDMS measured by the 13-interval sequence at different temperatures (with increasing slope, $T = 285.5, 296.5, 304,$ and 310 K, respectively). In all measurements, the pulsed gradient amplitude was increased from zero to a maximum value of $|g_{max}| = 35$ T/m [25]

The slope observed at the highest temperature (310 K) corresponds to a self-diffusion coefficient of the PEE–PDMS of $(1.36 \pm 0.04) \times 10^{-14}$ m^2 s^{-1}. The diffusivity decreases to $(3.0 \pm 0.1) \times 10^{-15}$ m^2 s^{-1} at 285.5 K, which corresponds to a root mean square displacement of $\sqrt{\langle z^2 \rangle} = 55$ nm. These values belong to the smallest self-diffusion data measured by PFG NMR so far. They represent the current limits of sensitivity for diffusion studies by this technique.

Clearly, the sensitivity of PFG NMR depends on the hardware available but also on the system studied. It is mainly influenced by the nuclear magnetic relaxation times, since they limit the maximum accessible diffusion time Δ and the maximum possible pulsed-gradient duration δ. Usually, Δ can be varied over a range of a few ms up to ~ 1 s. For example, for the PFG NMR sequences in Fig. 1a and Fig. 1b+c, the upper limit for Δ is given by the transverse (T_2) and longitudinal (T_1) nuclear magnetic relaxation times, respectively. Additionally, δ is always limited by T_2, which restricts the application of this technique to systems with transverse relaxation times exceeding ~ 1 ms.

The PFG NMR hardware introduced offers a further advantage. The extremely short rise and fall times of the pulsed gradients (e.g. $g = 10$ T/m is achieved after 30 μs) allow diffusion studies with the 13-interval PFG NMR sequence (Fig. 1c) by maintaining diffusion times Δ as short as 3 ms but still providing sufficient spin echo attenuation for accurate diffusion studies. This feature is used in Sect. 4 to study the diffusion of liquids in sediments, where large internal magnetic field gradients require the use of pulse sequences with alternating pulsed field gradients.

3 Structure-Related Diffusion in Nanoporous Materials

3.1 The Different Regimes of Diffusion Measurement

Owing to their particular way of production [34], nanoporous materials (and in particular zeolites, as their most important representatives [9,35]) are generally available as assemblages of small particles. Depending on the relation between the mean particle diameter and the mean diffusion path covered during the PFG NMR experiments, one may distinguish between three different regimes of measurement. As schematically indicated by Fig. 3, one may thus specify the regimes of (i) intracrystalline diffusion (the root mean square displacements covered during the observation time are small in comparison with the particle diameters), (ii) long-range diffusion (the root mean square displacements covered during the observation time are large in comparison with the particle diameters), and (iii) the transition range, which provides information about the interface between the intracrystalline space and the surrounding fluid (in general, the gas atmosphere of the guest molecules) as well as about the exchange rate between them. The following subsections will provide examples of experimental studies for all of these special cases.

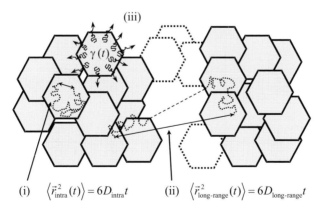

Fig. 3. The different regimes of diffusion measurement in beds of nanoporous particles

3.2 Intracrystalline Diffusion

Intracrystalline Transport Barriers

With the introduction of PFG NMR to zeolite science and technology [36, 37], in numerous cases the coefficients of intracrystalline diffusion obtained and generally accepted so far have turned out to be orders of magnitude below the correct values. Subsequent careful studies have revealed that very often this discrepancy could be attributed to shortcomings in the interpretation of the "classical", nonequilibrium techniques of diffusion measurement [9, 10]. However, even today, in very careful studies of zeolitic diffusion, the rate of molecular adsorption or desorption may appear to be much smaller than would result if it was estimated on the basis of the intracrystalline diffusivities observed by PFG NMR [38,39]. Owing to the most recent progress in PFG NMR instrumentation [25,29], an old hypothesis of explanation [40] has recently found confirmation. Figure 4 displays the coefficients of intracrystalline zeolitic diffusion for *n*-butane in silicalite-1 measured at three different temperatures, as a function of the mean distance over which the molecules have traveled during the experiments [41]. Most astonishingly, at the smallest temperature, the observed diffusivity (defined according to (7)) is found to decrease with increasing mean diffusion path. As a most direct explanation, one may speculate, therefore, that on their diffusion paths the molecules have to overcome internal barriers. Obviously, with increasing temperature, the thermal energy of the diffusants eventually becomes large enough to overcome these resistances. It is demonstrated in [42] that such a behavior may be predicted with rather simple assumptions, which may be based, for example, on a uniform barrier separation of 3 μm. The simulation results are indicated by the solid lines in Fig. 4.

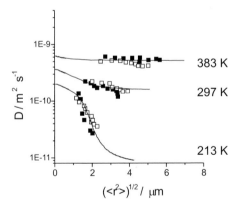

Fig. 4. Intracrystalline diffusion of *n*-butane in two samples of silicalite-1 (open and full symbols) measured at three different temperatures as a function of the mean distance over which the molecules traveled during the observation time. The solid lines represent the results of Monte Carlo simulation of restricted diffusion with a barrier separation of 3 μm [42]

Diffusion Anisotropy

Deviations from cubic structure in ordered porous materials have to lead to deviations from isotropic diffusion. Information of this type is provided by PFG NMR if one is analyzing the total curve rather than the first steep decay of the spin echo attenuation. Such experiments have been carried out with microporous crystalline materials (zeolites) of type MFI [43] and chabazite [44]. As a particular challenge, most recently a similar study has successfully been performed with ordered mesoporous materials of type MCM-41, which are known to consist of a hexagonal arrangement of long channels with diameters of the order of several nanometers [45]. In these studies [46], the PFG NMR spin echo attenuation was found to be satisfactorily approached by the theoretical dependence

$$\Psi(\delta g, \Delta) = \frac{1}{2} \int_0^\pi e^{-(\gamma\delta g)^2 \Delta \left(D_{\mathrm{par}} \cos^2 \Theta + D_{\mathrm{perp}} \sin^2 \Theta \right)} \sin \Theta \, d\Theta \qquad (20)$$

if one assumed a diffusion tensor with rotational symmetry, where the elements of the tensor (D_{par} and D_{perp}) correspond to the diffusion parallel and perpendicular, respectively, to the direction of the channels. Figure 5a shows examples of the fits of this model to the spin echo attenuations observed experimentally.

For the particular sample under study [47] with the data on water diffusivity shown in Fig. 5b, the following diffusion-based information about the channel architecture of MCM-41 has been acquired for the first time. (i) The mean diffusivity in the channel direction (D_{par}) is about one order of magnitude smaller than in the bulk phase of free water. Since the channel diameter exceeds the diameters of the diffusants by one order of magnitude, this experimental finding strongly suggests the existence of constrictions. Their separation should be notably below 1 μm. (ii) Though it is further reduced in comparison with transport in the direction of the channel axis,

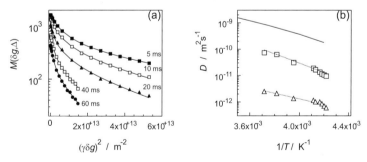

Fig. 5. PFG NMR diffusion studies for water in MCM-41. (a) Examples of the fits of the model of a diffusion tensor with rotational symmetry (20) to the experimentally observed signal intensities ($M(g\delta, \Delta)$) for 5 different diffusion times at $T = 263$ K. (b) Dependence of the parallel (□) and perpendicular (△) components of the diffusion tensor obtained, on the inverse temperature. For comparison, the full line represents the self-diffusion coefficients of supercooled bulk liquid water. Redrawn from Stallmach et al. [46].

there is also the possibility of water propagation in the direction perpendicular to the main channel axis. So far, one may only speculate whether this behavior is caused by the finite length of the channel elements, by some channel "reptation", or by a finite permeability of the channel walls.

Multicomponent Diffusion

Being able to selectively measure the diffusivities of the individual components in multicomponent systems is one of the great virtues of PFG NMR. In fact, in most of their practical applications, nanoporous materials are applied in contact with fluid mixtures rather than with individual components. Among a number of alternative possibilities for selective diffusion measurement in nanoporous materials [38, 39], the potentials of PFG NMR are particularly convincing if applied in the frequency rather than in the time domain, i.e. if applied under Fourier transformation [24, 48, 49]. Unfortunately, as a consequence of the increased line widths when applied to nanoporous host–guest systems, the conditions for applying high-intensity field gradients in Fourier transform (FT) NMR are rather limited. In particular, for studies in the regime of intracrystalline diffusion, i.e. for the measurement of root mean square displacements as small as possible, such powerful field gradient pulses are of extreme importance. By transferring the conception of field gradient pulse adjustment as introduced in [25] to FT PFG NMR [50], the range of applicability of PFG NMR to multicomponent systems was notably enlarged very recently.

As an example, Fig. 6 displays the results of a PFG NMR study of the mixture diffusion of n-butane and benzene adsorbed in zeolite NaX. These spectra were measured using the 13-interval sequence (Fig. 1c) with $|g_{max}| = 20$ T/m, $\delta = 500$ μs, $\tau = 1.5$ ms, and $\Delta = 25$ ms. From the intensity of the two lines in the NMR spectra in dependence on the applied pulsed field gradients (Fig. 6a), the attenuation plots of each of the two components (Fig. 6b) are easily found. Their

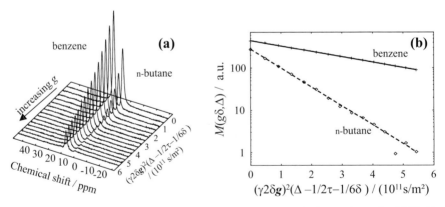

Fig. 6. FT PFG NMR diffusion measurements for benzene and n-butane in zeolite NaX. (a) NMR spectra as a function of applied pulsed gradient amplitude g. (b) Spin echo attenuation plot for each individual species

slopes correspond to diffusion coefficients of $(1.03 \pm 0.04) \times 10^{-11}$ m^2/s and $(2.94 \pm 0.10) \times 10^{-12}$ m^2/s for n-butane and benzene, respectively.

FT PFG NMR diffusion studies are well established [48, 51–54] and are also presented in Chap. 3 of this book. What is really new is the quality in attaining low diffusivities on the basis of the instrumental improvements described in Sect. 2.3 and [50]. In fact, to our knowledge, such small diffusivities have so far not been observed by Fourier transform PFG NMR with comparably short gradient widths and observation times.

3.3 Long-Range Diffusion

Diffusion in Particle Agglomerates

In their technological use as selective adsorbents or catalysts, nanoporous materials are generally used in the form of compacted agglomerates rather than as individual crystallites or particles. In such cases, the rate of molecular transport relevant to the efficiency of a given technological processes is generally determined by the rate of transport through the agglomerate rather than by the diffusivity in any of the individual particles. As an example, Fig. 7 shows a scanning electron micrograph of an industrial sample of MCM-41 [55]. It is clearly visible that the sample consists of secondary particles of some 100 μm diameter.

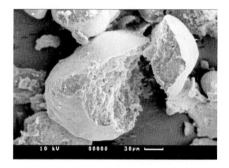

Fig. 7. Scanning electron micrograph of an industrial sample of MCM-41 [55,56]

The typical diffusion paths observable by PFG NMR are thus much smaller than these particles. On the other side, however, they are much larger than the extensions of the primary particles of genuine MCM-41 structure, which can scarcely be distinguished from each other in the micrograph. Therefore, over the relevant diffusion paths, the adsorbent must be expected to act as a quasi-isotropic host system. As a consequence, the curvature of the PFG NMR attenuation plots for benzene shown in Fig. 8 cannot be attributed to the occurrence of diffusion anisotropy as discussed in Sect. 3.2. It is most likely an expression of the heterogeneity of the compacted material. Such a tendency is expected to be reduced with increasing observation time, since then the averaging procedure embraces increasingly larger regions. In complete agreement with this expectation, the echo attenuation in the master plot of Fig. 8 is in fact found to approach an exponential dependence (i.e. a linear dependence in a semilogarithmic representation) with increasing observation time.

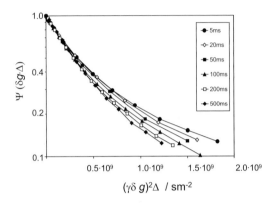

Fig. 8. PFG NMR spin echo attenuation for benzene in the MCM-41 particles shown in Fig. 7 as function of observation time Δ, which is given in the figure legend [56]

Figure 9 shows the effective diffusivities as a function of the loading for different temperatures, where the pore filling factor Θ is referred to the pore volume of the genuine MCM-41 structure within the particles. There are at least two remarkable features resulting from the displayed data. First, over most of the considered concentration range, the diffusivities are notably larger than in the bulk liquid. This behavior may be referred to the fact that the effective diffusivity in the regime of long-range diffusion is generally given by the relation [1,9]

$$D_{\text{long-range}} = p_{\text{inter}} D_{\text{inter}} , \tag{21}$$

with p_{inter} and D_{inter} denoting the relative amount of molecules outside of the nanoporous particles and their diffusivity. Though under the typical conditions of molecular adsorption, p_{inter} is clearly much smaller than 1, owing to the considerably enhanced mean free path in the space between the adsorbent particles, D_{inter} may attain values which are much larger than the bulk diffusivities. Hence, in the given case, obviously the increase in D_{inter} overcompensates the decrease due to the factor p_{inter}, so that the product as provided by (21) exceeds the diffusivity in the bulk liquid. As a consequence, the dramatic decrease of the effective diffusivity around pore filling factors close to one half of the saturation capacity of the MCM-41 phase might be attributed to the formation of a liquid phase in the secondary pore system between the primary MCM-41 particles, which reduces the mean free path. In fact, comparison of the hystereses of the adsorption isotherms with those of ordinary MCM-41 materials (i.e. of samples without a secondary pore system) confirms this conclusion [56].

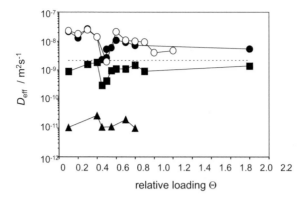

Fig. 9. Effective diffusivities of benzene in the MCM-41 particles shown in Fig. 7 as a function of the loading Θ for temperatures of 298 K ($\circ$, $\bullet$), 258 K ($\blacksquare$) and 208 K ($\blacktriangle$). The dashed line indicates the diffusivity of liquid benzene at 298 K [56]

Tortuosity Factor of Long-Range Diffusion

Figure 10 shows the temperature dependence of the long-range diffusivities obtained for ethane in a bed of zeolite NaX [57] and a theoretical estimate of the temperature dependence based on (21) using the approximation

$$ D_{\text{inter}} = \frac{1}{3} u \lambda_{\text{inter}} \frac{1}{\tau}, \tag{22} $$

where u, λ_{inter}, and τ denote the mean thermal velocity, the effective mean free path, and the tortuosity factor as defined via (9), respectively. The effective mean free path results by reciprocal addition of the mean free path in the gas phase and of the mean crystal diameter, which – for the given packing density of zeolite crystallites – serves as a good estimate of the mean pore diameter in the bed [57, 58].

At high gas phase concentrations, which in the closed NMR sample tubes are attained at high temperatures, bulk-phase diffusion is the dominating mechanism for D_{inter}. At sufficiently low temperatures, the so-called Knudsen (pore) diffusion prevails, i.e. mutual collisions of the diffusants are negligible in comparison with the particle–wall collisions. Most interestingly, satisfactory agreement is only obtained by assuming vastly differing tortuosity factors in these two regimes. For identical tortuosity factors, satisfactory agreement between the experimental data and the theoretical estimates would only result from assuming that the effective mean free path under the Knudsen regime is 2.5 μm, which would be rather difficult to rationalize for packings of crystallites with diameters one order of magnitude larger. In fact, we consider these findings from PFG NMR as a first experimental confirmation of some very recent theoretical papers claiming the existing of similarly vast differences in the tortuosities in dependence on the mechanism of molecular propagation [59, 60].

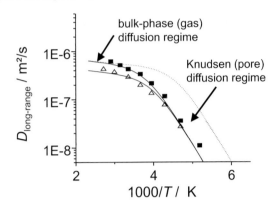

Fig. 10. Temperature dependence of the long-range diffusion coefficients of ethane in a bed of NaX zeolite for two different loadings corresponding to gas phase pressures of 50 mbar (■) and 80 mbar (△) at $T = 295$ K. The lines which fit the experimental data well represent the theoretical estimates of the temperature dependence based on (21) and (22) with different tortuosities in the bulk phase ($\tau_b = 1.6$) and Knudsen ($\tau_K = 16$) regimes. The dashed line assumes only one tortuosity factor of $\tau_b = \tau_K = 1.6$ for both regimes [57]

3.4 Transition Range

Using a mesoscopic perspective, molecular exchange at the boundary between intracrystalline and intercrystalline space, i.e. at the crystal surface, may be quantified to a much better extent using the transmission probability ρ defined by an appropriately chosen boundary condition for intracrystalline diffusion in (1).

The physical significance of the transmission probability may appear in essentially two different ways. As a first option, (1) and (2) may be considered to describe an experiment in which the NMR signal contribution of particles reaching the interfaces is lost as a consequence of nuclear magnetic relaxation at the boundaries. In this case, ρ denotes the surface relaxivity. This is the usual situation that one is confronted with when considering fluids within macroporous materials (e.g. sedimentary rocks). We shall refer to exactly this situation in Sect. 4.1. References [18–20] provide an excellent theoretical framework for their interpretation. In the short-time limit, up to second order in the molecular displacements, the effective diffusivities for diffusion in isolated spherical pores of radius R_s are found to be given by the relations [20]

$$\frac{D_{\text{eff}}(t)}{D_0} = 1 - \frac{4}{3\sqrt{\pi}} \frac{1}{R_s} \sqrt{D_0 t} - \frac{1}{2R_s^2} D_0 t + \dots , \tag{23}$$

$$\frac{D_{\text{eff}}(t)}{D_0} = 1 - \frac{2}{3\sqrt{\pi}} \frac{1}{R_s} \sqrt{D_0 t} - \frac{1}{R_s^2} D_0 t + \dots , \tag{24}$$

for the cases of reflecting ($\rho = 0$, (23)) and absorbing ($\rho = \infty$, (24)) boundaries, respectively. Applied to our case, the NaX zeolite crystals are approximated as spheres with radius R_s, and D_0 denotes the genuine (intracrystalline) diffusivity.

A second option, which follows exactly the same theoretical framework, is related to the interrelation of the rates of intracrystalline diffusion (D_{intra}, see Sect. 3.2) and long-range diffusion ($D_{\text{long}-\text{range}}$, see Sect. 3.3). The latter quantity has been shown to be represented by the product of the relative amount of molecules in the intercrystalline space and their diffusivity (see (21)). Let us first consider the condition $D_{\text{intra}} \ll D_{\text{long}-\text{range}}$. Since generally the activation energy of intracrystalline diffusion is much smaller than the heat of adsorption, for most systems this condition may be easily fulfilled by choosing sufficiently high temperatures. If there are no additional transport resistances at the outer surface of the particles, molecules encountering the particle surface will soon dissipate over the whole space. In the PFG NMR experiment their contribution to the signal may be easily separated from that of the molecules which have not encountered the surface and are still within the crystals. Considering only these molecules, we do have exactly the same situation as with absorbing boundaries [61]. If – on the other side – molecular exchange with the surroundings is excluded by either additional transport resistances at the surface ("surface barriers") or by an extremely low rate of molecular transport through the bed of particles (generally ensured at sufficiently low temperature), diffusion is essentially confined to the intraparticle space, so that now the particle surface acts as a reflecting wall.

Both limiting cases have been considered in PFG NMR experiments with beds of zeolite crystallites. In the literature, the case of completely restricted diffusion has been repeatedly used for an independent estimate of the mean crystal radius R via (10) [9]. Equation (10) holds under the condition that the molecular displacements during the diffusion time in an infinitely large crystal are much larger than the diameter of the given crystal.

The case of absorbing boundaries is considered in the so-called NMR tracer desorption technique [9,62]. In this technique, owing to the sensitivity of PFG NMR with respect to molecular displacements, one determines the relative part of the molecules which during the time of the PFG NMR experiment are able to leave their individual crystallites. In the absence of additional transport resistances at the crystallite surface, this fraction is a sole function of the intracrystalline diffusivity and the crystal size.

In [63], these conventional routes to determine the size of zeolite crystallites have been complemented for the first time by an analysis on the basis of (23) and (24). Figure 11 represents the attained effective diffusivities in a plot, which has been obtained by the best fitting of (23) and (24) to the experimental data with D_0 and R_s as fitting parameters. The systems under study were n-hexane (C_6H_{14}) with single-component adsorption, and n-hexane and tetrafluoromethane (CF_4) with two-component adsorption in zeolite NaX. The temperatures were chosen to yield values of $\sqrt{D_0 t}$ as small as possible. The measuring conditions for n-hexane correspond to reflecting boundary conditions and for tetrafluoromethane to absorbing boundary conditions.

The values of the genuine intracrystalline diffusivities D_0 and the crystallite radii obtained by the fitting procedure are summarized in Table 1. For comparison, the

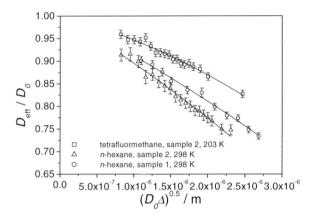

Fig. 11. Relative effective diffusivities for *n*-hexane ($\triangle$, sample 2) and tetrafluoromethane ($\square$, sample 2) for two-component adsorption, and for *n*-hexane for single-component adsorption ($\bigcirc$, sample 1). The lines represent the appropriate fits of (23) and (24), respectively [63]

crystallite extensions resulting from the "conventional" methods, i.e. from the long-time limit of totally restricted diffusion and from NMR tracer desorption, as well as from microscopic sample analysis, are also given. In these studies, for the first time the influence of the finite size of the zeolite crystallites has been simultaneously considered for one and the same sample (i) in the long-time limit of completely restricted diffusion, (ii) by NMR tracer exchange, and (iii) in the short-time limit of diffusion. It is remarkable that all these mutually independent techniques lead to satisfactorily agreeing data on the crystal size. Note that this good agreement could be obtained only by considering the second-order terms as given in (23) and (24) instead of confining the analysis to the first-order approach of (8).

So far, the measurements have been carried out under the limiting conditions of either reflecting or absorbing boundaries, for which analytical expressions are readily available. The consideration of the intermediate case will be a challenging task of future investigations, with respect to both the experimental procedure and the theoretical analysis.

4 Diffusion and Relaxation in Sediments

In NMR studies of natural porous sediments, where the pore fluids such as water or oil surround the solid pore matrix, the grain surface represents a boundary for the diffusing molecules which cannot be penetrated. However – owing to the reduced molecular mobility and the natural content of paramagnetic species – nuclear relaxation at the grain surface ("surface relaxation") may be important [64–66]. Clearly, the surface relaxation affects only a few molecular layers of the fluid covering the grain surface, but – if the pore size is not too large – diffusion between the molecular

Table 1. Values of the genuine intracrystalline diffusivities D_0 and the crystallite radii R_s obtained by fitting the relative effective diffusivities plotted in Fig. 11 with the models presented by (23) and (24). For comparison, the average radii $\langle R_c \rangle$ obtained from NMR tracer desorption, from the long-time limit of restricted diffusion (10), and from analysis of the size distribution measured with optical microscopy are also included [63]

Loading and temperature	Method	$D_0/\text{m}^2\,\text{s}^{-1}$	$R_s/\mu\text{m}$	$\langle R_c \rangle/\mu\text{m}$
2 n-C$_6$H$_{14}$/cav., 298 K	(23)	3.53×10^{-10}	9.1	
1 n-C$_6$H$_{14}$/cav., 298 K	(23)	5.54×10^{-10}	7.8	
1 CF$_4$/cav., 203 K	(24)	4.17×10^{-10}	9.1	
1 CF$_4$ /cav., 203 K	Tracer desorption			10.8
1 n-C$_6$H$_{14}$/cav., 298 K	(10)			11.9
Unloaded zeolites	Optical microscopy			10.5–10.9

surface layer and the pore body leads to a fast exchange so that the decay of the total nuclear magnetization is characterized by a single relaxation time. This relaxation time of the diffusion-mediated relaxation process represents an estimate of the pore size.

Solving the diffusion equation (1) for the nuclear magnetization in an isolated pore under the condition of fast diffusional averaging (see e.g. [68]) as described above leads to a well-established and practically important relation between the relaxation rates ($1/T_1$ and $1/T_2$) and the surface-to-volume ratio S_V of the pore [64, 67, 68]:

$$\frac{1}{T_{1,2}} = \rho_{1,2}S_V + \frac{1}{T_{1,2}^0}. \tag{25}$$

In (25), the surface relaxivity $\rho_{1,2}$ describes the efficiency of relaxation at the pore/grain interface and $T_{1,2}^0$ denotes the transverse (index 2) and longitudinal (index 1) bulk relaxation times.

4.1 Relaxation Time Studies in Sediments

Generally, in natural consolidated and unconsolidated sediments, the pores are heterogenous in shape and size and the surface relaxivity $\rho_{1,2}$ is not known. This leads to a total magnetization decay for the fluid in the pore space which is given by the superposition of the decays from each individual pore. Consequently, the resulting complex magnetization decay is often treated as a superposition of exponential decays, yielding the so-called relaxation time distribution $p(T_{1,2})$ as a result of data inversion [69, 70]. Despite the difficulties associated with the physics and the data analysis of such complex magnetization decays, NMR relaxation time studies are well established for characterizing porous rocks since they provide information on fluid–matrix interaction nondestructively, i.e. with the natural pore fluid occupying the pore space (see e.g. [64,70,71]). A collection of articles dealing with the history and the state of the art of the technical application of such petrophysical NMR investigations may be found in [72]. In the following, we present two examples where NMR relaxation time studies were used for the first time to investigate pore space properties of sediments originating from near-surface sources in central Germany.

Carbonate Rocks from an Aquifer

As the first example, Fig. 12a represents the distribution of longitudinal relaxation times ($p(T_1)$) for water in porous carbonate rocks (reef chalk, Werra series $CaC1$) originating from a Zechstein aquifer 50 km southwest of Leipzig (Profen area, central Germany, well UFZ1 [73]). The samples shown in Fig. 12b were cut from 10 cm diameter cores obtained during drilling the well. They were water-saturated, and longitudinal relaxation curves were measured at a 30 MHz ^{1}H resonance frequency by varying the delay d_1 in the inversion recovery ($\pi/2 - d_1 - \pi$) NMR pulse sequence.

The relaxation time distributions extend over roughly two orders of magnitude in T_1, reflecting the heterogeneity of the pore space in the samples. The geometrically averaged relaxation times, which serve as a measure of the mean pore sizes via (25), are included in Fig. 12a. At long relaxation times the distributions extend up to values which are identical to the longitudinal relaxation time of the bulk pore water ($T_1^0 \approx 2.5$ s) measured independently. Thus, the corresponding components of the water in the rock samples do not show significant surface relaxation. According to (25), this means that they are most likely trapped in large pores. The relative amount of water in such large pores (s_w) may be determined by introducing a cutoff

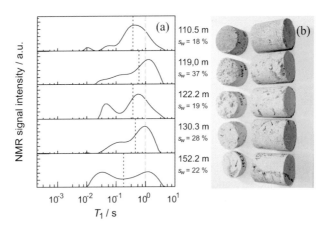

Fig. 12. (a) Distribution of longitudinal relaxation times (T_1) of five water-saturated carbonate rock samples originating from a Zechstein aquifer in central Germany. The depth is indicated in the figure. The dotted lines represent the arithmetic mean relaxation times T_1. The gray dashed line at $T_1 = 1$ s indicates the cutoff for the determination of the relative amount (s_w) of water not showing significantly reduced relaxation times. (b) Photographs of the samples of 4 cm diameter and 5 cm length prepared for the NMR studies [73]

relaxation time T_1^c. In Fig. 12a, such relative amounts are calculated for a value of $T_1^c = 1$ s. Assuming that water which has T_1 values larger than T_1^c did not reach the pore/grain interface during diffusion times $T_1^0 \geq t \geq T_1^c$, the associated pore radii R_1 may be estimated from the corresponding root mean square displacement via (6). For $T_1^c = 1$ s one obtains $R_1 \gtrsim 0.1$ mm as the pore radius.

Visual inspection of the sample surface (Fig. 12b) shows that such large pores are in fact present in these carbonate rocks. That they continue into the interior of the samples is demonstrated by measurements of the spatial distribution of pore water using 3d spin echo magnetic resonance imaging (MRI). Figure 13 represents examples of these investigations, showing that large areas of high water content are inhomogeneously distributed across the samples. The NMR relaxation time studies yield the relative amount of water contained in these pores, which, owing to their large size, are dominant for the water transport in the aquifer.

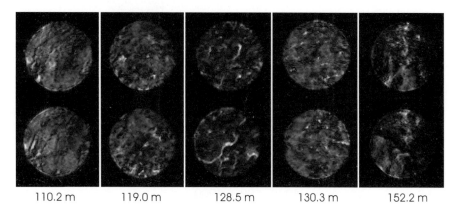

| 110.2 m | 119.0 m | 128.5 m | 130.3 m | 152.2 m |

Fig. 13. MRI studies of the spatial distribution of water in the five carbonate rock samples shown in Fig. 12. Two successive slices (slice thickness ~ 3 mm, in-plane resolution 0.5 mm) from the center of each sample are displayed [73]

Grain Size Fractions of a Glacial Sand Deposit

As a second example, Fig. 14 shows how the relaxation time distributions depend on the grain size in unconsolidated sands. The sand studied originates from a glacial sand deposit 10 km northwest of Leipzig (Rückmarsdorf area, central Germany). It was dry-sieved for a grain size analysis. The four grain size fractions, with diameters between 0.1 mm and 1.0 mm (see Fig. 14b), as well as the original sand (Fig. 14c), were saturated with water and introduced into sample containers of 4 cm diameter and 5 cm length. The transverse relaxation decay was measured at a 30 MHz ^{1}H resonance frequency using the Carr–Purcell–Meiboom–Gill (CPMG) NMR pulse sequence $(\pi/2 - [d_2 - \pi - d_2 - \text{echo}-]^n)$ with $n = 4096$ spin echoes and $d_2 = 300$ µs.

With increasing grain size, the relaxation time distributions and the geometric-mean transverse relaxation times plotted in in Fig. 14a shift to longer times. Qualitatively, this observation corresponds well to the behavior expected according to (25). However, the unknown surface relaxivity and the pore geometry, which is expected to be well connected in such beds of sand grains, do not allow a simple analysis of these relaxation time distributions with respect to surface-to-volume ratios and averaged pore and grain sizes [75]. This disadvantage may be circumvented by using time-dependent PFG NMR diffusion studies as demonstrated in the following two sections.

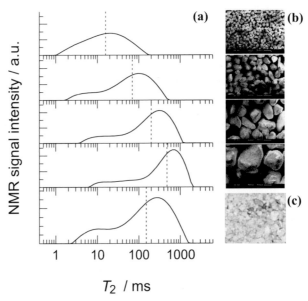

Fig. 14. (a) Distribution of transverse relaxation times (T_2) of five water-saturated sand samples originating from a glacial sand deposit in central Germany. The topmost four samples were obtained by sieving the original sand (bottom). The dotted lines represent the arithmetic-mean relaxation times T_2. (b) Scanning electron micrographs of the four grain size fractions obtained by sieving the original sand. The bar corresponds to a length of 300 µm [74]. (c) Photograph of the original sand

4.2 Determining Surface-to-Volume Ratios and Genuine Diffusivities

For arbitrary pore/grain interface shapes, the early time dependence of the effective diffusivity (8) is expected to depend only on the surface-to-volume ratio [18,20]. The special case of diffusion within spherical boundaries ((23) and (24), [20]) was used in Sect. 3.4 to analyze the time dependence of the effective intracrystalline diffusivities in zeolites. In order to prove the validity of this short-time diffusion approach for natural sediments, PFG NMR diffusion studies were performed with a bed of glass spheres, with a consolidated sandstone (Buntsandstein), and with unconsolidated sands using water and hexadecane as pore fluids.

In such samples, internal magnetic field gradients are known to affect the spin echo attenuation if measured with unipolar pulsed field gradients. Therefore, the 13-interval NMR pulse sequence (see Fig. 1c) with very short but intense pulsed field gradients ($\delta = 150\,\mu$s, $|g_{\max}| = 10\,$T/m) was applied, which enabled investigation with short transverse relaxation periods ($2\tau = 700\,\mu$s) and with variable observation times (Δ) as short as 3 ms [75]. Examples of the effective diffusivities obtained are presented in Fig. 15a. The results of the application of the short-diffusion-time model for S_V and D_0 are shown in Fig. 15b.

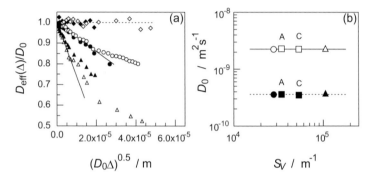

Fig. 15. (a) Relative effective self-diffusion coefficients $D_{\mathrm{eff}}(\Delta)/D_0$ as a function of $(D_0\Delta)^{0.5}$ for water (open symbols) and hexadecane (full symbols) in the bulk ($\lozenge, \blacklozenge$), in a bed of glass spheres ($\circ, \bullet$), and in a sandstone (Buntsandstein: $\triangle, \blacktriangle$). The solid lines represent the results of the fits of (8) to the early time dependence of these data. (b) Genuine diffusivities (D_0) for water (open symbols) and hexadecane (full symbols) as function of the the surface-to-volume ratios (S_V) obtained by fitting (8) to the effective diffusivity measured for the glass spheres ($\circ, \bullet$), the sandstone ($\triangle, \blacktriangle$), and two other unconsolidated sands denoted by A and C ($\square, \blacksquare$) [75]

Although the bulk diffusivities of water ($2.3 \times 10^{-9}\,$m^2/s) and hexadecane ($3.6 \times 10^{-10}\,$m^2/s) at the measurement temperature of 298 K differ by almost one order of magnitude, the dependences of the relative effective diffusivities on the diffusion lengths ($\sqrt{D_0\Delta}$) follow the same trends in the Buntsandstein and the glass spheres (see Fig. 15a). The same measurements with the bulk fluids show that D_{eff} is independent of $\sqrt{D_0\Delta}$. Thus, the decay of the relative effective diffusivities in the heterogeneous samples must be caused by the restriction of the diffusion path at the pore/grain interface. Since water and hexadecane show the same dependence, surface relaxation – which depends on the nature of the pore–grain interaction – cannot significantly influence the measured values of the effective diffusivities.

The application of the short-time diffusion model (8) yields the surface-to-volume ratio and the genuine diffusivities, which are plotted in Fig. 15b. It is interesting to note that all D_0 values obtained by fitting of the effective diffusivities coincide within the experimental uncertainty with the bulk diffusivities of the two fluids measured independently. This coincidence is necessary to prove that the short-time diffusion approach yields correct values for the surface-to-volume ratio [75]. Additionally, it could be demonstrated that the surface-to-volume ratio measured for the bed of glass spheres, $(27.7 \pm 3.0) \times 10^3$ m^{-1}, agrees very well with the value of $(27.9 \pm 2.7) \times 10^3$ m^{-1} which can be calculated from the model of random packing of spheres using a radius of the spheres of 200 µm and a porosity of the bed of 0.35 ± 0.02 (see e.g. (5) in [75]). Similar good agreement between the PFG NMR S_V ratios and the theoretical estimate on the basis of porosity and sphere radius was reported for a number of different beds of spheres in [74, 76], demonstrating the reliability of the S_V ratios obtained by this method.

4.3 Studying Fractal Surfaces of Sand Grains

Using the PFG NMR approach to measure surface-to-volume ratios described in Sects. 3.4 and 4.2, the surface properties of the unconsolidated sediment used for the relaxation time studies in Sect. 4.1 were investigated in more detail. Figure 16 displays the effective diffusivities of water in the four different grain size fractions of this quartz sand (compare Fig. 14b) [74, 77].

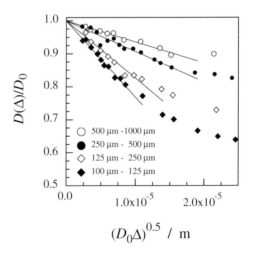

Fig. 16. Effective self-diffusion coefficients $D(\Delta)/D_0$ as a function of $(D_0\Delta)^{0.5}$ for water in the four grain size fractions of the unconsolidated quartz sand. The solid lines represent the results of the fits of (8) to the early time dependence of these data [74]

Via (8), the distinct increase in the initial slopes of the $D(\Delta)/D_0$ representations with decreasing grain diameter may easily be attributed to the corresponding increase in the surface-to-volume ratio. With the known grain density and porosity, these surface-to-volume ratios may be transferred into specific surface areas $S_\mathrm{m} = S/m_\mathrm{g}$, this means, into surface areas per mass of the grains (m_g). Their representation in Fig. 17 versus the averaged diameter (d_g) of the grain size fractions reveals an interesting feature of these natural sand grains: their specific surface area decreases less than linearly with increasing average grain diameter [74, 77].

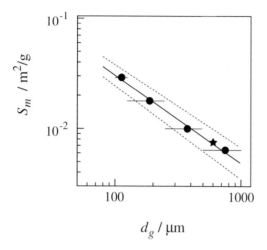

d_g / µm

Fig. 17. Specific surface areas S_m as a function of the averaged grain diameter d_g of the grain size fractions (•) and the original sand (⋆). The full and dotted lines represent the $\log S_\mathrm{m}$ vs. $\log d_\mathrm{g}$ fit and its confidence interval (slope -0.80 ± 0.05), respectively. The horizontal error bars show the width of the screen intervals used for sieving analysis [74]

Following the concept of Avnir et al. [78], who proposed an approach to determine the fractal dimension (D_s) of the surface area of granulated porous media by analyzing the scaling behavior of the measured specific surface in dependence on the grain diameter [74, 78],

$$S_\mathrm{m} \equiv \frac{S}{m_\mathrm{g}} \propto \frac{d_\mathrm{g}^{D_\mathrm{s}}}{d_\mathrm{g}^3} = d_\mathrm{g}^{D_\mathrm{s}-3}, \qquad (26)$$

the deviation in the slope of the log–log plot from -1 in Fig. 17 may be attributed to a fractal geometry of the grain surfaces. The $\log S_\mathrm{m}$ vs. $\log d_\mathrm{g}$ fit yields a slope of -0.80 ± 0.05 clearly deviating from -1. According to (26), it refers to a fractal dimension of the surface area of the sand grains of $D_\mathrm{s} = 2.20 \pm 0.05$.

Thus, even in samples with irregular pore space geometries such as natural sand grains, PFG NMR self-diffusion studies of the confined pore fluids are suitable for revealing geometric properties of the pore walls. However, for the validity of a fractal analysis of the measured surface areas as performed above, one has to keep in mind that a hierarchy of length scales determines its applicability (see Fig. 18 and [77, 79]).

The length scale of the observed diffusion process ($r \approx \sqrt{D_0 \Delta}$), which is generally of the order of a few μm, determines the lower limit for surface curvature radii (R_s) which contribute to the measured surface-to-volume ratio. Possible smaller features on the surface, which may be explored by adsorption studies, where the radius of the adsorbate molecule (R_m) determines the resolution of the surface area measurements, are averaged over the diffusion length. On the other hand, the surface curvature radii cannot significantly exceed the radii of the grains, which in the present case are on the order of $0.1 - -1$ mm. This situation is illustrated in Fig. 18.

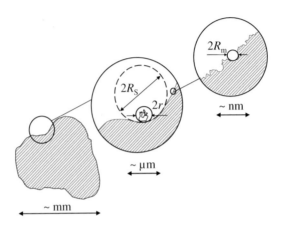

Fig. 18. Hierarchy of length scales involved in surface area measurements by PFG NMR self-diffusion studies in granulated porous media

5 Diffusion Under Internal Confinement in Multicomponent Polymer Systems

5.1 Ternary Polymer Blends

Multicomponent polymer systems often self-organize into structures on a mesoscopic length scale because of the repulsive interaction present between different polymers [80–82]. This leads to materials combining the mechanical properties of the constituents or having even superior properties. The diffusion of the different species through these structures is expected to be influenced by the thermodynamic barriers caused by the repulsive interactions. This corresponds to case c) in Sect. 2.1.

A very common system is a binary blend of homopolymers A and B. Their demixing on a macroscopic length scale can be prevented by adding a certain amount of diblock copolymers A–B which segregate at the A/B interfaces and prevent domain growth and coalescence [80–82]. These ternary blends have been found to form a variety of structures, for instance a common disordered phase (usually at high temperatures), a two-phase structure consisting of large A- and B-rich domains with the diblock copolymers decorating and stabilizing the interfaces, and a lamellar structure with the homopolymers dissolved in the lamellae formed by the diblock copolymers [83–91]. Additionally, a bicontinuous microemulsion (BμE) has been identified for a narrow range of compositions between the two-phase and the lamellar state [87–93]. In this structure, both homopolymer-rich domains are continuous, with most of the diblock copolymers at the (continuous) interface. The interfacial tensions are very small and the spontaneous curvature close to zero. Demixing takes place on a local scale, but no long-range order persists.

The dynamics in this and the other phases have so far been studied using dynamic mechanical measurements and dynamic light scattering [94, 95]. The results from these methods, however, are often difficult to explain, because a whole range of dynamic processes (local segment dynamics, diffusion, collective dynamics of the structure, etc.) contribute to the quantity measured. PFG NMR monitors long-range diffusion only and therefore allows studies of the influence of the BμE structure on the diffusivities of the polymers blended [33]. It can thus contribute to the understanding of the complex dynamics in self-organized mesoscopically structured polymer blends.

5.2 Homopolymer Diffusion Through a Bicontinuous Structure

Temperature-Dependent Behavior

The sample under study consisted of equal volumes of the homopolymers poly-(ethylethylene) (PEE) and poly(dimethylsiloxane) (PDMS) (molar masses 1.71 and 2.13 kg/mol, respectively), and 10 volume-% of compositionally nearly symmetric PEE–PDMS diblock copolymers (10.4 kg/mol). It forms a BμE with a repeat distance of the domains of $\sim$ 75 nm below $\sim$ 90 °C and a common disordered phase above [94].

The echo attenuations in the BμE state (Fig. 19a) are markedly nonexponential, and curves measured at different diffusion times Δ do not coincide, especially at high values of $(\gamma \delta g)^2 \Delta$. Single-exponential curves which are independent of Δ are expected for the case of normal diffusion of a single species, i.e. if the system is homogeneous on the length scale probed. As the blend consists of as many as three species, all echo attenuation curves were analyzed in a model-free way, namely by numerical inverse Laplace transformation (ILT) using the routine REPES [96]. In this way, distribution functions of inverse diffusion coefficients, $A(D^{-1})D^{-1}$, vs. $\log((\gamma \delta g)^2 \Delta)$ were obtained (equal-area representation, Fig. 19b). In the BμE state, they show two peaks for all values of Δ. The corresponding diffusivities D, derived from the center of mass of each peak, do not depend on Δ (Fig. 19c), and thus both diffusion processes are normal; only the relative amplitude of the slow decay decreases with increasing Δ (Fig. 19a). The smallest mean square displacement measured in the ternary blend is $\sim$ 190 nm, which is larger than the domain size [94,95]. Thus, the diffusion processes observed are isotropic, i.e. the interfaces in the

BµE state are buckled on a length scale smaller than the mean square displacements measured, and spatial heterogeneities are averaged out in the PFG NMR experiment.

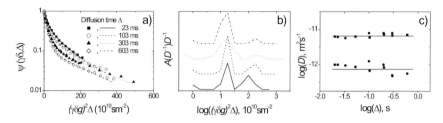

Fig. 19. (a) Semilogarithmic representation of the echo attenuation, $\Psi(\gamma\delta, \Delta)$, of the ternary blend as a function of $(\gamma\delta g)^2\Delta$ at 50 °C for different values of the diffusion time Δ. The lines are fits using REPES. (b) Corresponding distribution functions of inverse diffusion coefficients (same line types as in (a)). The curves are shifted vertically. (c) Diffusion coefficients of the two processes vs. Δ. The horizontal lines indicate the values averaged over all Δ

Identification of the Diffusing Species

An Arrhenius representation of the resulting average diffusion coefficients (Fig. 20a) allows the determination of the activation energies and a comparison with the average diffusivities of the parent components. The activation energies of the fast and the slow process in the BµE state are similar to the ones of bulk PDMS and PEE homopolymer, respectively, but the diffusion coefficient of the fast process is a factor of ~ 10 smaller than that of PDMS in the bulk, whereas the diffusion coefficient of the slow process is slightly higher than that of bulk PEE. The diffusion coefficients from the two processes in the ternary blend are both higher than the average diffusion coefficients of the PEE–PDMS diblock copolymer melt.

In order to identify which of the three species present in the sample gives rise to the fast process in the BµE phase, we made use of the large difference in the transverse nuclear magnetic relaxation times T_2 of PDMS (~ 200 ms at 25 °C) and PEE ($\sim 1 - 20$ ms at 25 °C) in the BµE. In the stimulated-spin-echo PFG NMR pulse sequence (Fig. 1b) for the measurement of $\Psi(\gamma\delta, \Delta)$, T_2-relaxation takes place during the τ-intervals after the first and the third $\pi/2$-pulse. When τ is increased from 3 ms to 25 ms, the contribution of PEE to the NMR signal vanishes, whereas the contribution from the PDMS remains almost unchanged (Fig. 20b). The remaining NMR signal shows the fast decay, which thus represents the diffusion of PDMS homopolymers through the BµE. We attribute the slow decay to the diffusion of PEE; the diblock copolymer is not expected to give rise to a significant signal, because it is present in a small amount only.

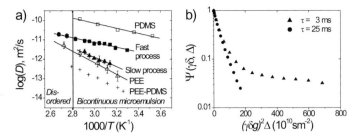

Fig. 20. (a) Arrhenius representation of the self-diffusion coefficients obtained from PFG NMR: fast (■) and slow process (▲) in the ternary blend. Self-diffusion coefficients of the PDMS (□) and PEE homopolymers (△). (+) Average diffusivities of the PEE–PDMS diblock copolymer in the lamellar state. The lines are linear fits. The vertical line denotes the phase transition from the BµE to the disordered state as identified using dynamic light scattering [94]. (b) Echo attenuation curves of the ternary blend measured at 25 °C for different values of τ

Influence of Tortuosity and Domain Viscosities

We ascribe the change of the diffusivities of the two homopolymers with respect to the bulk to two factors: the tortuosity of the BµE structure and the difference in the viscosities of the PEE and PDMS domains. Tortuosity alone can explain the reduction of the diffusivity of the PDMS homopolymers (Fig. 20a); however, it cannot explain why the diffusivity of the PEE homopolymers is higher in the BµE than in the bulk. We therefore have to assume that there is a dynamic equilibrium between the PDMS and PEE homopolymers. The exchange of the two homopolymers between the PEE and the PDMS domains is assumed to be faster than the shortest diffusion time in the experiment (23 ms). The probability p for a homopolymer to be in its own domain, which is equal to the fraction of time spent in this domain, is thus smaller than unity. In order to correlate the PFG NMR diffusivities of the two observable components with the structural and microdynamic parameters of the system, we make use of the Einstein relation (6) together with the principle of detailed balance. The average diffusivity of PDMS as a diffusant then reads

$$\langle D_{\mathrm{PDMS}} \rangle = \frac{D_{\mathrm{PDMS}}^{\mathrm{PDMS}}}{\tau_{\mathrm{PDMS}}} \times \left(1 - p_{\mathrm{PDMS}}^{\mathrm{PEE}}\right) + \frac{D_{\mathrm{PDMS}}^{\mathrm{PEE}}}{\tau_{\mathrm{PEE}}} \times p_{\mathrm{PDMS}}^{\mathrm{PEE}}, \qquad (27)$$

where $D_{\mathrm{PDMS}}^{\mathrm{PDMS}}$ and $D_{\mathrm{PDMS}}^{\mathrm{PEE}}$ are the diffusivities of PDMS in the PDMS- and PEE-rich domains, τ_{PDMS} and τ_{PEE} are the tortuosity factors of the PDMS and PEE domains, and $p_{\mathrm{PDMS}}^{\mathrm{PEE}}$ is the probability that a PDMS homopolymer is located in the PEE phase. The equivalent for PEE as a diffusant reads

$$\langle D_{\mathrm{PEE}} \rangle = \frac{D_{\mathrm{PEE}}^{\mathrm{PEE}}}{\tau_{\mathrm{PEE}}} \times \left(1 - p_{\mathrm{PEE}}^{\mathrm{PDMS}}\right) + \frac{D_{\mathrm{PEE}}^{\mathrm{PDMS}}}{\tau_{\mathrm{PDMS}}} \times p_{\mathrm{PEE}}^{\mathrm{PDMS}}, \qquad (28)$$

with analogous definitions. In the case of PDMS, both the small value of $p_{\mathrm{PDMS}}^{\mathrm{PEE}}$ and the higher viscosity of PEE in comparison with PDMS (leading to smaller values of

$D_{\mathrm{PDMS}}^{\mathrm{PEE}}$ in comparison with $D_{\mathrm{PDMS}}^{\mathrm{PDMS}}$) make the second term on the right-hand side of (27) negligibly small in comparison with the first one. Thus, the PDMS diffusivity in the BµE is found to be comparable to that in the bulk phase, reduced by the tortuosity. Analogously, in (28), $p_{\mathrm{PEE}}^{\mathrm{PDMS}}$ is expected to be much lower than $1 - p_{\mathrm{PEE}}^{\mathrm{PDMS}}$. However, this effect may be (over-)compensated by the fact that $D_{\mathrm{PEE}}^{\mathrm{PDMS}}$ is much larger than $D_{\mathrm{PEE}}^{\mathrm{PEE}}$, because PDMS has a lower viscosity than PEE. The experimental data show that this effect is so pronounced that the PEE mobility in the BµE exceeds that in the bulk (Fig. 20a). Assuming that, at each temperature, $\tau_{\mathrm{PDMS}} = \tau_{\mathrm{PEE}} \equiv \tau$ and $p_{\mathrm{PDMS}}^{\mathrm{PEE}} = p_{\mathrm{PEE}}^{\mathrm{PDMS}} \equiv p$ (no asymmetry in the shape of the two domains has so far been observed [94]) and using values of $D_{\mathrm{PDMS}}^{\mathrm{PEE}}$ and $D_{\mathrm{PEE}}^{\mathrm{PDMS}}$ obtained using dynamic light scattering on binary PEE/PDMS blends, we can estimate τ to be ~ 6 at 30 °C and ~ 3 at 80 °C from (27) and (28) (p decreases with decreasing temperature). This variation of the tortuosity values with temperature is probably related to the structural changes of the BµE: small-angle neutron scattering shows that the absolute value of the curvature increases with temperature [95]. However, the values of the tortuosity found are higher than the values calculated for water diffusion in various bicontinuous phases in water/lipid systems ($\sim 1.6 - 1.7$ in [97] and [98]). This discrepancy may be due to the presence of the block copolymers, which has not been taken into account in the above discussion, and to the contamination of the domains by the other type of homopolymer. This contamination alters the viscosities of the domains, and thus the diffusivities $D_{\mathrm{PDMS}}^{\mathrm{PDMS}}$ and $D_{\mathrm{PEE}}^{\mathrm{PEE}}$ should be replaced by modified diffusion coefficients, which, however are not straightforward to quantify because the viscosity of a polymer blend (which determines the diffusivity) can be a nonlinear function of the viscosities of the blend components.

The case of a very slow exchange of the homopolymers between the PEE and PDMS domains (i.e. slower than our highest observation time, 603ms) cannot be completely excluded, but the fact that in the temperature range studied the blend is close to the disordered state makes this assumption unlikely.

To summarize, in a ternary PDMS/PEE/PEE–PDMS diblock copolymer blend in the bicontinuous microemulsion state, we have observed two diffusional processes in the BµE state. They are attributed to the diffusion of PDMS and PEE homopolymers in the limit of fast exchange between domains. With respect to the bulk, the diffusivities are influenced by the tortuosity of the bicontinuous structure and by the changes in domain viscosity related to the exchange of homopolymers between different domains. PFG NMR thus proves to be useful for studies of confinement in self-organized, mesoscopically structured multicomponent polymer systems. Selective labeling by deuteration of the polymer of interest offers the possibility to study the components separately.

6 Conclusion

We have highlighted the potential of PFG NMR for elucidating the internal dynamics of complex systems, characterized by the existence of internal interfaces. Since very often the functionality of such systems depends on their internal matter exchange, in addition to their relevance for fundamental research, such investigations are of great value for the characterization of such systems with respect to their technological application and exploitation.

The span of the systems considered in this chapter illustrates the diversity of phenomena and problems, including typical soft-matter issues, as also discussed in Chaps. 7 and 8 as well as the application of tailored nanoporous solids and the special features of their guest molecules, which are the focus of the Chaps. 5 and 6. Moreover, the present chapter has also involved the discussion of transport phenomena in geological objects, which – in parallel to rather specific items – reveal a wealth of phenomena related to those in the other systems. The spectrum of systems most successfully investigated by PFG NMR is completed by the presentation of diffusion studies on biomaterials in Chaps. 12 and 13.

Owing to its versatility and efficiency, PFG NMR has proven to be a most helpful technique for comparison between the physical reality in nature and the results of molecular simulation and modeling as considered, for example, in Chaps. 1 and 3. Equally important, on the other hand, are the challenging issues for experimentalists in the field exerted by the numerous suggestions and stimulations of the theoreticians. The most prominent examples are the prediction of the possibility of decreasing (!) intracrystalline diffusivities in zeolites with increasing temperature (Chap. 1) and the occurrence of single-file diffusion in zeolites with a one-dimensional channel structure (Chap. 3). Thus, in view of both the most recent progress in PFG NMR instrumentation, as specified in Sect. 2.3 of this chapter and in Chap. 12, for example, and the spectacular development of novel materials of complex structure over the last few years, PFG NMR is most likely to continue its remarkable development as a technique of choice for numerous problems in fundamental and applied research dedicated to the internal dynamics of complex systems.

Acknowledgements

We wish to thank all our students and colleagues from the University of Leipzig who – during the 9 years of financial support by the DFG through the research grant SFB 294 – performed most of the experimental work presented in this chapter, provided the materials studied, or helped in the interpretation of the measurement results.

We are especially grateful to K. Almdal (Risø National Laboratory, Roskilde, Denmark), J. Caro (University of Hannover), H.-R. Gläser (Environmental Research Institute Halle/Leipzig GmbH), H. Jobic (CNRS, Villeurbanne, France), J. McKendry (Resonance Instruments Ltd., Witney, UK), D.M. Ruthven (university of Maine, USA), F. Schüth (MPI für Kohlenforschung, Mülheim/Ruhr), S. Spange (University of Chemnitz), P. Štěpánek (Institute of Macromolecular Chemistry, Prague, Czech Republic), D. Theodorou (National Technical University of Athens, Greece), and J. Weitkamp (University of Stuttgart) who did not participate in the SFB 294 but nevertheless supported our research with their scientific problems, ideas, and experiences.

References

1. J. Kärger, P. Heitjans, R. Haberlandt (Eds.): *Diffusion in Condensed Matter* (Vieweg, Braunschweig, 1998)
2. P. Heitjans, J. Kärger (Eds.): *Diffusion in Condensed Matter: from the Elementary Step to Transport in Complex Systems* (Springer, Berlin, Heidelberg, 2003)
3. T.Y. Tan, U. Gösele: "Electronic structure and properties of semiconductors", in *Handbook of Semiconductor Technology*, Vol. 1, ed. by K.A. Jackson, W. Schröter (Wiley-VCH, New York, 2000), pp. 231–290
4. T.Y. Tan, U. Gösele: "Diffusion in semiconductors", in *Diffusion in Condensed Matter: from the Elementary Step to Transport in Complex Systems*, ed. by P. Heitjans, J. Kärger (Springer, Berlin, Heidelberg, 2003)
5. A.R. Allnatt, A.B. Lidiard: *Atomic Transport in Solids* (Cambridge University Press, Cambridge, 1993)
6. H. Mehrer, F. Wenwer: "Diffusion in metals", in *Diffusion in Condensed Matter*, ed. by J. Kärger, P. Heitjans, R. Haberlandt (Vieweg, Braunschweig, 1998), pp. 1–35
7. H.C. Berg: *Random Walks in Biology* (Princeton University Press, Princeton, 1999)
8. M. Doi, S.F. Ewards: *The Theory of Polymer Dynamics* (Oxford Science Publications, Oxford, 1986)
9. J. Kärger, D.M. Ruthven: *Diffusion in Zeolites and Other Microporous Solids* (Wiley, New York, 1992)
10. N.Y. Chen, T.F. Degnan, C.M. Smith: *Molecular Transport and Reaction in Zeolites* (VCH, New York, 1994)
11. P.T. Callaghan: *Principles of NMR Microscopy* (Clarendon Press, Oxford, 1991)
12. R. Kimmich: *NMR Tomography, Diffusometry, Relaxometry* (Springer, Berlin, Heidelberg, 1997), p. 526
13. B. Blümich: *NMR Imaging of Materials* (Clarendon Press, Oxford, 2000)
14. F. Stallmach, J. Kärger: Adsorption **5**, 117 (1999)
15. J. Kärger, F. Stallmach: "Anomalous diffusion", in *Diffusion in Condensed Matter: from the Elementary Step to Transport in Complex Systems*, ed. by P. Heitjans, J. Kärger (Springer, Berlin, Heidelberg, 2003)
16. J. Kärger, W. Heink: J. Magn. Reson. **51**, 1 (1983)
17. W. Feller, *An Introduction to Probability Theory* (Wiley, New York, 1970)
18. P.P. Mitra, P.N. Sen, L.M. Schwartz, P. LeDoussal: Phys. Rev. Lett. **68**, 3555 (1992)
19. P.P. Mitra, P.N. Sen: Phys. Rev. B **45**, 143 (1992)
20. P.P. Mitra, P.N. Sen, L.M. Schwartz: Phys. Rev. B **47**, 8565 (1993)
21. J. Bear: *Dynamics of Fluids in Porous Media* (American Elsevier, New York, 1972)
22. O.E. Stejeskal, J.E. Tanner: J. Chem. Phys. **42**, 288 (1965)
23. R.M. Cotts, M.J.R. Hoch, T. Sun, J.T. Markert: J. Magn. Reson. **83**, 252 (1989)
24. J. Kärger, H. Pfeifer, W. Heink: Adv. Magn. Reson. **12**, 2 (1988)
25. P. Galvosas, F. Stallmach, G. Seiffert, J. Kärger, U. Kaess, G. Majer: J. Magn. Reson. **151**, 260 (2001)
26. J. Kärger, N.-K. Bär, W. Heink, H. Pfeifer, G. Seiffert: Z. Naturforsch. **50a**, 186 (1995)
27. *Technical Manual for Gradient Amplifier 8604* (Crown International Inc., Elkhart, 1991)
28. *Operator's Manual: TEC 7700 Power Supply Amplifiers* (Crown International Inc., Elkhart, 1996)
29. P. Galvosas, F. Stallmach, G. Seiffert: *Verfahren zur Steuerung von Verstärkern*, Deutsches Patent- und Markenamt, Aktenzeichen 102 16 493.2 (2002)
30. M.I. Hrovat, C.G. Wade: J. Magn. Reson. **45**, 67 (1981)

31. P.T. Callaghan: J. Magn. Reson. **88**, 493 (1990)
32. W.S. Price, K. Hayamizu, H. Ide, Y. Arata: J. Magn. Reson. **139**, 205 (1999)
33. S. Gröger, F. Rittig, F. Stallmach, K. Almdal, P. Štepánek, C.M. Papadakis: J. Chem. Phys. **117**, 396 (2002)
34. F. Schüth, K.S.W. Sing, J. Weitkamp (Eds.): *Handbook of Porous Materials* (Wiley-VCH, Weinheim, 2002)
35. C. Baerlocher, W.M. Meier, D.H. Olson: *Atlas of Zeolite Framework Types*, 5th edn (Elsevier, Amsterdam, 2001), p. 302
36. J. Kärger, J. Caro: J. Chem. Soc. Faraday Trans. I **73**, 1363 (1977)
37. J. Kärger, H. Pfeifer: Zeolites **7**, 90 (1987)
38. J. Kärger, D.M. Ruthven: "Diffusion and adsorption in porous solids", in *Handbook of Porous Materials*, ed. by F. Schüth, K.S.W. Sing, J. Weitkamp (Wiley-VCH, Weinheim, 2002), p. 2089
39. J. Kärger, S. Vasenkov, S.M. Auerbach: "Diffusion in zeolites", in *Handbook of Zeolite Catalysts and Microporous Materials*, ed. by S.M. Auerbach, K.A. Carrado, P.K. Dutta, (Marcel Dekker Inc., New York, 2003), p. 341
40. J. Kärger, D.M. Ruthven: Zeolites **9**, 267 (1989)
41. S. Vasenkov, W. Böhlmann, P. Galvosas, O. Geier, H. Liu, J. Kärger: J. Phys. Chem. B. **105**, 5922 (2001)
42. S. Vasenkov, J. Kärger: Micropor. Mesopor. Mater. **55**, 139 (2002)
43. U. Hong, J. Kärger, H. Pfeifer, U. Müller, K.K. Unger: Z. Phys. Chem. **173**, 225 (1991)
44. N.K. Bär, J. Kärger, H. Pfeifer, H. Schäfer, W. Schmitz: Micropor. Mesopor. Mater. **22**, 289 (1998)
45. J.S. Beck, J.C. Vartuli, W.J. Roth, M.E. Leonowicz, C.T. Kresge, K.D. Schmitt, C.T.W. Chu, D.H. Olson, E.W. Sheppard, S.B. McCullen, J.B. Higgins, J.L. Schlenker: J. Am. Chem. Soc. **114**, 10834 (1992)
46. F. Stallmach, J. Kärger, C. Krause, M. Jeschke, U. Oberhagemann: J. Am. Chem. Soc. **122**, 9237 (2000)
47. U. Oberhagemann, M. Jeschke, H. Papp: Micropor. Mesopor. Mater. **33**, 165 (1999)
48. P. Stilbs: Prog. NMR Spectrosc. **19**, 1 (1987)
49. R.R. Ernst, G. Bodenhausen, A. Wokaun: *Principles of Nuclear Magnetic Resonance in One and Two Dimensions* (Oxford Science Publications, Oxford, 1991)
50. P. Galvosas, S. Gröger, F. Stallmach, in prep.
51. U. Hong, J. Kärger, H. Pfeifer: J. Am. Chem. Soc. **113**, 4812 (1991)
52. E.J. Cabrita, S. Berger, P. Bräuer, J. Kärger: J. Magn. Reson. **157**, 124 (2002)
53. S.S. Nirvarthi, A.V. McCormick: J. Phys. Chem. **99**, 4661 (1995)
54. W.S. Price, K. Hayamizu, H. Ide, Y. Arata: J. Magn. Reson. **139**, 205 (1999)
55. S. Spange, A. Gräser, H. Müller, Y. Zimmermann, P. Rehak, C. Jäger, H. Fuess, C. Baehtz: Chem. Mater. **13**, 3698 (2001)
56. C. Krause, F. Stallmach, D. Hönicke, S. Spange, J. Kärger: Adsorption **9**, 235 (2003)
57. O. Geier, S. Vasenkov, J. Kärger: J. Chem. Phys. **117**, 1935 (2002)
58. P. Levitz: J. Phys. Chem. **97**, 3813 (1993)
59. V.N. Burganos: J. Chem. Phys. **109**, 6772 (1998)
60. K. Malek, M.-O. Coppens: Phys. Rev. Lett. **87**, 125505 (2001)
61. S. Frey, J. Kärger, H. Pfeifer, P. Walther: J. Magn. Res. **79**, 336 (1988)
62. J. Kärger: AIChE J. **28** 417 (1982)
63. O. Geier, R. Snurr, F. Stallmach, J. Kärger: J. Chem. Phys. **120**, 367 (2004)
64. W.E. Kenyon: Nucl. Geophys. **6**, 153 (1992)
65. R.L. Kleinberg, W.E. Kenyon, P.P. Mitra: J. Magn. Reson. **A 108**, 206 (1994)

66. Y.-Q. Song, S. Ruy, P.N. Sen: Nature **406**, 178 (2000)
67. D. Michel, H. Pfeifer: Z. Naturforsch. **20a**, 220 (1965)
68. K.R. Brownstein, C.E. Tarr: Phys. Rev. A **19**, 2446 (1979)
69. D.P. Gallegos, D.M. Smith: J. Colloid Interface Sci. **122**, 143 (1988)
70. Y.-Q. Song, L. Venkataramanan, M.D. Hürliman, M. Flaum, P. Frulla, C. Straley: J. Magn. Reson. **154**, 261 (2002)
71. F. Stallmach: "Möglichkeiten und Grenzen der NMR zur petrophysikalischen Charakterisierung von Lagerstättengesteinen", in *DGMK Tagungsbericht 9901* (Deutsche Wissenschaftliche Gesellschaft für Erdöl, Erdgas und Kohle e.V., Celle, 1999)
72. J.A. Jackson: *Concepts in Magnetic Resonance*, Vol. 13/6 (Wiley Interscience, New York, 2001)
73. H.-R. Gläser, F. Stallmach, W. Watzauer: Grundwasser, in prep.
74. F. Stallmach, C. Vogt, J. Kärger, K. Helbig, F. Jacobs: Phys. Rev. Lett. **88**, 105505 (2002)
75. C. Vogt, P. Galvosas, N. Klitzsch, F. Stallmach: J. Appl. Geophys. **50**, 455 (2002)
76. L.L. Latour, P.P. Mitra, R.L. Kleinberg, C.H. Sotak: J. Magn. Reson. A **101**, 342 (1993)
77. F. Stallmach, J. Kärger: Phys. Rev. Lett. **90**, 039602 (2003)
78. D. Avnir, D. Farin, P. Pfeifer: J. Colloid Interface Sci. **103**, 112 (1985)
79. D. Candela, P.-Z. Wong: Phys. Rev. Lett. **90**, 039601 (2003)
80. F.S. Bates: Science **251**, 898 (1991)
81. I.W. Hamley: *The Physics of Block Copolymers* (Oxford University Press, Oxford, 1998)
82. C.M. Papadakis, K. Almdal, K. Mortensen, F. Rittig, P. Štěpánek: Macromolecular Symposia, **162**, 275 (2000)
83. D. Broseta, G.H. Fredrickson: J. Chem. Phys. **93**, 2927 (1990)
84. R. Holyst, M. Schick: J. Chem. Phys. **96**, 7728 (1992)
85. P.K. Janert, M. Schick: Macromolecules **30**, 137 (1997)
86. P.K. Janert, M. Schick: Macromolecules **30**, 3916 (1997)
87. F.S. Bates, W.W. Maurer, P.M. Lipic, M.A. Hillmyer, K. Almdal, K. Mortensen, G.H. Fredrickson, T.P. Lodge: Phys. Rev. Lett. **79**, 849 (1997)
88. M.A. Hillmyer, W.W. Maurer, T.P. Lodge, F.S. Bates, K. Almdal: J. Phys. Chem. B **103**, 4814 (1999)
89. D. Schwahn, K. Mortensen, H. Frielinghaus, K. Almdal: Phys. Rev. Lett. **82**, 5056 (1999)
90. D. Schwahn, K. Mortensen, H. Frielinghaus, K. Almdal, L. Kielhorn: J. Chem. Phys. **112**, 5454 (2000)
91. L. Kielhorn, M. Muthukumar: J. Chem. Phys. **107**, 5588 (1997)
92. W.W. Maurer, F.S. Bates, T.P. Lodge, K. Almdal, K. Mortensen, G.H. Fredrickson: J. Chem. Phys. **108**, 2989 (1998)
93. H. Kodama, S. Komura, K. Tamura: Europhys. Lett. **53**, 46 (2001)
94. T.L. Morkved, B.R. Chapman, F.S. Bates, T.P. Lodge, P. Stepánek, K. Almdal: Faraday Discuss. **112**, 335 (1999)
95. T.L. Morkved, P. Štepánek, K. Krishnan, F.S. Bates, T.P. Lodge: J. Chem. Phys. **114**, 7247 (2001)
96. J. Jakeš: Czech J. Phys. **38**, 1305 (1988)
97. T. Feiweier, B. Geil, E.-M. Pospiech, F. Fujara, R. Winter: Phys. Rev. E **62**, 8182 (2000)
98. D.M. Anderson, H. Wennerström: J. Phys. Chem. **94**, 8683 (1990)

^{17}O NMR Studies of Zeolites

Dieter Freude and Thomas Loeser

Universität Leipzig, Institut für Experimentelle Physik I,
freude@uni-leipzig.de

Abstract. Multiple-quantum magic-angle spinning (MQMAS) and double rotation (DOR) techniques were applied in the fields of 17.6 T and 11.7 T to the study the ^{17}O NMR of oxygen-17 enriched zeolites A, LSX and sodalite. In addition, some ^{29}Si and ^{1}H MAS NMR experiments were performed. Zeolites with the ratio Si/Al = 1 were chosen, in order to have an alternating distribution of silicon and aluminum atoms. A linear correlation between the isotropic value of the chemical shift and the Si–O–Al bond angle α (taken from X-ray data) could be found for the zeolites A and LSX, but not for sodalites. Hydration of the zeolites causes a downfield ^{17}O NMR chemical shift of about 8 ppm with respect to the dehydrated zeolites. Ion exchange of the hydrated zeolites generates stronger chemical shift effects. The increase of the basicity of the oxygen framework of the zeolite LSX is reflected by a downfield shift of ca. 10 ppm going from the lithium to the cesium form, and the substitution of sodium by thallium in the zeolite A causes a shift of 34 ppm for the O3 site. ^{17}O DOR NMR spectra are superior to ^{17}O 3QMAS NMR spectra, featuring a resolution increase by a factor of two and are about equal with respect to the sensitivity. The residual linewidths of the signals in the ^{17}O DOR and ^{17}O 5QMAS NMR spectra can be explained by a distribution of the Si-O-Al angles in the zeolites.

1 Introduction

Oxygen is the most abundant element in the earth's crust, but ^{16}O has no nuclear spin. The ^{17}O isotope, with nuclear spin $I = 5/2$, has the disadvantage of a very low natural abundance (0.037%) and the high cost of isotopic enrichment. Solid-state NMR spectroscopy is mostly applied to the spin-1/2-nuclei ^{1}H, ^{13}C, ^{31}P and ^{29}Si. The demand for characterization of inorganic material and some recently developed experimental techniques have led to a growing interest in high-resolution solid-state NMR of quadrupole nuclei with half-integer spin, e.g. ^{27}Al and ^{17}O, for which the electric quadrupole interaction strongly broadens the NMR signal in the powder spectra, see [1]. The ^{27}Al NMR spectroscopy (nuclear spin $I = 5/2$, nuclear quadrupole moment $Q = 0.15 \times 10^{+28}$ m^{2}) of inorganic materials is already well established. The applications of solid-state ^{17}O NMR spectroscopy should be extended, in order to obtain an additional tool for structural probing of the oxygen framework of inorganic materials and the basic properties of porous catalysts as well.

Relatively few solid-state ^{17}O NMR studies have been published since the first investigation on MnO$_2$ was performed by J.A. Jackson [2] in 1963. The fact that the nuclear quadrupole moment of ^{17}O is much smaller than that of the ^{27}Al nuclei

D. Freude, T. Loeser, ^{17}O NMR Studies of Zeolites, Lect. Notes Phys. **634**, 163–183 (2004)
http://www.springerlink.com/

leads one to assume that the quadrupole broadening of the NMR signal is relatively small. However, the very anisotropic bonding of the oxygen atoms in many solids produces strong electric field gradients at the oxygen nuclei and quadrupole coupling constants C_{qcc} of several MHz. The mean values and the standard deviations for about 300 references to values of C_{qcc} in inorganic materials [1] are 4.2 ± 1.5 MHz and 4.8 ± 3.5 MHz for ^{17}O and ^{27}Al, respectively.

New NMR techniques, e.g. dynamic angle spinning (DAS) [3, 4], double rotation (DOR) [5], multiple-quantum excitation in combination with magic-angle spinning (MQMAS) [6], and satellite transition in combination with magic-angle spinning (CTMAS) [7] have recently been developed for quadrupole nuclei with half-integer spins. In addition, the perturbing effect of the electric quadrupole interaction is re-duced at the higher magnetic fields, which are now available. Numerous ^{17}O NMR investigations applying these techniques have been performed, in order to correlate the NMR parameters of resolved oxygen signals obtained with structure data ob-tained by diffraction methods: Grandinetti et al. [8] investigated the SiO_2 polymorph coesite, Mueller et al. [9] measured the ^{17}O signals of diopside, forsterite, clinoen-statite, wollastonite, and larnite by DAS NMR, and Bull et al. [10, 11] investigated the silicon-rich zeolites Sil-Y and ferrierite by ^{17}O DOR NMR. We investigated sev-eral hydrated zeolites by ^{17}O 3Q MAS and DOR NMR and demonstrated reasonably good resolution in a field of 17.6 T, whereas we obtained insufficient resolution in the lower field of 11.7 T [12, 13]. Pingel et al. [13] proposed a correlation between Si–O–Al angles and the ^{17}O chemical shift, whereas Bull et al. [10, 11] claimed that no simple correlation appears to exist between the zeolite bond angles and the ^{17}O NMR parameters. Freude et al. [14] argued that the influence of the adsorbed molecules and the various cations cannot be described by a linear correlation. Loeser et al. [15] found no such correlation for sodalites. Also, Profeta et al. [16] claimed that no simple correlation between the chemical shift, C_{qcc} NMR parameters, and the Si–O–Si angle exists.

Whereas for ^{29}Si NMR a linear correlation between the chemical shift $\delta(^{29}Si)$ and the mean value of $\rho(^{29}Si)$ is well established [11, 17-22], a similar correlation for ^{17}O has been observed for zeolites A and LSX [13, 14] but not confirmed in several other studies [8, 10, 11].

In earlier empirical correlations, it was found that the ^{17}O quadrupole coupling constant C_{qcc} increases and the asymmetry parameter η decreases with increas-ing bond angle α. But no monotonic correlation was found to exist between the ^{17}O chemical shift $\delta(^{17}O)$ and the Si–O–Si angle α [8]. The same holds for *ab initio* calculations of the isotropic ^{17}O chemical shifts [10, 11], which yield a rough quali-tative agreement with the experimental shifts (up to 3 ppm deviation). No monotonic $C_{qcc}(\alpha)$-dependence as in [8] was found in the calculations [10, 11]. Other *ab initio* calculations [21] reproduced well small differences in $\delta(^{17}O)$ for oxygen sites with similar local structures, but showed deviations up to approximately 10 ppm in the relative difference for oxygen sites with different local structures. There was also no correlation found between $\delta(^{17}O)$ and α, but the calculated ^{17}O electric field gra-dient (EFG) related parameters were in agreement with the experimental findings.

A correlation between $\delta(^{17}\text{O})$ and the Si–O bond length for nonbridging OH groups was reported in [23].

Various correlations between C_{qcc} and the bond angle α have been determined: C_{qcc} as a function of $\cos\alpha/(\cos\alpha - 1)$ [8,24]; C_{qcc} as a linear function of the bond angle α [25]; C_{qcc} as a linear function of $\cos\alpha$ [26]; and C_{qcc} linearly decreasing with increasing Si–O distance [27]. [10, 11] show that larger values of C_{qcc} are obtained for peaks at lower chemical shift $\delta(^{17}\text{O})$. For the correlation between the asymmetry parameter η and the bond angle α, $\eta = -3(\cos\alpha + 1)/(3\cos\alpha - 1)$ [28], $\eta = 1 + \cos\alpha$ [8], $\eta \propto (\alpha + \text{const.})^{-1}$ [25], and η as a function of $\cos\alpha/(\cos\alpha - 1)$ [24] have been obtained.

Another topic is how the isotropic value of the ^{17}O NMR chemical shift can monitor molecule–framework interactions and basic properties of the zeolite framework. From the viewpoint of the methodology, the residual linewidth of MQMAS and DOR spectra is an important subject of interest. It is well known that multiple-quantum and double-rotation techniques are complementary tools for getting highly-resolved ^{17}O NMR spectra of solids, see [13]. But it is not clear yet why the DOR technique (compared with 3QMAS or 5QMAS) yields a better resolution of the signals. Sources of line broadening and their different manifestations in DOR and MQMAS NMR will be discussed.

2 Materials and Methods

Zeolites A and LSX were donated by Wilhelm Schwieger and by Tricat Zeolites GmbH. Na,K-LSX denotes the zeolites $\text{Na}_{0.7}\text{K}_{0.3}$-LSX. The samples were dehydrated in a nitrogen stream for more than 12 hours at 220 °C and then treated at this temperature with H_2O (22%–43% ^{17}O enriched) for 3 hours. After the enrichment procedure, some samples were dehydrated again at 400 °C under vacuum, cooled and sealed, or loaded after activation with formic acid (40 HCOOH molecules per unit cell) and sealed.

The sodalite samples were provided by Prof. Dr. G. Engelhardt. They were prepared from kaolinite and sodium hydroxide solution by hydrothermal synthesis following Route B described in [29]. The products were characterized by XRD, thermogravimetry (TG) and ^{29}Si and ^{1}H MAS NMR [29].

The framework of the sodalites under study (Si/Al = 1) consists of a perfectly periodic array of all-space-filling $\{4^6 6^8\}$ polyhedra (β-cages) formed by a network of alternating corner-sharing SiO_4 and AlO_4 tetrahedra with a unit cell content $\{SiAlO_4\}_6^{6-}$ (the equivalent of two β-cages). The β-cages of the sodalites contain Na^+ cations for the compensation of the negative charge and may host OH^- anions and H_2O molecules. The general composition of hydro- and hydroxosodalites is expressed by $\text{Na}_{6+x}\{SiAlO_4\}_6(OH)_x \cdot nH_2O$. The abbreviation $(6+x){:}x{:}n$ denote the number of sodium ions, hydroxyl anions, and water molecules, respectively, per unit cell. One unit cell equals two β-cages. The structure of these sodalites, the dynamics of the guest molecules in the β-cages, and the dehydration and rehydration behavior were subject of extensive research by X-ray and neutron diffraction, thermal analysis,

FT IR, and ^{1}H, ^{23}Na, and ^{29}Si MAS NMR [29–37]. Hydrated 8:2:2 and dehydrated 8:2:0 hydroxosodalites as well as hydrated 6:0:8, dehydrated 6:0:0 and intermediate 6:0:4 hydrosodalites forms were found [29].

Hydrosodalites exhibit a large volume expansion upon dehydration from 6:0:8 to 6:0:0. The Si–O–Al bond angle increases from 136.2° to 156.3°, and the crystal volume increases by 8% (X-ray results [32]). The hydroxosodalite modifications 8:2:2 and 8:2:0 have bond angles of 138.7° (X-ray diffraction [30]) and 132.9° (neutron diffraction [34]). In sodalite 8:2:2 a hydrogen dihydroxide anion, $O_2H_3^-$, is formed [35], whereas in 8:2:0 the oxygen atom of a hydroxyl group is centered in a tetrahedron consisting of 4 Na^+ cations inside the β-cage, with the OH group performing vibrational and jump motions [36, 37].

For the ^{17}O enrichment the 6:0:8 and 8:2:2 sodalite samples were dehydrated in a nitrogen stream for 12 hours at 220 °C and then treated at this temperature with H_2O (22%–43% $H_2^{17}O$) for 3 hours. Fractions of the ^{17}O-enriched sodalites 6:0:8 and 8:0:2 were dehydrated under vacuum at 200 °C and 600 °C, respectively. The resulting 6:0:0 sample was cooled and sealed in glass ampoules fitting the 4 mm MAS rotors; the 8:2:0 sodalite is not rehydrated under these conditions [29] and hence not was sealed.

^{17}O 3QMAS experiments and ^{17}O DOR experiments were performed in an external magnetic field of 17.6 T (Bruker DMX 750 with narrow-bore magnet) at 101.7 MHz. Additional DOR experiments were done in a field of 11.7 T (Bruker MSL 500 with wide-bore magnet). Probes constructed at the Institute of Chemical Physics in Tallinn and Bruker MAS probes were used for the DOR and MQMAS experiments, respectively. Multiple-quantum experiments were performed by means of two adjusted strong pulses (or a second pulse as in FAM2 [38]) and an additional weak z-filter pulse [39]. The ^{17}O NMR scales are referenced to $H_2^{17}O$. A total ring-down delay of 7 µs after the z-filter pulse was used. The repetition time was in the range from 200 ms to 10 s corresponding to the measured longitudinal relaxation times T_1, which vary from 100 ms to 500 ms for the various zeolite samples under study. The dehydrated zeolites were loaded with 10 kPa gaseous oxygen, in order to shorten the value of T_1. A 24 phase cycle was applied twice with an additional 30° phase shift in between, in order to acquire the real and imaginary parts for the second Fourier transform. A nonselective nutation frequency of 120 kHz was determined for a $H_2^{17}O$ sample at the Larmor frequency of 101.7 MHz. For this rf power, the widths of the first and the second pulse in the 3QMAS (5QMAS) experiments were adjusted for maximum ^{17}O 3Q MAS NMR signal to 3.5 (3.8) µs and 1.2 (2.0) µs, respectively. The selective $\pi/2$-pulse length of the z-filter was adjusted to 32 µs. The FAM2 pulse [40], which was used in a few 3Q experiments, consists of a phase-alternating four-pulse sandwich with relative durations taken from [40] and an adjusted total duration of 1.2 µs. The time step t_1 in the multiple-quantum dimension was increased in steps of the reciprocal spinning frequency, in order to avoid spinning sidebands. The time domain consisted of 64 and 1024 steps in t_1 and t_2, respectively. 480 scans were accumulated for each t_1 step. The program XfShear [41] was used for the shearing of the 2D spectra. The simulation of the ^{17}O spectra was carried out with dmfit [42], which was set up to vary the isotropic chemical shift $\delta(^{17}O)$, the quadrupole coupling constant C_{qcc}, and the amplitude until the best approximation of the spectrum was yielded. DOR experiments were performed typically in a synchronized manner [43], with an outer rotor speed of about 1400 Hz and an inner rotor speed of 6000–6500 Hz.

3 Results and Discussion

^{17}O 3QMAS and DOR NMR spectra of the hydrated zeolite Na,K-LSX (low silica X type with 70% Na and 30% K) are shown in Figs. 1 and 3, respectively. Only three signals are resolved in the isotropic projection of the two-dimensional 3QMAS spectrum (Figs. 1 and 2), but four signals can be observed in the 5QMAS and DOR spectra (Fig. 2). X-ray data from Porcher et al. [44] prove the existence of four different oxygen sites in the faujasite-type zeolite LSX. Therefore, the peak in the middle position of the 3QMAS spectrum was fitted by two signals, and four slices were used for the determination of the quadrupole parameter from the spectra obtained (Fig. 1). The deconvolutions of the spectra give intensities of 25%:21%:26%:28% and 24%:30%:25%:21% for MAS and DOR, respectively. These intensities does not differ significantly from the ratio 1:1:1:1, as expected from the faujasite structure for the oxygen sites 1, 2, 3 and 4. The intensities in a preliminary study of the zeolite Na-LSX [13] deviated from this ratio by an enhanced intensity (34%) in one position, which was explained by a preferred enrichment of the oxygen atoms in the O1 position. In this study, the zeolite Na,K-LSX, which is the originally synthesized form, shows the best agreement with the expected 1:1:1:1 ratio. Other cation exchanged forms (Li, Na, K, and Rb) of LSX show ^{17}O signals with slightly different intensities. A significant deviation can be observed for the zeolite Cs-LSX: The 3QMAS spectrum gives a signal intensity of only 7% for the O4 site, whereas the corresponding signal disappears in the DOR spectrum. The numbering of the four signals in Fig. 1 refers to the usual numbering of the four oxygen positions in the faujasite, see [44], e.g. oxygen site 1 denotes bridges between adjacent cubooctahedra. The assignment is based on the assumption that the chemical shift decreases with increasing bond angle. This assumption will be discussed afterward in detail.

Figure 2 shows spectra obtained by the MQMAS and DOR techniques in the field of 17.6 T. The DOR spectra monitor the superposition of the isotropic chemical shift and the isotropic second-order quadrupole shift, whereas the MQMAS spectra exhibit the isotropic chemical shift only. The spectral range of the MQMAS spectra (104 ppm for I = 5/2 and 3Q and only 20.4 ppm for 5Q) is limited by the MAS spinning frequency of 15 kHz, cf. [45]. It can be concluded from Fig. 2 that the resolution of the 5Q spectra exceeds the resolution of the 3Q spectra, but the signal-to-noise ratio of the free induction decay of the 5Q experiment is lower by a factor of 10 (although a weekend run was made for each 5QMAS NMR spectrum). The sensitivity of the 3Q experiment per scan exceeds that of the 5Q experiment by a factor of about 40 for an equal number of scans. The DOR spectra are superior in resolution compared with the 3QMAS spectra. The signal-to-noise ratio of the free induction decay (FID) of one DOR acquisition is about equal to that for the FID of one 3QMAS

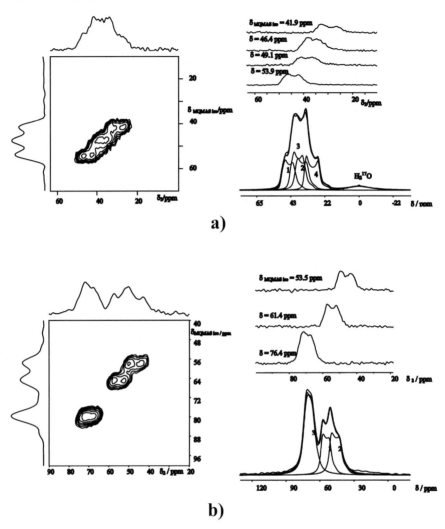

Fig. 1. ^{17}O MAS NMR spectra of the hydrated zeolites Na,K-LSX (a) and Tl-A (b). The sheared 3QMAS spectrum is presented with anisotropic projection on the top and with isotropic projection on the left hand side. Anisotropic slices of the 2D spectrum for four (LSX type) or three (A type) isotropic values are given on the right-hand side above. The spectrum below is a usual MAS spectrum with a fit, which uses the results of the 2D spectrum. The fit of the MAS spectrum gives quantitative results in opposite to the intensities of the 2D spectrum.

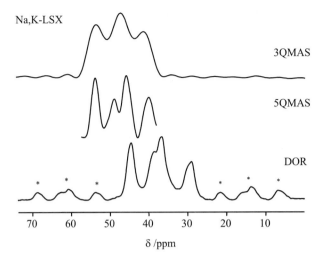

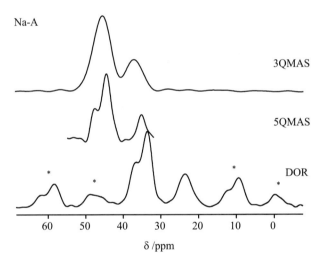

Fig. 2. Isotropic projections of the ^{17}O 3QMAS, 5QMAS NMR and ^{17}O DOR NMR spectra of the zeolites Na,K-LSX and Na-A.

acquisition, since the relatively low filling factor of the DOR coil (compared with MQMAS) is compensated by the higher excitation efficiency of the single-transition. Therefore, we used DOR in external fields of 17.6 T and 11.7 T for the determination of the isotropic NMR parameters $\delta_{CS\,iso}$ and $P_Q = C_{qcc}(1 + \eta^2/3)^{1/2}$, whereas 3QMAS slices were used for the determination of the asymmetry parameter η of the quadrupole interaction. The latter parameter could not be obtained from the DOR spectra, since the resolution of the spinning sidebands was not sufficient for a sideband analysis.

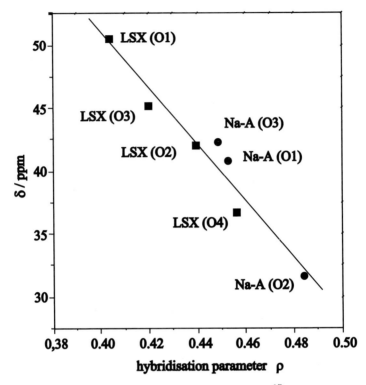

Fig. 3. Correlation between the isotropic chemical shift of the ^{17}O DOR NMR and the s-character of the oxygen hybrid orbitals for the oxygen sites of the zeolites Na-A (circles) and Na,K-LSX (rectangles). The straight line represents the best linear fit.

Table 1 shows ^{17}O NMR parameters, Si–O–Al bond angles α, and values for the s-character of the oxygen hybrid orbitals determined by the equation $\rho = \cos\alpha/(\cos\alpha - 1)$. The upper part of the table describes four oxygen sites in the zeolite Na,K-LSX and three oxygen sites in the zeolite Na-A. The data for the zeolite Na-A are slightly revised with respect to those in a previous paper [13], since new 3QMAS and DOR experiments have been performed. Two averages are given at the bottom of the table. The crucial point of the data for zeolites LSX and A in Table 1 is that the assignments of the values obtained for the ^{17}O NMR chemical shift to the oxygen site numbers and angles, which were taken from the literature, is problematic. Only a few lines in Table 1 seem to be free of doubt: TlA (O3), and Na-A (O3), Na,K-LSX (mean O1, O2, O3, O4), Tl-A (mean O1, O2, O3), Na-A (mean O1, O2, O3), Tl-A (mean O1, O2), Na-A (mean O1, O2). The signal of the site O3 in the zeolites A has about a twofold intensity owing to the twofold occupation of the site O3 with respect to O1 or O2. Also, the averages of the isotropic values of the chemical shift and averaged values for the bond angles or s-characters are unproblematic.

Table 1. Values of the isotropic chemical shift δ of the ^{17}O NMR, the quadrupole parameters $P_Q = C_{qcc}(1 + \eta^2/3)^{1/2}$ (taken from fits of 3QMAS slices and MAS spectra), the asymmetry parameters η taken from fits of 3QMAS slices in the case of zeolites A and LSX, and from $\eta = 1 + \cos\alpha$ [8] in case of the sodalites), and the SiO–Al bond angles α taken from the literature.

Zeolite	Site	δ / ppm	P_Q/ MHz	η	α/ °	for α
Na, K-LSX hydrated	O1	50.6	3.3	0.3	132.5	[50]
	O3	45.2	3.4	0.3	136.3	
	O2	42.1	3.3	0.2	141.5	
	O4	36.8	3.6	0.15	146.8	
	mean(O1,O2,O3,O4)	44.8	3.4	0.25	138.1	
Na, K-LSX dehydrated	O1	42.5	3.2	0.3	134.4	[51]
	O3	38.7	3.3	0.3	140.7	
	O2	37.9	3.3	0.3	145.5	
	O4	33.1	3.3	0.3	146.9	
Na-A hydrated	O1	60.7	3.3	0.15	148	[52]
	O2	53.4	3.6	0.05	161	
	O3	75.5	3.2	0.2	144	
	mean(O1,O2,O3)	39.9	3.5	0.15	147.7	
	mean(O1,O2)	37	3.5	0.1	151.5	
Tl-A hydrated	O1	40.9	3.4	0.15	145.6	[53]
	O2	31.7	3.6	0.05	159.5	
	O3	42.4	3.4	0.2	144.3	
	mean(O1,O2,O3)	67	3.3	0.15	148.6	
	mean(O1,O2)	57.4	3.4	0.1	153.8	
SOD 6:0:8	O	39.1	3.4	0.30	136.2	[32]
SOD 6:0:0	O	36.3	4.3	0.10	156.3	[32]
SOD 8:2:2	O	36.0	3.4	0.25	138.7	[30]
SOD 8:2:0	O	39.2	3.5	0.30	132.9	[34]

Additional information about the assignment of ^{17}O NMR signals to oxygen sites can be obtained from the asymmetry parameter η of the electric field gradient tensor. Equation (2) describes the plausible facts that η decreases from the value one to zero as α goes from 90° to 180°. The signals with the lowest values of η should be assigned to the largest angles in this range. Unfortunately, the accuracy of the asymmetry parameter, which has been obtained from the fit of the slices in the 2D 3QMAS spectra, is not very high (±0.05). But the value $\eta = 0.05$ for the signals, which has been connected in Table 1 to site O2 of the hydrated zeolites Tl-A and Na-A, proves that the assignment of all three signals for the zeolite A is justified. The differences in the values of the asymmetry parameters of the four signals of the zeolite Na,K-LSX ($\eta = 0.3, 0.3, 0.2, 0.15$ for O1, O2, O3, O4, respectively) are less significant, but also in agreement with the assignment of the ^{17}O NMR signals to the sites. Therefore, the rule that the chemical shift decreases with increasing bond angle seems to be valid also for the zeolite Na,K-LSX. Figure 3 plots the seven isotropic chemical-shift values (calculated from the ^{17}O DOR NMR spectra) of the zeolites

Na-A and Na,K-LSX against the values of the s-character of the oxygen bonds in the various Si–O–Al sites. The solid line in Fig. 4 corresponds to the correlation $\delta(^{17}O)$ /ppm $= -214\rho + 136$. The best fit for the correlation of chemical shift with bond angle gives δ /ppm $= -0.65\alpha$ /$°$ + 134. This deviates only slightly from the relation δ /ppm $= -0.71\alpha$ /$°$ + 143.7, which was obtained by previous 3QMAS experiments [13]. The corresponding correlation coefficients are 0.924 and 0.918, if the chemical shift is plotted against α or ρ, respectively. This result supports the existence of a correlation between chemical shift and bond angle in the hydrated zeolites LSX and A. But two facts of our own experimental findings, which will be discussed afterwards, argue against a general correlation. The dehydration of the zeolite and adsorption of molecules other than water changes the Si–O–Al angles and the values of the corresponding chemical shift. The functions $\delta(\alpha)$ obtained vary upon dehydration or adsorption of other molecules. In addition, the ^{17}O chemical shift depends on the type of the zeolitic cation. For example, a downfield shift of 34.1 ppm can be observed for the O3 signal in the ^{17}O NMR spectrum going from the sodium to the thallium form of the zeolite A.

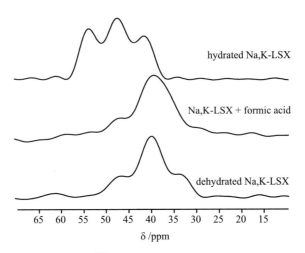

Fig. 4. Isotropic projections of the ^{17}O 3QMAS spectra of the zeolite Na,K-LSX, dehydrated, loaded with formic acid and hydrated.

It is wellknown that the interaction of the framework with adsorbed molecules influences the structure of zeolites. For example, different X-ray patterns exist for the hydrated and the dehydrated faujasites and zeolites A [44, 46, 47]; and the symmetry changes from monoclinic to orthorhombic in the zeolite ZSM-5 upon adsorption of organic molecules [48]. One effect is that the framework bond angles are changed by the adsorbed molecules. This obviously goes along with a change of the ^{17}O NMR chemical shift. However, the ^{17}O NMR shift upon dehydration (compared with the hydrated zeolites) cannot be explained by the changes of bond angles only. The fit for the chemical-shift values obtained from the 3QMAS spectra of the dehydrated zeolites A and LSX in fused glass ampoules (upon Si-O-Al bond angles taken from X-ray data) gives δ /ppm= $-0.65\alpha/°$ + 130, whereas the function for the hydrated zeolites is $\delta/ppm = -0.65\alpha/°$ + 134. This indicates a low field shift of about 4 ppm due to the interaction of the water molecules with the framework oxygen atoms in addition to the shift that is caused by the change of the bond angles.

Figure 4 shows the spectra of the dehydrated zeolite Na,K-LSX, the zeolite after loading with 40 HCOOH molecules per unit cell and the hydrated zeolite, which contains 60 water molecules per unit cell. The most significant differences in these ^{17}O spectra and in corresponding spectra of the zeolite A [49] exist between the hydrated and dehydrated zeolites. The centers of gravity are 47.7 ppm and 40.2 ppm in the spectra of the zeolite Na-LSX, see Fig. 4, and 40.6 ppm and 32.1 ppm in the spectra of zeolite Na-A [49], for the hydrated and dehydrated samples, respectively. From X-ray data [46] it can be concluded that the Si–O–Al angles vary upon dehydration. The average bond angles increase upon dehydration only slightly by 2.5° and 0.5° for the LSX and A types, respectively. The downfield shift in the ^{17}O spectra of zeolites upon hydration is influenced mainly by the interaction of the polar water molecules with the oxygen framework of the zeolite. The simple explanation is that water molecules attract charge from the electronic shell of the oxygen framework, which causes a paramagnetic ^{17}O NMR shift. The relatively weaker interaction between the molecules of formic acid and the LSX framework causes a chemical shift (center of gravity) of 42.4 ppm, which is only 2.2 ppm downfield with respect to that of the dehydrated sample (Fig. 4).

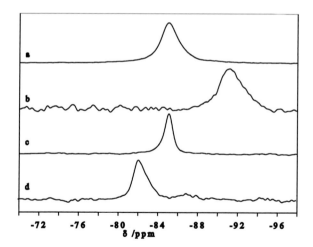

Fig. 5. ^{29}Si spectra of the hydrosodalites 6:0:8 (a), 6:0:0 (b), and the hydroxosodalites 8:2:2 (c), 8:2:0 (d). The spectra are scaled to equal maximum amplitude. The dehydrated sample 6:0:0 was measured in a glass tube in which it had been sealed after dehydration. All spectra were measured in the field of 11.7 T at a spinning frequency of 5 kHz.

Figure 5 shows the ^{29}Si MAS NMR spectra of the ^{17}O-enriched hydrosodalites 6:0:8, 6:0:0 and hydroxosodalites 8:2:2, 8:2:0. Engelhardt et. al. [29] found, for the hydrosodalites 6:0:8, 6:0:0 and hydroxosodalites 8:2:2 and 8:2:0, single ^{29}Si lines at −82.5 ppm, −90.5 ppm, −84.3 ppm, and −81.2 ppm, respectively. For the corresponding ^{17}O-enriched samples, we observed slightly different line positions, at −85.0 ppm, −91.3 ppm, −85.1 ppm, and −82.2 ppm, respectively.

The broad, less intense line in Fig. 5d at −86.9 ppm can be explained by silanol groups in defect sites of the crystal structure. The intensity ratio of this line and that of ordinary framework silicon of the hydroxosodalite 8:2:0 at −82.2 ppm is strongly influenced by the different spin–lattice relaxation times. An increase of the repetition time from 150 s to 600 s decreases the relative intensity of the peak of the defect sites by a factor of two. Hence the fraction of defect sites in 8:2:0 is smaller than the intensity ratio in Fig. 5d suggests, owing to the very high T_1 of the framework silicon.

The Si–O–Al bond angles α of the sodalite species 6:0:8, 6:0:0, 8:2:2 and 8:2:0 were taken from the references given in Table 1.

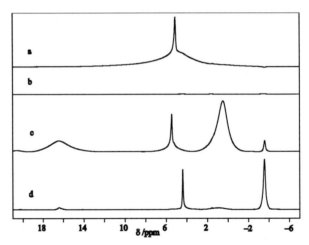

Fig. 6. ^{1}H spectra of the ^{17}O-enriched hydrosodalites 6:0:8 (a), 6:0:0 (b), and hydroxosodalites 8:2:2 (c), 8:2:0 (d) measured in the field of 11.7 T at a spinning frequency of 10 kHz. The spectra (apart from (b)) are scaled to equal maximum amplitude. The dehydrated sample 6:0:0 was measured in a glass tubes in which it had been sealed after dehydration.

The results of the ^{1}H MAS NMR measurements are shown in Fig. 6. The spectrum of the hydrosodalite 6:0:8 (Fig. 6a) shows a superposition of a broad line at 5.0 ppm and a narrow line at 5.1 ppm, the relative intensities of which are 90% and 10%, respectively. The peak at 5.0 ppm can be assigned to the water molecules in the sodalite 6:0:8. A value of 4.3 ppm has been reported for the 6:0:8 phase [29]. The peak at 5.1 ppm is not observed in the non-^{17}O-enriched sample. It can be assumed that the hydrothermal treatment during the ^{17}O exchange produced silanol groups in defect sites.

The spectrum of the hydroxosodalite 8:2:2 (Fig. 6c) shows two broad lines at the positions 0.5 ppm and 16.5 ppm and two sharp resonances at 5.4 ppm and −3.6 ppm, the relative intensities of which are 60%, 25%, 13% and 2%, respectively. The lines at 16.5 ppm and 0.5 ppm can be associated with the central and terminal protons, respectively, of the cage-filling hydrogen dihydroxide anion H–O· · ·H· · ·O–H$^-$ [29, 31]. The resonances are broadened by dynamic exchange of the two protons, which also accounts for the small deviation from the intensity ratio of 2:1 [54]. The peak at 5.4 ppm can be identified with impurities, e.g. hydrosodalite 6:0:8 or NaOH on the surface [29]. The resonance at −3.6 ppm indicates traces of the hydroxosodalite 8:2:0 which have been produced by the dehydration of the sample prior to the ^{17}O isotopic enrichment in combination with the inability of 8:2:0 to rehydrate.

The ^{1}H spectrum of the dehydrated hydrosodalite (Fig. 6b) indicates that practically no water remains in the cages, i.e. the form 6:0:0 is produced. In contrast, there remain two narrow lines and one broad line in the ^{1}H-spectrum of the dehydrated hydroxosodalite at −3.6 ppm, 4.4 ppm, and 0.8 ppm. The relative intensities are 64%, 27%, and 9%. The more intense narrow peak originates from OH-groups in the cages of the hydroxosodalite 8:2:0 [29], and the broad line from the terminal protons of the dihydroxide anion of the species 8:2:2 owing to incomplete dehydration. The broad resonance of the central proton is too weak to be detected and disappears into the baseline. The intensity ratio of the broad and narrow resonances discussed proves that the content of hydrated hydroxosodalite is less than 7%. The less intense narrow line at 4.4 ppm can be explained by silanol groups in defect sites, in agreement with the ^{29}Si MAS NMR measurements.

There exists only a single oxygen site in the sodalite framework, and thus the line assignment is obvious. Zeolites A and LSX contain 3 and 4 crystallographically different oxygen sites, respectively.

Figure 7 displays the ^{17}O MAS NMR spectra of the sodalite samples. The spectrum of the hydrosodalite 6:0:8 (Fig. 7a) shows two lines. The line at 39.1 ppm can be identified as a quadrupole powder pattern and can hence be assigned to the framework oxygen. The second-order quadrupolar broadening is superimposed on a distribution of the chemical shifts. The asymmetry parameter η was determined from the equation $\eta = 1 + \cos \alpha$ [8], the quadrupole parameter P_Q from $P_Q = C_{qcc}(1 + \eta^2/3)^{1/2}$, and C_{qcc} from a fit of the ^{17}O MAS NMR spectrum. The results are given in Table 1. The second line arises from the H$_2$O molecules in the sodalite cages and the small fraction of silanol groups. It has its center of gravity at −24 ppm.

The ^{17}O MAS NMR spectrum of the hydrosodalite 6:0:0 (Fig. 7b) shows a line at 36.3 ppm, which can be fitted in the way described above.

The ^{17}O spectrum of the hydroxysodalite 8:2:2 (Fig. 7c) consists of two signals. The line at 36.0 ppm (relative intensity 83%) exhibits a quadrupole powder pattern and is assigned to the framework oxygen. The other line, at −40.3 ppm, with a relative intensity of 17% is assigned to the oxygen of the H$_3$O$_2$ anion.

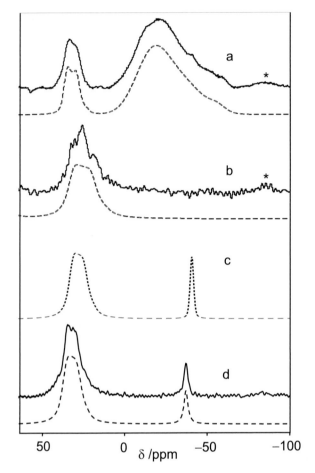

Fig. 7. ^{17}O spectra of the hydrosodalites 6:0:8 (a), 6:0:0 (b), and the hydroxosodalites 8:2:2 (c), 8:2:0 (d) (solid lines) and the corresponding simulations (dashed lines). The intensities of the spectra are scaled to equal maximum amplitude. (The acquisition time was 5 ms and hence the displayed frequency of the thermal noise is in the range of 200 Hz.) The dehydrated sample 6:0:0 (b) was measured in a sealed glass tube. All spectra were measured in the field of 17.6 T at a spinning frequency of 12 kHz.

Upon dehydration of the 8:2:2 sodalite the two peaks are slightly shifted. In addition to the quadrupole powder pattern of the framework oxygen at 39.2 ppm, one line is observed at −37.1 ppm. The latter can be assigned to the OH groups in the hydroxosodalite 8:2:0. The relative intensities of the two signals are 87% and 13%, respectively.

Plots of the isotropic chemical shift $\delta(^{17}O)$ as a function of the bond angle α of the framework oxygen are given in Fig. 8. For every zeolite species a separate linear fit was applied to the data. It becomes obvious that there exist individual linear

Table 2. Best linear fits of the isotropic ^{17}O chemical shift $\delta(^{17}O)$ with respect to the T–O–T bond angle α (taken from Fig. 8).

Zeolite	Slope in the $\delta - \alpha$ plot / ppm	$\delta(\alpha = 0)$ in the $\delta - \alpha$ plot / ppm	Correlation coefficient
Na,K-LSX hydrated	−0.92	171	0.989
Na,K-LSX dehydrated	−0.62	126	0.909
Na-A hydrated	−0.69	141	0.999
Sodalites	−0.11	53	0.668

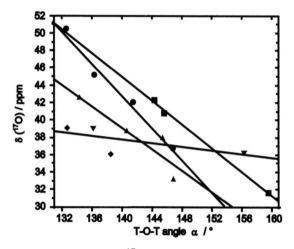

Fig. 8. Correlation between the isotropic ^{17}O chemical shift and the T-O-T bond angle a for the oxygen sites of the hydrated zeolites Na-A (squares), Na,K-LSX (circles), dehydrated Na,K-LSX (up triangles), and for the oxygen sites in the sodalites 6:0:8, 6:0:0 (down triagles) and 8:2:2, 8:2:0 (diamonds). The straight lines represent the corresponding best linear fits for the zeolites LSX, A and the sodalites.

dependences between α and $\delta(^{17}O)$ for the zeolites Na-A and Na,K-LSX (correlation coefficients greater than 0.9). For the sodalites the correlation is significantly worse (correlation coefficient 0.668). In contrast to the correlations for Na-A and Na,K-LSX different states of hydration were considered in a single fit in case of the sodalites. The influence of adsorbed water molecules on the ^{17}O chemical shift of the framework oxygen (cf. [14]) and the different cation configurations could account for the deviation from a linear correlation.

The slopes of the linear-fit functions (Table 2) cover a range from −0.11 ppm/° for the sodalites to −0.92 ppm/° for the hydrated zeolite Na,K-LSX.

Figure 9 shows the mean values (center of gravity) of the isotropic ^{17}O NMR chemical shift in the spectra of several cation-exchanged zeolites, and the ionic radius

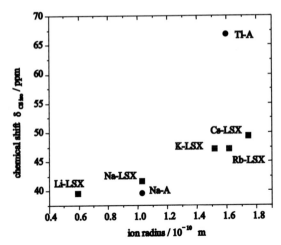

Fig. 9. Plot of the centers of gravities of the isotropic chemical shifts (^{17}O 3QMAS NMR) of the hydrated zeolites A (circles) and LSX (rectangles) against the ionic radius of the cation.

of the cations taken from [55]. The downfield shift increases with increasing ionic radius if zeolites of type A and LSX are considered separately. The cation exchange influences also the Si–O–Al angles in the zeolites. X-ray data for the zeolite Na-A [52] and Tl-A [53] give mean Si–O–Al angles of 148.4° and 149.2° for the hydrated sodium and thallium forms, respectively. ^{29}Si MAS NMR spectra of zeolites A show changes of the signal position from −89.7 ppm (Na-A) to −92.2 ppm (Tl-A). The change of the chemical shift (2.5 ppm) is considerably higher than that calculated from the change of the mean bond angle (0.8°), which gives ca. 0.5 ppm by means of the equation $\delta(^{29}\text{Si})/\text{ppm} = -223.9 \langle \cos\alpha/(\cos\alpha - 1)\rangle + 12.8$ [19]. We could not find any X-ray data in the literature which give the bond angles in hydrated zeolites LSX if the sodium ions are fully exchanged by cesium ions. The ^{29}Si MAS NMR spectra show changes of the signal position from −82.8 ppm for Li-LSX and −84.6 ppm for Na, K-LSX up to −89.7 ppm for Cs-LSX. The shift of about 7 ppm would give a change of about 9° in the mean bond angle. A value of 139.3° was determined by Olson [50]) for the hydrated zeolite Na-X. However, the application the equation above, which was derived for sodium zeolites [19], to zeolites exchanged with larger cations seems to be questionable, and X-ray studies should be performed, in order to determine the bond angles for all cation forms of the zeolite LSX.

The large shift of the center of gravity on going from the sodium to the thallium form of the zeolite A (see Fig. 9), in connection with the small variation of the mean Si–O–Al bond angle of only 0.8°, proves that the ^{17}O NMR chemical shift strongly depends on the radius of the cation. A nearly linear dependence was found for the five cations in the zeolite LSX. The tendency to an increasing chemical shift with increasing cationic radius can be explained in the following way. A decreasing distance from the cation to the oxygen framework attracts an increasing charge from the electronic shell of the oxygen framework which causes a paramagnetic contribution to the ^{17}O NMR shift. A base is an electron pair donor. The electron density of the oxygen framework increases with increasing cation radius. This leads to the increase of the basic properties of the zeolite.

The last aspect of this study concerns the resolution of the NMR spectra. It is wellknown that both distributions of the quadrupole parameter P_Q and distributions of the isotropic value of the chemical shift $\delta_{CS\ iso}$ limit the resolution of the MQMAS and DOR NMR spectra of quadrupole nuclei, whereas the resolution for spin-1/2 nuclei depends only on the latter effect. The ^{29}Si MAS NMR spectra show a single signal at -89.7 or -84.6 ppm, which is 1.0 ppm or 0.7 ppm broad (FWHM = full width at half maximum), for the hydrated zeolite Na-A or Na,K-LSX, respectively. The broadening of the spectra can be explained by a distribution of the mean Si–O–Al bond angles, which is caused by non-regularly distributed cations, since the positions of some cations in the unit cell are ill-defined. Some other effects (field inhomogeneity, incorrect adjustment of the magic-angle, susceptibility broadening, mobility, and a short T_1 of heteronuclei) that could cause a line broadening were excluded. Assuming this to be the dominant effect, the width of the distribution of angles can be calculated from the equation which describes the bond angular dependece of the ^{29}Si MAS NMR. We obtain 1.5° and 0.9° for the zeolites Na-A and Na,K-LSX, respectively (by using the mean angle for all sites, since the calculated value is slightly different for each site). These values characterize the FWHM for the distribution of the mean Si–O–Al angle, which is an average over four angles. Doubled values for the distribution widths of *one* Si–O–Al angle are expected if the angles are independently distributed. Thus, the residual linewidths of the ^{29}Si MAS NMR yield distribution widths (FWHM)of the (non-averaged) Si–O–Al angles of 3.0° and 1.8° for the zeolites Na-A and Na,K-LSX, respectively.

This distribution of angles can be inserted into the equation $\delta(^{17}O)$ /ppm = $-0.65\alpha° + 134$ which was obtained above. It gives a distribution of the ^{17}O chemical shift with an FWHM of 1.8 ppm and 1.5 ppm for the hydrated zeolites Na-A and Na,K-LSX, respectively. The angularly dependent distribution of the quadrupole coupling constant can be estimated by equation $C_{qcc} = C_{qcc}(180°) 2 \cos \alpha / (\cos \alpha - 1)$ [8,24]. For the sake of simplicity, we use mean quadrupole coupling constants of 3.4 MHz and 3.2 MHz, and we use again the mean angle of 148.4° and 139.3° for the hydrated zeolites Na-A and Na,K-LSX, respectively. The values 0.06 MHz and 0.05 MHz for the FWHM of the distributions of the quadrupole coupling constants are obtained for the hydrated zeolites Na-A and Na,K-LSX, respectively.

With the Larmor frequency ν_L and the quadrupole coupling parameter $P_Q = C_{qcc}(1 + \eta^2/3)^{1/2}$, we have in the case of a nuclear spin $I = 5/2$ and multiple- or single-quantum transitions for MQ MAS or DOR, respectively, the following equations for the isotropic value of the total shift:

$$\delta_{MQMAS\ iso} / \text{ppm} = \delta_{CS\ iso} / \text{ppm} + \frac{3 \times 10^6}{850} \frac{P_Q^2}{\nu_L^2}$$

and

$$\delta_{DOR\ iso} / \text{ppm} = \delta_{CS\ iso} / \text{ppm} - \frac{3 \times 10^6}{500} \frac{P_Q^2}{\nu_L^2}.$$

The isotropic quadrupole shifts on the right hand side of the equations, which are of opposite sign (factor $-10/17$) for the multiple-quantum transitions compared with the single-quantum transitions, are distributed owing to the distribution of the quadrupole coupling constants. Inserting the distributions estimated above, we obtain quadrupole shifts of the order of magnitude of 0.1 ppm (the exact values are 0.08 ppm and 0.06 ppm in the case of 3QMAS and -0.13 ppm and -0.1 ppm in the case of DOR for the hydrated zeolites Na-A and Na,K-LSX, respectively).

Thus, the influence of an angular distribution (Si–O–Al angle) on the isotropic quadrupole shift (about ± 0.1 ppm) can be neglected in comparison with the influence on the isotropic chemical shift (about 2 ppm). Figure 2 demonstrates the different resolutions in the ^{17}O spectra for the 3QMAS, 5QMAS and DOR techniques. The deconvolution gives linewidths of about 2 ppm for the single signals in the 5QMAS and DOR spectra. This value does not differ significantly from the values which were estimated for the ^{17}O NMR spectra by means of the ^{29}Si NMR spectra (1.8 ppm and 1.5 ppm, for the hydrated zeolites Na-A and Na,K-LSX, respectively). Thus, the residual linewidths of the 5QMAS and DOR spectra can be explained by a distribution of the Si–O–Al angles, which is caused by a nonhomogeneous distribution of cations in the zeolite. This is a natural broadening independent of the NMR techniques. Now the question arises of why the residual linewidth of the 3QMAS NMR spectra is about twice as broad as that for 5QMAS and DOR NMR. The excitation procedure does not influence the residual linewidth of the 3QMAS spectra: application of FAM2 gives the same resolution (and also the same intensity oppositely to the original work [38]). Also the lowering of the hf power by a factor of two does not influence the residual linewidth of the spectra. The improved resolution of 5Q compared with 3QMAS NMR was first reported by Sarv et al. [56]. Also, Mildner et al. [57] claimed that the resolution of 3QMAS experiments was less than the resolution obtained by a 5QMAS technique similar to the RIACT technique. Pike et al. [58] calculated the relative scaling factors and obtained a better resolution by a factor of 4.19 for 5Q compared with 3Q MAS (spin = 5/2), but they could not find experimental evidence for this. Nonsecular dipolar couplings between ^{17}O and ^{27}Al nuclei can also influence the resolution of ^{17}O NMR spectra in the lower field [59]. This can be neglected in this study, since the Larmor frequency of the aluminum nuclei is more than two orders of magnitude higher than the corresponding quadrupole coupling constant. Further work seems to be necessary, in order to explain the lower resolution of 3Q MAS NMR compared with DOR NMR.

4 Conclusions

It was confirmed that in principle, the isotropic ^{17}O chemical shift $\delta(^{17}O)$ decreases with increasing T–O–T bond angle α for the zeolites under study. The experimental values can be described by the equations $\delta(^{17}O) = 141 - 0.69\,\alpha$, $\delta(^{17}O) = 171 - 0.92\,\alpha$, and $\delta(^{17}O) = 126 - 0.62\,\alpha$ (α in degrees, $\delta(^{17}O)$ in ppm) for the zeolites hydrated Na-A and hydrated and dehydrated Na,K-LSX with correlation coefficients of 0.999, 0.989 and 0.909, respectively. For the sodalites, a linear correlation was not found. This behavior could be explained by the effect of varying water content and cation configuration in the hydro- and hydroxosodalite samples.

The dehydration of the zeolites causes ^{17}O NMR chemical shift changes by the superimposed effects of the well-known changes of the Si–O–Al bond angles and the effect of polarization of the framework by the adsorbed water molecules. The total effect is about 8 ppm, whereas the angularly corrected effect amounts about 4 ppm. The low-field shift due to the adsorption interaction is relatively small (ca. 2.2 ppm) for formic acid.

A downfield shift of ca. 10 ppm on going from the lithium to the cesium form of the zeolite LSX and the shift of ca. 34 ppm for the O3 signal after the substitution of sodium by thallium cations in the zeolite A reflect the increase of the basicity of the oxygen framework of the zeolite caused by ion exchange with larger cations.

^{17}O DOR NMR spectra are superior to ^{17}O 3QMAS NMR spectra with respect to the resolution by a factor of two. The application of the FAM2 excitation [38] does not improve the resolution or intensity in the ^{17}O 3QMAS NMR spectra. The signal-to-noise ratios of the DOR and 3QMAS NMR spectra are comparable, whereas that of the 5QMAS NMR spectra is lower by more than one order of magnitude, and the spectral window is lower by a factor of five. This limits the application of the 5QMAS technique to ^{17}O NMR. The residual linewidths of the signals in ^{17}O DOR and ^{17}O 5QMAS NMR are caused by a distribution of the Si–O–Al angles in the zeolites.

Acknowledgments

We acknowledge the contributions to this work from Dr. U. Pingel, Dr. D. Prochnow. Dr. H. Ernst, Dr. A. Samoson, B. Knorr, and D. Prager. We are grateful to Prof. Dr. D. Michel, Prof. Dr. R. Stannarius, Prof. Dr. Pöppl, Dr. S. Mulla-Osman, DP Denis Schneider, Prof. Dr. G. Engelhardt, and Prof. Dr. W. Schwieger for advice and help. This work was supported by the Deutsche Forschungsgemeinschaft under the projects Fr 902/9 and SFB 294 and by the Max-Buchner-Stiftung.

References

1. D. Freude: "Quadrupole nuclei in solid-state NMR", in *Encyclopedia of Analytical Chemistry*, ed. by R.A. Meyers (Wiley, Chichester, 2000), p. 12188
2. J.A. Jackson: J. Phys. Chem. Solids **24**, 591 (1963)
3. A. Llor, J. Virlet: Chem. Phys. Lett. **152**, 248 (1988)
4. B.F. Chmelka, K.T. Mueller, A. Pines, J. Stebbins, Y. Wu, J.W. Zwanziger: Nature **339**, 42 (1989)
5. A. Samoson, E. Lippmaa, A. Pines: Mol. Phys. **65**, 1013 (1988)
6. L. Frydman, J.S. Harwood: J. Am. Chem. Soc. **117**, 5367 (1995)
7. Z.H. Gan: J. Chem. Phys. **114**, 10845 (2001)

8. P.J. Grandinetti, J.H. Baltisberger, I. Farnan, J.F. Stebbins, U. Werner, A. Pines: J. Phys. Chem. **99**, 12341 (1995)
9. K. T. Mueller, J.H. Baltisberger, E. W. Woote, A. Pines.: J. Phys. Chem. **96**, 7001 (1992)
10. L. Bull, A. Cheetham, T. Anupold, A. Reinhold, A. Samoson, J. Sauer, B. Bussemer, Y. Lee, S. Gann, J. Shore, A. Pines, R. Dupree: J. Am. Chem. Soc. **120**, 3510 (1998)
11. L.M. Bull, B. Bussemer, T. Anupold, A. Reinhold, A. Samoson, J. Sauer, A.K. Cheetham, R. Dupree: J. Am. Chem. Soc. **122**, 4948 (2000)
12. J.P. Amoureux, F. Bauer, H. Ernst, C. Fernandez, D. Freude, D. Michel, U.T. Pingel: Chem. Phys. Lett. **285**, 10 (1998)
13. U.T. Pingel, J.P. Amoureux, T. Anupold, F. Bauer, H. Ernst, C. Fernandez, D. Freude, A. Samoson: Chem. Phys. Lett. **194**, 345 (1998)
14. D. Freude, T. Loeser, D. Michel, U. Pingel, D. Prochnow: Solid State Nucl. Magn. Reson. **20**, 46 (2001)
15. T. Loeser, D. Freude, G.T.P. Mabande, W. Schwieger: Chem. Phys. Lett. **370**, 32 (2003)
16. M. Profeta, F. Mauri, C. J. Pickard: J. Am. Chem. Soc. **125**, 541 (2003)
17. J. Klinowski: Anal. Chim. Acta. **283**, 929 (1993)
18. G. Engelhardt: Stud. Surf. Sci. Catal. **52**, 151 (1989)
19. R. Radeglia, G. Engelhardt: Chem. Phys. Lett. **114**, 28 (1985)
20. C.A. Fyfe, Y. Feng, H. Grondey: Microporous Mater. **1**, 393 (1993)
21. X.Y. Xue, M. Kanzaki: Solid State Nucl. Magn. Reson. **16**, 245 (2000)
22. J.A. Tossell: J. Non-Cryst. Solids **120**, 13 (1990)
23. S.E. Ashbrook, A.J. Berry, S. Wimperis: J. Am. Chem. Soc. **123**, 6360 (2001)
24. K. Vermillion, P. Florian, P. Grandinetti: J. Chem. Phys. **108**, 7274 (1998)
25. A.V. Larin, D.P. Vercauteren: Int. J. Quantum Chem. **82**, 182 (2001)
26. H. Maekawa, P. Florian, D. Massiot, H. Kiyono, M. Nakamura: J. Phys. Chem. **100**, 5525 (1996)
27. T.M. Clark, P.J. Grandinetti: Solid State Nucl. Magn. Reson. **16**, 55 (2000)
28. U. Sternberg: Solid State Nucl. Magn. Reson. **2**, 181 (1993)
29. G. Engelhardt, J. Felsche, P. Sieger: J. Am. Chem. Soc. **114**, 1173 (1992)
30. I. Hassan, H.D. Grundy: Acta Cryst. **C39**, 3 (1983)
31. G. Engelhardt, P. Sieger, J. Felsche: Anal. Chim. Acta **283**, 967 (1993)
32. J. Felsche, S. Luger: Zeolites **6**, 367 (1986)
33. J. Felsche, S. Luger: Ber. Bunsenges. Phys. Chem. **90**, 731 (1986)
34. S. Luger, J. Felsche, P. Fischer: Acta Cryst. **C43**, 1 (1987)
35. M. Wiebcke, G. Engelhardt, J. Felsche, P. B. Kempa, P. Sieger, J. Schefer, P. Fischer: J. Phys. Chem. **96**, 392 (1992)
36. W. Buehrer, J. Felsche, S. Luger: J. Chem. Phys. **87**, 2316 (1987)
37. O. Elsenhans, W. Bührer, I. Anderson, J. Nicol, T. Udovic, F. Rieutord, J. Felsche, P. Sieger, G. Engelhardt: Physica B **180 & 181**, 661 (1992)
38. P.K. Madhu, A. Goldbourt, L. Frydman, S. Vega: J. Chem. Phys. **112**, 2377 (2000)
39. J. P. Amoureux, C. Fernandez, S. Steuernagel: J. Magn. Reson. A **123**, 116 (1996)
40. A. Goldbourt, P.K. Madhu, S. Vega: Chem. Phys. Lett. **320**, 448 (2000)
41. J.P. Amoureux, C. Fernandez, Y. Dumazy: J. Chim. Phys. **92**, 1943 (1995)
42. D. Massiot, F. Fayon, M. Capron, I. King, S. Le Calve, B. Alonso, J.O. Durand, B. Bujoli, Z.H. Gan, G. Hoatson: Magn. Reson. Chem. **40**, 70 (2002)
43. A. Samoson, E. Lippmaa: J. Magn. Reson. **84**, 410 (1989)
44. F. Porcher, M. Souhassou, Y. Dusausoy, C. Lecomte: Eur. J. Mineral. **11**, 333 (1999)
45. J.P. Amoureux, C. Fernandez: Solid State Nucl. Magn. Reson. **10**, 211 (1998)
46. M.M.J. Treacy, J.B. Higgins, R. von Ballmoos: Zeolites **16**, 323 (1996)

47. J.J. Pluth, J.V. Smith: J. Am. Chem. Soc. **102**, 4704 (1980)
48. C.A. Fyfe, J.H. O'Brien, H. Strobl: Nature **326**, 281 (1987)
49. U. Pingel: "^{17}O NMR -Spektroskopie von porösen Materialien", Ph.D. Thesis, Universität Leipzig (2000)
50. D. H. Olson: J. Phys. Chem. **74**, 2758 (1970)
51. F. Porcher, M. Souhassou, Y. Dusausoy, C. Lecomte: Eur. J. Mineral **11**, 333 (1999)
52. V. Gramlich, W.M. Meier: Z. Kristallogr. **133**, 134 (1970)
53. P.E. Riley, K. Seff, D.P. Shoemaker: J. Phys. Chem. **76**, 2593 (1972)
54. G. Engelhardt, P. Sieger, J. Felsche: Angew. Chem. Int. Ed. Engl. **31**, 1210 (1992)
55. D.R. Lide: *CRC Handbook of Chemistry and Physics* (CRC Press, Boca Raton, FL, 1996)
56. P. Sarv, C. Fernandez, J.P. Amoureux, K. Keskinen: J. Phys. Chem. **100**, 19223 (1996)
57. T. Mildner, M. E. Smith, R. Dupree: Chem. Phys. Lett. **306**, 297 (1999)
58. K.J. Pike, R.P. Malde, S.E. Ashbrook, J. McManus, S. Wimperis: Solid State Nucl. Magn. Reson. **16**, 203 (2000)
59. S. Wi, L. Frydman: J. Chem. Phys. **112**, 3248 (2000)

Paramagnetic Adsorption Complexes in Zeolites as Studied by Advanced Electron Paramagnetic Resonance Techniques

Andreas Pöppl[1], Marlen Gutjahr[2], and Thomas Rudolf[3]

[1] Universität Leipzig, Institut für Experimentalphysik II,
   poeppl@physik.uni-leipzig.de
[2] Universität Leipzig, Institut für Experimentalphysik II,
   marlen@physik.uni-leipzig.de
[3] Universität Leipzig, Institut für Experimentalphysik II,
   rudof@rz.uni-leipzig.de, trudolf@gmx.de

Abstract. One major challenge in the research of microporous materials is the spectroscopic characterization of surface adsorption complexes. If the adsorbed molecules are paramagnetic the structure of the formed complexes, the dynamics of the adsorbed molecules, and the specific adsorption sites at the metaloxide surface can be accessed by electron spin resonance spectroscopy. Such studies may provide a unique understanding of the various physical processes giving rise to the observed adsorption phenomena but also allow a detailed characterization of the nature of the respective adsorption sites and their chemical properties. In that way, adsorption of nitric oxide and di-tert-butyl nitroxide in ZSM-5 and Y zeolites is employed to study the electron pair acceptor properties, commonly denoted as Lewis acidity, of alkali metal cations and aluminum defect centers in these molecular sieve systems. Strategies are presented on the basis of continuous wave electron spin resonance spectroscopy to deduce the electron pair acceptor strength of the surface sites from the spin density distribution in the formed adsorption complexes, the electric surface field at the adsorption site, and the adsorption-desorption behavior of the probe molecules. However an unambiguous evidence of the direct coordination of the probe molecules to the specific surface sites is a necessary requirement of the proposed methods. Therefore, particular emphasis is given to the determination of the structure of the adsorption complexes by pulsed electron double resonance and electron spin echo envelope modulation spectroscopy.

1 EPR Spectroscopy of Adsorbed Molecules in Zeolites

The spectroscopic characterization of the interaction of adsorbate molecules with surface sites is a major topic in the study of microporous materials such as zeolites. Such studies provide valuable information about the adsorptive and catalytic properties of the surfaces sites on a microscopic scale. Among the various kinds of adsorption centers on the inner surface of zeolites, acid sites have attracted special interest because they give rise to the unique acid properties of these crystalline solids, which these materials attractive for specially tailored heterogeneous catalytic applications [1,2]. For catalytic applications of microporous materials, the determination of the structure and concentration of the acid sites, as well as their acidity, are of the utmost importance.

A. Pöppl, M. Gutjahr, T. Rudolf, Paramagnetic Adsorption Complexes in Zeolites as Studied by Advanced Electron Paramagnetic Resonance Techniques, Lect. Notes Phys. **634**, 185–215 (2004)
http://www.springerlink.com/   © Springer-Verlag Berlin Heidelberg 2004

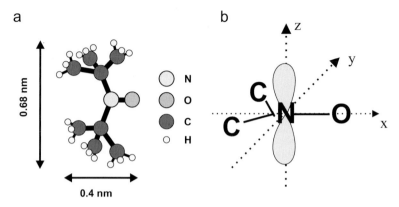

Fig. 1. Structure of the di-tert-butyl nitroxide (DTBN) molecule (a), and principal-axes frame of the **g** and $\mathbf{A}^N$ tensors (b)

Various spectroscopic tools have been successfully developed on the basis of ^{1}H nuclear magnetic resonance (NMR) and infrared (IR) spectroscopy to characterize proton-donating Brønsted acid sites [4–6], which are generally formed by surface OH groups. Here, the acid strength of the Brønsted centers can be accessed by either the proton chemical shift or the frequency of the vibrational modes of the OH moieties.

Electron pair acceptor centers, so-called Lewis acid sites, constitute the second major group of acid surface sites in zeolites. Lewis sites show a large diversity with respect to their structure and electron pair acceptor strength, which makes their spectroscopic characterization by NMR or IR methods a much more demanding task than that for Brønsted sites. Whereas alkali metal cations in zeolites are known to be only weak electron pair acceptors, aluminum defect centers, commonly denoted as true Lewis sites, are believed to have acidities comparable to well-defined strong Lewis acids such as aluminum trichloride and boron trifluoride. Various extra-framework species such as Al^{3+}, AlO_3^{3-}, AlO^+, $(-O-Al)^{2+}$, and AlOOH have been suggested to act as such true Lewis sites in zeolites [5, 7, 8]. However, unambiguous information about the nature of these centers and their electron pair acceptor strength is still missing.

The natural approach to characterize Lewis acid surface sites in microporous materials is to study their interaction with probe molecules having electron pair donor properties. Suitable probes are, especially, paramagnetic nitroxide radicals such as di-tert-butyl nitroxide (DTBN, $(C_4H_9)_2NO$)), shown in Fig. 1, and nitric oxide (NO) molecules, which both form adsorption complexes with electron pair acceptor centers that can subsequently be investigated by electron paramagnetic resonance (EPR) techniques. In that way we can take advantage of the superior sensitivity and information content of advanced EPR methods, such as multifrequency EPR, pulsed electron nuclear double resonance (ENDOR), one-dimensional (1D) electron spin echo envelope modulation (ESEEM), and two-dimensional (2D) hyperfine sublevel correlation (HYSCORE) spectroscopy, to explore the Lewis acid surface centers. The EPR spectra of the adsorption complexes formed can be described by a standard spin Hamiltonian in angular-frequency units,

$$\mathcal{H} = \frac{2\pi\beta_e}{h}\mathbf{B}\,\mathbf{g}\,\mathbf{S} + 2\pi\mathbf{S}\,\mathbf{A}^N\,\mathbf{I}^N + 2\pi\sum_i^N \mathbf{S}\,\mathbf{A}^i\,\mathbf{I}^i$$

$$+ 2\pi\sum_i^N \mathbf{I}^i\,\mathbf{Q}^i\,\mathbf{I}^i - \frac{2\pi\beta_n}{h}\sum_i^N g_{n,i}\mathbf{B}\,\mathbf{I}^i. \tag{1}$$

The first term describes the electron Zeeman interaction of the unpaired electron spin (quantum number $S = 1/2$) of the paramagnetic adsorption complex with the applied external magnetic field, where $\mathbf{S}$ and $\mathbf{B}$ are the vector representations of the electron spin operator and the external magnetic field, and $\mathbf{g}$ is the Zeeman splitting or so-called $\mathbf{g}$ tensor. The hyperfine (hf) coupling between the unpaired electron spin and the nuclear spin of the ^{14}N nucleus (quantum number $I^N = 1$) of the nitrogen atom is measured by the hf interaction tensor $\mathbf{A}^N$ in the second term. The third term, with superhyperfine (shf) coupling tensors $\mathbf{A}^i$ defines the shf interaction between the unpaired electron spin $\mathbf{S}$ and the nuclear spins $\mathbf{I}^i$ of the ligand atoms, here in particular those of the adsorption sites formed by alkali metal cations or extra-framework aluminum species. The $\mathbf{I}^N$ and $\mathbf{I}^i$ designate the spin operators of the respective nuclear spins. Finally, the last two terms of (1) collect the nuclear quadrupole (nq) and Zeeman interactions of all coupled nuclear spins $\mathbf{I}^i$ including that of the ^{14}N atom. The quantities $\mathbf{Q}^i$ and $g_{n,i}$ are the nq tensors and the nuclear g factors. The $\mathbf{g}$ tensor and the nitrogen hf and metal ion shf couplings each contain valuable information about the structure and the electronic properties of the adsorption complexes formed, which offers the opportunity to characterize the Lewis acid adsorption site in a very detailed manner.

The first two terms in the spin Hamiltonian of (1), with the electron Zeeman and the nitrogen hf interaction, determine the anisotropic EPR powder spectra of the adsorption complexes and can easily be accessed by continuous-wave (cw) EPR spectroscopy. Though the ligand shf coupling in the third term is usually one order of magnitude smaller in comparison with the ^{14}N hf interaction and contributes only to the inhomogeneous line broadening of the EPR spectrum it is crucial for the elucidation of the geometrical and electronic structure of the adsorption complexes formed. Today, pulsed ENDOR and HYSCORE experiments offer the opportunity to determine such ligand shf and nq interactions even in disordered systems, as outlined in Sect. 2. It seems to be worth noting that the analysis of the ligand shf coupling of the unpaired electron spin at the DTBN or NO molecule with the nuclear spin $\mathbf{I}^i$ at the aluminum or alkali metal cation nucleus gives the only direct experimental evidence that the paramagnetic probe molecule actually interacts with the surface site studied.

Once the direct coordination of the nitroxide or NO probes with the Lewis acid surface site is established, several approaches can be employed to characterize quantitatively its electron pair acceptor strength:

– The unpaired electron of the nitroxide radical is located in an antibonding $2p\pi^*$ molecular orbital (MO) at the NO group, which is formed by nitrogen and oxygen $2p$ atomic orbitals (AOs). It has been proposed that the electronic structure of the

N–O moiety is composed of a mixture of two resonance structures [9]

$$
\begin{array}{cc}
\text{I} & \text{II} \\
> \dot{\text{N}}^+ - \overline{\text{O}}\,|^- & > \ddot{\text{N}} - \dot{\text{O}}\,|.
\end{array}
\qquad \longleftrightarrow \qquad (2)
$$

Coordination to an electron pair acceptor leads to a redistribution in the N–O π bonding system in comparison with the isolated probe molecule. Obviously, the accompanying change in the unpaired spin density $\rho^{\text{N}}_{2p\pi}$ at the nitrogen atom in the $2p\pi$ MO can be taken as a quantitative measure of the electron pair acceptor strength of the Lewis site and can be monitored by the ^{14}N hf coupling tensor $\mathbf{A}^{\text{N}}$ in a cw EPR spectrum. Cohen and Hoffman [10] demonstrated the potential of this approach in the study of a series of group IV Lewis acids in solution by employing DTBN radicals as paramagnetic probes. However, this method is only applicable if the adsorption sites do not significantly influence the spin polarization effects in the DTBN probe. Fortunately, this can easily be verified by a linear relationship between the isotropic and dipolar nitrogen hf couplings $A^{\text{N}}_{\text{iso}}$ and $B^{\text{N}}_{xx/yy}$ [9, 11]. Otherwise, only qualitative information about the electron pair acceptor strength of the surface sites can be gained from the ^{14}N hf interaction. Later, DTBN probes were used to study qualitatively the properties of electron pair acceptor sites on various catalyst surfaces, among them silica, faujasite-type zeolites [11–13], alumina [14, 15], and Al$_2$O$_3$ films [16]. We shall show in Sect. 4 that the approach is likewise applicable to NO probes.

- Whereas for nitroxide probes the ^{14}N hf is the key parameter for the characterization of Lewis centers, the popularity of nitric oxide as a probe molecule is mainly based on the sensitivity of its g tensor parameters with respect to the specific adsorption site [17–19, 21, 22]. In gaseous NO, the $^2\Pi_{1/2}$ ground state of the molecule is diamagnetic as the spin and orbital angular momentum cancel each other [23, 24]. Note that the paramagnetism of gaseous NO results from the first excited $^2\Pi_{3/2}$ state, which gives a characteristic nine-line EPR spectrum at $g = 0.8$ that can be observed at moderate NO gas pressures [23, 25]. However, upon adsorption, the orbital angular momentum is quenched by the electric fields at the adsorption site. The $^2\Pi_{1/2}$ ground state becomes paramagnetic [26–28] and accessible to EPR spectroscopy. This quenching of the orbital angular momentum is accompanied by the lifting of the degeneracy between the antibonding $2p\pi^*_x$ and $2p\pi^*_y$ MOs, where the unpaired electron resides in the latter [17]. For the NO adsorption complexes, the principal values of the $\mathbf{g}$ tensor are given by [19, 29]

$$
\begin{aligned}
g_{xx} &= g_{\text{e}} \frac{\Delta}{\sqrt{\lambda^2 + \Delta^2}} - \frac{\lambda}{E}\left(\frac{\Delta - \lambda}{\sqrt{\lambda^2 + \Delta^2}} - 1\right), \\
g_{yy} &= g_{\text{e}} \frac{\Delta}{\sqrt{\lambda^2 + \Delta^2}} - \frac{\lambda}{E}\left(\frac{\Delta - \lambda}{\sqrt{\lambda^2 + \Delta^2}} + 1\right), \\
g_{zz} &= g_{\text{e}} - \frac{2l\lambda}{\sqrt{\lambda^2 + \Delta^2}},
\end{aligned}
\qquad (3)
$$

with the NO spin–orbit coupling constant $\lambda = 121$ cm^{-1}. Here l is a covalency factor and measures the effective g factor of the orbital contribution, which equals 1 for the free NO molecule and changes slightly on adsorption. The parameters E and Δ define the energy splitting between the $2p\pi_y^*$ and $2p\sigma^*$ MOs and between the $2p\pi_y^*$ and $2p\pi_x^*$ MOs, respectively. The g_{zz} principal axis points along the NO bond, and g_{yy} defines the symmetry axis of the $2p\pi_y^*$ MO. The splitting Δ can be used as a measure of the electric surface field at the NO adsorption site [17, 19–22, 26, 27, 30]. However, all three principal values of the **g** tensor must be known, according to (3), to determine precisely the splitting Δ together with the parameters E and l, which are given by the solution of a fourth-order polynomial equation. The orthorhombic distortion $\mid g_{xx} - g_{yy} \mid$ of the **g** tensor of the NO adsorption complexes is in the order of 10^{-3} or even less but can be resolved by the application of multifrequency EPR spectroscopy at low temperatures [22]. Another, more serious peculiarity in using the energy splitting Δ for the characterization of a surface site is that it depends not only on the electrical surface field but also on the coordination geometry of the NO adsorption complex. Recently, it has been shown by density functional methods [31] that, especially, the bond angle in $M^+ - NO$ ($M^+ = Na^+, Cu^+$) species has a significant influence on the parameter Δ. Since the bond angle can be expected to depend on various structural properties of the adsorption site in a complicated manner and can only be accessed experimentally by a detailed analysis of the ligand shf interaction, the energy splitting Δ provides only restricted information for a comparison of different adsorption sites.

– EPR measurements at low temperatures are useful for characterizing the adsorbed state of the probe molecule at the solid surface. Otherwise, experiments at temperatures above 80 K allow the study of the adsorption and desorption processes of the probes from the Lewis acid sites. In the case of DTBN radicals, the desorption process is difficult to monitor because it is usually accompanied by anisotropic rotations of the nitroxide, which already start below the desorption temperature and can, further, be observed at elevated temperatures. Otherwise, NO has been successfully employed to study desorption phenomena [25, 32, 33] by taking advantage of its unique magnetic properties. There, the key feature is that the NO $^2\Pi_{1/2}$ ground state is paramagnetic only in the adsorbed state of the molecule, where the orbital angular momentum is quenched by the surface field. Thus, the desorption of the NO molecules from the adsorption sites leads to a drastic decrease in the EPR signal intensity of the NO adsorption complexes with rising temperature and eventually to the total disappearance of their EPR signal. In a study of aluminum defect centers in a series of H-ZSM-5 zeolites, Witzel et al. [32] defined the maximum in the rate of change with temperature of the EPR signal intensity of the NO adsorption complex as the desorption temperature of the probe molecules. This desorption temperature was taken as a measure of the acid strength of the true Lewis acid sites. Later, the desorption energies of the NO probes, as a measure of the electron pair acceptor strength of the adsorption sites, could be deduced directly from line-broadening effects in the EPR spectra of the adsorption complexes [25, 33], as described in Sect. 4. The approach was justified by the observation of the EPR signal of the $^2\Pi_{3/2}$ state of the desorbed NO molecules in the gas phase at temperatures above the desorption temperature.

Naturally, the dimensions of the channels and cage openings of the various zeolite structures determine the choice of the probe molecule. Because of its small size, the diatomic nitric oxide can be employed to investigate electron pair acceptor centers in almost all microporous materials. Otherwise, the size of the nitroxide radicals prevents their use for the study of surface sites in many zeolites that have small pore diameters, such as zeolites A (Fig. 2a) and ZSM-5 (Fig. 2c).

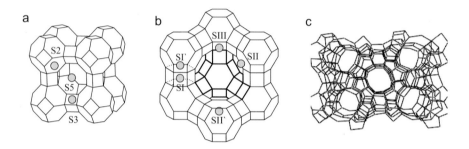

Fig. 2. Schematic representations of the structure and cation positions of zeolites (a) A, (b) Y, and (c) ZSM-5

In the case of the relatively small nitroxide molecule DTBN, the X and Y zeolites (Fig. 2b), with their 0.74 nm entrance window into the α-cage [3], already sets the lower application limit for nitroxide radicals. Therefore, DTBN has been used in this work to study the electron pair acceptor properties of alkali metal cations in Y zeolites, as discussed in Sect. 3. Finally, Sect. 4 summarizes the characterization of sodium cations and aluminum defect centers in zeolites A and ZSM-5, which have pore openings below 0.68 nm [3], by NO probes. However, as outlined above, a necessary requirement for such investigations is the elucidation of the adsorption site of the paramagnetic probe molecule and the determination of the structure of the adsorption complex foremd by pulsed ENDOR or HYSCORE techniques. Thus we shall give a brief introduction to ENDOR and HYSCORE spectroscopy in disordered systems in the following Sect. 2, first.

2 Pulsed ENDOR and HYSCORE Spectroscopy of Paramagnetic Adsorption Complexes

For the characterization of solid surfaces by probe molecules, the identification of their adsorption sites on the basis of a specific interaction between the adsorbate molecule and the adsorbate site is a major part of all investigations. For instance, in adsorption complexes of paramagnetic probe molecules with alkali metal cations or aluminum defect centers, the shf coupling between the unpaired electron in the adsorbate molecule and the metal ion ligand nucleus is the key interaction for identifying the adsorption site. It carries important information about the coupled nuclear spin, the structure of the adsorption complex, and the distribution of the electron spin density within the complex. Unfortunately, such ligand shf couplings are often very weak and contribute only to inhomogeneous broadening of the EPR signal. However, they can be resolved by ENDOR [34–36] or ESEEM [36–38] techniques. In single crystals most investigations of paramagnetic complexes have been performed by cw ENDOR spectroscopy [35]. Pulsed EPR experiments such as pulsed ENDOR and ESEEM spectroscopy have found their merits in the investigation of disordered materials [36–38]. Especially, the Davies pulsed ENDOR [39] and the 2D HYSCORE sequence [40] have been proven to be powerful tools to determine weak shf and nq interactions of ligand nuclei in powders. The methodology of pulsed ENDOR and ESEEM methods has been described in detail in a recent monograph by Schweiger and Jeschke [36]. Therefore, it seems to be sufficient to give only a short description of the working principles of the Davies and HYSCORE sequences here which is necessary for the understanding of the experiments presented. We shall focus rather on specific implications of their application to exploring the structure of NO and DTBN adsorption complexes in zeolites.

For the explanation of the Davies ENDOR experiment and the HYSCORE experiment, we shall refer to the energy level diagram of an $S = 1/2, I = 3/2$ spin system in Fig. 3, which illustrates the case where, for instance, the unpaired electron

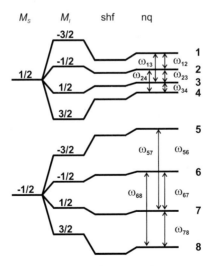

Fig. 3. Energy level scheme for an $S = 1/2, I = 3/2$ spin system in the case $|Q_{zz}| < |A_{zz}| < |\omega_I/\pi|$. M_S and M_I are the magnetic electron and nuclear spin quantum numbers, and the parameters ω_{ij} denote the nuclear transition frequencies between the spin states i and j

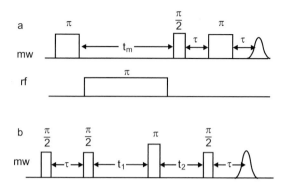

Fig. 4. Pulse sequences of the Davies ENDOR (a) and HYSCORE (b) experiment

spin of an adsorbed NO or DTBN molecule couples to the nuclear spin of a sodium or lithium cation. Figure 4a shows the pulse sequence of the Davies ENDOR experiment. It belongs to the group of spin polarization transfer experiments. In the Davies ENDOR sequence, the applied microwave (mw) pulses are selective with respect to one allowed EPR transition, such as the transition $6 \leftrightarrow 2$. The first mwπ-pulse inverts the polarization across this EPR transition. If the succeeding radio frequency (rf) π-pulse is in resonance with, for instance, the nuclear spin transition $2 \leftrightarrow 1$, the polarization across the $6 \leftrightarrow 2$ EPR transition disappears and the electron spin echo amplitude detected by the mw $\pi/2-\tau-\pi-\tau-$echo detection sequence is zero. In that way the ENDOR spectrum of the nuclear transition frequencies can be measured by sweeping the frequency of the rf π pulse.

A peculiarity of the pulsed ENDOR experiment, but also of ESEEM experiments, arises in disordered systems, since the excitation range of the mw pulses ($\leq 0.5\,\text{mT}$) amounts only to a small fraction of the total width of the anisotropic EPR powder pattern ($\geq 10\,\text{mT}$). Therefore, only a limited number of paramagnetic complexes with selected orientations with respect to the external magnetic field contribute to the ENDOR spectrum measured at a given observer position within the EPR powder spectrum. This is well-known by the term "orientation-selective spectroscopy". The technique offers the oppurtunity to determine the orientation of the principal-axis frame of the ligand shf and nq tensors with respect to the $\mathbf{g}$ and $\mathbf{A}^{\text{N}}$ tensors which are defined by the molecular axes frame of the NO or DTBN molecule (Fig. 1b). Orientation-selective ENDOR spectroscopy has already been discussed in a series of publications [36,42,43]. Therefore, we shall give only a brief outline of the simulation procedure for pulsed ENDOR spectra used in this work [46].

In a first step of the calculation, the $\mathbf{g}$ tensor orientations that contribute to the EPR resonance positions at the selected magnetic field B_0 in the orientation-selective ENDOR experiment have to be determined. The $\mathbf{g}$ tensor orientations are defined by pairs (θ_i, ϕ_i) of horizontal and azimuthal angles θ and ϕ as usual. In our case of NO or DTBN adsorption complexes, the EPR powder pattern is governed by the anisotropic electron Zeeman and nitrogen hf interactions. The tensors of both, the interactions $\mathbf{g}$ and $\mathbf{A}^{\text{N}}$ are coaxial [20, 22]. The EPR resonance positions at a given observer field B_0 in the ENDOR experiment are then expressed by the well-known relations [42]

$$B_0 = \frac{h\nu_{\mathrm{mw}} - \sum\limits_{i=1}^{3} A^{\mathrm{N}}(\theta, \phi) M_i^{\mathrm{N}}}{\beta_e g(\theta, \phi)}, \tag{4}$$

where ν_{mw} is the mw frequency, $M_I^{\mathrm{N}} = -1, 0, 1$ are the magnetic spin quantum numbers of the ^{14}N nucleus ($I^{\mathrm{N}} = 1$), and all other symbols have their usual meaning. The effective g value and hf coupling A^{N} are given by

$$g(\theta, \phi) = \left(\sum_j g_{jj}^2 l_j^2\right)^{1/2},$$

$$A^{\mathrm{N}}(\theta, \phi) = \left(\frac{\sum\limits_j \left(A_{jj}^{\mathrm{N}} g_j l_j\right)^2}{g(\theta, \phi)}\right)^{1/2}, \quad j = x, y, z, \tag{5}$$

with the direction cosines

$$l_x = \cos\phi \sin\theta, \quad l_y = \sin\phi \cos\theta, \quad l_z = \cos\theta \tag{6}$$

defining the orientation of the external magnetic field B_0 with respect to the g tensor frame. The selected orientations (θ_i, ϕ_i) at the observer field B_0 are then determined by calculating the EPR spectrum according to (4)–(6) and searching for all the g tensor orientations giving rise to a resonance position within the field interval $B_0 \pm \Delta B_{12}^{\mathrm{inh}}$. Here it is assumed that the bandwidth of the mw pulses in the pulsed EN-DOR experiments is smaller than the inhomogeneous EPR linewidth $\Delta B_{12}^{\mathrm{inh}}$ and can consequently be neglected. For the analysis of the ENDOR spectra, the spin Hamiltonian of the total spin system in (1) is reduced to an effective two-spin Hamiltonian of an electron spin coupled to one ligand nuclear spin [44]. Then the ENDOR spectra are calculated from the selected orientations (θ_i, ϕ_i) by an exact diagonalization of the effective spin Hamiltonian given in the g tensor principal-axis system [45],

$$\mathcal{H} = \frac{2\pi\beta_e}{h} B_0(0,0,1)\tilde{\mathbf{R}}(\phi_i, \theta_i, 0)\mathbf{g}\mathbf{S} - \frac{2\pi\beta_n}{h} g_l B_0(0,0,1)\tilde{\mathbf{R}}(\phi_i, \theta_i, 0)\mathbf{I}^l$$
$$+ 2\pi\mathbf{S}\tilde{\mathbf{R}}(\alpha, \beta, \gamma)\mathbf{A}^l\mathbf{R}(\alpha, \beta, \gamma)\mathbf{I}^l + 2\pi\mathbf{I}^l\tilde{\mathbf{R}}(\alpha', \beta', \gamma')\mathbf{Q}^l\mathbf{R}(\alpha', \beta', \gamma')\mathbf{I}^l \tag{7}$$

where g_l is the nuclear g factor and $\mathbf{A}^l$ and $\mathbf{Q}^l$ are the hf and nuclear quadrupole (nq) interaction tensors of the ligand nucleus, both given in their principal coordinate system. The quantity $\mathbf{R}$ is the Euler matrix. The two sets of Euler angles α, β, γ and α', β', γ' define the transformation of the diagonal $\mathbf{A}^l$ and $\mathbf{Q}^l$ tensors into the g tensor coordinate frame. The ENDOR transition frequencies ν_{ij} are calculated from the differences between the eigenvalues $\nu_{ij} = (\omega_i - \omega_j)/2$ obtained within the two electron spin manifolds $M_S = \pm 1/2$. For each frequency ν_{ij}, the relative ENDOR intensities P_{ij} were obtained from the matrix elements

$$P_{ij} = \left\langle \psi_i \left| B_{\mathrm{rf}}(0,1,0) \, \tilde{\mathbf{R}}(\phi_i, \theta_i, 0) \frac{2\pi}{h} \left(-\beta_n g_l \mathbf{I}^{\mathbf{l}} + \beta_e g \mathbf{S} \right) \right| \psi_j \right\rangle^2, \qquad (8)$$

where ψ_i and ψ_j are the corresponding eigenvectors of the spin states i and j. Equation (8) takes into account the interaction between the rf field B_{rf} and the electron spin that gives rise to the so-called hyperfine enhancement [45]. This interaction may lead to pronounced differences in the ENDOR intensities between signals from different M_S states in experiments at conventional mw frequencies, such as in X-band measurements [35]. However, in the W-band also, the hyperfine enhancement effect can not be neglected [46]. For the simulation of the orientation-selective ENDOR spectra, the above procedure is repeated for each set of contributing $\mathbf{g}$ tensor orientations (θ_i, ϕ_i). The final ENDOR spectrum is then obtained by adding all individual ENDOR spectra using a Gaussian weighting function with a variance $\Delta B_{12}^{\mathrm{inh}}$.

The HYSCORE experiment [40] in Fig. 4b is based on the well-known three-pulse ESEEM experiment [41]. It uses nonselective mw pulses. If the ligand shf interaction is anisotropic, the nonselective mw pulses cause a branching of all allowed and forbidden EPR transitions in Fig. 3 [36]. The pulse sequence starts with a $\pi/2$-pulse, which generates allowed and forbidden electron spin coherences. The second $\pi/2$-pulse transforms the electron spin coherences to nuclear spin coherences. After this generator sequence nuclear spin coherences with frequencies ω_{ij} $(i, j = 1, 2, 3, 4)$ and ω_{kl} $(k, l = 5, 6, 7, 8)$ in the two electron spin manifolds $M_S = \pm 1/2$ (Fig. 3) evolve during the evolution period t_1. Then the mixing π-pulse exchanges the populations in the two M_S manifolds and creates correlations between the nuclear spin coherences of the two electron spin manifolds with frequencies ω_{ij} and ω_{kl}. After a second evolution period t_2, the nuclear spin coherences are finally transferred back to observable electron spin coherence by the fourth $\pi/2$ mw pulse, which is detected by an electron spin echo. The amplitude of the echo is modulated by the nuclear spin transition frequencies ω_{ij} and ω_{kl}. The symmetric nuclear coherence transfer pathways $\omega_{ij} \leftrightarrow \omega_{kl}$ and $\omega_{kl} \leftrightarrow \omega_{ij}$ during the evolution period $t_1 - \pi - t_2$ lead to cross-peaks $(\omega_{ij}, \pm \omega_{kl})$, $(\omega_{kl}, \pm \omega_{ij})$ in the 2D frequency domain spectra that correlate the nuclear transition frequencies of the two M_S manifolds [36]. With the exception of ^{14}N ligand nuclei that have large nq couplings [50], in general cross-peaks exclusively correlating nuclear spin coherences with $\Delta M_I = \pm 1$ will appear with considerably higher intensity in the 2D spectra in comparison with cross-peaks caused by nuclear spin coherences with $\Delta M_I \geq \pm 2$ [47–49]. In addition, the cross peak intensities in the two quadrants $(+, \pm)$ of the 2D spectra will be different for related features $(\omega_{ij}, \omega_{kl})$, $(\omega_{ij}, -\omega_{kl})$. This can be used to discriminate between the case of strong $(A_{zz} > \omega_I/\pi)$ and weak $(A_{zz} < \omega_I/\pi)$ coupling where ω_I is the nuclear Larmor frequency [36]. In the case of $S = 1/2$, $I = 1/2$ spin systems or systems with small nq couplings, cross-peaks $(\omega_{ij}, \omega_{kl})$, $(\omega_{kl}, \omega_{ij})$ in the $(+, +)$ quadrant will dominate the 2D spectrum for $(A_{zz} < \omega_I/\pi)$. Otherwise, for $(A_{zz} > \omega_I/\pi)$, cross-peaks $(\omega_{ij}, -\omega_{kl})$, $(\omega_{kl}, -\omega_{ij})$ in the $(+, -)$ quadrant will have considerably higher intensities. The full power of the HYSCORE experiment emerges in disordered systems where the cross-peaks transform into well-separated ridges in the 2D spectra [51]. The spread of the spectrum into two dimensions considerably improves the spectral resolution in comparison with 1D ESEEM or pulsed ENDOR methods, so that second-order effects in the ligand shf interaction [52, 53] and weak nq splittings can be resolved even in spectra of disordered materials [54, 55]. In this work, all simulations of HYSCORE spectra were performed in the time domain. Subsequently, the frequency domain spectra were obtained by 2D Fourier transformation (FT) and displayed as magnitude spectra, in analogy to the experimental HYSCORE spectra [54]. The time domain spectra were calculated by using the general expression for the modulation of the four-pulse ESEEM

$$
\begin{aligned}
E_{\mathrm{mod}}^{\mathrm{4p}}\left(\tau, t_1, t_2\right) = {} & \frac{1}{4\left(2I+1\right)} \sum_{ijklmn} M_{il} M_{lj}^{+} M_{jm} M_{mk}^{+} M_{kn} M_{nl}^{+} \\
& \times \exp\left\{-i\left(\omega_{ij}\tau + \omega_{ik}t_2 + \omega_{ln}\tau + \omega_{lm}t_1\right)\right\} \\
& + \frac{1}{4\left(2I+1\right)} \sum_{ijklmn} M_{il} M_{lj}^{+} M_{jm} M_{mk}^{+} M_{kn} M_{nl}^{+} \\
& \times \exp\left\{-i\left(\omega_{ik}\tau + \omega_{jk}t_1 + \omega_{mn}\tau + \omega_{ln}t_2\right)\right\} \\
& + c.c.,
\end{aligned}
\tag{9}
$$

derived by Reijerse et al. [56]. Here the i, j, k run over the $M_S = 1/2$ and the l, m, n over the $M_S = -1/2$ manifold. The elements of the transition element matrix $\mathbf{M}$ are evaluated from the EPR transition probabilities

$$
M_{il} = \left\langle \psi_i \left| B_{\mathrm{mw}}(1,0,0)\tilde{\mathbf{R}}(\phi_i, \theta_i, 0)\frac{2\pi\beta_e}{h}\mathbf{S} \right| \psi_l \right\rangle
\tag{10}
$$

where B_{mw} is the mw field. Again, the nuclear transition frequencies ω_{ij}, ω_{kl} between the nuclear spin states i, j and k, l, as well as the eigenvectors ψ_i and ψ_l of the corresponding spin states i and l, are calculated by an exact diagonalization of the effective spin Hamiltonian in (7). As in the pulsed ENDOR experiment, the angles θ_i, ϕ_i are determined by the orientationselectivity of the HYSCORE experiment and must be calculated as described above.

3 Di-Tert-Butyl Nitroxide (DTBN) Adsorption Complexes

3.1 Structure of DTBN Adsorption Complexes

DTBN probes were employed to study the electron pair acceptor properties of alkali metal cations in zeolite Y. In general, for the determination of the structure of adsorption complexes by EPR methods, the experiments must be performed at temperatures sufficiently low to avoid any motional averaging of the anisotropic magnetic interactions. Figure 5 illustrates low-temperature X-band (9.7 GHz) EPR spectra of DTBN adsorbed on a series of alkali-metal-ion-exchanged Y zeolites. Furthermore, EPR spectra of DTBN in a frozen solution in toluene and adsorbed on zeolite HY with Brønsted acid surface sites are presented for comparison. All EPR spectra show the

Table 1. Nitrogen hf coupling parameters of DTBN in Y zeolites and toluene [11]

	LiY	NaY	KY	RbY	CsY	HY	Toluene
$A_\parallel^N$ (mT)	4.00	3.93	3.8	3.72	3.55	3.98	3.40
$A_\perp^N$ (mT)	0.60	0.56	0.57	0.54	0.5	0.7	0.49
$\rho_{2p\pi}^N$	0.67	0.66	0.63	0.624	0.60	0.64	0.57

typical spectral features of immobilized nitroxide radicals, with a ^{14}N hf splitting into three lines due to the interaction of the unpaired electron spin ($S = 1/2$) with the nuclear spin of the ^{14}N ($I = 1$) nucleus. The principal values of the axially symmetric ^{14}N hf coupling tensor $\mathbf{A}^N$ and the deduced unpaired spin density $\rho_{2p\pi}^N$ at the nitrogen in the $2p\pi$ MO are summarized in Table 1. For all samples, the **g** tensor was found to be axially symmetric with, $g_\perp = 2.0072$, $g_\parallel = 2.0023$.

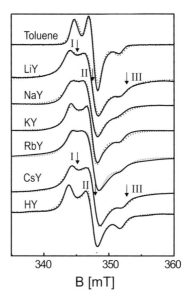

Fig. 5. Experimental (solid lines) and simulated (dashed lines) X-band EPR spectra of DTBN adsorbed on various Y zeolites at 20 K. Simulations were done using the parameters given in Table 1. The arrows indicate the spectral positions I, II, and III, where orientation-selective HYSCORE measurements were performed on Li$^+$–DTBN and Cs$^+$–DTBN adsorption complexes [11]

The weak shf interaction between the unpaired electron spin at the DTBN molecule and the nuclear spin of the alkali metal cation could not be resolved in the cw EPR spectra of the adsorption complexes (Fig. 5). Therefore orientation-selective HYSCORE experiments needed to be employed to prove the direct coordination of the paramagnetic probe molecule and to determine the coordination geometry of the adsorption complexes formed. The following procedure was applied to study the structure in DTBN adsorption complexes in zeolites LiY and CsNaY [57, 58]. Primarily, HYSCORE spectra at the so-called powder position in the EPR spectrum (position II in Fig. 5) were measured to determine the principal values of the alkali metal ion shf coupling tensor $\mathbf{A}^l$. At spectral position II, a complete HYSCORE powder spectrum can be measured because all crystallites with $\mathbf{g}$ tensors whose angle θ between the $g_\parallel$ axis and the external magnetic field $\mathbf{B}$ is in the range $0° \leq \theta \leq 90°$ contribute to the EPR powder pattern. Then orientation-selective HYSCORE experiments were performed at positions I and III, corresponding to the $g_\perp$ and $g_\parallel$ spectral positions in the EPR powder pattern. For position III, only $\mathbf{g}$ tensor orientations with $\theta = 0°$ are selected, whereas at position I, all crystallites with $\mathbf{g}$ tensor orientations $\theta = 90°, 0° \leq \phi \leq 180°$ contribute to the HYSCORE spectrum. Finally, simulations of these orientation-selective spectra with the principal values of the shf tensor $\mathbf{A}^l$ already deduced reveal the orientation of its principal-axis system with respect to the $\mathbf{g}$ tensor axis frame, expressed in terms of a set of Euler angles α, β, γ.

Figure 6 illustrates experimental and simulated orientation-selective HYSCORE spectra of DTBN adsorbed on zeolite LiY. ^{1}H cross-peaks from the protons of the methyl groups of DTBN and intense cross-peak features from the lithium isotopes ^{7}Li $(I = 3/2)$ and ^{6}Li $(I = 1)$ of a single species of DTBN adsorption complexes with Li cations (species Li:A) can be observed in the 2D spectrum. All correlation features appear in the (+,+) quadrant in accordance with the weak-coupling case $(A^{Li}_{zz} < \omega_{Li}/\pi)$. Especially, the dominant $(\Delta M_I = \pm 1, \Delta M_I = \pm 1)$ cross-peaks at about (6 MHz, 6 MHz) of the ^{7}Li isotope show the characteristic ridge-type shape which is typical for HYSCORE spectra of disordered systems. As expected for a nuclear spin with $I > 1/2$ [48, 49] additional correlation features of the ^{7}Li isotope of type $(\Delta M_I = \pm 1, \Delta M_I = \pm 2), (\Delta M_I = \pm 2, \Delta M_I = \pm 1)$, and $(\Delta M_I = \pm 2, \Delta M_I = \pm 2)$ are present in the HYSCORE spectrum besides the dominant $(\Delta M_I = \pm 1, \Delta M_I = \pm 1)$ cross-peaks at about (6 MHz, 6 MHz). The experimental spectra gave no indication of a splitting of the ^{7}Li cross-peaks due to the nq interaction. Therefore, only an upper limit for the ^{7}Li nq coupling $Q^{Li}_{zz} \leq 0.06$ MHz can be deduced from the linewidth of the cross-peak ridges parallel to the $\omega_1 = \omega_2$ frequency axis in the HYSCORE spectra [57].

Orientation-selective HYSCORE spectra of DTBN adsorbed on zeolite CsNaY are presented in Fig. 7. Intense cross-peaks of the cesium isotope ^{133}Cs $(I = 7/2)$ are observed in both quadrants $(++), (+-)$ of the 2D spectra. The spectral features in the $(++)$ quadrant are assigned to species Cs:A of the DTBN adsorption complex with $A^{Cs}_{zz} < \omega_{Cs}/\pi$. The additional cross peaks in the second quadrant $(+-)$ indicate the formation of a second type of adsorption complex of DTBN with Cs cations in CsNaY (species Cs:B), with $A^{Cs}_{zz} > \omega_{Cs}/\pi$. Though no nq splitting was observed, an upper limit of $Q^{Cs}_{zz} \leq 0.06$ MHz has been deduced from the width of the Cs cross-peak ridges [58]. The principal values of the shf coupling tensors $\mathbf{A}^{Li}, \mathbf{A}^{Cs}$ of the DTBN adsorption complexes determined, their Euler angles α, β, γ, and the isotropic and dipolar shf couplings $A^l_{iso}, B^l_{xx,yy}$ are summarized in Table 2. The angle α could not be determined experimentally, because of the axially symmetric $\mathbf{g}$ tensor of the DTBN radical, but was assumed to be $\alpha = 0$ on the basis of the

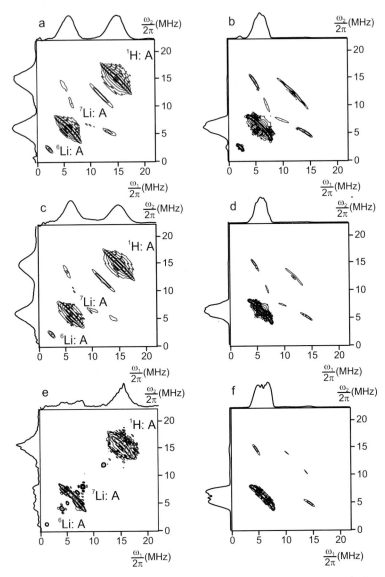

Fig. 6. Experimental (a, c, e) and simulated (b, d, f) HYSCORE spectra of the Li^{+}– DTBN adsorption complex in zeolite LiY at the spectral positions II (a, b), I (c, d), and III (e, f) [57]. Spectra were recorded with $\tau = 96$ ns at $T = 15$ K

symmetry of the DTBN molecule. From the orientation of the tensors $\mathbf{A}^{Li}$, $\mathbf{A}^{Cs}$, we deduce a linear complex structure for the DTBN adsorption complexes Li:A, Cs:A (Fig. 8a), which are formed at the SII cation positions in the supercage of the Y zeolite [57, 58] (Fig. 2b). The second set of cesium shf coupling data Cs:B obtained for CsNaY zeolites revealed a bent complex structure as depicted in Fig.

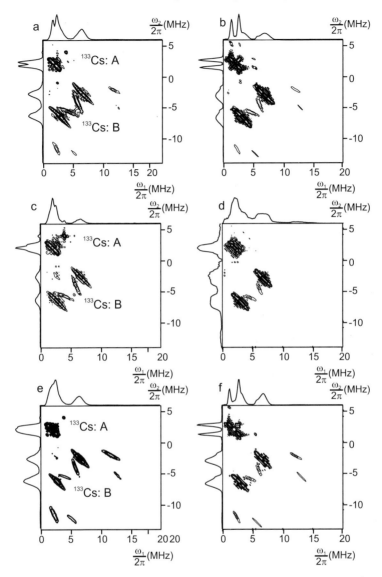

Fig. 7. Experimental (a, c, e) and simulated (b, d, f) HYSCORE spectra of the Cs$^+$– DTBN adsorption complex in zeolite CsNaY at the spectral positions II (a, b), I (c, d), and III (e, f) [58]. Spectra were recorded with $\tau = 96$ ns at $T = 15$ K

8b. This DTBN adsorption complex is formed at the SIII cation positions in the supercage [58]. In contrast to site SII, opposite to each SIII site there is another SIII site. The latter prevents the coordination of the DTBN molecule to the Cs$^+$ cation at the SIII position in a linear complex geometry owing to spatial constraints. It seems worth noting that the SIII cation positions are not occupied in LiY zeolites [59, 60],

Table 2. Alkali metal ion shf coupling parameters, bond lengths ($r_{\mathrm{O-M+}}$), and M$^+$–O–N bond angles (β_{bond}) of M$^+$:DTBN adsorption complexes in zeolites LiY and CsNaY [57]

Zeolite	Species	A^l_{xx} (MHz)	A^l_{yy} (MHz)	A^l_{zz} (MHz)	α ($^\circ$)	β ($^\circ$)	γ ($^\circ$)	A^l_{iso} (MHz)	$B^l_{xx/yy}$ (MHz)	$r_{\mathrm{O-M+}}$ (nm)	β_{bond} ($^\circ$)
LiY	Li:A	−4.3	−4.3	2.6	0	90	0	−2.0	2.3	0.19	180
CsNaY	Cs:A	−1.9	−1.9	0.7	0	90	0	−1.5	0.4	0.25	180
CsNaY	Cs:B	−7.2	−7.2	12.6	0	50	0	9.0	1.8	0.15	130

explaining the absence of such a bent DTBN adsorption complex in the LiY system. For all three adsorption complexes Li:A, Cs:A, Cs:B, a variance in the angle β of $\Delta\beta = \pm25^\circ$ was found, indicating a substantial structural disorder of the adsorption complexes formed.

In the case of the linear geometries of the DTBN adsorption complexes Li:A and Cs:A reasonable bond lengths $r_{\mathrm{O-Li+}} = 0.19$ nm [57] and $r_{\mathrm{O-Cs+}} = 0.25$ nm [58] were found by using a simple point dipole approximation. The negative sign of the isotropic shf interactions $A^{\mathrm{Li}}_{\mathrm{iso}}$ and $A^{\mathrm{Cs}}_{\mathrm{iso}}$ indicates a substantial spin polarization of s-type AOs of the cation. In a linear complex geometry no mixing between the $2p\pi^*$

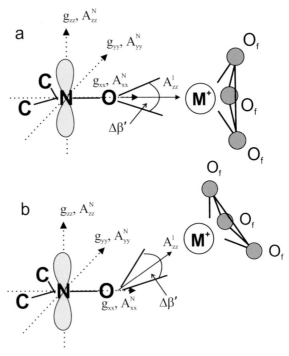

Fig. 8. Structural model of Li$^+$-DTBN and Cs$^+$-DTBN adsorption complexes in LiY and CsNaY zeolites: complex structures (a) Li:A, Cs:A and (b) Cs:B. The alkali metal cations and framework oxygens are indicated by M$^+$ and O$_f$, respectively.

MO at the N–O group of the DTBN molecule which hosts the unpaired electron spin and the AOs of the cation is possible, preventing a positive spin density at the cation. Alternatively, for the bent complex structure Cs:B of the DTBN adsorption complexes in zeolite CsNaY an admixture of the $6s$ cesium AO in the ground state of the unpaired electron located in a $2p\pi^*$ MO is likely. The latter would lead to a significant contribution of spin density at the cesium ion. Indeed, a positive A_{iso}^{Cs} value was found for the structure Cs:B (Table 2), which translates into an unpaired electron spin density in the Cs $6s$ orbital of $\rho_{6s}^{Cs} = 0.0036$. The point dipole approximation provides an N–O–Cs bond angle of $\beta_{bond} = 137°$ and a slightly smaller bond length $r_{O-Cs^+} = 0.21$ nm for the complex Cs:B [58].

3.2 Characterization of the Electron Pair Acceptor Properties of the Adsorption Sites

In the previous section we have confirmed the direct coordination of the DTBN molecules with the alkali metal cations in Y zeolites in the example of zeolites LiY and CsNaY as a necessary precondition for further studies of their Lewis acidity by DTBN probes. But if the unpaired spin density $\rho_{2p\pi}^N$ at the nitrogen in the $2p\pi^*$ MO of the N–O group of the DTBN molecule is taken as a quantitative measure of the electron pair acceptor strength of the surface sites studied the internal structure of the DTBN probe must not be changed by the formation of the adsorption complex. Such structural changes of the nitroxide upon adsorption can easily be excluded by an analysis of its spin polarization effects. It is well known that the isotropic nitrogen hf coupling A_{iso}^N of the nitroxide radicals results from a spin polarization of the σ electrons by the unpaired spin density in the $2p\pi^*$ MO of the N–O group [61],

$$A_{iso}^N = Q_{NN}^N \rho_{2p\pi}^N + Q_{NN}^O \rho_{2p\pi}^O, \tag{11}$$

where Q_{NN}^N, Q_{NN}^O are the σ–π interaction parameters. The spin polarization parameters Q_{NN}^N, Q_{NN}^O are very sensitive to structural changes of the DTBN molecule [61]. The quantities $\rho_{2p\pi}^N$ and $\rho_{2p\pi}^O$ denote the spin densities in the $2p\pi^*$ MO at the nitrogen and oxygen atoms. Under the assumption that the overlap spin density and the loss of spin density from the N–O π system to the zeolite cation can be neglected so that $\rho_{2p\pi}^N + \rho_{2p\pi}^O = 1$, a linear relation between A_{iso}^N and the dipolar nitrogen hf coupling $B_{xx/yy}^N$ holds [61]:

$$A_{iso}^N = \frac{Q_{NN}^N - Q_{NN}^O}{B_0^N} B_{xx/yy}^N + Q_{NN}^O. \tag{12}$$

Here, B_0^N is the dipolar hf coupling of a single electron in a $2p$ orbital at the nitrogen atom. Indeed, a linear relation between the isotropic and dipolar nitrogen hf interaction parameters could be found for M^+–DTBN adsorption complexes in zeolites and DTBN dissolved in toluene (Fig. 9a), indicating no substantial structural changes of the probe molecule upon complex formation with alkali metal cations in Y zeolites. The σ–π interaction parameters determined $Q_{NN}^N = 2.6$ mT, $Q_{NN}^O = -0.1$ mT [11], are

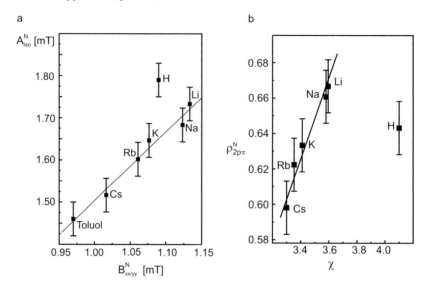

Fig. 9. DTBN adsorbed in various Y zeolites and in deuterated tolueneat 20 K. (a) Isotropic nitrogen hf coupling A_{iso}^N versus dipolar hf coupling $B_{xx/yy}^N$. (b) Nitrogen spin density $\rho_{2p\pi}^N$ in dependence on the electronegativity of the alkali metal ions and protons in zeolite Y [11]

in satisfactory accordance with previous studies of DTBN in aprotic solutions [61]. In contrast, the data for DTBN adsorbed on the protonated zeolite HY having only Brønsted acid sites deviate significantly from the linear relationship found, which suggests structural changes of the probe molecule.

For the characterization of the strength of the electron pair acceptors the electronegativity of the zeolite cations was related to the electron spin density $\rho_{2p\pi}^N$ at the nitrogen atom of the adsorbed spin probes. Figure 9b shows that $\rho_{2p\pi}^N$ increases linearly with ascending electronegativity of the cations, which has been taken from [62]. Note that the DTBN molecules interact with the cations via the oxygen of their N–O group (Fig. 8). Thus the zeolite cations, acting as an electron pair acceptor site, meaning a Lewis acid center, cause a shift of a nonbonding electron pair towards the oxygen atom and consequently a shift of spin density towards the nitrogen atom ($> \dot{N}^+ - \overline{\underline{O}}\,|^- - M^+$). As a result we observe an enhanced contribution of structure I in (2) to the total electronic structure of the NO goup. The larger the electronegativity of the zeolite cation is, the more pronounced is this effect. Therefore, the electron pair acceptor strength of the alkali metal cations in Y zeolites follows the relation $Li^+ > Na^+ > K^+ > Rb^+ > Cs^+$. A similar result was found for DTBN in X zeolites after ion-exchange with a series of alkali metal ions [13]. Obviously, the approach presented offers a straightforward experimental method to quantify the electron pair acceptor strength of adsorption sites in zeolites by the π-orbital spin density ρ_π^N of adsorbed DTBN probe molecules. Moreover, the electron pair acceptor strength of the alkali metal cations in Y zeolites can be compared directly with that of well-defined Lewis acids in aprotic solutions studied likewise by DTBN probes [61].

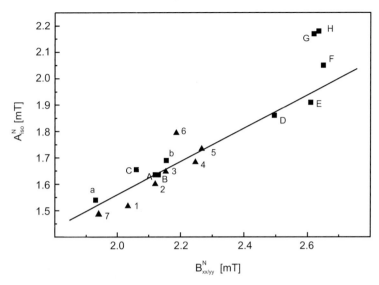

Fig. 10. Isotropic nitrogen hf coupling A_{iso}^{N} versus dipolar hf coupling $B_{xx/yy}^{N}$ of DTBN in solution and adsorbed in Y zeolites: (a) toluene, (b) phenol, (A) Ti(OPh)$_4$, (B) GeCl$_4$, (C) SiCl$_4$, (D) SnCl$_4$, (E) TiCl$_4$, (F) AlCl$_3$, (G) H$_2$O: TiCl$_4$, (F) H$_2$O: TiBr$_4$ (data taken from [61]). (1) CsY, (2) RbY, (3) KY, (4) NaY, (5) LiY, (6) HY, (7) deuterated toluene (data taken from [11])

Figure 10 shows that the alkali metal cations in Y zeolites have an electron pair acceptor strength similar to that of weak Lewis acids such as GeCl$_4$, Ti(OPh)$_4$, and SiCl$_4$ but definitely smaller than that of the strong Lewis acids SnCl$_4$, TiCl$_4$, and AlCl$_3$.

Of course, such a comparison is only justified if the $\sigma-\pi$ interaction parameters Q_{NN}^{N}, Q_{NN}^{O} of the nitroxide are the same for adsorption on solid surfaces and in solution. This is the case for the alkali metal ions in Y zeolites but is not fulfilled for the protonated HY material. Consequently, the data on the HY zeolite, together with those obtained for protonic solutions (Fig. 10), differ significantly from the linear relationship given by (12).

4 Nitric Oxide (NO) Adsorption Complexes

4.1 Structure of NO Adsorption Complexes

In the previous section we have successfully applied the nitroxide radical DTBN for the study of the electron pair acceptor properties of the surface sites located in the supercages of Y zeolites. Unfortunately, the diameter of the micropores of a number of zeolites which are of specific interest for catalytic and adsorptive applications is to small to allow the application of DTBN probes. Important examples of this group of microporous materials are, for instance, the ZSM-5 and A zeolites. To characterize

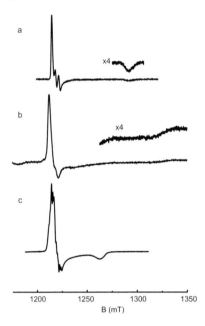

Fig. 11. Q-band EPR spectra of NO adsorbed on (a) NaA, (b) NaZSM-5, and (c) HZSM-5 zeolites at 10 K

the electron pair acceptor properties of sodium cations and aluminum defect centers (true Lewis acid sites) in NaA, NaZSM-5, and HZSM-5 zeolites, we have employed the small nitric oxide molecule as a paramagnetic probe.

Naturally, in a first step we again have to establish the direct coordination of the NO probe molecules with the adsorption sites of interest by cw and pulsed EPR experiments at low temperatures. For that purpose Na^+–NO and Al–NO adsorption complexes in dehydrated NaA, NaZSM-5, and HZSM-5 zeolites were studied by multifrequency EPR spectroscopy at X-, Q-, and W-band frequencies [22,33]. As an example, Fig. 11 illustrates the Q-band (34 GHz) EPR spectra of the NO adsorption complexes in these materials at $T = 10$ K, which are governed by an orthorhombic **g** tensor and a pronounced nitrogen hf splitting into three lines along the y principal axis of **g**. The spin Hamiltonian parameters of the adsorption complexes determined by spectral simulations of their X-, Q-, and W-band spectra, together with the calculated energy splitting Δ between the antibonding $2p\pi_y^*$ and $2p\pi_x^*$ NO MOs, are summarized in Table 3. It is worth noting that the principal values of the nitrogen hf coupling tensor $\mathbf{A}^N$ could not be determined completely because the ^{14}N hf splitting is resolved only in the EPR spectra along the y principal axis of the **g** tensor. The spectrum of Al–NO complexes in HZSM-5 zeolites features an additional shf splitting into six fairly well-resolved lines, which is presumably assigned to the interaction of the unpaired electron spin at the NO molecule with one ^{27}Al ($I^{Al} = 5/2$) nuclear spin from an aluminum defect site. Comparable spectra have been reported in the literature [18,28] and are typical of NO coordinated to aluminum defect centers. The nature of these aluminum defect centers in HZSM-5 zeolites is either Al^{3+} or AlO^+ and has been discussed elsewhere [33,63]. No shf structures have been observed in the EPR spectra of Na^+–NO adsorption complexes.

Table 3. Spin Hamiltonian parameters and $2p\pi_y^*$–$2p\pi_x^*$ energy splitting Δ of NO adsorption complexes in zeolites NaA, NaZSM-5, and HZSM-5 at $T = 10$ K [22,33]

	g_{xx}	g_{yy}	g_{zz}	A_{xx}^N (MHz)	A_{yy}^N (MHz)	A_{zz}^N (MHz)	A_{yy}^{Al} (MHz)	Δ (eV)
NaA	1.999	1.994	1.884	16	92	–	–	0.272
NaZSM-5	1.994	1.991	1.846	32	102	–	–	0.165
H-ZSM-5	1.999	1.999	1.927	–	84	–	31	0.137

Therefore, pulsed ENDOR experiments have been employed to resolve the alkali metal ion shf interaction and in that way to prove the direct coordination of the NO probes with the surface sites. A detailed analysis of the geometrical and electronic structure has been performed for Na^+–NO complexes in NaA zeolites by orientation-selective pulsed ENDOR spectroscopy at W- and X-band frequencies [46]. As the deviation of the **g** tensor of the Na^+–NO complex from axial symmetry is small (Table 3), the analysis of the orientation-selective ^{23}Na ENDOR experiments benefits especially from the high spectral resolution at W-band (94 GHz) frequencies, since the g_{xx} spectral position can readily be distinguished from the g_{yy} region. This in turn allows one to unambiguously determine the principal values and the orientation of the principal axes of the sodium shf coupling tensor with respect to the **g** tensor principal-axis frame. The superior orientationselectivity of the W-band experiment is nicely demonstrated by two-pulse field-swept electron spin echo (ESE) experiments. The W- and X-band field-swept ESE spectra of the Na^+–NO complexes in NaA together with calculated EPR resonance fields in the **g** tensor xz and xy planes as a function of the angle θ between the external magnetic field and the z axis of the **g** tensor, are illustrated in Fig. 12. Such θ-versus-resonance-field plots are very useful for the selection of proper field positions for orientation-selective ENDOR measurements.

A series of orientation-selective W- and X-band pulsed ENDOR spectra of the Na^+–NO adsorption complex in zeolite NaA is illustrated in Fig. 13. The various magnetic field-positions where the ENDOR spectra were taken are indicated by arrows in Fig. 12. All W-band spectra (Fig. 13a) reveal two ^{23}Na ENDOR signals from the $M_S = \pm 1/2$ electron spin manifolds with an average spacing of about 10 MHz, which are symmetrically situated with respect to the sodium nuclear Larmor frequency ν_{Na} (weak-coupling case, $A^{Na} < 2\nu_{Na}$). Furthermore, ^{27}Al ENDOR signals from weakly coupled distant aluminum nuclei can be observed. In the spectra recorded near the g_{zz} spectral region of the EPR powder pattern (spectra a–d), additional shoulders are observed at each sodium ENDOR line. We assume that these shoulders are caused by the nq interaction of the ^{23}Na nucleus, which has a nuclear spin $I^{Na} = 3/2$. This interpretation is supported by the splitting of the sodium ENDOR signals into three lines observed for the spectrum taken at the g_{xx} position

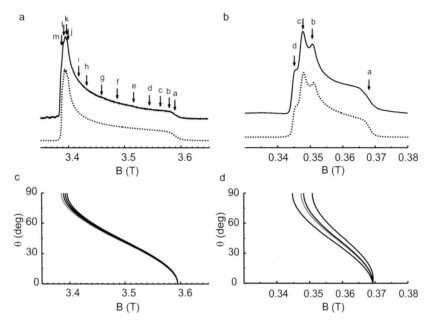

Fig. 12. W- and X-band two-pulse field-swept ESE spectra of the Na$^+$–NO adsorption complex in zeolite NaA: experimental (solid line) and simulated (dashed line) (a) W- and (b) X-band spectra. Calculated magnetic resonance fields in the **g** tensor xz (dashed line) and yz (solid lines) plane for (c) W- and (d) X-band [46]

at the low-field edge of the EPR powder spectrum (spectrum m), although baseline distortions somewhat obscure the nq triplets. The X-band ENDOR spectra (Fig. 13b) show only ^{23}Na ENDOR signals from one M_S state in the frequency range between 7 and 10 MHz. The ^{23}Na ENDOR signals from the second M_S state are expected to appear at frequencies below 2 MHz (intermediate-coupling case, $A^{\mathrm{Na}} \approx 2\nu_{\mathrm{Na}}$), a spectral range which is not accessible to our experimental setup. The two weak signals at about 12 MHz and 14 MHz (spectra a–c) are due to the nitrogen central-nucleus hf interaction in the Na$^+$–NO complex. The additional ENDOR signal at about 4 MHz is again assigned to weakly coupled distant ^{27}Al nuclei. On the basis of the spectral simulation of the orientation-selective ENDOR spectra presented, we obtain an axially symmetric sodium shf coupling tensor $\mathbf{A}^{\mathrm{Na}}$ with $A^{\mathrm{Na}}_{xx} = A^{\mathrm{Na}}_{yy} = 6.3$ MHz, $A^{\mathrm{Na}}_{zz} = 10.9$ MHz, $\alpha = 90°$, and $\beta = 35°$ [46]. The data translate into isotropic and dipolar ^{23}Na shf couplings of $A^{\mathrm{Na}}_{\mathrm{iso}} = 7.8$ MHz and $B^{\mathrm{Na}}_{xx/yy} = -1.55$ MHz. A later inclusion of the ^{23}Na nq interaction in the spectral analysis (not shown) yields the sodium nq parameters $Q^{\mathrm{Na}}_{xx} = -0.41$ MHz, $Q^{\mathrm{Na}}_{yy} = -0.07$ MHz, $Q^{\mathrm{Na}}_{zz} = 0.48$ MHz, $\alpha' = 90°$, $\beta' = 35°$, and $\gamma' = -90°$ [46]. The ^{14}N ENDOR signals observed in the X-band allowed us to determine the principal values $A^{\mathrm{N}}_{xx} = 25.3$ MHz and $A^{\mathrm{N}}_{zz} = 26.3$ MHz of the nitrogen hf interaction tensor, which are not accessible by cw EPR

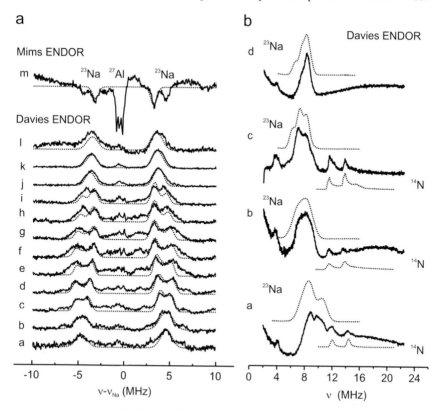

Fig. 13. Experimental (solid lines) and simulated (dashed lines) orientation-selective (a) W- and (b) X-band pulsed ENDOR spectra of the Na^+–NO adsorption complex in zeolite NaA. The simulations took into account only the ^{23}Na shf and ^{14}N hf couplings. The ^{23}Na and ^{14}N nq interactions were neglected [46]

spectroscopy (Table 3). Together with the principal value $A^N_{yy} = 92$ MHz determined by EPR, A^N_{xx} and A^N_{yy} provide $A^N_{iso} = 47.7$ MHz and $B^N_{xx/yy} \approx -21.9$ MHz.

The sodium and nitrogen hf coupling data obtained allow us to specify the distribution of the unpaired spin density over the Na^+–NO adsorption complex. We determine $\rho^{Na}_{3s} = 0.009$, $\rho^N_{2s} = 0.031$, $\rho^N_{2p\pi} = 0.458$, and $\rho^O_{2p\pi} = 0.502$ because $\rho^{Na}_{3s} + \rho^N_{2s} + \rho^N_{2p\pi} + \rho^O_{2p\pi} = 1$. The wave function of the unpaired electron is then given by

$$\psi = (0.46)^{1/2}\psi_{2p\pi}(N) + (0.03)^{1/2}\psi_{2s}(N)$$
$$+ (0.50)^{1/2}\psi_{2p\pi}(O) + (0.01)^{1/2}\psi_{3s}(Na). \qquad (13)$$

The spin density distribution gives information about the change in electronic structure of the NO molecule upon adsorption at the cation. In analogy to the DTBN molecule, the electronic structure of NO is composed of a mixture of two resonance structures [24],

$$\text{I} \qquad\qquad\qquad \text{II}$$
$$: \dot{N} = \ddot{O} : \qquad \longleftrightarrow \qquad : \ddot{N}^- = \dot{O} :^+ \qquad\qquad (14)$$

The unpaired spin densities $\rho_{2p\pi}^N$, $\rho_{2p\pi}^O$ measure the contributions of the resonance structures I and II, respectively, in (14). If we compare the $\rho_{2p\pi}^N$ and $\rho_{2p\pi}^O$ values of the Na$^+$–NO adsorption complex with those obtained for free NO molecules $\rho_{2p\pi}^N =$ 0.65, $\rho_{2p\pi}^O = 0.35$ [24], we recognize an enhanced contribution of structure II to the total electronic structure of the NO molecule in the adsorption complex, indicated by a shift of the unpaired spin density in the $2p\pi_y^*$ MO towards the oxygen atom. This shift is caused by the electron pair acceptor or Lewis acid property of the sodium cation, resulting in a net negative charge at the nitrogen atom in the adsorption complex (Na$^+$– $: \ddot{N}^- = \dot{O} :^+$). As a result, the redistribution of the unpaired spin density in the NO probe molecule upon complex formation with a Lewis acid surface site offers again the opportunity to characterize the electron pair acceptor strength of the adsorption site by the nitrogen hf coupling, in analogy to adsorbed nitroxide radicals. The proposed electronic structure of the NO molecule in the adsorption complex indicates a substantial ionic bond contribution to the metal ion – NO bond and is supported by the small unpaired spin density at the sodium ion. Covalent bond contributions are expected to be small, which implies a weak bonding of the NO to the Na$^+$ cation. Actually, the NO probes are only physisorbed at the cations, as shown below.

A bent coordination geometry of the Na$^+$–NO complex can be deduced from the orientation of the $\mathbf{A}^{Na}$ shf coupling tensor with respect to the $\mathbf{g}$ tensor coordinate frame, as shown in Fig. 14. The $2p\pi_y^*$ MO with the unpaired electron spin lies within the Na$^+$–NO complex plane. Note that the z-axis of the $\mathbf{g}$ tensor is aligned parallel to the N–O bond direction and the y-axes of $\mathbf{g}$ and $\mathbf{A}^N$ tensors point along the symmetry axis of the $2p\pi_y^*$ MO [17]. Using a point dipole approximation, we calculate a bond length $r_{N-Na^+} = 0.21$ nm and a bond angle of $\beta_{bond} = 142°$ from the ^{23}Na dipolar shf coupling tensor $\mathbf{B}^{Na}$ [46]. On the basis of a comparison of the measured ^{23}Na nq parameters of the Na$^+$–NO adsorption complex with those of sodium cations in dehydrated NaA zeolites obtained by NMR [64], it has been concluded that the NO adsorption complex is formed at cation site S2 (Fig. 2b) in the NaA zeolite [46].

In the case of Na$^+$–NO and Al–NO adsorption complexes in zeolites NaZSM-5 and HZSM-5, the direct coordination of the NO probe molecules could be established by Davies ENDOR experiments in the X-band (Fig. 15). It is worth noting that for the Al–NO complex, the ^{27}Al ENDOR spectrum verified the former assignment of the additional splitting in the EPR spectrum at the g_{yy} spectral position to a ^{27}Al shf coupling [63].

4.2 Electron Pair Acceptor Properties of Adsorption Sites Characterized by the Desorption Behavior of NO Probes

NO adsorption complexes are not stable in zeolites at higher temperatures because nitric oxide desorbs at $T \geq 100$ K. As outlined in Sect. 1, the study of the desorption

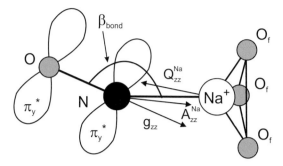

Fig. 14. A schematic drawing of the bent structure of the Na^+–NO adsorption complex in zeolite NaA. The z principal axes of the $\mathbf{g}$, $\mathbf{A}^{Na}$, and $\mathbf{Q}^{Na}$ tensors lie within the Na–N–O complex plane and form angles with the Na^+–NO bond direction of 38°, 3°, and 8°, respectively. The cation at site S2 is coordinated to the framework oxygens, O_f, in the six-membered rings in a trigonal symmetry. Only the three oxygens in the first coordination sphere are shown [46]

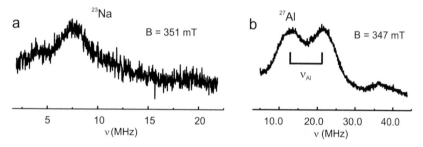

Fig. 15. X-band Davies ENDOR spectra of (a) Na^+-NO and (b) Al-NO adsorption com- plexes in zeolites NaZSM-5 and H-ZSM-5.

process of the basic NO probe molecules from their adsorption sites offers the opportunity to characterize the electron pair acceptor strength of Lewis acid surface sites. There are two possibilities for studying this desorption process by EPR, either by measurement of the adsorbed state or by monitoring the paramagnetic $^2\Pi_{3/2}$ molecular state of the desorbed NO molecules in the gas phase [25,32,33]. As an example, Fig. 16 illustrates the temperature dependences obtained for the EPR intensity of the adsorbed state of NO at $g \approx 2$ and of the rotational ground state ($J = 3/2$) of the $^2\Pi_{3/2}$ state of the desorbed molecules with its characteristic nine-line EPR pattern at $g \approx 0.8$ [23], for zeolite NaA. The intensities of the EPR signal of the adsorbed state are found at the low-temperature side and were obtained by double integration of its first-derivative EPR signal. The high-temperature side shows the temperature dependence of the EPR intensity of the desorbed NO molecules, which is normalized with respect to the intensity obtained at $T = 300$ K. To obtain relative concentrations c_{NO} of desorbed NO molecules, the doubly integrated intensities of the gas phase signals have been normalized with respect to their intensity at $T =$

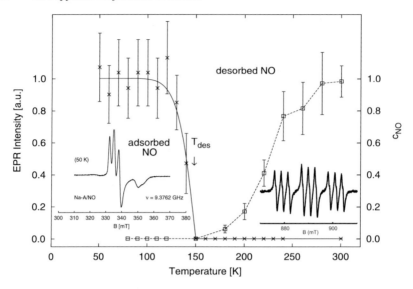

Fig. 16. Adsorption and desorption behavior of NO on NaA studied in the X-band with relative EPR intensities of the adsorbed state (crosses) and relative concentrations c_{NO} of desorbed NO molecules (squares). The inserts show the X-band EPR signal of the adsorbed (Na^+–NO complex) and desorbed ($^2\Pi_{3/2}$ state) NO molecules [33]

300 K and weighted by the temperature-dependent population of the $^2\Pi_{3/2}$ state of NO [33]. The EPR intensities of the NO adsorption complexes decrease continuously with rising temperature for 120 K $\leq T \leq$ 140 K and disappear at $T \geq$ 140 K. It should be noted that the relative EPR intensities provide only a rough measure of the concentration of the NO adsorption complexes, since they also depend on the temperature-dependent paramagnetic susceptibility and quality factor of the cavity. But the disappearance of the EPR signals at higher temperatures is a strong indication of a substantial decrease in the concentration of the Na^+–NO adsorption complexes. At slightly higher temperatures $T >$ 150 K, desorbed molecules appear in the gas phase above the zeolite material and their relative concentration continuously increases with rising temperature. Obviously, the temperature-dependent EPR experiments monitor the desorption of the NO probe molecules from their cation adsorption sites into the gas phase. According to Fig. 16, we can define a characteristic desorption temperature T_{des} from the highest temperature where gas phase EPR signals cannot be detected. The experimentally found desorption temperatures for NO adsorbed on NaA, NaZSM-5, and H-ZSM-5 are given in Table 4. It is not surprising that zeolite HZSM-5, with its aluminum defect centers known to act as strong true Lewis acid sites, yields the highest T_{des} value.

At this stage of the analysis, the EPR experiment yields the same information as conventional temperature-programmed desorption (TPD) methods. But alternatively, it offers, moreover, the opportunity to determine microscopically the energy E_{des} of the desorption of the NO probe molecules from their adsorption sites. Here the key

Table 4. Desorption temperatures T_{des}, desorption energies E_{des}, and nitrogen hf couplings A_{yy}^{N} of NO adsorption complexes [33]

Zeolite	Adsorption complex	T_{des} (K)	E_{des} (kJ/mol)	A_{yy}^{N} (MHz)
NaA	Na^+–NO	150	7.1	92
NaZSM-5	Na^+–NO	190	4.1	102
H-ZSM-5	Al–NO	240	20.2	84

information lies in the temperature-dependent line shape of the EPR signals of the NO adsorption complexes at $T < T_{des}$ [25,33]. The decrease in the EPR signal intensity of the NO adsorption complexes with proceeding desorption is accompanied by a strong increase in the homogeneous linewidth δB^{hom} of their EPR signal, which measures the reciprocal lifetime of the adsorbed state. The homogeneous linewidth follows an Arrhenius behavior

$$\delta B^{hom}(T) = b_1 \exp -\frac{E_{des}}{k_B T}, \tag{15}$$

where the preexponential factor is b_1 and the Boltzmann constant is k_B, which can be used to evaluate an activation energy E_{des} from the temperature dependence of $\delta B^{hom}(T)$. Both the homogeneous line broadening of the EPR signal of the adsorbed NO and the Arrhenius behavior of the homogeneous linewidth are illustrated in Fig. 17 for the example of the Na^+–NO complex in NaA zeolites.

Above a certain energy E_{des}, the dissociation probability is appreciable. One expects a dissociation rate proportional to $\exp(-E_{des}/k_B T)$ which is involved as

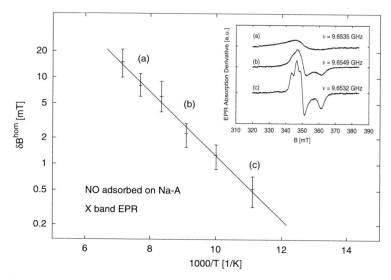

Fig. 17. Arrhenius plot of the homogeneous EPR linewidth δB^{hom} of adsorbed NO on NaA. The insert shows the corresponding X-band EPR spectra of the Na^+–NO adsorption complex at (a) $T = 130$ K, (b) $T = 110$ K, and (c) $T = 90$ K [33]

the Arrhenius factor [65]. The specific values of this energy E_{des} are characteristic of different adsorption sites in microporous materials and have the microscopic character of a desorption energy per NO adsorption complex in molar units. The desorption energies E_{des} determined for NO adsorbed at sodium cations and aluminum defect centers in zeolites NaA, NaZSM-5, and HZSM-5 are likewise summarized in Table 4. Again the highest desorption energy is found for the strong-Lewis-acid aluminum defect centers. The low desorption energies of $E_{des} <10$ kJ/mol obtained for the Na$^+$–NO centers indicate that the nitric oxide is only weakly physisorbed at the sodium cations, which have only a weak Lewis acidity in comparison with aluminum defect centers.

In the previous section we have shown that the electron pair acceptor properties of the Lewis acid sites lead to a redistribution of the electron density within the adsorbed nitric oxide molecules, where the unpaired spin density in the $2p\pi_y^*$ NO molecular orbital is shifted toward the oxygen atom. Consequently, the nitrogen hf coupling decreases with rising electron pair acceptor strength of the Lewis acid site. The nitrogen hf coupling data in Table 4 support this model. The resulting relation A_{yy}^N(HZSM-5)$<$ A_{yy}^N(NaA)$<$ A_{yy}^N(NaZSM-5) shows that the aluminum defect centers (true Lewis acid sites) in zeolite HZSM-5 possesses a higher electron pair acceptor strength or Lewis acidity than the sodium cations in the NaA and NaZSM-5 in accordance with the determined desorption energies E_{des}. We should note that the nitrogen hf coupling paramters A_{yy}^N can easily be deduced from the low temperature EPR spectra and in that way provide a simple tool to characterize the electron pair acceptor strength of the adsorption site.

If we compare the parameters obtained in Table 4, it seems to be surprising that a lower desorption temperature T_{des} was found for NaA than for NaZSM-5, because E_{des}(NaA)$>$ E_{des}(NaZSM-5). This indicates that the local desorption behavior of the probe molecules at a specific Lewis acid site may deviate from the overall adsorption/desorption properties of the zeolite. Therefore, if the acidity of different Lewis sites is characterized by means of the measurement of the desorption temperatures the same zeolite host materials should be used. Otherwise, the desorption energies or nitrogen hf couplings should be determined for the NO adsorption complexes formed because these methods probe microscopically the actual local properties at the adsorption site.

5 Conclusions

EPR spectroscopy of adsorbed paramagnetic probe molecules is a versatile tool to characterize the electron pair acceptor properties of surface sites in microporous materials and thereby opens up an opportunity to study the acidity of Lewis sites in these materials. The advantage of the EPR method is that it allows a microscopic characterization of the structure of the NO adsorption complex formed and of the nature of the specific Lewis acid adsorption sites. ESEEM or pulsed ENDOR spectroscopy can be applied to prove unambiguously the direct coordination of the probe molecules with the surface site of interest.

In our studies we employed two paramagnetic probe molecules with Lewis basic properties, NO and DTBN. Information about the electron pair acceptor strength of Lewis acid adsorption sites is provided by either the spin Hamiltonian parameters of the adsorption complexes formed or the study of desorption processes of the paramagnetic probe molecules. Nitric oxide has the advantage that, because of its small size, most of the catalytically relevant surface sites are accessible to this probe molecule. Three alternative methods can be suggested to explore the acidity of the adsorption sites by NO probes: (i) determination of the desorption temperature T_{des} by EPR spectroscopy of the paramagnetic $^2\Pi_{3/2}$ molecular state of desorbed NO at $T > 110$ K, (ii) evaluation of the desorption energy E_{des} of NO from the temperature-dependent homogeneous line broadening of the EPR signal of the NO adsorption complex in the range 60 K $< T < 180$ K, and (iii) measurement of the principal value A_{yy}^{N} of the nitrogen hf coupling tensor $\mathbf{A}^{\mathrm{N}}$ by EPR at low temperatures as a measure of the electron pair acceptor strength of the adsorption site. It is worth noting that A_{yy}^{N} is taken as a measure of the spin density $\rho_{2p\pi}^{\mathrm{N}}$ in the $2p\pi$ MO of NO at the nitrogen atom. Therefore, an increasing Lewis acidity results in a decreasing principal value A_{yy}^{N}. The last two approaches yield information about the electron pair acceptor strength of the Lewis acid sites in a true microscopic sense.

Although the application of the nitroxide radical DTBN is restricted to zeolites with large pore openings, it offers, in contrast with nitric oxide, the unique opportunity to compare directly the electron pair acceptor strengths of adsorption sites on solid surfaces with those of well-defined Lewis acids in solutions. Here again, the key parameter is the nitrogen hf coupling, which measures the spin density $\rho_{2p\pi}^{\mathrm{N}}$. However, the DTBN coordinates via the oxygen atom of its NO group to the Lewis acid site. As a result, the nitrogen hf coupling increases with the Lewis acidity of the adsorption sites.

Acknowledgments

This research was supported by the priority program "Hochfeld EPR-Spektroskopie" and the Fonds der Chemischen Industrie. The authors are gratefully indebted to Prof. D. Goldfarb and Dr. P. Manikandan, Weizmann Institute of Science, Rehovot, for help with the W-band ENDOR measurements and to Prof. R. Böttcher and Prof. D. Michel, Universität Leipzig, for helpful discussions.

References

1. R.P. Townsend: *Properties and Applications of Zeolites* (Society of Chemical Industry, London, 1980)
2. J. Weitkamp: "New directions in zeolite catalysis", in *Catalysis and Adsorption by Zeolites*, ed. by G. Ohlmann, H. Pfeifer, R. Fricke (Elsevier, Amsterdam, 1991), p. 21
3. N. Herron, D.R. Corbin: *Inclusion Chemistry in Zeolites: Nanoscale Materials by Design* (Kluwer Academic, Dordrecht, 1995)

4. E. Brunner, H.G. Karge, H. Pfeifer: Z. Phys. Chem. **176**, 173 (1992)
5. H. Pfeifer: NMR Basic Principles and Progress **31**, 31 (1994)
6. G. Ertl, H. Knözinger, J. Weitkamp: *Handbook of Heterogeneous Catalysis* (VCH, Weinheim, 1997)
7. G.H. Kühl: J. Phys. Chem. Solids **38**, 1259 (1977)
8. H.G. Karge: Stud. Surf. Sci. Catal. **65**, 133 (1991)
9. A.H. Cohen, B.M. Hoffman: J. Am. Chem. Soc. **95**, 2061 (1973)
10. A.H. Cohen, B.M. Hoffman: Inorg. Chem. **13**, 1484 (1973)
11. M. Gutjahr, A. Pöppl, W. Böhlmann, R. Böttcher: Colloids Surf. A: Physiochem. Eng. Aspects **189**, 93 (2001)
12. G.P. Lozos, B.M. Hoffman: J. Phys. Chem. **2**, 2110 (1974)
13. H. Ulbricht, P. Köhler: Z. Chem. **25**, 253 (1985)
14. E.V. Lunina, G.L. Markaryan, A.V. Fionov, A.V. Astashin, R.I. Samoilova, M.T. Zdravkova: Appl. Magn. Res. **2**, 675 (1991)
15. E.V. Lunina: Appl. Spec. **50**, 1413 (1996)
16. U.J. Katter, T. Hill, T. Risse, H. Schlienz, M. Beckendorf, T. Klijner, M. Hamann, H.-J. Freud: J. Phys. Chem. **101**, 552 (1997)
17. P.H. Kasai, R.J. Bishop, Jr.: "Electron Spin resonance studies of zeolites" in: *Zeolite Chemistry and Catalysis*, ed. by J.A. Rabo, ACS Monograph 171 (American Chemical Society, Washington, DC, 1976) pp. 351–391
18. J.H. Lunsford: J. Phys. Chem. **72**, 4163 (1968)
19. B.M. Hoffman, N.J. Nelson: J. Chem. Phys. **50**, 2598 (1969)
20. D. Biglino, H. Li, R. Erickson, A. Lund, H. Yahiro, M. Shiotani: Phys. Chem. Chem. Phys. **1**, 2887 (1999).
21. H. Yahiro, A. Lund, R. Aasa, N.P. Benetics, M. Shiotani: J. Phys. Chem. A **104**, 7950 (2000)
22. T. Rudolf, A. Pöppl, W. Hofbauer, D. Michel: Phys. Chem. Chem. Phys. **3**, 2176 (2001)
23. R. Beringer, J.G. Castle, Jr.: Phys. Rev. **78**, 581 (1950)
24. G.C. Dousmanis: Phys. Rev. **97**, 967 (1955)
25. T. Rudolf, A. Pöppl, W. Brunner, D. Michel: Magn. Reson. Chem. **77**, 93 (1999)
26. J.H. Lunsford: J. Chem. Phys. **46**, 4347 (1967)
27. J.H. Lunsford: J. Phys. Chem. **72**, 2141 (1968)
28. A. Gutsze, M. Plato, H.G. Karge, F. Witzel: J. Chem. Soc. Faraday Trans. **92**, 2495 (1996)
29. C.L. Gardner, W.A. Weinberger: Can. J. Chem. **48**, 1317 (1970)
30. Z. Sojka, M. Che, E. Giamello: J. Phys. Chem. **101**, 4831 (1997)
31. K. Neyman, D.I. Ganyushin, V.A. Nasluzov, R. Rösch A. Pöppl, M. Hartmann: Phys. Chem. Chem. Phys. **5**, 2429 (2003)
32. F. Witzel, H.G. Karge, A. Gutsze: "ESR measurements for the characterization of acidic Lewis sites in zeolites", in *Proceedings of the 9th International Zeolite Conference*, ed. by R. von Ballmoos et al. (Butterworth-Heinemann, Montreal, 1993), pp. 283–295
33. T. Rudolf, A. Pöppl: J. Magn. Reson. **155**, 45 (2002)
34. L. Kevan, L. Kispert: *Electron Spin Double Resonance Spectroscopy* (Wiley, New York, 1976)
35. A. Schweiger: *Structure and Bonding 51* (Springer, Berlin, Heidelberg, 1982)
36. A. Schweiger, G. Jeschke: *Principles of Pulse Electron Paramagnetic Resonance* (Oxford University Press, Oxford, 2001)
37. L. Kevan: "Developments in electron spin echo modulation analysis", in *Modern Pulsed and Continuous-Wave Electron Spin Resonance*, ed. by L. Kevan, M.K. Bowman (Wiley, New York, 1990), pp. 43–118

38. S.A. Dikanov, Yu.D. Tsvetkov: *Electron Spin Modulation (ESEEM) Spectroscopy*, (CRC Press, Boca Raton 1992)
39. E.R. Davies: Phys. Lett. A **47**, 1 (1974)
40. P. Höfer, A. Grupp, H. Nebenführ, M. Mehring: Chem. Phys. Lett. **132**, 279 (1986)
41. W.B. Mims: Phys. Rev. B **6**, 3543 (1972)
42. G.C. Hurst, T.A. Henderson, R.W. Kreilick: J. Am. Chem. Soc. **107**, 7294 (1985)
43. B.M. Hoffman, J. Martinsen, R.A. Venters: J. Magn. Reson. **59**, 110 (1984)
44. E.J. Reijerse:"Electron Spin Echo Spectroscopy on Transition Metal Compounds", Ph.D. Thesis, Katholieke Universiteit Nijmegen (1986)
45. A. Schweiger, H. Günthard: Chem. Phys. **70**, 1 (1982)
46. A. Pöppl, T. Rudolf, P. Manikandan, D. Goldfarb: J. Am. Chem. Soc. **122**, 10194 (2000)
47. A. Pöppl, R. Böttcher: Chem. Phys. **221**, 53 (1997)
48. P.J. Carl, D.E.W. Vaughan, D. Goldfarb: J. Phys. Chem. B **106**, 5428 (2002)
49. M. Gutjahr, R. Böttcher, A. Pöppl: Appl. Magn. Reson. **22**, 401 (2002)
50. A.M. Tyryshkin, S.A. Dikanov, E.J. Reijerse, C. Burgard, J. Hüttermann: J. Am. Chem. Soc. **121**, 3396 (1999)
51. P. Höfer: J. Magn. Reson. A **111**, 77 (1994)
52. S.A. Dikanov, M.K. Bowman: J. Magn. Reson. A **116**, 125 (1995)
53. A. Pöppl, L. Kevan: J. Phys. Chem. B **100**, 3387 (1996)
54. A. Pöppl, M. Hartmann, W. Böhlmann, R. Böttcher: J. Phys. Chem. A **102**, 3559 (1998)
55. A. Pöppl, M. Gutjahr, M. Hartmann, W. Böhlmann, R. Böttcher: J. Phys. Chem. A **102**, 7752 (1998)
56. E.J. Reijerse, J.J. Shane, E. de Boer, P. Höfer, D. Collison: "One and two dimensional ESEEM of nitrogen coordinated oxo-vanadium(IV) complexes", in *Electron Paramagnetic Resonance of Disordered Systems*, ed. by N.D. Yordanov (World Scientific, Singapore, 1991), pp. 253–271
57. M. Gutjahr, R. Böttcher, A. Pöppl: J. Phys. Chem. B **106**, 1345 (2002)
58. M. Gutjahr, R. Böttcher, A. Pöppl: J. Phys. Chem. B **107**, 13117 (2003)
59. D. Forano, R.C.T. Slade, E.K. Anderson, I.G.K. Anderson, E. Prince: J. Solid State Chem. **82**, 95 (1989)
60. R. Jelinek, A. Malek, G.A. Ozin: J. Phys. Chem. **99**, 9236 (1995)
61. A.H. Cohen, B.M. Hoffman: J. Phys. Chem. **78**, 1313 (1974)
62. M.J. Mortier: J. Catal. **55**, 138 (1978)
63. A. Pöppl, T. Rudolf, D. Michel: J. Am. Chem. Soc. **120**, 4879 (1998)
64. G.A.H. Tijink, R. Janssen, W.S. Veeman: J. Am. Chem. Soc. **109**, 7301 (1987)
65. N.G. van Kampen: *Stochastic Processes in Physics and Chemistry* (North-Holland, Amsterdam 1992)

Study of Conformation and Dynamics of Molecules Adsorbed in Zeolites by ^{1}H NMR

Dieter Michel[1], Winfried Böhlmann[2], Jörg Roland[3], and Samir Mulla-Osman[4]

[1] Universität Leipzig, Institut für Experimentalphysik II,
   michel@physik.uni-leipzig.de
[2] Universität Leipzig, Institut für Experimentalphysik II,
   bohlmann@physik.uni-leipzig.de
[3] Universität Leipzig, Institut für Experimentalphysik II,
   joerg.roland@siemens.com
[4] Universität Leipzig, Institut für Experimentalphysik II,
   samir@physik.uni-leipzig.de

Abstract. The chapter "Study of Conformation and Dynamics of Molecules Adsorbed in Zeolites by ^{1}H NMR" is concerned with the application of high-resolution (HR) solid-state NMR techniques to study the behavior of molecules adsorbed on surfaces of nanoporous solids, such as zeolitic molecular sieves. This includes a combined or alternative application of conventional high-resolution NMR methods and of high-resolution solid-state NMR techniques, including magic-angle sample spinning (MAS), cross-polarization (CP), high-power decoupling and appropriate multiple-pulse sequences for two- or higher dimensional NMR and multiple-quantum spectroscopy. The interaction of adsorbed molecules with adsorption centers in the internal surfaces of porous solids does not only lead to changes in the reorientational and translational mobility of the molecular species but influences also the molecular conformation. Examples will be given for simple olefins in interaction with inner zeolite surfaces. Conclusions about the correlation times of the internal reorientational and translational dynamics are derived in complete agreement with the conclusion obtained from diffusion coefficients by means of PFG NMR (second chapter). Since the methodical approach of HR MAS NMR in heterogeneous systems presented here is also valuable for the investigation of lyotropic crystalline phases using HR MAS NMR (in Chap. 12) And for the NMR studies of cartilage (in Chap. 13) it was also the aim of this chapter to elucidate also the methodical background of these measurements in some more detail.

1 NMR Spectroscopy in Heterogeneous Systems

1.1 General Introduction

Background and Aim

High-resolution nuclear magnetic resonance (NMR) spectroscopy has become an important tool for the investigation of structure and dynamics in many fields of physics and chemistry. Many studies focus on the investigation of molecules associated with interfaces, showing properties that belong neither to the liquid nor to the solid state. Typical examples are membranes [1, 2], peptides bound to a resin [3], and organic molecules sorbed in zeolites [4], in relation to the other work presented in this lecture note.

D. Michel, W. Böhlmann, J. Roland, S. Mulla-Osman, Study of Conformation and Dynamics of Molecules Adsorbed in Zeolites by ^{1}H NMR, Lect. Notes Phys. **634**, 217–274 (2004)
http://www.springerlink.com/

The restricted mobility of these molecules and the occurrence of additional local magnetic fields lead to broad resonance lines which limit, especially, the applicability of proton NMR studies. Since the details of the proton resonance spectra due to chemical shifts and indirect spin couplings are obscured by these line broadenings in many cases, only the overall longitudinal and transverse relaxation times (or linewidths) are available for a study of adsorbate–adsorbent interactions in most applications using the techniques of broad-line proton NMR spectroscopy. For a relatively long period these measurements provided the main source of information about the state and dynamics of adsorbed molecules. The applications of the above-mentioned NMR methods to the study of the mobility of adsorbed species were reviewed from both a theoretical and a practical point of view by Packer [5] and Pfeifer [6], and with regard to heterogeneous catalytic systems by Resing [7] and Derouane et al. [8]. Of particular interest was the systematic analysis of nuclear spin relaxation [6, 9, 10] for studying subtle details of the reorientational and translational motion of molecules adsorbed on solid interfaces. Since conventional relaxation time studies are outside of the scope of this contribution, they are included in the following sections only as far as they are related to high-resolution NMR measurements. In 1972, the first highly resolved ^{13}C NMR spectra of adsorbed molecules could be measured [9]. Later, also ^{15}N NMR measurements were carried out for molecules adsorbed on solid interfaces [10–12]. As is well known, the development of these methods is closely related to the application of Fourier transform NMR spectroscopy [12] and to the possibility of studying less abundant nuclei such as ^{13}C and ^{15}N with large ranges of chemical shifts in order to circumvent at least partially the problems in spectral resolution mentioned.

With these experiments, it became possible to study in detail the change in the electronic state of the molecules due to their interaction with the surface. Moreover, from the ^{13}C relaxation times and the nuclear Overhauser factors, additional information about the geometry and the mobility of the adsorbed molecules can be derived. The rapid increase in the number of papers in this field was related to the possibility of identifying molecular species and of studying, for instance, reacting adsorbates and examining product formation and reaction kinetics. These studies consequently also permit the identification of the surface-active sites in the presence of adsorbed molecules, and the investigation of catalytic transformations. The application of high-resolution ^{13}C and ^{15}N NMR spectroscopy of adsorbed molecules has been reviewed by several authors (see [13–19]). In particular, questions of sensitivity and resolution important in the study of adsorbed species were addressed in more detail in [16] and will therefore not be treated in detail here.

In order to receive highly resolved spectra for molecules adsorbed on solid surfaces, a combined or alternative application of conventional high-resolution NMR methods and of high-resolution solid-state NMR techniques is necessary, including magic-angle spinning (MAS) of the sample, cross-polarization (CP), high-power decoupling, and appropriate multiple-pulse sequences for two- or higher-dimensional NMR and multiple-quantum spectroscopy. The first application of solid-state ^{13}C CP NMR spectroscopy to adsorbed species was published by Kaplan et al. [20]. A broader application of this technique was achieved by combining it with the technique of magic-angle spinning [21]. Valuable information can be obtained about chemically bound surface species and about strongly adsorbed species in which the molecular motion is greatly reduced. Important examples are hydrocarbon molecules which are trapped inside the zeolite channels as a result of pore plugging, and carbon-containing residues fixed in zeolitic structures after catalytic reactions. A comprehensive survey of the various experimental techniques available in multinuclear solid-state NMR, of the type of physical and chemical information retrievable from the spectra and of a wide range of applications of solid-state NMR in the various branches of chemistry is given in the book by Fyfe [22].

Scope and Organization

In spite of the suitable measuring conditions given by the good sensitivity, high-resolution ^{1}H NMR spectroscopy of adsorbed molecules has seldom been used successfully. For rigid systems, the primary reason for the lack of good resolution is the presence of homonuclear dipolar interaction, which usually dominates the NMR linewidth. The degree of restriction of the mobility of adsorbed molecules, however, is in many cases comparable to the internal thermal mobility occurring for macromolecular systems of biological origin such as proteins. In the latter case two-dimensional (2D) ^{1}H NMR spectroscopy, typical of liquid systems, and other sophisticated techniques have been applied with great success [23]. In contrast to the apparently sufficient magnitude of the thermal mobility, the modern 2D ^{1}H NMR techniques have been applied only very seldom to molecules adsorbed on solid interfaces or similar heterogeneous systems [24]. The reason for this situation is the limited resolution in the ^{1}H NMR spectra of the adsorbed molecules. Owing to the still sufficient mobility of adsorbed molecules in many cases, the limitation arises mainly from interactions with paramagnetic impurities and from magnetic-susceptibility effects in polycrystalline adsorbent materials. In our previous work [4], it could be demonstrated that a better spectral resolution in the proton NMR spectra of adsorbed molecules may be achieved by using the magic-angle spinning technique in order to reduce the influence of the local magnetic fields on the linewidths due to magnetic-susceptibility effects.

Therefore, it seems necessary to discuss the possibilities of and limitations on improving the spectral resolution in ^{1}H MAS NMR spectroscopy in heterogeneous systems (section 1.2). Since the adsorbed molecules, for instance in zeolitic molecular sieves, possess a restricted but still appreciable mobility as mentioned above, the combined influence of magic-angle spinning and thermal motion will be treated in detail in this section. In particular, it will be shown that for the heterogeneous systems considered also, high-field ^{1}H MAS NMR spectroscopy leads to an essential improvement in the resolution of chemical shifts. On this basis it becomes possible to study subtle details in the change of the ^{1}H NMR chemical shift of the molecules upon adsorption. In Sect. 2 it is shown for simple olefins adsorbed in NaX zeolites that this advantageous situation allows one to derive conclusions about changes in the conformation of the molecules in the course of the adsorption. It is well known from previous work [16] on ^{13}C and ^{15}N high-resolution NMR studies that typical changes in the chemical shifts of adsorbed molecules occur with respect to molecules in the liquid or gaseous state, which also play an important role in the ^{1}H MAS NMR studies presented here. Since the chemical shift may change with temperature and pressure as a result of intermolecular effects such as hydrogen bonds, van der Waals bonds, local electric and magnetic fields, and particle exchange, ^{1}H NMR spectra should be even more sensitive to adsorption interactions. Since the changes in the chemical shifts measured for nuclei in adsorbate molecules may be of the same order of magnitude as the influence of the bulk susceptibility of the sample under study and the contribution of the van der Waals interactions, a central problem is to determine the specific contribution to the shifts which is due to adsorption interactions. With respect to the necessary shielding correction of the chemical shift, we refer here to a previous monograph [16]. A further important problem is the averaging of NMR spectra by exchange processes between physisorbed molecules and molecules bonded at the adsorption sites.

A main object of this contribution is to show how the methods of two-dimensional (2D) ^{1}H MAS NMR spectroscopy and related techniques can be used for the study of molecule–surface interactions and to analyze structural changes in the course of the adsorption. Examples are given for allyl alcohol and 1-butene sorbed in NaX zeolite. In Sect. 4 it will be shown that homonuclear cross-relaxation studies using ^{1}H NOESY NMR can be used not only to derive distance information but also to study the thermal mobility and to contribute to a deeper understanding of the dynamics of molecules sorbed in zeolites with a definite internal structure. Since it has been shown (section 3.2) that homonuclear, ^{1}H–^{1}H cross-relaxation is mainly sensitive to the slower processes in the restricted motion, besides ^{1}H NOESY NMR experiments (section 4.1) also the determination of ^{1}H–^{13}C cross-relaxation rates (section 4.2) and the estimation of the relaxation rate differences of protons bonded to ^{12}C and ^{13}C nuclei (section 4.3) are necessary in order to enable a more detailed analysis of the reorientation and translational motion of molecules in zeolites. In this context, the role of higher spin orders in the cross-relaxation is discussed also (section 4.4). Finally, in Sect. 4.5 a model for the motion of adsorbed molecules is derived which contains two different correlation times in the spectral density function and which is formally similar to the model-free approach discussed by Lipari and Szabo. The correlation times are discussed in comparison with pulsed field gradient (PFG) NMR measurements (see Chap. 1) of the self-diffusion coefficients of the adsorbed molecules. This enables us to compare the correlation time for the slower motion with the mean residence time of the molecules in a large cavity (supercage) derived from the self-diffusion coefficients, and allows also an interpretation of the shorter correlation time in terms of a (faster but not completely isotropic) reorientation of the molecules within the supercages. In conclusion, the ^{1}H NMR studies performed allow conclusions about the change in the conformation of the molecules during interaction with adsorption sites and enable selective measurements of the reorientation and translation motions of the adsorbed molecules.

1.2 High-Resolution ^{1}H NMR of Adsorbed Molecules

Susceptibility Effects: Static Local Fields

It is well known that when a polycrystalline adsorbent, such as a zeolite, is placed in a strong magnetic field, the microcrystals produce a nonuniform local magnetic field that is roughly six orders of magnitude smaller. Consequently, the resultant magnetic field experienced by an adsorbed molecule depends on its position and on the orientation and distribution of crystallites, giving rise to susceptibility broadening of the NMR lines [25, 26].

Different methods have been proposed to alleviate this loss in resolution. Mank et al. [27] suggested a method that relies upon susceptibility matching, i.e. filling the intercrystalline space between the adsorbent grains with a liquid possessing a susceptibility close to that of the crystallites. A more suitable approach includes magic-angle spinning, as shown in [28, 29]. Local field components that show the transformation properties of traceless second-rank tensors under changes of sample orientation are averaged to zero by MAS in a way similar to the anisotropic magnetic shielding and dipolar couplings between nuclear spins. Although the effect of MAS on the local magnetic fields has been previously realized and discussed [26, 30, 31], a detailed analysis of the line narrowing that it produces is still outstanding. Vander Hart et al. [26] suggested that the magnetic-susceptibility broadening of NMR lines in polycrystalline organic solids can be completely removed by MAS in the case of a material with an isotropic susceptibility tensor and substantially reduced when the tensor is anisotropic. Garroway [30] indicated that MAS not only eliminates the broadening effect due to magnetic susceptibility but also the shift itself. On the basis of theoretical considerations, Alla and Lippmaa [31] found that the width of the susceptibility-broadened spectral lines scales under MAS by a factor determined by the anisotropy of the magnetic-susceptibility tensor.

In the following, a theoretical approach is presented [32] that allows a more detailed look at the local magnetic field distribution in a polycrystalline sample exposed to a strong external magnetic field. This approach leads to a classification of different contributions to the NMR linewidth caused by susceptibility broadening and a description of their behavior under MAS. The magnetic field gradients associated with the local fields can also be characterized within the same mathematical framework. Throughout the paper, we consider exclusively magnetic induction, but the term "magnetic field" is used synonymously.

Crystallites are described as bodies consisting of continuous matter. We first need to properly describe the shielding magnetic field produced by a single crystallite in a homogeneous external field B. This field depends on the material, as well as the shape and orientation of the crystallite. For typical crystallite shapes, e.g. cubes and tetrahedrons, only numerical solutions are available. However, analytical descriptions exist for a body of spherical shape with an arbitrary susceptibility tensor, and for an ellipsoidal body with an isotropic susceptibility [33, 34].

For both cases, the shielding field within the crystallite c is homogeneous and the magnetic induction can be described by the formula $\mathbf{B}_c = \frac{1}{3}\mathbf{X}_c\mathbf{B} = \frac{1}{3}\chi_c^{\mathrm{iso}}\chi_c\mathbf{B}$, where χ_c^{iso} denotes the isotropic value and χ_c is the geometric part of the tensor $\mathbf{X}_c$. For simplicity, we assume $\mathbf{X}_c$ to be axially symmetric, i.e. to be fully described by χ_c^{iso}, the anisotropy $\Delta\chi_c$, and the orientation. $\Delta\chi_c$ is defined by the representation of the principal-axis values of χ_c as $1 - \Delta\chi_c/3$, $1 - \Delta\chi_c/3$, and $1 + 2\Delta\chi_c/3$. For a spherical crystallite, $\mathbf{X}_c$ coincides with the susceptibility tensor of the material. In the case of an ellipsoid with an isotropic susceptibility, $\mathbf{X}_c$ describes demagnetization effects: χ_c can be written as $\chi_c = 3\mathbf{N}_c$, where $\mathbf{N}_c$ is a demagnetization tensor that is characterized by three demagnetization factors and for which $\mathrm{Tr}(\mathbf{N}_c) = 1$ [31]. Since the internal fields inside an ellipsoidal crystallite with an isotropic susceptibility and inside a spherical crystallite with an arbitrary susceptibility tensor differ only in their geometric part χ_c, the shape effect on the internal field can be incorporated into the following theory at any time using the appropriate χ_c. Outside the crystallite, the magnetic field is analogous to that produced by a magnetic dipole $\mu_c = (1/\mu_0)V_c\mathbf{X}_c\mathbf{B}$ placed at the center of the crystallite of volume V. For real crystallites, the above description serves as an approximation that is well justified. Any inhomogeneties of the internal field are primarily due to surface effects and cannot significantly affect the bulk of a crystallite. The external field of an arbitrarily shaped particle, when expanded in terms of a multipole series, yields a dipolar field as the first nonvanishing contribution. Higher-order approximations scale with higher powers of the inverse distance and can be easily neglected.

Hence, for the local field inside a crystallite, we assume

$$\mathbf{B}_c^{in} = \frac{1}{3}\mathbf{X}_c\mathbf{B}. \tag{1}$$

The field $\mathbf{B}_c$ at a position $\mathbf{r}$ outside of the crystallite is given by

$$\mathbf{B}_c^{out}(\mathbf{r}) = \frac{\mu_0}{4\pi}\frac{1}{r_c^3}\left[\mu_c - 3\left(\mu_c\mathbf{n}_c\right)\mathbf{n}_c\right] \tag{2}$$

where

$$\mu_c = \frac{1}{\mu_0}V_c\mathbf{X}_c\mathbf{B}, \quad \mathbf{r}_c = r_c\mathbf{n}_c.$$

The vector $\mathbf{r}_c$ connects $\mathbf{r}$ with the position $\mathbf{R}_c$ of the center of crystallite c, i.e. $\mathbf{r}_c = \mathbf{R}_c - \mathbf{r}$. Later, the index c will be used to distinguish between different crystallites. In further calculations, we represent the external field $\mathbf{B}$ by its magnitude B and by a direction $\mathbf{b}$ that is assumed to coincide with the z-axis of the laboratory frame. For NMR considerations, we are interested only in the component b_c of the local magnetic field that is parallel to $\mathbf{B}$. Using (1) and (2), we obtain

$$b_c^{in} = \mathbf{b}\mathbf{B}_c^{in} = \frac{1}{3}B\chi_c^{iso}\mathbf{b}\chi_c\mathbf{b} \tag{3}$$

and

$$b_c^{out} = \mathbf{b}\mathbf{B}_c^{out} = \frac{1}{4\pi}B\chi_c^{iso}\frac{V_c}{r_c^3}\left[\mathbf{b}\chi_c\mathbf{b} - 3\left(\mathbf{b}\chi_c\mathbf{n}_c\right)\mathbf{n}_c\mathbf{b}\right] \tag{4}$$

for the local field inside and outside, respectively, of a crystallite c. The above representation is convenient for the calculation of the MAS average, which is simply accomplished by expressing $\mathbf{b}$ in a coordinate frame fixed to the rotating sample,

$$\mathbf{b} = \left(\sqrt{\frac{2}{3}} \cos \varphi, \sqrt{\frac{2}{3}} \sin \varphi, \sqrt{\frac{1}{3}} \right). \tag{5}$$

Complete MAS averaging, which usually can be obtained with moderate spinning speeds, corresponds now to an average over φ. For any symmetric tensor χ, we find that the MAS average of $\mathbf{b}\chi\mathbf{b}$ yields its isotropic value. Also, for a pair of arbitrary vectors $\mathbf{u}$ and $\mathbf{v}$, the geometric term $(\mathbf{bu})\,(\mathbf{vb})$ leads to an average of $\mathbf{uv}/3$. Hence, the local field under MAS, b_c^{M}, is

$$b_c^{\text{in M}} = \frac{1}{3} B \chi_c^{\text{iso}} \tag{6}$$

for positions inside the crystallite, and

$$b_c^{\text{out M}} = \frac{1}{4\pi} B \chi_c^{\text{iso}} \frac{V_c}{r_c^{\,3}} \left(1 - \mathbf{n}_c \chi_c \mathbf{n}_c \right) \tag{7}$$

for those outside the crystallite. Since the distinction is clear enough from the context, we shall drop the superscripts used to distinguish inside and outside fields from now on.

a)

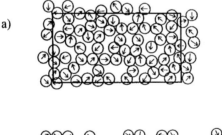

b)

Fig. 1. Models for the calculation of the overall local field distribution. The crystallites are marked by a circle with an arrow showing the orientation. The polycrystalline sample is approximated by the following arrangements. (a) Real sample: no pattern in the distribution of crystallites exists. The field distribution can only be obtained by a calculation of the field at each point r in the sample. (b) Model sample: crystallites are randomly oriented; their positions are restricted to a regular lattice. Each lattice position is occupied with a probability p. Three possible configurations of crystallites are given. The field must be calculated for all possible configurations and all points r inside a primitive unit cell (indicated by a box). Taken from [32]

The *total magnetic field* b_{loc} produced by a polycrystalline adsorbent at a position **r** within a crystallite $c = 0$ can be simply described by adding the shielding field caused by crystallite 0 and the field due to all neighboring crystallites $c \neq 0$, $b_{\text{loc}} = \sum_{i=0}^{N} b_i$. In the simpler case where **r** refers to a position in the intercrystalline space, no shielding field is present and all particles can be treated in the same way. Thus, for any given set of positions and orientations of the crystallites, b_{loc} can be calculated for an arbitrary position **r** by means of (3)–(5) or (6) and (7). The calculation of the distribution of b_{loc} due to all possible positions **r** in this configuration can only be accomplished by a numerical treatment. Thus, rather than considering the whole sample volume for a given configuration, we analyze a restricted volume, comparable to the size of a crystallite, and assume a distribution of configurations (Fig. 1). Details of the numerical calculations are given in [32].

The total local field can be written as a sum of contributions b_ν from all lattice positions ν. Since the quantities b_ν represent independent random variables, the distribution of b_{loc} can be approximated by using the central limit theorem. Thus, b_{loc} has a Gaussian distribution with a center $\langle b_{\text{loc}} \rangle = \sum_\nu \langle b_\nu \rangle$ and a mean quadratic deviation $\langle\langle b_{\text{loc}}^2 \rangle\rangle = \sum_\nu \langle\langle b_\nu^2 \rangle\rangle$, where the double bracket is defined by $\langle\langle b^2 \rangle\rangle = \langle b^2 - \langle b \rangle^2 \rangle$. The values $\langle b_\nu \rangle$ and $\langle\langle b_\nu^2 \rangle\rangle$ can be calculated by averaging (3)–(5) and (6) and (7) over all orientations. For the internal field of a crystallite inside the primitive unit cell (which has by convention the number 0) surrounding **r**, we obtain

$$\langle b_0 \rangle = \frac{1}{3} B \chi_0^{\text{iso}}, \tag{8}$$

$$\langle\langle b_0^2 \rangle\rangle = \frac{4}{405} \left(B \chi_0^{\text{iso}} \Delta\chi_0 \right)^2 \tag{9}$$

in the static case and

$$\langle b_0^{\text{M}} \rangle = \frac{1}{3} B \chi_0^{\text{iso}}, \tag{10}$$

$$\langle\langle b_0^{\text{M2}} \rangle\rangle = 0 \tag{11}$$

in the case of MAS. For other lattice positions ($\nu \neq 0$), we must additionally account for the probability p of finding a crystallite in the position $\mathbf{r}_\nu$, and arrive at

$$\langle b_\nu \rangle = \frac{1}{4\pi} p B \chi_\nu^{\text{iso}} \frac{V_\nu}{r_\nu^3} \left(1 - 3\cos^2 \gamma_\nu \right), \tag{12}$$

$$\langle\langle b_\nu^2 \rangle\rangle = p \left(\frac{1}{4\pi} B \chi_\nu^{\text{iso}} \Delta\chi_\nu V_\nu / r_\nu^3 \right)^2 \left(\frac{4}{45} + \frac{1}{15} \cos^2 \gamma_\nu + \frac{1}{5} \cos^4 \gamma_\nu \right)$$
$$+ p\left(1 - p\right) \left(\frac{1}{4\pi} B \chi_\nu^{\text{iso}} \frac{V_\nu}{r_\nu^3} \right)^2 \left(1 - 3\cos^2 \gamma_\nu \right)^2 \tag{13}$$

and

$$\langle b_\nu^{\mathrm{M}} \rangle = 0, \tag{14}$$

$$\langle\langle b_\nu^{\mathrm{M2}} \rangle\rangle = p\frac{4}{45}\left(\frac{1}{4\pi}B\chi_\nu^{\mathrm{iso}}\,\Delta\chi_\nu\,\frac{V_\nu}{r_\nu^3}\right)^2. \tag{15}$$

Here γ_ν denotes the angle between the external magnetic field and the vector $\mathbf{r}_\nu$, given by $\mathbf{n}_\nu\mathbf{b} = \cos\gamma_\nu$. Contributions from all lattice positions ν give the center and width of the total local field distribution at a position $\mathbf{r}$ (Fig. 2).

In conclusion to this part, both the shielding field and the field produced by neighboring crystallites can contribute to the linewidth in a static sample. The non-vanishing contribution from the shielding field occurs when the susceptibility tensor is anisotropic or the crystallite has a nonspherical shape ($\Delta\chi \neq 0$). The mean shielding field derived by averaging over all orientations, however, is independent of $\Delta\chi$ (**??**). The field originating from a neighboring crystallite depends on the orientation of this crystallite in the external magnetic field and its position relative to the site of the resonating spin (4). Under MAS conditions, the orientation of the crystallite has no influence on its shielding field (6); all orientations lead to the same shielding field, which is identical to the average shielding field in the nonspinning case. Thus, with and without MAS, we find the same center of the distribution (see (8) and (10)). Examination of the contribution of the shielding field to the distribution width shows that it is completely eliminated by MAS (11). For a neighboring crystallite, a similar conclusion results only if $\Delta\chi = 0$ (see (7) and (13)). However, a nonvanishing contribution to the distribution width remains under MAS for neighboring crystallites with an anisotropic magnetic susceptibility. The overall local field distribution is characterized by an FWHM of $0.153\,\chi^{\mathrm{iso}}\,\Delta\chi\sqrt{pB}$. If the distribution width of the local fields is proportional to the observed NMR linewidth, the line narrowing achieved by MAS can be indicated by a line-narrowing factor, which is the ratio of the FWHM values without and with MAS (Fig. 3). The continuous decrease of the line-narrowing factor with increasing p is a result of the $\sqrt{p}$ dependence of the FWHM with MAS. A general analytical formula for the MAS line-narrowing factor cannot be obtained in our model.

The ability of MAS to reduce NMR susceptibility broadening was also studied earlier by Alla and Lippmaa [31]. Susceptibility and shape effects were included for the shielding field, while dipolar fields were caused by neighboring crystallites that were assumed to be of spherical shape. The formulas derived on the basis of a single crystallite correspond to ((4) (assuming $\Delta\chi = 0$) and (7) in our model. Assuming that the angle-dependent terms are statistically independent, a line-narrowing factor $3/\Delta\chi$ is derived from the ratio of both equations. The factor $3/\Delta\chi$ given in [31] agrees well with the results presented here for $0.2 \leq \Delta\chi \leq 1$ and reasonable values of p (Fig. 3).

With respect to the experimental situation, the discussion of the static local inhomogeneous magnetic fields may be summarized as follows:

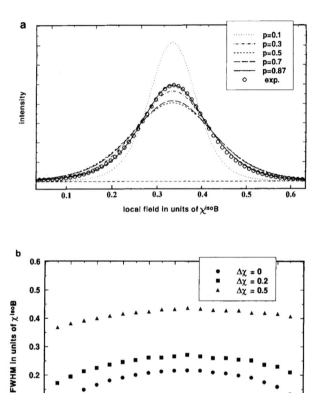

Fig. 2. Overall local field distributions in the static case. (a) Normalized overall local field distributions for $\Delta\chi = 0$ and various values of p. The distributions are symmetrical and have the same center. With a probability p close to 0 or 1, the distribution widths are smaller owing to a narrower variation range in the local environment of each crystallite. (b) Full width at half maximum (FWHM) of the overall local field distribution for various values of $\Delta\chi$ and p. Increasing anisotropy $\Delta\chi$ leads to an increase in the FWHM of the distribution. The FWHM maximum for $p = 1/2$ is less distinct for higher $\Delta\chi$. Taken from [32]

(i) There is an appreciable reduction of the quadratic deviation in the MAS case, which explains, at least qualitatively, the strong line narrowing observed. The residual linewidth in the MAS NMR spectra of molecules embedded in a certain crystallite of the adsorbent material is given by the influence of the local magnetic fields due to the neighboring crystallites.

(ii) A Gaussian NMR line shape has to be expected in the static case without thermal motion, with a second moment $M_2 = \gamma^2 \langle\langle b_{loc}^2 \rangle\rangle \propto B^2$. This means that the line broadening may increase strongly with rising magnetic field B.

(iii) In the experiment which is still to be discussed, both a Lorentzian and a Gaussian contribution to the line shape have been observed, even in the static case. In the case of MAS with higher rotational frequencies, the line shape was found to be Lorentzian. The two effects are not reflected in this treatment and have still to be explained in terms of motional influences. In this way, we also hope to understand the limits in spectral resolution.

Susceptibility Effects: Influence of MAS and Thermal Motion

The theoretical approach in this case is based on a paper by Fenzke et al. [35], where it was assumed that the spectra are dominated either by dipole–dipole interaction or by chemical shift anisotropy. This treatment can be generalized to the present situation, where the susceptibility influence dominates, because, as shown above, the essential contribution to the local magnetic fields acting on a molecule adsorbed in a certain crystallite (e.g. a grain of a polycrystalline zeolite material) is supplied by the neighboring crystallites, i.e. it is produced by magnetic dipole moments.

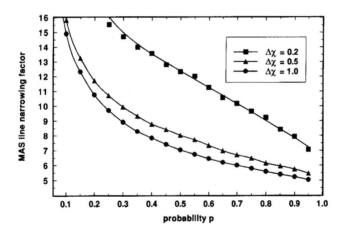

Fig. 3. Ratio of FWHM values for the local field distribution with and without application of MAS (MAS line-narrowing factor) for various $\Delta\chi$ and p.Teken from [32]

In terms of the paper [35], the free induction decay $D(t)$ can be described by a product of a time-dependent envelope function $\Phi(t)$ and a part which is responsible for the position of the center of the lines in the spectra with the Larmor frequency ω_0:

$$D(t) = \Phi(t)\exp\{i\omega_0 t\}, \tag{16}$$

where

$$\Phi(t) = \left\langle \exp\left(\int_0^t i\omega(t')\,dt'\right)\right\rangle.$$

The angular frequency $\omega(t)$ is proportional to the component of the local magnetic field along the direction z of the external magnetic field. Hence, for a predominant dipole contribution, one can write

$$\omega(t) \propto 3\cos^2\theta(t) - 1,$$

similarly as in [10]. The expression for $\Phi(t)$ can be expanded into a Taylor series,

$$\Phi(t) \approx 1 - \frac{1}{2}M_2 \int_0^t \int_0^t G(t' - t'')\,dt'\,dt'',$$

where $M_2 = \langle\omega^2\rangle$ is the second moment of the induced static field and $G(t' - t'')$ is the correlation function. In our case the correlation function G is influenced by the thermal motion of the molecules (G_{therm}) in the zeolite, as well as by the macroscopic rotation of the sample (G_r). Since these motions are statistically independent of each other, we can write

$$G(t' - t'') = G_{\text{therm}}(t' - t'')G_r(t' - t''). \tag{17}$$

In the simplest case, which has still to be analyzed critically, we assume that the thermal motion is characterized by one correlation time τ_c:

$$G_{\text{therm}} = \exp(-|t' - t''|/\tau_c). \tag{18}$$

The correlation function for MAS, $G_r(t' - t'')$, is given by

$$G_r(t' - t'') = \frac{5}{4}\langle(3\cos^2\theta(t') - 1)(3\cos^2\theta(t'') - 1)\rangle$$
$$= \frac{5}{6}\cos[\omega_r(t' - t'')] + \frac{5}{12}\cos[2\omega_r(t' - t'')] \tag{19}$$

using the definition of the magic angle, viz. $\cos^2\theta_m = 1/3$; ω_r is the angular frequency for MAS. These equations then lead to an envelope function

$$\Phi(t) = \exp\left\{-\frac{M_2}{3}(J(2\omega_r, t) + 2J(\omega_r, t))\right\}. \tag{20}$$

The spectral density function $J(\omega_r)$, defined as usual by the Fourier transform of the correlation function $G(t' - t'')$, is given by

$$J(\omega_r, t) = \frac{t\tau_c}{1 + (\omega_r\tau_c)^2} - \frac{\tau_c^2(1 - (\omega_r\tau_c)^2)}{(1 + (\omega_r\tau_c)^2)^2}\left[1 - \exp\left(-\frac{t}{\tau_c}\right)\cos(\omega_r t)\right]$$
$$- \frac{\omega_r\tau_c^3}{(1 + (\omega_r\tau_c)^2)^2}\exp\left(-\frac{t}{\tau_c}\right)\sin(\omega_r t). \tag{21}$$

The Fourier transform of the envelope function $\Phi(t)$ yields the shape for each line in the NMR spectrum, including the central line and the sideband pattern, as will be discussed in detail in the simulation of the spectra. In the absence of MAS ($\omega_r = 0$), the envelope function $\Phi(t)$ simplifies to the well known formula (Anderson–Weiss theory)

$$\Phi(t) = \exp\left\{ -M_2\tau_c \left[t + \tau_c \exp\left(-\frac{t}{\tau_c} \right) - \tau_c \right] \right\}, \qquad (22)$$

which is used in order to analyze the static spectra and to estimate M_2 and the correlation time τ_c by means of a line shape analysis, as is shown now. Typical ^{1}H NMR spectra of 1-butene molecules adsorbed in NaX zeolites are shown in Fig. 4, which were measured at a proton Larmor frequency of 600 MHz. Four strongly broadened lines are superimposed, which represent contributions from the chemically inequivalent protons in the groups =CH$_2$, =CH–, –CH$_2$–, and –CH$_3$ of the adsorbed 1-butene molecule. The influence of the indirect spin-spin interactions is hidden and the difference in the chemical shifts of the chemically inequivalent protons in =CH$_2$ is not visible. If one decomposes the spectra in Fig. 4, the line shape of each of the four lines deviates from a purely Gaussian shape, which is to be expected if the local fields are not influenced by thermal motions as assumed in [32]. If the envelope function of the free induction is computed numerically according to (22), we may estimate the second moment M_2 of the static field distribution and the order of magnitude of the correlation time. After Fourier transformation of these time domain functions $\Phi(t)$ for a variety of M_2 and τ_c values, frequency domain signals were obtained (the "theoretical" NMR spectra). Formally, the line shape can be approximated by a superposition of a Gaussian (G) and Lorentzian (L) contribution with a G/L ratio of

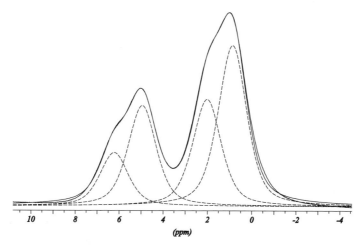

Fig. 4. Deconvolution of the static NMR spectra of 1-butene taken at 600 MHz (293 K), sorbed in NaX. The linewidth is 910 Hz for all four lines, which show the expected ratio of intensities of 1:2:2:3. Taken from [63]

0.45 ± 0.15. We then extracted those pairs of values (τ_c, M_2) that lead to a linewidth comparable to the experimental results. It is clear that such a procedure enables only a rough estimation of the second moment and the order of magnitude of the correlation time. The best fits are obtained for $1.5 \times 10^{-4}\text{s} \leq \tau_c \leq 2.5 \times 10^{-4}\text{s}$ and $2.1 \times 10^7\text{s}^{-2} \leq M_2 \leq 4 \times 10^7\text{s}^{-2}$.

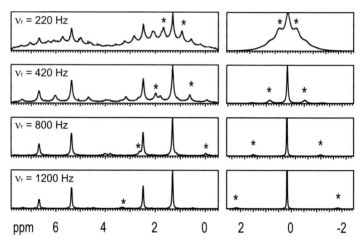

Fig. 5. ^{1}H MAS NMR spectra for 1-butene (taken at 600 MHz, $T = 293$ K) at various MAS frequencies (left), in comparison with the Fourier transform of the envelope function $\Phi(t)$ (see text) computed for the same rotational frequencies and taking the values $\tau_c = 5 \times 10^{-3}\text{s}$ and $M_2 = 4.7 \times 10^6\text{s}^{-2}$ (right). The first sideband of the $-CH_3$ group and the corresponding peaks in the simulation are marked by asterisks (*). Sideband patterns appear at a MAS frequency of about 200 Hz. The linewidths (full widths at half maximum, FWHM) in the simulated (FWHM$_{sim}$) and the experimental (FWHM$_{exp}$) spectra agree up to MAS frequencies of about 800 Hz. For higher MAS frequencies, the simulations lead to smaller linewidths (FWHM$_{exp} = 25$ Hz, FWHM$_{sim} = 4$ Hz, at $\nu_r = \omega_r/2\pi = 1200$ Hz) if the same τ_c values are used. Taken from [63]

A more detailed analysis relies on the study of the MAS NMR shape of the spectra at different MAS frequencies. It is necessary to treat the ranges of lower and higher rotational frequencies separately in order to check the applicability of the model used, because the computations lead to similar but not fully identical results in the two cases. The computations based on (20) and (21) show that the rotational sidebands at rotational frequencies of about 200 Hz are reflected in an appropriate way (Fig. 5) if a correlation time of $\tau_c = 4 \times 10^{-3}$ s and a second moment of $M_2 = 4.8 \times 10^6$ s^{-2} are used for NMR measurements at 600 MHz. At higher rotational frequencies (MAS frequencies above ca. 4 kHz), the line shape is purely of Lorentzian type, in good agreement with the experimental data. For higher MAS frequencies (4 kHz), the calculated linewidth thus amounts to 1.4 Hz, in contrast to the experimental values of FWHM$_{exp} \cong 8$ Hz. The Lorentzian line shape in the spectra found for higher MAS frequencies is confirmed by the calculations. From the values for the ^{1}H NMR linewidths (FWHM of ca. 8–10 Hz at a Larmor frequency of 600 MHz and MAS frequencies of 4 to 6 kHz), correlation times in the range of 3×10^{-4} s $\leq \tau_c \leq 5 \times 10^{-4}$ s were derived. The correlation times thus determined differ by a factor of at least 10 from those values which lead to a good fit of the rotational sideband spectra. It should also be mentioned that mainly the magnitude of the correlation time is essential, and that changes of the second moment do not allow a better fit of both the sideband structure and the residual linewidths.

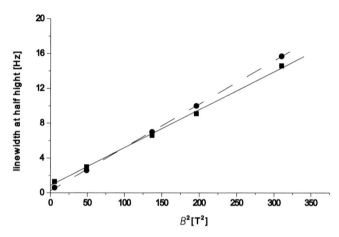

Fig. 6. Dependence of the residual linewidths in ^{1}H MAS NMR experiments on molecules sorbed in porous materials on the dc magnetic field B. Solid circles: results from simulations with M_2 values (proportional to B^2) taken from the analysis of the static linewidth and the linewidth for high MAS frequencies. Solid squares: experimental linewidth (also taking into account the indirect spin–spin coupling) after subtraction of a residual linewidth of 1.3 Hz due to other field-independent influences. The segmented and straight lines show respective best fits. The quadratic dependence of the linewidth on the external field can be verified. Taken from [63]

This behavior can only be understood if the averaging process is considered in more detail. The motional averaging mechanism is related to translational motions of the adsorbed species over distances which are comparable to the size of the grains of the adsorbent. This is consistent with the fact, already explained above, that the distribution of the local magnetic fields is essentially due to contributions from neighboring particles. Thus, the model of a single correlation time can only be valid if the grains are equal in size (no distribution of the "correlation times") and/or if the translational motion is accompanied by jumps over long distances (viz., where the "correlation time" reflects the mean lifetime at the adsorption sites and where the flight time is comparatively small). The results illustrate the necessity to use different values for τ_c to explain the calculation of MAS sideband patterns at lower MAS frequencies and the residual linewidths at higher MAS frequencies. A more detailed model for such a distribution of correlation times has not yet been treated in the present line shape analysis. This situation does not essentially limit the reliability of this line shape analysis, because of several reasons:

(i) The main features of the line shape are reflected in an appropriate way, i. e. a mixture of Gaussian and Lorentzian contributions at low rotational frequencies and a purely Lorentzian line shape at MAS frequencies above ca. 1 kHz were found, since the second moment M_2 of the local field distribution is proportional to B^2, which has been checked for a range of **B** fields between 2.36 and 17.6 T (Fig. 6).

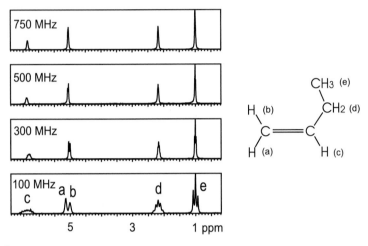

Fig. 7. ^{1}H NMR spectra for 1-butene sorbed in an NaX zeolite, measured at resonance frequencies of 750, 500, 300, and 100 MHz. The increase of the residual total linewidth for higher external magnetic fields leads to the fact that the indirect spin–spin interaction is only resolved at lower fields. The better chemical-shift resolution for measurements at high B fields can be clearly inferred from the spectra. The letters a, b, c, d, and e characterize the assignment of the various resonance lines. Taken from [63]

(ii) The estimated range of correlation times is consistent with the dimensions of the zeolite crystallites (grains) with respect to the thermal averaging process. Using a coefficient for self-diffusion of $D = 2.5 \times 10^{-9}$ m^{2}s^{-1} from independent PFG NMR measurements on the same systems at room temperature, then according to the well known equation $(\langle l^2 \rangle)^{1/2} = (6D\tau_c)^{1/2}$ a mean square displacement in the range $(\langle l^2 \rangle)^{1/2} = 1.5$ μm to 8.6 μm can be found using correlation times of $\tau_c = 1.5 \times 10^{-4}$ s and $\tau_c = 5 \times 10^{-3}$ s. This mean displacement is comparable to the mean diameters of the zeolite crystallites, which are specified to be 5–8 μm in our experiments.

(iii) The application of MAS techniques leads to an appreciable line narrowing in the ^{1}H NMR spectra, which is obvious from Figs. 7 and 8, where measurements for various adsorbed molecules at different resonance frequencies are shown. Since the remaining linewidth increases proportionally to B^2, the question is which B fields

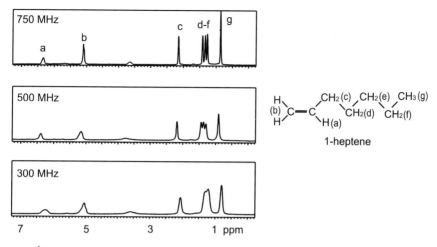

Fig. 8. ^{1}H NMR spectra for 1-heptene sorbed in an NaX zeolite with a loading of 4 molecules per supercage, at different magnetic fields. A gain in chemical-shift resolution is observed for high magnetic fields (see also Fig. 7). Taken from [63]

are still useful in order to receive a better resolution. The measurements show that even at proton resonance frequencies of 750 MHz, the resolution of chemical-shifts for the adsorbed molecules is better than in lower fields (Figs. 7 and 8).

Thus, the nonlinear increase of residual linewidths is still of smaller influence than the increase in chemical-shift resolution. This behavior will be demonstrated on the basis of a simulation by means of a commercial program. The results of the simulations for resonance frequencies of 100 and 500 MHz are shown in Figs. 9 and 10, respectively. The simulations confirm the statements about the limit of resolution obtained by using high magnetic fields for NMR studies of adsorbed molecules under MAS conditions.

2 Analysis of Chemical Shifts

2.1 Introduction

The aim of this section is to investigate the behavior of simple hydrocarbons, such as olefins, in zeolites in relation to the adsorption dynamics and to structural changes of the molecules which are relevant to catalytic transformations. These systems are characterized by a relatively high internal mobility of the embedded molecules, with thermal reorientation times between 10^{-9} s and 10^{-11} s. They are suitable for measurements of highly resolved ^{1}H NMR spectra based on the combined application of MAS techniques, measurements at high external magnetic fields, and the use of pure solid adsorbents (e.g. zeolites of type NaX) with a very small concentration of paramagnetic impurities. ^{1}H MAS NMR measurements were run at Larmor frequencies

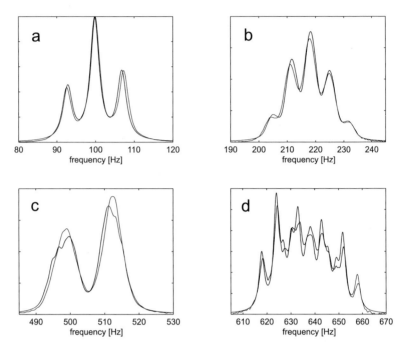

Fig. 9. Comparison between simulated and experimental ^{1}H MAS NMR spectra for 1-butene in an NaX zeolite measured at room temperature (with $\nu_{\mathrm{r}} = 3.5$ kHz, resonance frequency 100 MHz). Simulations were carried out by calculating the resonance frequencies and intensities, followed by the application of an additional line broadening with a single Lorentzian-shaped line with a half width of 2.6 Hz for all groups. Figures (a) to (d) show the four separated CH$_n$ groups of 1-butene ($-$CH$_3$, $-$CH$_2-$, $=$CH$_2$, $=$CH$-$, respectively), with strongly developed patterns due to the indirect couplings of the protons. A fairly good agreement between the simulations and the experimental spectra can be seen. Taken from [63]

between 100 and 750 MHz and at temperatures between 130 K and ambient temperatures, using moderate spinning frequencies of ca. 4–5 kHz. The use of MAS for adsorbate–adsorbent systems in general becomes possible if the techniques of sealed samples is applied [36]. This allows a pretreatment of the samples under defined conditions and a loading of the materials with a definite amount of molecules through the gas phase. Commercial NaY zeolites, from UOP Molecular Sieve Adsorbents, and an ultrapure zeolite, NaX, prepared by Shdanov et al. [37], were very suitable [38]. Because of the relatively high internal dynamics, which is also expressed by the values of the self-diffusion coefficients of the adsorbed molecules measured by means of PFG NMR techniques (see Chap. 1 of this book), it is important that the experimentally measured NMR parameters are always mean values between molecules on the adsorption sites and the residual molecules in the zeolite cages. As will be discussed in this section, the thermodynamic equilibrium between the adsorption complexes formed and the other adsorbed molecules sorbed in the large cavities can be charac-

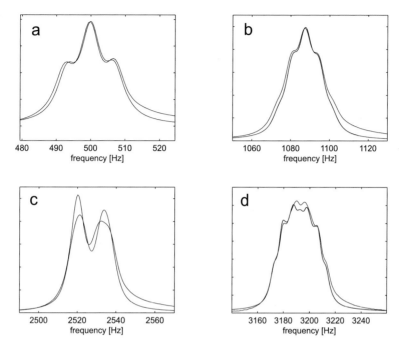

Fig. 10. Comparison between simulated and experimental ^{1}H MAS NMR spectra for 1-butene in an NaX zeolite measured at room temperature (with $\nu_r = 3.5$ kHz, resonance frequency 500 MHz). The line shape is simulated by a Lorentzian curve with a linewidth of 6.6 Hz. Figures (a) to (d) show the four separated CH$_n$ groups of 1-butene (–CH$_3$, –CH$_2$–, =CH$_2$, =CH–, respectively). The splitting due to the indirect spin–spin coupling vanishes as expected. Taken from [63]

terized in more detail if the temperature is varied over a wide range, thus covering different equilibrium states. The detailed investigation of this dynamics allows one to derive conclusions about changes of the electronic structure of the molecules in contact with adsorption sites and conformational changes, in spite of the fact that the mean lifetimes of the molecules on the adsorption sites may be very short.

In the following section it will be shown that ^{1}H NMR studies (see [39] for more details) are much more sensitive to adsorption interactions than those for ^{13}C nuclei forming the skeleton of the molecule.

2.2 ^{1}H NMR Chemical Shifts and Line Shape Analysis

The appreciable increase in spectral resolution achieved in ^{1}H NMR for 1-butene molecules sorbed in NaX is demonstrated in Fig. 11 in comparison with spectra for 1-butene molecules dissolved in liquid CDCl$_3$. All proton resonance lines for 1-butene molecules can be resolved in the adsorbed state but only a few line splittings due to indirect spin–spin couplings between the various protons (coupling constants

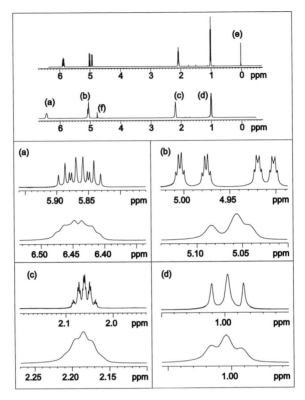

Fig. 11. Comparison of ^{1}H NMR spectra of 1-butene ($T = 298$ K, $\nu_H = 600.13$ MHz, $\nu_r = 5$ kHz) dissolved in CDCl$_3$ (top) and sorbed in NaX ($\nu_r = 5$ kHz, second spectrum from top). The inserts show the line shape for different molecular groups ((a) =CH–, (b) =CH$_2$, (c) –CH$_2$–, and (d) –CH$_3$) in the adsorbed and in the solution state. Peak (e) is due to TMS in solution. In the spectrum of the adsorbed molecules, the –CH$_3$ group chemical shift was calibrated to 1 ppm. Peak (f) belongs to the remaining H$_2$O or DHO in D$_2$O, which was used for locking but could not be taken for the calibration of the spectra. Taken from [64]

J_{HH}) are visible. J-coupling patterns with J_{HH} smaller than about 6 Hz cannot be resolved. The ^{1}H NMR spectra show clearly the changes of the chemical shifts in the adsorbed state with respect to the molecules in solution. The simulation reveals that the same J_{HH} coupling constants as found for 1-butene in solution may be used as a reasonable approximation. In the temperature range from 295 K to 343 K there are almost no changes in the residual linewidth of ca. 5 Hz, which shows that the homogeneous contributions to the linewidths from the dipolar spin–spin interactions are not dominant, because they are expected to change from group to group. Hence, in the simulations one may use the same inhomogeneous line shape function for each line in accordance to the still-dominant influence of the averaged local fields. The chemical equivalence of the two –CH$_2$– and the three –CH$_3$ group protons in the adsorbed species can be clearly inferred from the spectra.

Compared with the liquid/gaseous state, the mean position of the proton NMR signal for the group =CH– of the molecule in the adsorbed state is appreciably shifted to lower magnetic fields. This shift also depends on loading (Fig. 12) and temperature (Fig. 13). It can be interpreted as changes of the electronic densities as a result of interactions of the adsorbed molecules, with the exchangeable Na^+ cations in the NaX zeolites acting as adsorption sites. The situation is similar to that found in former ^{13}C NMR chemical-shift studies [16], where this conclusion was drawn by using zeolites with various cations (e.g. Li^+, Na^+, K^+, Rb^+, Cs^+, and Tl^+) [16]. Furthermore, the verification of specific interactions of the π electrons of the 1-butene molecules with the Na^+ cations at certain crystallographic sites in the zeolite cage was the subject of quantum-chemical MO calculations, where also the conformation of the 1-butene molecules in interaction with the cations was investigated [40, 41]. Owing to changes in the charge distribution at the carbon atoms, the ^{13}C NMR lines in the group =CH– are also much more shifted relative to the position in the "free" (gaseous or solved) 1-butene molecule than was found for the lines of the groups =CH_2 and –CH_2–. It seems to be well understood from the previous work that the geometry in the sp^2 hybridized part of the adsorbed molecule remains practically unchanged. In quantum mechanical calculations [41], not only the resulting changes of the electron density in the adsorbed 1-butene molecules were treated as the result of interactions with Na^+ adsorption sites. The calculations [41] also point out that changes in the conformation of the molecules relative to the gaseous state may occur.

Two chemically nonequivalent protons in the terminal =CH_2 group appear, as expected from the sp^2 hybridization in the olefinic part of the molecule. This leads to different chemical shifts of the two protons in the group =CH_2 and to different $^3J_{HH}$ coupling constants for these protons and the proton in the group =CH– (viz. 10.23 Hz and 17.32 Hz for the *cis* and the *trans* proton positions, respectively), which can be detected in solution as well as in the adsorbed state. Hence, the proton NMR spectra of the =CH_2 group for adsorbed molecules are characterized by a superposition of two doublets; the weaker J-couplings to more distant protons are not resolved. In contrast to the NMR spectrum of gaseous 1-butene, the chemical shifts of the *cis* and *trans* proton positions are not constant but change with temperature as well as with loading, i.e. with the mean total number of molecules adsorbed in a large cavity (supercage). This effect will be discussed in detail below. As mentioned, a rather good agreement of the ^1H NMR signals between the simulated spectra and the experimental ones is achieved for the aliphatic groups –CH_2–, –CH_3 and for the =CH– group using the J-coupling constants for the molecule in the gaseous phase and taking into account the line shifts observed for the adsorbed molecules. Of special interest is a distinct change in the lineshape for the terminal molecular group =CH_2 with temperature and loading. It is obvious that the measured difference between the chemical shifts of the two protons in the group =CH_2 for molecules embedded in NaX zeolites decreases at lower temperature (Fig. 13) and becomes approximately equal to zero at about 270 K. In the ^1H NMR measurements at a still higher Larmor frequency of 750 MHz, which have now became possible, sufficiently resolved proton NMR spectra could also be measured at lower temperatures (230 K). It is clearly seen that the chemical

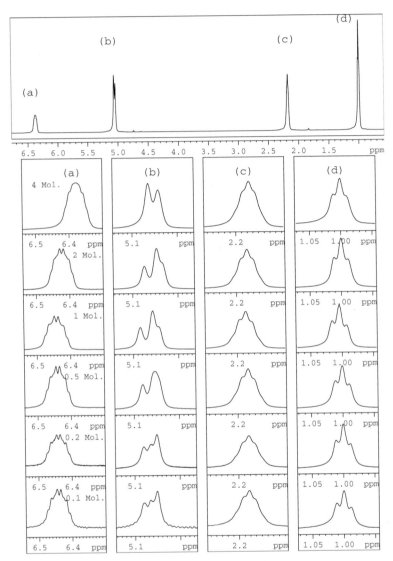

Fig. 12. 1-butene sorbed in NaX ($T = 298$ K, $\nu_H = 600.13$ MHz, $\nu_r = 5$ kHz) for different loadings. The upper part shows the complete ^{1}H HR MAS NMR spectrum with a loading of 4 molecules per supercage. In the inserts, the dependence of the line shape and line position for each molecular group ((a) =CH, (b) =CH$_2$, (c) –CH$_2$– and (d) –CH$_3$) is illustrated for various loadings, viz. 4, 2, 1, 0.5, 0.2, and 0.1 molecules per supercage. The line splitting due to the coupling constants is partially resolved. Taken from [64]

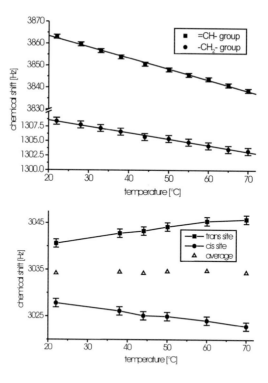

Fig. 13. Chemical shifts (in Hz) for 1-butene sorbed in NaX in dependence on temperature for the =CH– and –CH₂– groups (top) and for the protons of the =CH₂ group (bottom) of the molecule (measurements at 600.13 MHz). The arithmetic average ($\triangle$ (bottom)) of the cis-proton ($\bullet$ (bottom)) and trans-proton ($\blacksquare$ (bottom)) of the =CH₂ group is independent of temperature in the range investigated. Taken from [64]

shifts for the two protons in the group =CH₂ again become different from zero at still lower temperatures (Fig. 14). The spectra then show two doublets again with about the same ${}^3J_{HH}$ coupling constants as described above, but clearly indicate a crossing of the two lines at ca. 270 K, as may be inferred from the inversion in the ${}^3J_{HH}$ coupling pattern. It is important to note that no such changes can be found for 1-butene in solution. The same typical changes in the ${}^1H$ NMR line shape for the group =CH₂ may be observed if the total loading is varied between 4 and 0.1 molecules per supercage (Fig. 12). Hence, in order to prove which conclusions may be derived from the ${}^1H$ NMR spectra, the influence of the dynamics of adsorption has to be treated in more detail.

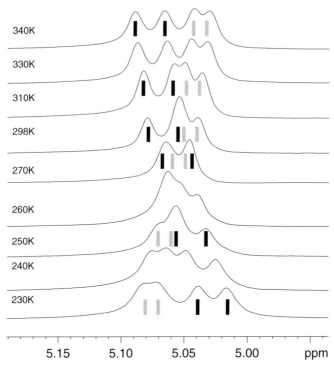

Fig. 14. ^{1}H MAS NMR spectra of 1-butene in NaX: spectral range of the group =CH$_2$ in measurements run at $\nu_{\mathrm{H}} = 749.98$ MHz and $\nu_{\mathrm{r}} = 5$ kHz. Drastic changes in the line shape are seen in the measurements at different temperatures. At the highest and lowest temperatures shown, the line shape is given by a superposition of two doublets with $^3 J_{\mathrm{HH}}$ coupling constants of ca. 10 Hz and 17 Hz and chemical-shift differences corresponding to $(+)$ 0.35 ppm and $(-)$ 0.5 ppm for a loading of 1 molecule per supercage (see text). The positions of the two peak doublets underlying the complex signals are sketched as guides for the eye. Taken from [39]

2.3 Dynamics of Adsorption. Exchange Between Different Adsorption Sites

To understand the influence of the adsorption dynamics on the ^{1}H NMR spectra, we start with the analysis of the ^{13}C NMR chemical shifts with varying total number of adsorbed molecules. The change in the chemical shifts can be explained if we consider an exchange between molecules adsorbed on the cationic sites and those that are physisorbed in the zeolite cages. This exchange has already been described in the interpretation of former ^{13}C NMR chemical-shift measurements [16], which were checked for comparison in this work also. In order to understand the ^{1}H NMR spectra, we interpret first the results that can be derived from the ^{13}C spectra under comparable conditions. One may suggest [16] that the molecules exist in a dynamic equilibrium between two states, viz. temporarily "bonded" at the adsorption sites (fraction p_{C}) and in a physisorbed state within the zeolite cavities (fraction p_{M}).

In accordance with the experimental situation, it is further assumed that there is a fast exchange between both states. If we denote by $p_C = N_C/N$ the fraction of the complexes formed (N is the total number of adsorbed molecules), then we have to admit that the number N_C of the complexes formed with the adsorption sites is in general not identical to the number N_A of adsorption sites ($N_C \leq N_A$). Because of the fast exchange, an averaged ^{13}C chemical shift,

$$\delta = p_C \delta_C + p_M \delta_M \,, \tag{23}$$

is measured for the various molecular moieties and we assume that the changes found in the ^{13}C chemical shifts are simply due to a change of the fractions p_C and p_M for both types of molecules. In a small interval we find an approximately linear variation with temperature. For proton-decoupled ^{13}C NMR spectra, where we find a single line for each group, we may simplify the interpretation of the chemical shifts δ: if we denote by δ the observed resonance shift with respect to the state M (i.e. if we put $\delta_M = 0$), we find $\delta = \delta_C(N_C/N)$ for the resonance shift of adsorbed molecules exchanging with those bound in a complex MA.

We consider here only one type of adsorption site A (number N_A) and assume that the probability of the decomposition of the complex can be described by its equilibrium constant K. As is well known [16,42], the equilibrium may be described by the equation

$$M + A \Leftrightarrow MA \tag{24}$$

from which the relation

$$K = N_C/[(N - N_C)(N_A - N_C)] \tag{25}$$

may be derived in order to evaluate the ratio N_C/N from the measured chemical shift δ. Here, N_C denotes the number of molecules bound to adsorption sites, and $N - N_C$ is the number of physisorbed molecules that are not bound to adsorption sites. In the extrapolation of δ to very small loading, $N \to 0$, a maximum value $\delta_m = \delta_C[KN_A/(KN_A + 1)]$ can be estimated, and from the analysis of the quantity $x = \delta/\delta_m$ versus the total number N of adsorbed molecules,

$$\frac{Nx^2}{1 - x} = \frac{1}{x_m} \frac{Nx}{1 - x} - \frac{N_A}{x_m^2} \,, \tag{26}$$

where $x_m = \delta_m/\delta_C$, the quantity KN_A may be determined [16]. In previous ^{13}C NMR chemical-shift measurements at room temperature [16] ($T \approx 295$ K) this data treatment allowed the conclusion that we have a "weak interaction" between simple olefins and Na$^+$ adsorption sites in NaX zeolites (Si/Al ratio of 1.35), characterized by a value $KN_A = 1.7 \pm 0.3$. "Weak interaction" means that even for a small number N of adsorbed molecules no predominant, exclusive bonding of the adsorbed molecules to the adsorption sites occurs, but always a "distribution" between the types M and MA according to the reaction equation proposed. Since we have about $N_A = 5.5$ Na$^+$ cations (preferentially on crystallographic SIII sites) acting as adsorption sites in NaX zeolites (Si/Al $\approx$ 1.35), a value $K = 0.3 \pm 0.1$ (at room temperature) is found [16].

These considerations can be adopted to understand the proton NMR spectra. At decreasing temperatures the fraction p_C of molecules at the adsorption sites increases and, thus, the measured average value of the chemical shift is determined more and more by the value for the bonded state. For the group =CH– and for the group –CH$_2$–, this value is shifted to lower magnetic fields.

To understand the temperature dependence of the spectra for the group =CH$_2$, we consider only the apparent superposition of two doublets which are observed in the case of adsorbed molecules if the splitting is mainly caused by the $^3J_{HH}$ coupling between the =CH$_2$ protons and the proton in the group =CH–, because the couplings to the protons in –CH$_2$– group (–1.66 Hz in the gaseous phase) cannot be resolved. At decreasing temperatures the chemical shift is determined more and more by the value for the bonded state. This is the reason why the apparent difference of the chemical shifts for the two protons in the group =CH$_2$ first becomes zero and then changes in the opposite direction. At the lowest temperature (230 K) accessible so far for the ^{1}H MAS NMR measurements at 750 MHz, a difference in the chemical shifts of the two =CH$_2$ group protons of ca. 0.05 ppm was measured (37.8 Hz). Here, the doublet with the larger chemical shift (relative to the chemical shift of the -CH$_3$ group) shows about the same $^3J_{HH}$ coupling as that doublet which appears with the smaller chemical shift for the butene/NaX system at ca. 340 K or for the molecules in the gaseous state. A similar change in the position appears for the other doublet.

The line shape for the group =CH$_2$ at the different temperatures can be evaluated by means of simulations of the NMR spectra, considering an exchange of the molecules between both states (characterized by different chemical shifts and the $^3J_{HH}$ coupling constants) and different fractions of bonded and physisorbed molecules depending on the temperature range considered. The simulations were performed using the software platform GAMMA [43]. The influence of relaxation on the linewidth has been neglected, but a line broadening, caused mainly by susceptibility effects, was taken into account. It must be noted that the simulation represents the real behavior only in a semiquantitative manner, since parameters assumed for the spin systems cannot be measured directly because of the dynamic equilibrium mentioned above are of crucial importance for the interpretation. However, some of these parameters may be estimated in a plausible manner. For the physisorbed molecules, the parameters that were measured for 1-butene in the gaseous state were used, which is a quite reasonable assumption. For molecules bonded to the adsorption sites, the spin system parameters were assumed to be given by those measured at 230 K. The exchange rate was assumed to be 2×10^5 s^{-1}. This value represents a fast exchange in the sense of the NMR timescale of the chemical shifts. No significant changes in the calculated spectra are observed if higher exchange rates are used. The assumption of a much lower exchange rate, i.e. of the order of magnitude of the inverse of the chemical-shift difference, would result in a typical line broadening in the ^{1}H or ^{13}C spectra that was never observed in the experiments. The value for this exchange rate was concluded from ^{13}C spectra of 1-butene sorbed in NaX. Compared with ^{13}C

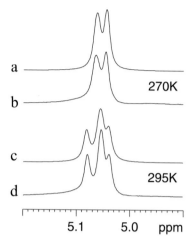

a

b

270K

c

d

295K

5.1   5.0   ppm

Fig. 15. Comparison of the region of the =CH$_2$ group of the ^{1}H spectra of 1-butene in NaX (a, c) with simulations (b, d), at two different temperatures. The simulations of the influence of exchange between molecules bonded at adsorption sites and physisorbed molecules were performed using constant spin system parameters (see text) and varying the fractional population of the two species. The fractional populations were 2.16 : 1 and 1.26 : 1 for the spectra measured at 270 K and 295 K, respectively. Taken from [39]

NMR spectra in chloroform, one observes changes of the chemical shift. The largest shift, of about 10 ppm (here 1500 Hz), occurs for the signal of the =CH– group. However, the linewidth of this signal, as for all other ^{13}C signals, is about 1 Hz, which is mainly caused by $\mathbf{B}_0$ inhomogeneities. The contribution of the chemical exchange to the linewidth [44, 45] is rather small. To calculate the rate, the chemical shift-difference of the ^{13}C signals in the physisorbed and the adsorbed state must be known. This difference must be larger than 1500 Hz, resulting in an exchange rate larger than $2 \times 10^4 \ \mathrm{s}^{-1}$.

The results of the simulations for two representative temperatures are shown in comparison with the experimental spectra in Fig. 15. The observed spectra could be reasonably reproduced by keeping the parameters of the spin system, as discussed, constant but varying the functions of adsorbed and free molecules, viz. the quantity $N_C/(N - N_C)$. Using (25), equilibrium constants can be calculated using the $N_C/(N - N_C)$ derived by simulating the spectra for the group =CH$_2$ measured at various temperatures. The values are shown in Table 1.

There is good agreement between the value of $K \approx 0.3$ (at $T \approx 295$ K) estimated above and the values obtained at 295 K and 270 K in Table 1. This shows that the spin system parameters chosen were quite reasonable.

From the observed and simulated spectra, we can draw the following conclusions. The changes in the signal structure of the =CH$_2$ group at varying temperature can be described by a dynamic equilibrium of bonded and physisorbed 1-butene in the NaX zeolite. In both states, the proton NMR line shape for the group =CH$_2$ is characterized by distinct changes in the chemical shifts for the two nonequivalent

Table 1. Calculation of the equilibrium constant K using (25). $N = 1$ is the number of 1-butene molecules per large cavity. N_C gives the number of bonded molecules per large cavity

T/K	295	270
$N_C/(N - N_C)$	1.26	2.16
N_C	0.557	0.68
K	0.255	0.448

protons but about the same values of the $^3J_{HH}$ coupling constants. This suggests that there are no changes in the sp^2 hybridization in the olefinic part caused by the adsorption. On the other hand, the distinct changes in the ^{1}H– and ^{13}C NMR spectra of the bonded molecule compared with those of the free one that were found for the groups =CH$_2$ and =CH–, suggest investigating whether there are any detectable changes in the conformation of the "bonded" adsorbed molecules in comparison with the "free" ones. For this reason, we have also measured intramolecular proton–proton cross–relaxation rates using ^{1}H nuclear Overhauser effect spectroscopy (NOESY).

3 Two-Dimensional MAS NMR Spectroscopy of Adsorbed Molecules

3.1 Introduction. New Possibilities for 2D NMR

Depending on the correlation time τ_c of molecular reorientation, different types of NMR experiments can be carried out. In the approximate range 10^{-10} s $< \tau_c < 10^{-7}$ s for the correlation times, favorable conditions exist for the 2D NMR experiments typical for liquids [38]. Typical examples of ^{1}H COSY NMR and ^{1}H NOESY NMR spectra are shown in Figs. 16 and 17, respectively. The 2D representation shown in Fig. 17 was recorded using TPPI (time-proportional phase incrementation) and the phase cycle given in [12] to suppress J cross-peaks in 512×512 data points. With a recycle delay of 2 s (the biggest proton T_1 value was about 400 ms), a mixing time of 800 ms and 16 scans for each free induction decay (FID) the overall recording time was about 6 h. It was transformed to a 2D spectrum of size 1024×1024 points and phase- and baseline-corrected in each dimension. To obtain quantitatively reliable cross-peak intensities, especially at shorter mixing times, measurements with increased signal-to-noise ratio are necessary.

The most important conclusion in this context is that the resolution is enhanced even at modest spinning rates (as low as 1 to 2 kHz) to such an extent that scalar interactions may dominate, as in high-resolution NMR spectroscopy of liquids. Hence, the utility of one and two-dimensional heteronuclear pulse sequences known from liquid-phase NMR may be demonstrated for structure investigations of adsorbed molecules. Also, without additional application of MAS techniques, it is well known (see [16]) that 1D ^{13}C NMR spectra without proton decoupling yield line splittings

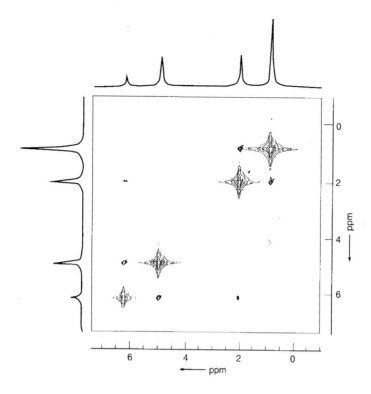

Fig. 16. Two-dimensional correlation spectroscopy (COSY): 2D ^{1}H NMR spectrum of 1-butene adsorbed in NaX. The magnitude spectrum clearly shows the three J-couplings between protons of neighboring molecular groups. Taken from [38]

due to scalar ^{1}H, ^{13}C coupling over one bond. Thus, it is not surprising that J-resolved 2D experiments can be performed rather easily for adsorbed molecules. The realization of better conditions for measuring highly resolved spectra discussed above led also to a considerable improvement of the resolution in the 2D heteronuclear ^{13}C, ^{1}H shift correlated spectra, as shown in Fig. 18, which is of general interest because of the importance of this experiment for spectral assignment when more complicated spectra of adsorbed species have to be interpreted. The advantage of the detection of the information inherent in the proton NMR spectra via heteronuclear 2D NMR spectra of adsorbed molecules is obvious, since in those cases where the ^{1}H NMR lines are not separated in a sufficient manner, a 2D representation of the proton part in a shift correlated spectrum leads also in fact to an increase in resolution. In a still better way, this advantage may be inferred from a 2D representation of ^{13}C, ^{1}H multiple-quantum NMR spectra (Fig. 19).

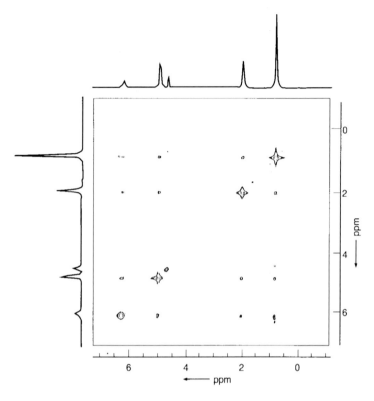

Fig. 17. Proton NOESY spectrum of 1-butene adsorbed in NaX. The additional diagonal peak at ca. 4.6 ppm is induced by a small amount of protons contained in deuterated water which was used to lock the spectrometer field. The spectrum has been phase-corrected in both dimensions. Positive cross-peaks indicate a thermal correlation time longer than 10^{-10} s. Taken from [38]

3.2 ^{1}H NOESY NMR on Adsorbed Molecules

The ^{1}H NOESY method is especially important because it provides the basis for studying structural changes in the course of the adsorption also. For instance, by applying the same experiments and strategies as applied to biomolecules, NOESY experiments can contribute to the study of the dynamics and conformation of adsorbed species. Since the molecules adsorbed in zeolites for example are of much smaller size, typically hydrocarbons with up to ten carbon atoms, the proton NMR spectra are much less crowded, leading to new possibilities for the determination and interpretation of cross-relaxation rates in comparison with the situation encountered for biomolecules in solution. Hence, adsorbed molecules are potentially suited as a model system for the refined interpretation of ^{1}H NOESY NMR measurements. For instance, variation of adsorbate, adsorbent, loading, and temperature leaves much choice to the experimentalist for methodical investigations of intra- and intermolecular cross-relaxation in simple molecules and of the influence of internal motions.

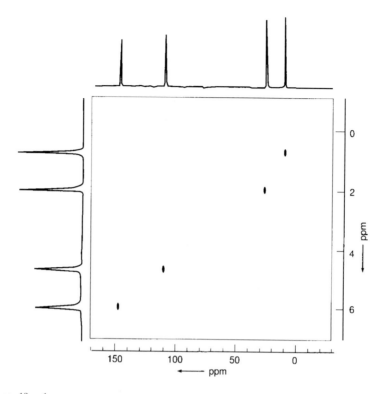

Fig. 18. ^{13}C, ^{1}H heteronuclear chemical shift correlated 2D spectrum of 1-butene absorbed in NaX. The pulse sequence and data processing have been chosen to obtain pure absorption peaks. Exponential multiplication with a line broadening of 50 Hz has been applied in the ^{13}C domain (top). The overall recording time was about 10 h. Taken from [38]

Cross-Relaxation

Cross-relaxation processes among nuclear spins may be described by a linear system of differential equations,

$$\frac{\mathrm{d}}{\mathrm{dt}}I_i(t) = -\sum_j R_{ij}I_j(t), \tag{27}$$

where I_i denotes the deviation of the magnetization of the proton species i from its thermal-equilibrium value and R_{ij} are elements of the dynamical matrix $\mathbf{R}$. The analytical expressions for R_{ij} in the case of dipolar relaxation among protons are based on the Solomon equations for a two-spin system [46], generalized to an arbitrary number of participating spins by neglecting cross-correlation effects [47,48]:

$$R_{ij} = \frac{1}{20}\left(\frac{\mu_0}{4\pi}\gamma^2\hbar\right)^2 [6J_{ij}(2\omega_0) - J_{ij}(0)] \quad \text{for} \quad i \neq j, \tag{28}$$

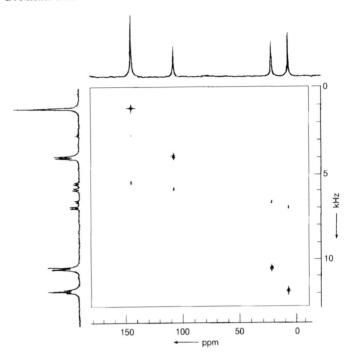

Fig. 19. Multiple-quantum 2D spectrum. The first dimension shows ^{13}C chemical shifts (top). In the second dimension, double-quantum (strong intensities) and zero-quantum (weak intensities) frequencies are obtained for each molecular group. Disregarding the line splittings due to J-coupling, the double- and zero-quantum frequencies are given by the sum $\omega_C + \omega_H$ and the difference $\omega_C - \omega_H$ of the offset frequencies for carbons (C) and protons (H), respectively. The corresponding ^{1}H spectrum can be constructed by evaluating the difference between the frequencies of the zero- and double-quantum peaks for each molecular group, which is equal to two times the proton resonance offset ω_H. Taken from [38]

$$R_{ii} = \sum_{j \neq i} \frac{1}{20} \left(\frac{\mu_0}{4\pi} \gamma^2 \hbar \right)^2 \left[6 J_{ij}(2\omega_0) + 3 J_{ij}(\omega_0) + J_{ij}(0) \right]. \qquad (29)$$

Here γ and ω_0 are the gyromagnetic ratio and the Larmor frequency, respectively, of the proton spin. The generalized spectral density $J_{ij}(\omega)$ characterizes the stochastic time dependence of the vector $\mathbf{r}_{ij}$ connecting spins i and j. For a fixed distance r_{ij} and an isotropic reorientation of $\mathbf{r}_{ij}$ with a correlation time τ_c, the generalized spectral density is given by

$$J_{ij}(\omega) = \frac{1}{r_{ij}^6} \frac{2\tau_c}{1 + \omega^2 \tau_c^2}. \qquad (30)$$

More complex motional models lead to different expressions for $J_{ij}(\omega)$ [47,49–51] but do not affect the expressions in (28) and (29). If cross-correlation effects play an essential role, additional intensities I_k should be included in (27), representing multiple-spin magnetization. They leave, however, cross-relaxation rates between single spin magnetizations ((28)) and (29) unchanged [48]. Cross-relaxation rates among all longitudinal (single- and multiple-) spin magnetizations of a three-spin system caused by dipolar interaction and chemical shift anisotropy have recently been reported by Chaudhry et al. [52]. The analytical expressions given in [52] indicate a dominance of cross-relaxation among single-spin magnetizations in a proton spin system for low molecular mobility ($\omega_0\tau_c > 1$) since (i) the corresponding rates R_{ij} are the only ones containing a generalized spectral density at zero frequency $J_{ij}(0)$, and (ii) chemical-shift anisotropy can be considered to be small compared with dipolar interactions. Therefore, we shall neglect cross-correlation henceforth. Note that dipolar interaction is the dominant interaction between proton spins (besides negligible scalar J-coupling) and is thus the only possible contribution to the cross-relaxation rates R_{ij}. Other relaxation mechanisms, such as those originating from paramagnetic impurities or proton chemical-shift anisotropy, affect only diagonal elements of $\mathbf{R}$. They can be taken into account in (28) and (29) by adding an "external" relaxation rate R_i^{ext} to each diagonal element R_{ii}. Magnetically equivalent protons, frequently occurring in methyl and methylene groups, can be adequately represented by a single index, in which case (28) and (29) are slightly modified [52].

The measurement of the matrix $\mathbf{R}$ is possible by ^{1}H NOESY NMR spectroscopy. The integrated peak volumes $N_{ij}(\tau)$ in a two-dimensional NOESY spectrum at a mixing time τ form the intensity matrix $\mathbf{N}(\tau)$, which is given by [53, 54]

$$\mathbf{N}(\tau) = [\exp(\mathbf{R}\tau)]\,\mathbf{N}(0). \qquad (31)$$

$\mathbf{N}(0)$ corresponds to the NOESY spectrum with vanishing mixing period: its diagonal elements $N_{ii}(0)$ contain the magnetizations I_i at the beginning of the mixing period, while all off-diagonal elements are zero. To extract R_{ij} from the NOESY spectra, different strategies are possible. If all peak intensities N_{ij} (including the diagonal peaks) for two different mixing times are known with sufficient accuracy, the matrix $\mathbf{R}$ can be calculated by using the matrix logarithm [54, 55]:

$$\mathbf{R} = \frac{1}{\tau_1 - \tau_0} \log\left[\mathbf{N}(\tau_1)\mathbf{N}^{-1}(\tau_0)\right]. \qquad (32)$$

If the intensity matrix $\mathbf{N}(\tau)$, has been determined for a series of different mixing times τ, (31) can be used to fit $\mathbf{R}$ to $N(\tau)$ in complete analogy to a single-exponential fit of longitudinal relaxation data. A less rigorous approach is needed if some intensities N_{ij} are inaccessible to quantitative measurements, as is often the case for diagonal peak intensities N_{ii} in the NOESY spectra of biomolecules. Then the experimental determination of $\mathbf{R}$ as described above is no longer possible, and the $N_{ij}(\tau)$ have to be interpreted directly. In contrast to the entry in the cross-relaxation matrix R_{ij}, each $N_{ij}(\tau)$ is, however, affected by all spins and not only by the pair of spins i and j. An interpretation of $N_{ij}(\tau)$ is usually performed by changing the parameters characterizing the conformation and dynamics of a hypothetical molecule until maximum agreement between the corresponding simulated intensities (27)–(30) and the experimentally obtained cross-peak intensities is reached ("full matrix approach") [56–58]. Thus, a model for the conformation and dynamics of the complete molecule is necessary in order to interpret the $N_{ij}(\tau)$.

From this point of view the interpretation of R_{ij} is much easier, since only a model of the time dependence of the distance vector connecting protons i and j is required. Hence, the determination of **R** amounts to a decomposition of the relaxation network and, therefore, serves as a convenient intermediate step between the determination of NOESY cross-peak intensities and their interpretation in terms of model parameters. For NOESY applications to adsorbed molecules, this procedure may be useful [59].

^{1}H NOESY NMR of Allyl Alcohol in NaX Zeolite

The experiments were carried out using an MSL 500 spectrometer (500 MHz proton resonance frequency; Bruker Biospin, Karlsruhe) under conditions of MAS with a rotor frequency of 4.1 kHz and at a variable temperature. The ^{1}H NOESY NMR spectra were recorded at 256×512 time domain points, transformed to 512×512 points

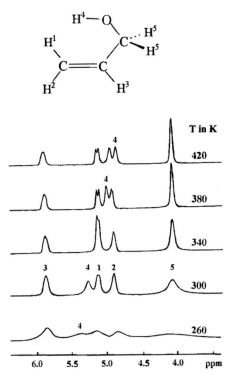

Fig. 20. Temperature dependence of ^{1}H NMR spectra of allyl alcohol adsorbed on NaX. One conformation (sp, sc) of allyl alcohol is sketched on the top. Peak assignment is indicated on the room temperature spectrum. Taken from [59]

in the 2D spectrum. Time-proportional phase incrementation was used to obtain pure absorption spectra. Suitable phase cycling schemes allowed only zero-quantum coherences to survive the mixing period [11]. 2D peak intensities (peak volumes) were determined by integration over the corresponding spectral region. ^{1}H NMR spectra of allyl alcohol adsorbed on NaX zeolite are shown in Fig. 20. At room temperature, all five magnetically inequivalent protons in the molecule are resolved. Individual peaks have been assigned by comparison with spectra of liquid allyl alcohol. The position

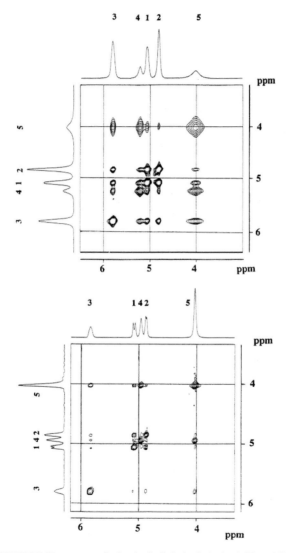

Fig. 21. ^{1}H NOESY NMR spectra of adsorbed allyl alcohol: (top) $T = 300$ K, $t = 80$ ms; (bottom) $T = 370$ K, $t = 400$ ms. Taken from [59]

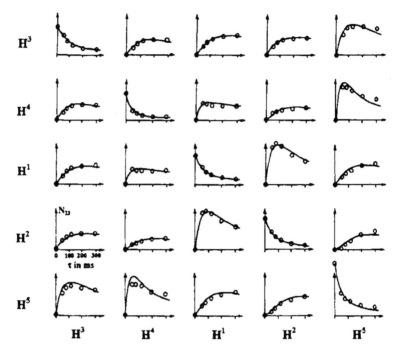

Fig. 22. Mixing-time dependence of NOESY peak intensities at 300 K. Proton labels are indicated at the bottom and left margins of the 5×5 matrix. Each entry represents the mixing-time dependence of the corresponding 2D NOESY intensity. A coordinate system has been sketched for $N_{23}(\tau)$. The vertical scale for diagonal entries N_{ii} is scaled by a factor of 4 compared with that for off-diagonal entries. The closest fit with a cross-relaxation rate matrix (which is $\mathbf{R}^{\text{ex}}$ (300 K) given in the text) is marked by a solid line. Taken from [59]

of the OH peak shifts upfield with increasing temperature. While strong line overlap dominates the appearance of the spectrum at 260 K, ^{1}H NMR linewidths are reduced with increasing molecular mobility at higher temperatures. Line splitting caused by the strongest proton J-couplings can already be observed at 300 K. NOESY spectra were taken at temperatures of 300 and 370 K (Fig. 21), where the hydroxyl group overlaps only to a small extent with other proton lines. The determination of 2D peak volumes is possible by integrating the intensity over appropriately chosen rectangular areas in the 2D spectrum. A mixing-time variation, including five (at $T = 300$ K) and three (at $T = 370$ K) mixing times τ, has been recorded. To obtain a convenient representation, all peak volumes have been arranged in a matrix, each entry of which is a plot of the time dependence of the corresponding 2D peak intensity (Fig. 22). In order to obtain their relative intensities, a one-dimensional single-pulse proton spectrum were been taken with the same repetition time as for the NOESY spectra. The integrated 1D intensities correspond to the intensities present at the beginning of the mixing period for a vanishing preparation time ($t_1 = 0$ in the F1 time domain), and hence to the 2D peak volumes for vanishing mixing time. To achieve the proper absolute intensity, a single scaling factor is necessary, which was adjusted visually to the data obtained from the 2D spectra.

Since all magnetically inequivalent protons are spectroscopically resolved, the mixing-time variation of the NOESY peak volumes can be interpreted directly with (31). A nonlinear fit procedure [60] has been used to find a matrix $\mathbf{R}^{\mathrm{ex}}$, the exponential evolution (31) of which shows the minimum quadratic deviation from the experimental data (Fig. 22). The symmetry condition $n_j R_{ij}^{\mathrm{ex}} = n_i R_{ji}^{\mathrm{ex}}$ (n_k is the number of magnetically equivalent spins in position k) [61] reduces the number of independent parameters from 25 to 15. For both temperatures, the nonlinear fit procedure led to a unique matrix $\mathbf{R}^{\mathrm{ex}}$ for a broad range of different starting matrices. The deviations of the experimental intensities and their best matrix fit mostly remain within the limits of experimental error (Fig. 22). The matrices obtained are

$$\mathbf{R}^{\mathrm{ex}}(300\,\mathrm{K}) = \begin{array}{ccccc} \mathrm{H}^3 & \mathrm{H}^4 & \mathrm{H}^1 & \mathrm{H}^2 & \mathrm{H}^5 \\ \left(\begin{array}{ccccc} -9.5 & 1.3 & 1.5 & 1.7 & 3.3 \\ 1.3 & -23.2 & 4.4 & 0.4 & 7.3 \\ 1.5 & 4.4 & -14.1 & 8.6 & 0.5 \\ 1.7 & 0.4 & 8.6 & -14.0 & 0.1 \\ 6.6 & 14.7 & 1.0 & 0.3 & -17.9 \end{array}\right) \end{array} \mathrm{s}^{-1} \begin{array}{c} \mathrm{H}^3 \\ \mathrm{H}^4 \\ \mathrm{H}^1 \\ \mathrm{H}^2 \\ \mathrm{H}^5 \end{array}$$

and

$$\mathbf{R}^{\mathrm{ex}}(370\,\mathrm{K}) = \begin{array}{ccccc} \mathrm{H}^3 & \mathrm{H}^1 & \mathrm{H}^4 & \mathrm{H}^2 & \mathrm{H}^5 \\ \left(\begin{array}{ccccc} -0.93 & 0.11 & 0.11 & 0.13 & 0.20 \\ 0.11 & -1.20 & 0.34 & 0.55 & 0.07 \\ 0.11 & 0.34 & -1.84 & 0.08 & 0.64 \\ 0.13 & 0.55 & 0.08 & -1.27 & 0.06 \\ 0.39 & 0.14 & 1.27 & 0.11 & -1.76 \end{array}\right) \end{array} \mathrm{s}^{-1} \begin{array}{c} \mathrm{H}^3 \\ \mathrm{H}^1 \\ \mathrm{H}^4 \\ \mathrm{H}^2 \\ \mathrm{H}^5 \end{array}$$

The rows and columns are labeled with the corresponding protons (compare Fig. 20). Note that the indices always refer to proton labels and not to the position in the matrices given here. The latter positions have been chosen to match the appearance of the 2D NOESY spectra.

Until now, no assumptions concerning the conformation and dynamics of the molecule have been necessary, except some rather general implications ensuring the validity of (27). However, models have to be used for the interpretation of entries in $\mathbf{R}^{\mathrm{ex}}$. Off-diagonal entries R_{ij}^{ex} are affected exclusively by the position and motion of spins i and j. The most suitable for interpretation are elements R_{ij}^{ex} for $i, j = 1$–3, since the allyl group can be approximated as rigid [61]. Further, we assume the reorientation of this group to be isotropic and therefore (30) to be applicable. The correlation time τ_{c} of thermal reorientation remains the only adjustable parameter. Table 2 lists experimental values R_{ij}^{ex} and model values R_{ij}^{mod}. Since the highest cross-relaxation rate is least influenced by errors, the correlation time τ_{c} has been determined by matching values for R_{12}^{ex} and R_{12}^{mod}. For room temperature, we find

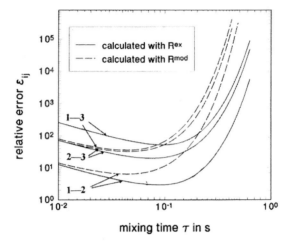

Fig. 23. Relative error propagator ϵ_{ij} for the proton pairs of the allyl group calculated using the experimental and the model values for **R**. The optimum mixing times, corresponding to the smallest experimental error, range from 50 to 100 ms. The presence of more protons (six compared with three for the model matrix) leads to an increase of the relative error rates calculated with the experimental matrix and a shortening of the optimum mixing time. Taken from [59]

$\tau_c = 6.1 \times 10^{-9}$ s. Thus, NOESY experiments at a single temperature allow an estimation of the molecular mobility of adsorbed species.

Conventionally, this information is obtained from *diagonal* elements in **R**, for instance as in the full relaxation analysis based on the temperature dependence of longitudinal relaxation times T_1 [6]. Although easier to measure, longitudinal relaxation times in many cases can be defined only by neglecting cross-relaxation. Additionally, they may be affected by all and not just a selected pair of spins (28) and (29) and by external relaxation such as relaxation due to paramagnetic impurities. Determination of *off-diagonal* elements in **R** is more sophisticated; their definition, however, is well founded and their physical interpretation straightforward. The latter

Table 2. Relaxation matrix entries and corresponding interproton distances for the allyl group, as determined experimentally for 300 K and evaluated by model calculations

Proton pair	$H^1 - H^2$	$H^1 - H^3$	$H^2 - H^3$
$R_{ij}^{\mathrm{mod}}(\tau_c = 6.1 \times 10^{-9}\ \mathrm{s})$	8.6 s^{-1}	0.41 s^{-1}	1.53 s^{-1}
r_{ij}^{mod}	1.86 Å	3.09 Å	2.48 Å
$R_{ij}^{\mathrm{ex}}(300\ \mathrm{K})$	8.6 s^{-1}	1.5 s^{-1}	1.7 s^{-1}
$\Delta R_{ij}^{\mathrm{ex}}$	1.1 s^{-1}	1.1 s^{-1}	1.1 s^{-1}
r_{ij}^{ex}	1.86 Å	2.66 Å	2.44 Å
$r_{ij}^{\mathrm{ex}} \pm \Delta r_{ij\pm}^{\mathrm{ex}}$	1.82–1.90 Å	2.27–3.10 Å	2.24–2.90 Å

is based on the fact that only direct interactions between two spins may give rise to direct magnetization transfer. Therefore, only internuclear dipolar interactions have to be taken into account, leading to a simple and unambiguous interpretation of cross-relaxation rates (see (28)). Therefore, we think the determination and interpretation of cross-relaxation rates to be a valuable extension of the capabilities of NMR to investigate molecular mobilities.

The two remaining experimentally obtained cross-relaxation rates within the allyl group have to follow the same model and correlation time. Hence, they can be translated into internuclear distances r_{ij}^{ex}, which can be compared with distances r_{ij}^{mod} taken from [62]. While the experimental cross-relaxation rate for proton pair 2–3 shows good agreement with the model value, for proton pair 1–3, with the longest internuclear distance, a deviation by a factor of almost four is observed (see Table 2). Owing to the inverse sixth-power dependence of the cross-relaxation rate on the internuclear distance r_{13} (see (30)), the experimental and model values for r_{13} differ only by less than 20%. To specify the accuracy of the experimental cross-relaxation rates R_{ij}, an analysis of the error propagation from the 2D peak volumes $N_{ij}(\tau)$ into the rates R_{ij} is necessary. An adequate treatment of this problem has recently been given by Macura [61]. He assumes equal and uncorrelated volume errors $\Delta N_{ij} = \Delta a$. Errors in the cross-relaxation rates are characterized by the relative error propagator ϵ_{ij} defined by $\epsilon_{ij}\Delta a = \Delta R_{ij}/R_{ij}$. On the basis of (32), with $\mathbf{N}(\tau_0)$ being the unity matrix, Macura derives an analytical expression for ϵ_{ij} which depends on the mixing time τ and the cross-relaxation network, characterized by the matrix $\mathbf{R}$. The relative error propagator ϵ_{ij} for the allyl group obtained with the experimental and model values of R_{ij} is depicted in Fig. 23 for varying mixing time. As has already been pointed out by Macura for model geometries, error propagation within the cross-relaxation network leads to highly inaccurate values for small cross-relaxation rates (such as R_{13}) owing to the presence of high rates (R_{12}). Absolute errors $\Delta R_{ij}^{\mathrm{ex}}$, determined according to Macura with the experimental matrix $\mathbf{R}^{\mathrm{ex}}$ (300 K) for a mixing time of $\tau = 50$ ms, and the corresponding error limits for the distance $r_{ij}^{\mathrm{ex}} \pm \Delta r_{ij}^{\mathrm{ex}}$ are included in Table 2. On the basis of R_{12}^{ex}, the error for the correlation time τ_{c} has been estimated to be $\Delta\tau_{\mathrm{c}} = 0.4 \times 10^{-9}$ s. Owing to the determination of $\mathbf{R}$ by a fit to a whole mixing-time dependence, the real experimental error of the relaxation rates R_{ij}^{ex} is expected to be smaller. Hence, the discrepancy between the R_{13} values obtained experimentally and with the model cannot be attributed only to experimental errors. A similar problem will arise in the study of 1-butene molecules adsorbed in zeolites of type NaX. Hence, reasons for this deviation will be still discussed in the next section.

^{1}H NOESY NMR of 1-Butene in NaX Zeolite

The experiments were carried out at proton resonance frequencies of 300, 500, and 600 MHz (MSL 300, MSL 500, and DRX 600 NMR spectrometers, Bruker Biospin Karlsruhe) under MAS conditions with rotor frequencies of 3 to 6 kHz and at variable temperatures [39, 63, 64]. A recycle delay of 10 s and a pulse width of ca. 12 μs which is a typical value for liquid-state NMR measurements, were used. The ^{1}H

MAS NOESY NMR spectra were recorded under similar conditions as mentioned in Sect. 3.2. The mixing times were varied from 0.8 to 2.1 s. Depending on the pore filling factor, 16 or 32 scans were taken for each measurement of a free induction decay (FID). The 2D NMR spectra were transformed to 1024×1024 points and were phase- and baseline-corrected in each dimension. In [42] it is shown that ^{1}H NOESY NMR spectra for the system 1-butene/NaX may be measured down to very low pore filling factors, such as 0.05 molecules per supercage. This allows, in particular, the study of the behavior of quasi-isolated molecules in zeolite cages. Over a wide range of coverage, positive cross-peaks were obtained. For the samples with 0.75 and 0.05 molecules per supercage, a mixing-time variation of the 2D peak volumes was recorded. A nonlinear fit procedure of the cross-relaxation matrix was used to find the relaxation rate matrices. The results were used for a more detailed analysis of the dynamics of reorientation and other motions of the adsorbed molecules, as outlined in more detail in Sect. 4.5.

For a further analysis of cross-relaxation experiments with adsorbed molecules to derive structural information, however, we have also to take into account that we observe average values for the relaxation parameters. In principle, under the assumption of a fast exchange (see Sect. 2.2 and 2.3), one observes an averaged cross-relaxation matrix [65], i.e. the cross-relaxation rates

$$R_{ij}^{\mathrm{obs}} = p_{\mathrm{C}} R_{ij}^{\mathrm{C}} + p_{\mathrm{M}} R_{ij}^{\mathrm{M}} \tag{33}$$

with the relative fractions being p_{C} and p_{M}. Here, the definition of "fast" is related to an exchange rate $k \gg \max(R_{ij}^{\mathrm{C}}, R_{ij}^{\mathrm{M}})$. The relation is fulfilled for the system investigated.

The dipole–dipole cross-relaxation rates R_{ij} between two protons i and j are defined in the usual way,

$$R_{ij} = \frac{1}{20} \left(\frac{\mu_0}{4\pi} \gamma^2 \hbar \right)^2 \frac{1}{r_{ij}^6} [6 J_{ij}(2\omega) - J_{ij}(\omega)] = \frac{F(\omega)}{r_{ij}^6}, \tag{34}$$

where a function $F(\omega)$ has been introduced to describe the influence of the thermal motion.

Although the fractions p_{C} and p_{M} can be estimated, the extraction of relevant distance information from (33) would be problematic, because the spectral density functions are still unknown. However, the discussion can be considerably simplified if we consider the real situation. First, the observed NOESY spectra have positive cross-peaks caused by negative cross-relaxation rates in the order of -0.04 s^{-1}. It is known that molecules in solution rotate freely with a correlation time of 10^{-12} s. For the purely physisorbed, i.e. free gaseous, 1-butene molecules in NaX, it can be assumed that the molecule definitely rotates not slower, but surely faster, as characterized by this correlation time. If one calculates a cross-relaxation rate using this correlation time and the standard distances (see below) of 1-butene molecules, the respective cross-relaxation should have a positive sign and the corresponding rates would be smaller than 0.001 s^{-1}. Therefore, it is clear that the influence of $p_{\mathrm{M}} R_{ij}^{\mathrm{M}}$ can be safely neglected in the discussion of the rates. The NOESY spectra and the cross-relaxation rates are dominated by the molecules that are bonded at the adsorption sites (fraction p_C).

For the further discussion, we consider only cross-relaxation rates of those proton spin pairs where we may assume that the shape of the reduced correlation function $J_{ij}(\omega)$ is the same. Hence, to arrive at this simplification we have to ensure that fast internal motions on the timescale of the correlation times for the reorientation of the whole molecule do not play any role. In particular, cross-peaks with methyl protons are not taken into account, because of the fast anisotropic rotation of the $-CH_3$ group. Moreover, there are no hints that the time constants for conformational changes are of comparable magnitude to the correlation times mentioned. We shall concentrate mainly on the dependence of the quantities R_{ij} on the factor $1/r_{ij}^6$. With respect to the distances r_{ij} we apply the structural data for 1-butene molecules in the gaseous phase according to microwave measurements by Kondo et al. [66].

The structure provided by these authors was used to measure the distances discussed, which we refer as the "standard" geometry. The other conformations of the 1-butene, which will be used for comparison, were produced by rotations of the C_2–C_3 bond without varying the distances. Of course, such an approach gives only qualitative results, but it may reasonably reproduce the experimental observation. An energy-optimized structure of the molecule shows only minor changes in the distances and the bond angles, but allows the same conclusions that are drawn below. The notation for the atoms follows the convention in [66].

Figure 24 shows the time dependence of the peak intensities for ^{1}H NOESY measurements at a temperature of 343 K. Similar measurements were also run at ambient temperature. In the matrix calculation, the peaks related to the protons in the group $=CH_2$ were not treated separately; only the mixing-time dependence of the total peak intensities has been considered so far. In particular, we are interested in the behavior of the terminal group $=CH_2$, i.e. we now try to derive separately the cross-relaxation rates R_{13} and R_{23} between the $=CH-$ proton (denoted in the following by (3), see Fig. 25) and the two different $=CH_2$ protons (denoted by (1) and (2)). In order to measure R_{23} and R_{13} separately, only the NOESY experiment at 343 K was suitable. At lower temperatures the cross-peaks overlap more strongly. As is well known, from the initial slope of the cross-peak volume versus mixing time, the correct cross-relaxation rates can estimated. In this case we would find $R_{23}/R_{13} = r_{13}^6/r_{23}^6$, since we may take the same spectral density function for the reorientation of both vectors. Using the distances $r_{23} = 2.406$ Å and $r_{13} = 3.07$ Å derived from the "standard" geometry of the sp^2-hybridized olefinic part of 1-butene, we would find a ratio value of $R_{23}/R_{13} \cong 4.3$. Note that there are no indications from the ^{13}C chemical shifts that structural changes in the olefinic part of the molecule occur. Therefore, the assumption of "standard" distances seems to be reasonable from this point of view. The experimental ratios of the cross-peaks between H^1, H^2, and H^3 were estimated via a deconvolution of the partially overlapping cross-peaks [39]. A Lorentzian deconvolution was performed on a sum of F2 rows that cover the region of the cross-peak. The result obtained for the cross-peak measured is shown in Fig. 26.

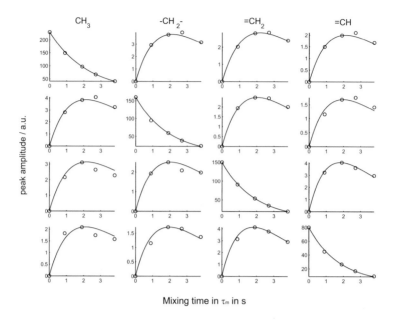

Mixing time in τ_m in s

Fig. 24. Plot of intensities of diagonal- and cross-peaks of the NOESY spectra measured at 343 K (o = experimental values, lines = calculated mixing-time dependences using the relaxation matrix **R** defined in (31)). Taken from [63]

At all mixing times $t_{\mathrm{mix}} \geq 0.9$ s, we obtained an experimental peak ratio close to one (see Fig. 27) in strong contrast to the value $\delta_{23}/\delta_{13} \cong 4.3$ mentioned above. Since we have measured at various mixing times clearly different from zero, however, it is not allowed to simply translate the ratios of the cross-peak intensities measured at a certain mixing time directly into ratios of relaxation rates and distance information, because the cross-peak intensities may grow with different rates depending on $F(\omega)$ and the distances. Therefore, to compare the theoretical and experimental ratios in a better way, we start again from the evolution of the total intensity of the =CH$_2$ group versus the mixing time, i.e. from the sum of the cross-relaxation rates originating from the pairs H^1–H^3 and H^1–H^2 as measured above. Taking the distances r_{23} and r_{13} as given above, we may estimate a value $F(\omega) \approx 12.63$ (Å)^6 s^{-1} at 343 K. This would correspond to a mean distance $r_{\mathrm{D}} = 2.6094$ Å, defined by

$$2/r_{\mathrm{D}}^6 = 1/r_{13}^6 + 1/r_{23}^6 \cdot \tag{35}$$

By means of $F(\omega)$, the intensities of the cross-peaks 1–3 and 2–3 were calculated for various mixing times in order to find the theoretical intensity ratios as shown in Fig. 27. In particular, for the mixing time of 0.9 s, close to the maximum of the peak volumes, we find a theoretical ratio of ca. 3, which clearly differs from the experimental values and indicates that the distances used cannot be applied for a further analysis. The experimental ratio (1.25 ± 0.2) practically does not depend on the mixing time.

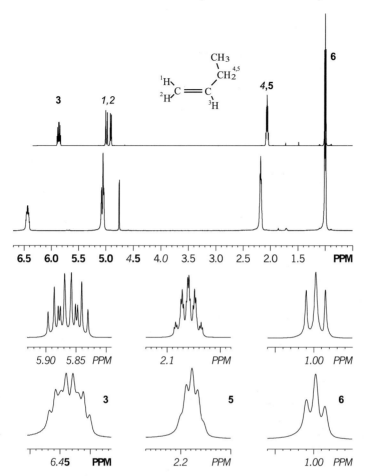

Fig. 25. Comparison of ^{1}H NMR spectra of 1-butene ($T = 298$ K, $\nu_H = 600.13$ MHz, $\nu_r = 5$ kHz) adsorbed (second from top) in NaX zeolite (1 molecule per large cavity) and dissolved in CDCl$_3$ (top). The expanded regions (bottom) show the line shapes for the groups 3 (=CH–), 4/5 (–CH$_2$–), and 6 (–CH$_3$) as indicated in the structure formula of 1-butene. In the spectrum of the adsorbed molecules, the –CH$_3$ group chemical shift was calibrated to 1 ppm. Since an appreciable influence of adsorption interactions on the ^{1}H chemical shift in the –CH$_3$ group was not found, this calibration also includes an approximation to the shielding correction. The clear changes in the chemical shifts for the other molecular groups are due to adsorption interactions. Because the ^{1}H NMR line shapes for the protons 1 and 2 are strongly temperature-dependent, details are shown only in the following figures. Taken from [63]

Therefore we may assume that it reflects the real ratio of averaged cross-relaxation rates $\langle R_{23}\rangle/\langle R_{13}\rangle$, which may be translated into distance information. This means that from the respective distance ratio $(\langle R_{23}\rangle/\langle R_{13}\rangle)^{1/6} = \langle r_{13}\rangle/\langle r_{23}\rangle \approx 1.04$, we might conclude that both protons (1, 2) are approximately equidistant from proton 3. These distances lie between the standard values of $r_{23} = 2.406$ Å and $r_{13} = 3.07$ Å

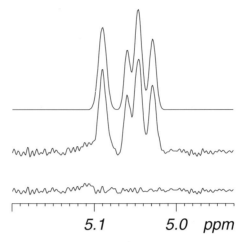

Fig. 26. Sum of slices in the F2 direction of a cross-peak in a ^{1}H NOESY spectrum ($T = 343$ K, mixing time 0.9 s) between the protons in the groups =CH– and =CH$_2$. A deconvolution using Gaussian line shapes was performed for the two line pairs to compare their integrals. No significant change was found when a Lorentzian line shape was used. The bottom trace shows the difference between the fitted curve (top) and the experimental spectra (middle). The integration leads to an amplitude ratio of the cross-peaks of about 1.2. Taken from [63]

used above and cannot be assumed to be equal to the mean distance r_D (see (35)). This also means that the estimation of the spectral density function $F(\omega)$ given above is poor. The finding may be the result of either an exchange of the molecules between those attached to adsorption sites and the remaining physisorbed ones (as clearly inferred from the spectra described in Sect. 2.3) and/or a (probably slow) proton delocalization within the group =CH$_2$ of the adsorbed molecule, as suggested in a previous paper [64]. In the first case the average experimental distance $\langle r_{13} \rangle \approx \langle r_{23} \rangle$ should depend on the population numbers and thus change with temperature. The accuracy achieved is not sufficient to allow further conclusions about the latter question, because of the poor estimation of the spectral density function $F(\omega)$. However, it should be mentioned again, that the general conclusion about the approximately equal mean values $\langle r_{13} \rangle \approx \langle r_{23} \rangle$ is independent of the estimation of $F(\omega)$.

We are interested, further, in estimating values for the intramolecular distances, especially the distances r_{13}, r_{23}, r_{24}, r_{25}, r_{34}, and r_{35}, to get information about the conformation of the adsorbed molecules. In the following, we discuss only the ratios between the cross-relaxation rates. Assuming equal spectral density functions $F(\omega)$, we find, for two different spin pairs i, j and k, l

$$\frac{R_{ij}}{R_{kl}} = \left(\frac{r_{kl}}{r_{ij}} \right)^6. \tag{36}$$

For 1-butene molecules in the gaseous phase the existence of rotational isomers has been investigated in detail [66,67]. It has been concluded [66] that the *skew* form

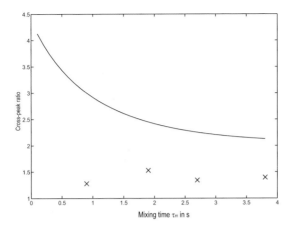

Fig. 27. Comparison of measured (x) and theoretical (line) intensity ratios of the cross-peaks between the protons of the $=CH_2$ and $=CH-$ groups at different mixing times. The statistical error of the measured ratios is less than about 0.1. An estimation of the systematic errors has not been performed. Since the experimental values do not deviate appreciably for all mixing times, the deviation between the experimental values and the calculated ones (see text) is obvious. Taken from [63]

is more stable in free 1-butene, though the energy difference is only ca. 630 J/mol. MO calculations of the interaction between 1-butene molecules and Na^+ cations [40,41] have shown that in this case the *skew* form is even more stable than the *cis* rotamer. These results should be applicable also to 1-butene molecules adsorbed in NaX zeolites, i.e. in adsorbed 1-butene the *skew* rotamer is more probable too. In this form the symmetry plane, (of the four carbons in the *cis* rotamer) is absent and, hence, the two protons in the group $-CH_2-$ should be chemically nonequivalent. This is not seen in the ^1H NMR spectra. The ^1H NMR spectra for adsorbed 1-butene molecules clearly reveal only one line for the two $-CH_2-$ protons. This could be understood in terms of an averaging process between the two *skew* configurations, although the potential barrier between the two *skew* positions in (free) 1-butene is 6.9 kJ/mol [66]. Hence, from the point of view of our ^1H NMR measurements, we might also take the *trans* configuration for adsorbed 1-butene as an average in a first approximation, in order to estimate distances derived from the ^1H NOESY NMR data. To eliminate the factor $F(\omega)$ in the cross-relaxation rates for various cross-peaks, we take the distances r_{34} and r_{45} between the proton in the group $=CH-$, position (1), and the two protons of the aliphatic $-CH_2-$ group, positions (4, 5), for the different rotamers and use the respective cross-relaxation rate as a "calibration" for a further discussion of distances estimated from the cross-relaxation rates.

We specify the following rates. The cross-relaxation rate R_{al} (aliphatic) is the sum of the two rates R_{34} and R_{35},

$$R_{\text{al}} = R_{34} + R_{35} = F\left(\omega\right)\left[\frac{1}{r_{34}^{6}} + \frac{1}{r_{35}^{6}}\right] = F\left(\omega\right)\frac{2}{\langle r_{\text{al}}\rangle^{6}}, \tag{37}$$

and we may use this relation to define an average distance r_{al} depending on the conformation considered. In the case of the cross-relaxation rates $R_{1\text{al}}$ and $R_{2\text{al}}$, we have a cross-peak with the two chemically identical protons in positions 4 and 5, i.e. we observe for instance a rate

$$R_{2\text{al}} = \frac{2F\left(\omega\right)}{\langle r_{2\text{al}}\rangle},$$

with

$$\frac{2}{\langle r_{2\text{al}}\rangle^{6}} = \frac{1}{r_{24}^{6}} + \frac{1}{r_{25}^{6}}$$

in analogy to (37). In case of the *skew* rotamers, we may even have different distances $r_{24} \neq r_{25}$. As already mentioned, the cross-relaxation rates R_{23} and R_{13} are approximately the same. In the numerical values of Table 3, we assume $R_{13} = R_{23}$ and $\langle r_{13}\rangle = \langle r_{23}\rangle$. From the ratios of the experimental cross-relaxation rates, $\langle R_{34}\rangle / (R_{13} + R_{23}) = \langle R_{34}\rangle / (2R_{13}) = (\langle r_{13}\rangle / \langle r_{34}\rangle)^{6}$ and $R_{34}/R_{24} = (\langle r_{24}\rangle / \langle r_{14}\rangle)^{6}$ (see also (33)), we obtain the following relations for the proton–proton distances,

$$r_{\text{D}} = 0.86\,\langle r_{24}\rangle \quad \text{and} \quad \langle r_{14}\rangle = \langle r_{24}\rangle = 1.06\,\langle r_{34}\rangle,$$

which can be compared with distance estimations for the conformations *cis*, *trans*, and *skew* of 1-butene molecules, as shown in Table 3. In the *trans* conformation, for instance, the distances in (32) are equal to $\langle r_{34}\rangle = 2.916$ Å. From the cross-relaxation rate $R_{34} = 2F(\omega)/\langle r_{34}\rangle^{6} = 0.031$ s^{-1} (at a resonance frequency of 600.13 MHz and $T = 343$ K), we may also derive values for $F(\omega)$. These values will not be discussed here further.

Apparently, the assumption of the *trans* conformation in the adsorbed state leads to the best adaptation to the experimental values. Mixed conformations as observed in the gaseous state or for molecules in solutions, i.e. *cis* and *skew* conformations seem not to be appropriate. This conclusion is in agreement with the discussion about the MO calculations mentioned above [40, 41]. Obviously, the conformation of the bonded adsorbed molecules is reflected mainly in the chemical shifts of the terminal group $=CH_2$ at 230 K. At higher temperatures, the relative fraction of the adsorbed molecules is much smaller with respect to that for the "physisorbed" ones, which have a similar conformation to that in solution or in the "free" (gaseous) state. This change in the population leads to drastic changes in the spectra measured for the $=CH_2$ protons at room temperatures, and higher, which now reflect the conformation of the "free" molecules. A similar conclusion can also be drawn when we inspect the ^{1}H NMR spectra for the group $=CH_2$ at a very low total number of adsorbed molecules, which are similar to those at higher loading and low temperatures. This clearly shows the strong influence of the population of the two different states of the adsorbed molecule on the adsorption dynamics and confirms the conclusions about changes in the conformation.

Table 3. Intramolecular proton distances for various rotamers and comparison with experimental ratios derived from ^{1}H NOESY NMR measurements on adsorbed 1-butene. The measurements were carried out at a resonance frequency of 600.13 MHz, at 343 K, and for a loading of about 1 molecule per large cavity. The distances r_{ik} and the factors $F(\omega)$ are given in units of Å (10^{-10} m) and (Å)^6 s^{-1}, respectively. In the numerical values, we have assumed $\langle r_{13} \rangle = \langle r_{23} \rangle$ for the position "D" and $\langle r_{14} \rangle = \langle r_{24} \rangle$ (for *cis*, *trans*) or used average distances according to (32)

	cis	*trans*	*skew*	Experiment
$r_{\mathrm{D}}/\langle r_{34} \rangle$	1.039	0.958	0.895	0.860 $r_{13} \approx r_{23}$
$\langle r_{24} \rangle/\langle r_{34} \rangle$	1.489	1.191	1.162	1.060 $r_{24} \approx r_{14}$
$\langle r_{34} \rangle$	2.511	2.724	2.916	
$F(\omega)$	3.80	6.15	9.2	

4 Thermal Mobility of Molecules in Zeolites

In the previous sections, high-resolution NMR methods were applied to elucidate the adsorption behavior and to study the structure of adsorbed molecules. Since the structures and spectral assignment of the test molecules are known, we concentrate in this section on NMR methods for the investigation of molecular dynamics, i.e. the observation and interpretation of various relaxation processes.

The conventional method for the investigation of the dynamical behavior of adsorbed molecules is the full relaxation analysis [6]. It is based on an interpretation of the temperature dependence of longitudinal and transverse relaxation times. Since some effort is necessary to quantify undesired paramagnetic contributions to relaxation, we try to avoid the interpretation of NMR relaxation times such as T_1 and T_2. Instead, various homonuclear and heteronuclear cross-relaxation rates are determined, which can be interpreted directly in terms of correlation times τ_c for the molecule or for a molecular group. Instead of extensive measurement of the easily accessible but difficult to interpret parameters, we try to use advanced NMR techniques for the determination of more directly interpretable ones. The methods outlined here are, in this sense, complementary to the full relaxation analysis. The results attained for the intracrystalline mobility of the hydrocarbons will be compared with conclusions from PFG NMR measurements of intracrystalline self-diffusion coefficients (see Chap. 1).

4.1 ^{1}H NOESY NMR Experiments

Cross-relaxation among proton spins detected by ^{1}H NOESY NMR was the basis of NMR conformation studies, as treated in the previous section. The experiments were run for allyl alcohol [4] and simple olefins, such as 1-butene, 1-pentene, and 1-heptene [42], adsorbed in NaX zeolites.

The cross-relaxation matrix $\mathbf{R}$ was derived from the matrices of intensities $\mathbf{N}(t)$ obtained experimentally for several mixing times. As is well known, the various cross-relaxation rates R_{ij} are determined only by the statistical time dependence of the vectors $\mathbf{r}_{ij}$ connecting the respective protons, which is described by the spectral density $J_{ij}(\omega)$. Thus we may arrive at an easy interpretation in the case of a fixed distance $\mathbf{r}_{ij}$ and isotropic reorientation. Assuming a molecular geometry as specified in [62], the best agreement between the experimentally obtained values R_{ij} and model calculations was found for a correlation time $\tau_c \approx 6 \times 10^{-9}$ s for the reorientation of the allyl group in NaX at a temperature of ca. 300 K [4].

^{1}H NOESY NMR spectra were also run on the system 1-butene/NaX to investigate the changes in mobility in dependence on the pore filling factor, varied in the range between 4 and about 0.05 molecules per supercage [42]. In all cases, positive cross-peaks were obtained. A nonlinear fit procedure [59] of the intensity matrix $\mathbf{N}(t)$ measured at various mixing times (0.8 s, 1.2 s, 1.5 s, 1.8 s, and 2.1 s) was used to find the relaxation rate matrices $\mathbf{R}$ for loadings of 0.75 and 0.05 molecules per supercage. For example, the matrices obtained at 290 K are given by the schemes

$$
\mathbf{R} \text{ (in s}^{-1}) =
\begin{array}{cccc}
-\text{CH}_3 & -\text{CH}_2- & =\text{CH}_2 & =\text{CH}- \\
\end{array}
\begin{pmatrix}
-0.41 & 0.14 & 0.08 & 0.13 \\
0.09 & -0.59 & 0.08 & 0.13 \\
0.05 & 0.08 & -0.50 & 0.25 \\
0.04 & 0.06 & 0.13 & -0.97
\end{pmatrix},
$$

$$
\mathbf{R} \text{ (in s}^{-1}) =
\begin{array}{cccc}
-\text{CH}_3 & -\text{CH}_2- & =\text{CH}_2 & =\text{CH}- \\
\end{array}
\begin{pmatrix}
-0.84 & 0.24 & 0.15 & 0.23 \\
0.16 & -0.99 & 0.13 & 0.20 \\
0.10 & 0.13 & -1.12 & 0.45 \\
0.08 & 0.10 & 0.22 & -1.45
\end{pmatrix}.
$$

In a first inspection of the data, a simplified model of an isotropic reorientation with a correlation time τ_c was applied, together with the assumption of a fixed distance r_{ij}. The results of this analysis are shown in Table 4 for 1-butene, 1-pentene, and 1-heptene adsorbed in NaX- type zeolites. The essential conclusion is that the correlation times τ_c are of the order of magnitude of ca. 10^{-9} s at a temperature of 290 K.

In the application of this method, several problems have to be taken into account:

(i) All magnetically nonequivalent protons have to be resolved in the [1] H NMR spectrum to justify a matrix fit to the experimental data. Otherwise, the dimension of the matrix $\mathbf{R}$ to be fitted becomes higher than the dimension of the experimentally obtained intensity matrices $\mathbf{N}(t)$. High inaccuracies in the relaxation rates would result from the increased number of fit parameters.

(ii) To interpret cross-relaxation rates, models are necessary to describe the time dependence of the positions of both protons relative to each other. This can be complicated for nonrigid molecules if the distance between the two protons under consideration is not fixed. Also, anisotropic motions may affect cross-relaxation rates.

(iii) A series of 2D NMR spectra is required. Hence, the method is in general unsuitable for a very low number of adsorbed molecules because of the low signal-to-noise ratio. At higher loadings, however, the presence of *intermolecular* cross-relaxation can complicate the interpretation of cross-relaxation rates.

(iv) For the analysis of cross-relaxation experiments with adsorbed molecules we have to take into account that we observe average values $\langle \mathbf{R} \rangle$ for the cross-relaxation matrix (see [39]), where in most cases a fast exchange between molecules bonded to adsorption sites (C) and the remaining physisorbed ones (M) may be assumed, i.e. $\langle R \rangle = p_C R_C + p_M R_M$. Although the fractions p_C and p_M might be estimated, the extraction of relevant distance information and correlation times could be problematic.

Table 4. Proton–proton cross-relaxation rates $\mathbf{R}$ between the =CH– and the =CH$_2$ groups of 1-butene, 1-pentene, and 1-heptene molecules adsorbed in NaX zeolite[a]

Olefin	Parameter	Molecules per supercage					
		0.04	0.25	0.50	0.75	1	4
1-butene	$R\,(\mathrm{s}^{-1})$	0.22	0.15		0.12	0.12	0.04
	$\tau_c\,(10^{-10}\,\mathrm{s})$	11	8.9		8.5	8.5	6.7
1-pentene	$R\,(\mathrm{s}^{-1})$		0.095	0.075	0.091		0.118[b]
	$\tau_c\,(10^{-10}\,\mathrm{s})$		7.9	7.5	7.8		6.2
1-heptene	$R\,(\mathrm{s}^{-1})$		0.64[c]				
	$\tau_c\,(10^{-10}\,\mathrm{s})$		21				

[a] The ^{1}H NOESY NMR measurements were run at ca. 290 K. The pore filling factors vary between 0.05 and 4 molecules per supercage.
[b] Measurements at a resonance frequency of 500 MHz.
[c] Measurements at a resonance frequency of 750 MHz.

In the real situation, the observed ^{1}H NMR NOESY spectra have positive cross-peaks, from which negative cross-relaxation rates in the order of -0.04 s^{-1} can be calculated. It is known that molecules in solution rotate freely with a correlation time of ca. 10^{-12} s. For purely physisorbed, i.e. for instance free gaseous, 1-butene molecules in NaX, it can be assumed that they rotate not much slower, but surely faster, as characterized by a correlation time of ca. 10^{-9} s. If one calculates a cross-relaxation rate, the respective cross-relaxation should have a positive sign and the corresponding rates would be smaller than 0.001 s^{-1}. Therefore it is clear that the influence of $p_M R_{ij}^M$ can be safely neglected in the discussion of the rates. Thus it might be expected that the NOESY spectra and the cross-relaxation rates are dominated only by those molecules that are bonded at the adsorption sites (fraction p_C) and are not sensitive to faster (possibly anisotropic) motions. Hence, since a more detailed measurement over a wide range of temperatures is not accessible so far, further information cannot be extracted from the cross-relaxation rates measured. Consequently, for a more detailed analysis of the dynamics of reorientation and other motions of the adsorbed molecules, the 2D ^{1}H NOESY NMR measurements should be compared with the results of measurements of the ^{1}H–^{13}C cross-relaxation rates and similar relaxation rates in order to derive conclusions about a motional model.

4.2 Determination of ^{1}H–^{13}C Cross-Relaxation Rates

With the inclusion of ^{13}C spins (in natural abundance), another access to the dynamics of individual molecular groups is possible if sufficient spectrometer time is available. An established method for the determination of mobility in high-resolution NMR in the fluid phase is the measurement of nuclear Overhauser enhancement (NOE) factors η [68, 69].

Under the condition of proton decoupling, the ^{13}C magnetization assumes an equilibrium value $I_S^{0,\text{NOE}}$ different from the equilibrium value I_S^0 without proton decoupling. Both equilibrium values can be measured by simple single-pulse experiments. In the case of a molecular group consisting of a ^{13}C spin and n directly coupled protons, the influence of other protons belonging to other groups in the same or other molecules on the NOE factor is negligible. The two equilibrium values are related by the well known simple formula

$$I_S^{0,\text{NOE}} = I_S \left(1 + \eta \right), \tag{38}$$

with the NOE factor

$$\eta = n_I \frac{\gamma_I}{\gamma_S} \frac{R_{IS}}{R_1^S}, \tag{39}$$

where R_{IS} and $R_1^S = 1/T_1^S$ are the proton-to-carbon cross-relaxation rate and the longitudinal carbon relaxation rate (denoted in the following as the autorelaxation rate), respectively. The latter can be determined by ^{13}C NMR inversion recovery experiments under the condition of continuous proton decoupling. Equations (38) and (39) then allow the calculation of the heteronuclear cross-relaxation rate R_{IS}.

Again, since R_{IS} is a cross-relaxation rate, it is independent of paramagnetic impurities and is affected only by the time-dependent orientation of the vector connecting the carbon and proton spins. In analogy to the proton–proton cross-relaxation rates (see Sect. 3), we find for the cross-relaxation rate R_{IS} at a fixed carbon–proton distance r_{IS}

$$R_{IS} = (1/20)(\mu_0 \gamma_I \gamma_S \hbar / 4\pi)^2 [6 J_{IS}(\omega_0^I + \omega_0^S) - J_{IS}(\omega_0^I - \omega_0^S)], \tag{40}$$

where γ_I and γ_S are the magnetogyric ratios for protons and carbon nuclei, respectively. Experimentally obtained values for η, R_{IS}, and R_1^S for 1-butene, 1-pentene, and 1-heptene molecules adsorbed in NaX zeolites are listed in Table 5. The correlation times given in Table 5 were obtained using a spectral density function

$$J_{IS}(\omega) = r_{IS}^{-6} 2\tau_c / (1 + \omega^2 \tau_c^2), \tag{41}$$

i.e. assuming a thermal motion with a single correlation time τ_c.

Table 5. ^{1}H–^{13}C cross-relaxations rates (R_S) at higher loading (4 molecules per supercage) of 1-butene, 1-pentene, and 1-heptene in NaX zeolite

Olefin	Parameter	=CH–	=CH$_2$	–CH$_2$–	–CH$_2$–	–CH$_2$–	–CH$_2$–	–CH$_3$
1-butene	R_S (s^{-1})	0.61	0.92	0.38				0.3
	η	0.94	1.12	1.50				1.57
	R_{IS} (s^{-1})	0.14	0.13	0.07				0.04
	τ_c (10^{-12} s)	13	12	6.5				3.7
1-pentene	R_S (s^{-1})	1.01	1.41	0.72	0.66			0.48
	η	0.78	1.15	1.12	1.23			1.09
	R_{IS} (s^{-1})	0.20	0.20	0.10	0.10			0.04
	τ_c (10^{-12} s)	19	19	9.3	9.3			3.7
1-heptene	R_S (s^{-1})	2.07	3.34	2.23	1.77	1.88	1.47	0.74
	η	1.81	1.11	0.98	0.97	1.07	1.05	1.36
	R_{IS} (s^{-1})	0.42	0.47	0.27	0.22	0.25	0.19	0.08
	τ_c (10^{-12} s)	39	44	25	21	23	18	7.4

As an essential conclusion from the analysis of these measurements, one can see that the correlation times derived are appreciably shorter than the values shown in Table 4, which were obtained from the ^{1}H NMR NOESY measurements. Moreover, slight differences occur between the various groups. Mainly, the values for the methyl group deviate from the other values, and the higher flexibility of the 1-heptene molecules towards the aliphatic end of the longer chains is also reflected in the correlation times observed. Since we considered only the intramolecular proton–proton vectors in the ^{1}H NMR NOESY measurements shown in Table 4, the change of the correlation times towards the methyl group in the aliphatic chains will not be discussed further here. It seems to be interesting, however, to understand the differences between the correlation times estimated from the NOESY measurements (Table 4) and from the latter measurements (Table 5), which deviate by about two orders of magnitude. Apparently, in the temperature range considered, the proton–carbon cross-relaxation rate is dominated by the process with the shortest correlation time. This finding is also seen in other heteronuclear relaxation time measurements, reported in the following section.

4.3 Relaxation Rate Difference of Protons Bound to ^{12}C and ^{13}C Nuclei

Protons in CH$_n$ groups which contain the rare ^{13}C nucleus have an additional relaxation path compared with the majority of protons, which are bound to the abundant ^{12}C atoms. Otherwise, they possess the same structural and dynamic properties. Consequently, the difference in the respective longitudinal relaxation rates $R_1(^{13}C)$ and $R_1(^{12}C)$ is an easily interpretable parameter characterizing the reorientation of the C–H bond. The difference is given by

$$R_1(^{13}C) - R_1(^{12}C)$$
$$= (1/20)(\mu_0\gamma_I\gamma_S\hbar/4\pi)^2[6J_{IS}(\omega_0^I + \omega_0^S) + 3J_{IS}(\omega_0^I) + J_{IS}(\omega_0^I - \omega_0^S)]. \quad (42)$$

It contains similar information about the molecular motion to the cross-relaxation rates R_{IS}. In previous work [4] it was shown that it is possible to determine the relaxation rate $R_1(^{13}C)$ in a proton inversion recovery experiment which is modified by carbon decoupling and indirect proton detection. To select the protons bound to ^{13}C nuclei from the rest, ^1H–^{13}C J-couplings were used in an INEPT sequence to transfer proton magnetization to ^{13}C spins. The ^{13}C spin system was decoupled during the recovery delay in order to prevent a magnetization recovery due to heteronuclear cross-polarization.

Table 6. Experimental values of proton relaxation rates, with $(R_1^{I'})$ and without (R_1^I) a directly bonded ^{13}C atom, for different molecular groups of 1-butene, measured at room temperature and at a loading of 4 molecules per supercage in NaX zeolite

Molecular group	=CH–	=CH$_2$	–CH$_2$–	–CH$_3$
$R_1^{I'}$ (s^{-1})	0.62	0.76	0.64	0.47
R_1^I (s^{-1})	0.35	0.47	1.48	0.38
τ_c (s)	1.3×10^{-11}	1.4×10^{-11}	0.75×10^{-11}	0.4×10^{-11}

Values for $R_1(^{13}C)$, and $R_1(^{12}C)$ together with the reorientation time τ_c determined by (41) and (42), are given in Table 6 for a sample with four molecules of 1-butene per supercage adsorbed in NaX. The agreement with values for τ_c determined on the basis of heteronuclear cross-relaxation rates is satisfying. Both methods work under similar conditions and give similar information. However, they might differ in the measuring time required. For the sample investigated here, the evaluation of the NOE factor required more spectrometer time. The whole situation suggests that one should discuss improved models for the spectral density function.

4.4 Cross-Relaxation Processes Including Multiple-Spin Order

As is known, single exponential decays of the longitudinal proton and carbon magnetization in a heteronuclear spin system can only be obtained if one of the spins is decoupled. Otherwise, cross-relaxation leads to magnetization exchange between the carbon and proton magnetizations $\langle S_z \rangle$ and $\langle I_z \rangle$, respectively, introducing more relaxation constants that are necessary for the description of longitudinal relaxation. Additionally, cross-correlation among various relaxation-active interactions, such as anisotropic chemical shielding of the carbon nucleus and dipolar interactions, leads to the involvement of different forms of multiple-spin order, e.g. $\langle 2S_z I_z \rangle$ or $\langle 4S_z I_z \rangle$ [70,71].

Consequently, the relaxation behavior of each molecular group becomes increasingly complex. On the other hand, cross-relaxation rates among all longitudinal components of the spin system can provide detailed information about the dynamics of the molecular group. For instance, the cross-relaxation rate from $\langle S_z \rangle$ to $\langle 4S_z I_z \rangle$ has been used to investigate side-chain motions in bipolymers [72].

For the experimental determination of various cross-relaxation rates, a systematic approach has been proposed [73], consisting of the concerted processing of NMR intensities obtained in a set of one-dimensional NMR experiments. Details are described in [73] and only a brief account of the conclusions will be given here.

It is particularly interesting that relaxation rates connecting single- and triple-spin order are not affected by chemical-shift interactions. Only the (completely cross-correlated) dipolar interaction between all three spins is relevant. Experiments were carried out, for instance, for the groups $-CH_2-$ and $=CH_2$ of 1-butene adsorbed in NaX zeolite [4, 73]. If both protons have the same distance r_{IS} to the carbon atom, if reasonable values for the angle between the two proton–carbon bonds and for the angle formed by a proton–carbon bond and the vector r_{II} connecting both protons are used, and if an isotropic reorientation is assumed (cf. the theoretical work in [52,72]), our observations contradict the theoretical values and suggest that a model assuming isotropic reorientation of a rigid, isolated molecular group is unable to explain cross-relaxation rates between longitudinal single- and triple-spin order [4, 73].

In conclusion, a consistent description of all relevant relaxation rates requires refined models for the motion of a molecular group. These may include rotational jumps and/or anisotropic reorientation.

4.5 A Model for the Motion of Adsorbed Olefin Molecules in NaX Zeolites

The problem in the interpretation of the results of the various relaxation time measurements reported above is to understand the differences in the behavior of the homonuclear and heteronuclear cross-relaxation rates, determined by means of ^{1}H NOESY experiments and ^{1}H–^{13}C NOE and related measurements, respectively. The measurements were performed for adsorbed simple olefins, such as 1-butene and 1-pentene, as very suitable model substances. Most probably, the reasons for these differences are the various frequency regions which are covered by the relaxation rates considered. Thus, on a rough inspection, using the simplified approach of spectral density functions with a single correlation time, the data seem to reflect that ^{1}H NOESY investigations are well suited to mean reorientation times of the molecules longer than the inverse proton Larmor frequencies. ^{1}H–^{13}C cross-relaxation and related measurements can conveniently be used for faster reorientation motions. A shortcoming of this work is that it was not achievable to change the measuring temperature over a wide range in order to study the dispersion of the relaxation times due to the variation of correlation times as a function of temperature. The experimental limitation is due to problems of spectral resolution in relation to the restricted mobility at lower temperatures.

In the context of the assumptions already mentioned, a further interpretation of the data is based on a spectral density function $J(\omega)$ with two correlation times τ_1 and τ_2, according to the formula

$$J(\omega) = \frac{2}{r_{ij}^6} \left[\frac{p\tau_1}{1 + (\omega\tau_1)^2} + \frac{(1-p)\,\tau_2}{1 + (\omega\tau_2)^2} \right], \qquad (43)$$

where $0 \leq p \leq 1$ is a probability factor and r_{ij} is the distance between the dipole-coupled nuclei. Since an unambiguous analytical solution of the equations including the NOESY and NOE data by using this expression for $J(\omega)$ is not possible, a graphical procedure was applied [42, 74].

This analysis leads to two correlation times, of the order of magnitude of $\tau_2 = 1 \times 10^{-9}$ s and $\tau_1 = 1 \times 10^{-11}$ s, and a weighting factor of $p \leq 0.23$. The value for τ_2 is approximately the same as the effective correlation time derived above from the ^{1}H NOESY NMR measurements under simplifying conditions.

The value of τ_1 possesses about the same magnitude as the values derived above from the NOE measurements and from the (proton) relaxation rates of those protons which are directly bound to the ^{13}C nuclei. Moreover, the effective correlation time τ_2 is approximately the same that been derived for the reorientation of the adsorbed molecule coupled with the jump (translation) motion of the molecules to another adsorption "site" and/or to a neighboring large cavity. This conclusion is confirmed by the fact that this correlation time is about the same as the mean residence time of a 1-butene molecule in a large cavity of the NaX zeolite, derived from the self-diffusion coefficient by assuming a mean jump length $(\langle l^2 \rangle)^{1/2} \approx 1.1$ nm (see Chaps. 1) and [75, 76]).

Here, the mean distance between two neighboring large cavities in a NaX zeolite is taken for $(\langle l^2 \rangle)^{1/2}$. Thus, τ_2 describes approximately the coupled reorientation/translation motion which may characterize the microscopic steps of the self-diffusion of the adsorbed molecules.

For the interpretation of the other motion, with correlation time τ_1, it is most probable that τ_1 is related to a fast but restricted reorientational and/or librational-type motion of the adsorbed molecules, e.g. a motion of the molecule when it is "localized" in a supercage in a time interval which is smaller than the mean lifetime in a supercage, which is comparable to τ_2. It is also very probable that the restriction is due to the "anchoring" of the molecules to the Na^+ cations near the zeolite surface. The factor p is a value describing the incomplete averaging of the dipole interaction. In this sense, the two-correlation-time model applied so far has some analogy with the model of Lipari and Szabo [51], introduced to improve the data from the structural analysis of macromolecular biological systems. Instead of the fraction p in our model, an "order parameter" S^2 was used which is proportional to the average $\langle 1 - 3\cos^2\theta \rangle$ over the Legendre polynomial, $S^2 \propto \frac{1}{2}\langle 1 - 3\cos^2\theta \rangle$, where the angle θ describes the anisotropic motion.

5 Conclusions

The resolution in the ^{1}H NMR spectra of adsorbed molecules may be improved by means of magic-angle sample spinning to such an extent that ^{1}H–^{1}H cross-relaxation rates between the various proton pairs may be studied. The techniques were applied to study the conformation and dynamics of adsorbed molecules, as well as their interaction with adsorption sites. Because of the large sensitivity in ^{1}H MAS NMR measurements in comparison with the investigation of less abundant nuclei (such as

^{13}C and ^{15}N), the number of adsorbed molecules can be varied over a wide range of loading. This includes also the possibility of studing quasi-isolated molecules in zeolite matrices.

The results of the studies may be summarized as follows:

(i) A theoretical model was presented to describe the effect of magnetic susceptibility on the position and width of the lines in ^{1}H MAS NMR spectra measured at different sample-spinning rates. This approach is suitable for explaining also the limits on spectral resolution at various flux densities B of the external magnetic field, including NMR studies at very high dc magnetic fields. The total width for the distribution of the local fields is mainly influenced by contributions resulting from the orientations of neighboring crystallites; it vanishes under the influence of MAS when the magnetic susceptibility is isotropic ($\Delta\chi = 0$) and cannot be completely eliminated if $\Delta\chi \neq 0$. At higher MAS frequencies (2 to 4 kHz), Lorentzian line shapes appear, with linewidths proportional to B^2. For a real powdered solid adsorbent, only numerical solutions are possible for a quantitative estimation of the linewidth.

(ii) The considerable improvement achieved in the chemical-shift resolution enables a study of subtle details in the variation of chemical shifts for simple olefins adsorbed in NaX zeolites. A relatively high spectral resolution is achieved in the ^{1}H MAS NMR spectra even for very low pore filling factors, where the molecules are quasi-isolated in the zeolite cages. The most interesting finding in these measurements is an appreciable change in the proton chemical shifts for the olefinic =CH$_2$ group for adsorbed molecules when the temperature is varied over a wider range between about 230 K and 340 K. This allows a clear differentiation between bonded and physisorbed 1-butene molecules, and enables a detailed treatment of the dynamics of adsorption and an interpretation in terms of changes in the conformation of the adsorbed molecules.

(iii) 2D ^{1}H NOESY MAS NMR measurements allow the study of intramolecular proton–proton distances in adsorbed simple olefin molecules (1-butene, and 1-pentene, 1-heptene). Their interpretation is in good agreement with the conclusion drawn from the chemical shifts, as mentioned. The problem here was to achieve an accuracy in the estimation of intramolecular distances that allows further conclusions about conformational changes. For this reason, the influence of the thermal motion was eliminated in the treatment of the cross-relaxation data. Furthermore, the analysis was based on the use of ratios of cross-relaxation rates for various cross-peaks in order to check the accuracy in determining different intramolecular proton–proton distances.

(iv) The change in the conformation of an adsorbed molecule due to an interaction with adsorption sites has also to be connected to changes in the electronic density at the various molecular groups of the adsorbed molecule due to the interaction with adsorption sites (here the structural Na$^+$ cations in the zeolite cages), derived from the ^{1}H or ^{13}C chemical shifts of the groups =CH– and =CH$_2$ of the adsorbed molecule. In this sense the spectroscopic data are very interesting for a deeper understanding of the "activation" of adsorbed molecules in relation to catalytic transformations.

(v) A model was derived for the reorientation of adsorbed olefin molecules in NaX zeolites. The problem in the interpretation of the results of the various relaxation time measurements was to understand the differences in the behavior of the homonuclear and heteronuclear cross-relaxation rates, determined by means of ^{1}H NOESY experiments and ^{1}H–^{13}C NOE and related measurements, respectively. The measurements were performed for adsorbed simple olefins, such as 1-butene and 1-pentene, as very suitable model substances. A consistent interpretation of the data was achieved on the basis of a spectral density function $J(\omega)$ with contributions with two different correlation times τ_1 and τ_2. The effective correlation time τ_2 is approximately the same as may be estimated, from PFG NMR diffusion measurements, for the reorientation of an adsorbed molecule coupled with the jump (translation) motion of the molecule to another adsorption "site" and/or to a neighboring large cavity. The quantity τ_1 is most probably related to a fast but restricted reorientational and/or librational-type motion of the adsorbed molecules, e.g. a motion of the molecule when it is "localized" in a supercage.

Acknowledgments

The authors are greatly indebted to Dr. Uwe Schwerk for his essential contribution to the development and application of high-resolution ^{1}H MAS NMR to the study of molecules attached to solid interfaces. Useful comments from Dr. André Pampel (Leipzig), and Mr. Gert Klotzsche's help in the NMR measurements at very high magnetic fields by means of an AVANCE 750 MHz spectrometer (Bruker Biospin) are kindly acknowledged. The authors are very glad that they could use the high-field NMR spectrometer DMX 750 at the Technical University of Munich (especially mentioning the valuable support by Professor Horst Kessler) at the beginning of this work. We would like to thank in particular Professor Sergey Petrovich Zhdanov (St. Petersburg) for providing high-quality zeolite materials in the initial phase of this work. Mr. Lutz Moschkowitz helped us very much in the preparation of vacuum-treated samples with a definite loading suitable for MAS measurements. In particular, we are very grateful to the Deutsche Forschungsgemeinschaft for the support of this work in relation to the high-field equipment initiative.

References

1. F. Volke, A. Pampel: Biophys. J. **68**, 1960 (1995)
2. D. Huster, K. Arnold, K. Gawrisch: J. Phys. Chem. B **103**, 243 (1999)
3. K. Elbayed, M. Bourdonneau, J. Furrer, T. Richert, J. Raya, J. Hirschinger, M. Piotto: J. Magn. Reson. **136**, 127 (1999)
4. U. Schwerk, D. Michel: Colloids Surf. **115**, 267 (1996)
5. K.J. Packer: *Nuclear Spin Relaxation Studies of Molecules Adsorbed on Surfaces*, in *Progress in NMR Spectroscopy*, Vol. 3, ed. by J.M. Emsley, J. Feeney, L.H. Sutcliffe (Pergamon, London 1967), p. 87

6. H. Pfeifer: *Nuclear Magnetic Resonance and Relaxation of Molecules Adsorbed on Solids*, in *NMR Basic Principles and Progress*, Vol. 7, ed. by P. Fiehl, E. Fluck, R. Kosfeld (Springer, Berlin, 1972), p. 53

7. H.A. Resing: Adv. Mol. Relax. Process. **1**, 109 (1967/68)

8. E.G. Derouane, J. Fraissard, J.J. Fripiat, W. E. E. Stone: Catal. Rev. **7**, 121 (1972)

9. D. Michel: Z. Phys. Chem. (Leipzig) **252**, 263 (1973)

10. D. Michel, A. Germanus, D. Scheller, B. Thomas: Z. Phys. Chem. (Leipzig) **262**, 113 (1981); D. Michel, A. Germanus, H. Pfeifer: J. Chem. Soc. Faraday Trans. 1, **78**, 237 (1982)

11. D. Michel, A. Germanus, H. Pfeifer: J. Chem. Soc. Faraday Trans. 1, **78**, 237 (1982)

12. R.R. Ernst, G. Bodenhausen, A. Wokaun: *Principles of Nuclear Magnetic Resonance in One and Two-Dimensions* (Clarendon Press, Oxford, 1987)

13. H. Pfeifer: Phys. Rep. C**26**, 293 (1976)

14. H. Pfeifer, W. Meiler, D. Deininger: Annu. Rep. NMR Spectrosc. **15**, 291 (1983)

15. J.B. Nagy, G. Engelhardt, D. Michel: Adv. Colloid Interface Sci. **23**, 67 (1985)

16. G. Engelhardt, D. Michel: *High-Resolution Solid-State NMR of Silicates and Zeolites* (Wiley, Chichester, 1987)

17. J.B. Nagy: "Multinuclear magnetic resonance in liquids and solids–Chemical applications", in NATO ASI Ser. C**322**, 371 (1990), ed. by P. Granger, R.K. Harris

18. H. Pfeifer, H. Ernst: Annu. Rep. NMR Spectrosc. **28**, 91 (1994)

19. J. Klinowski: Anal. Chim. Acta **283**, 929 (1993)

20. S. Kaplan, H.A. Resing, J.S. Waugh: J. Chem. Phys. **59**, 5681 (1973)

21. E.O. Stejskal, J. Schaefer, J.M.S. Hennis, M.K. Tripodi: J. Chem. Phys. **61**, 2351 (1974)

22. C.A. Fyfe: *Solid-State NMR for Chemists* (CFC Press, Guelph, Canada, 1983)

23. K. Wüthrich: *NMR of Proteins and Nucleic Acids* (Wiley, Chichester, 1986)

24. M. Ebener, G. von. Fircks, H. Günther: Helv. Chem. Acta **743**, 1296 (1991)

25. L.E. Drain: Proc. Phys. Soc. **80**, 1380 (1962)

26. D.L. Vander Hart, W.L. Earl, A.N. Garroway: J. Magn. Reson. **44**, 361 (1981); D.L. Vander Hart: *"Magnetic susceptibility & high-resolution NMR of liquids and solids"*, in *Encyclopedia of Nuclear Magnetic Resonance*, Vol. **5**, ed. by D.M. Grant, R.K. Harris (Wiley, Chichester, 1996), p. 2938

27. V.V. Mank, W.G. Vasilyev, F.D. Ovtcharenko, J.F. Zubenko: Dokl. Akad. Nauk SSSR **207**, 133 (1972) [In Russian]

28. D. Doskocilova, B. Schneider: Chem. Phys. Lett. **6**, 381 (1970); D. Doskocilova, D.T. Tao, B. Schneider: Czech. J. Phys. B**25**, 202 (1975)

29. E.P. Whipple, P.J. Green, M. Ruta, R.L. Bujalski: J. Phys. Chem. **80**, 1350 (1976)

30. A.N. Garroway: J. Magn. Reson. **49**, 168 (1982)

31. M. Alla, E. Lippmaa: Chem. Phys. Lett. **87**, 30 (1982)

32. U. Schwerk, D. Michel, M. Pruski: J. Magn. Reson. A **119**, 157 (1996)

33. L.D. Landau, E.M. Lifshitz: *Electrodynamics of Continuous Media* (Pergamon, Oxford 1984)

34. J.A. Osborn: Phys. Rev. **67**, 351 (1945)

35. D. Fenzke, B.C. Gerstein, H. Pfeifer: J. Magn. Reson. **98**, 469 (1992)

36. D. Freude, M. Hunger, H. Pfeifer: Chem. Phys. Lett. **91**, 307 (1982)

37. S.P. Shdanov, S.S. Kovoshchov, N.N. Samulevich: *Synthetic Zeolites* (Khimia, Moscow, 1981)

38. U. Schwerk, D. Michel: Z. Phys. Chem. **189**, 29 (1995)

39. D. Michel, A. Pampel, J. Roland: J. Chem. Phys. **119**, 9242 (2003)

40. W. Meiler, D. Deininger, D. Michel: Z. Phys. Chem. (Leipzig) **258**, 139 (1977)

41. R. Lochmann, W. Meiler: Z. Phys. Chemie (Leipzig) **258**, 1059 (1977)

42. W. Böhlmann, D. Michel, J. Roland: Magn. Res. Chem. **37**, 126 (1999)

43. S.A. Smith, T.O. Levante, B.H. Meier, R.R. Ernst: J. Magn. Reson. A **106**, 75 (1994)
44. H. Fribolin: *Ein- und zwei-dimensionale NMR-Spektroskopie* (VCH, Weinheim, 1992)
45. J.W. Emsley, J. Feeney, L.H. Sutcliffe: *High-Resolution NMR Spectroscopy*, Vol. 2, (Pergamon, Oxford, 1965)
46. I. Solomon: Phys. Rev. **99**, 559 (1955)
47. J. Tropp: J. Chem. Phys. **72**, 6035 (1980)
48. D. Canet, J.B. Robert: *Behaviour of the NMR Relaxation Parameters at High Fields*, in *NMR Basic Principles and Progress*, Vol. 25, ed. by P. Diehl, E. Fluck, H. Günther, R. Kosfeld, J. Seelig (Springer, Berlin, Heidelberg, 1990), p. 45
49. D.E. Woessner: J. Chem. Phys. **42**, 1855 (1965)
50. D.E. Woessner: J. Chem. Phys. **36**, 1 (1962)
51. G. Lipari, A. Szabo: J. Am. Chem. Soc. **104**, 546 (1982)
52. A. Chaudhry, J. Pereira, T.J. Norwood: J. Magn. Reson. A **111**, 215 (1994)
53. S. Macura, R.R. Ernst: Mol. Phys. **41**, 95 (1980)
54. E.T. Olejniczak, R.T. Gampe, S.W. Fesik: J. Magn. Reson. **67**, 28 (1986)
55. P.A. Mirau: J. Magn. Reson. **80**, 439 (1988)
56. B.A. Borgias, T.L. James: J. Magn. Reson. **79**, 493 (1988)
57. S.P. Edmondson: J. Magn. Reson. **98**, 283 (1992)
58. J.D. Baleja: J. Magn. Reson. **96**, 619 (1992)
59. U. Schwerk, D. Michel: J. Phys. Chem. **100**, 352 (1996)
60. W.H. Press: *Numerical Recipes in Pascal: "The Art of Scientific Computing"* (Cambridge University Press, Cambridge, 1989)
61. S. Macura: J. Magn. Reson. A **112**, 152 (1995)
62. Landolt-Börnstein New Series II/21 (Springer, Berlin, Heidelberg, 1992), p. 248
63. J. Roland, D. Michel: Magn. Reson. Chem. **38**, 587 (2000)
64. J. Roland, D. Michel, A. Pampel: *Investigation of conformational changes of organic molecules sorbed in zeolites by HR MAS NMR spectroscopy*, in *Magnetic Resonance in Colloid and Interface Science. Proceedings of the NATO Advanced Research Workshop*, ed. by J. Fraissard, O. Lapina (Kluwer, Dordrecht, 2002), p. 83
65. D. Neuhaus, M. Williamson: *The Nuclear Overhauser Effect in Structural and Conformational Analysis* (VCH, New York, 1989)
66. S. Kondo, E. Hirota, Y. Morino: J. Mol. Spectrosc. **28**, 471 (1968)
67. A.A. Bothner-By, C.N. Colin, H. Günther: J. Am. Chem. Soc. **84**, 2748 (1962)
68. J.H. Noggle, R.E. Schirmer: *The Nuclear Overhauser Effect. Chemical Applications* (Academic Press, New York, 1971)
69. A. Abragam: *Principles of Nuclear Magnetic Resonance* (Clarendon, Oxford, 1994)
70. T.E. Bull: J. Magn. Reson. **72**, 397 (1987)
71. L.G. Werbelow, D.M. Grant: *Intramolecular Dipolar Relaxation in Multispin Systems*, in *Advances in Magnetic Resonance*, Vol. 9, ed. by J.S. Waugh (Academic Press, New York, 1977), p. 189
72. M. Ernst, R.R. Ernst: J. Magn. Reson. A **110**, 202 (1994)
73. U. Schwerk: *High-Resolution NMR Methods for the Characterization of Conformation and Dynamics of Molecule Sorbed in Zeolites*, Ph.D. Thesis, University of Leipzig (1996)
74. J. Roland: *HR-MAS-NMR-Untersuchungen zur Strukturänderung und Dynamik von adsorbierten Molekülen an Zeolithen*, Ph.D. Thesis, University of Leipzig (2002)
75. J. Kärger, D. Michel: Z. Phys. Chem., (Leipzig) **257**, 983 (1976)
76. J. Kärger, D. Michel, A. Petzold, J. Caro, H. Pfeifer, R. Schöllner: Z. Phys. Chem. (Leipzig) **257**, 1009 (1976)

Molecular Dynamics of Liquids in Confinement

Friedrich Kremer[1] and Ralf Stannarius[2]

[1] Universität Leipzig, Institut für Experimentalphysik I,
   kremer@physik.uni-leipzig.de
[2] Otto-von-Guericke-Universität Magdeburg, Institut für Experimentalphysik,
   ralf.stannarius@physik.uni-magdeburg.de

Abstract. Surface and confinement effects may have substantial effects on the molecular and collective dynamics of liquids. Within this chapter, we describe the confinement effects on the dynamical modes of isotropic liquids. Characteristic for many systems is the formation of surface layers with a retarded dynamics. Proper treatment of the adsorber walls ("lubrication") can reduce or completely remove such layers. In glass-forming liquids, the limiting space can lead to the suppression of the formation of collectively reorienting clusters in the liquid, and consequently to an acceleration of the dynamics compared to the bulk liquid. The combination of retarding and accelerating effects in a system where molecular migration between the boundary layers and the core is effective is describes in a dynamic model that calculates the influences of molecular exchange on dielectric spectra.

1 Introduction

Molecular and collective dynamics in a confining space are determined by the counterbalance of surface and confinement effects [1]. The former results from interactions of the host system with guest molecules which take place at the interface between both; the latter originates from the inherent length scale on which the underlying molecular fluctuations take place. Surface effects cause a retardation, while confinement effects are characterized by an acceleration of the molecular dynamics with decreasing spatial dimensions of the confining space (Fig. 1). Hence in glass forming systems [2–7], an increase or decrease, respectively, of the calorimetric glass transition temperature is observed. It is evident that this counterbalance must depend sensitively on the type of confined molecules (glass-forming liquids, polymers, or liquid crystals), on the properties of the (inner) surfaces (wetting or nonwetting), and on the architecture of the molecules with respect to the walls (grafted, layered, or amorphous).

Confining geometries [8–24] can be realized in various ways: by containing the system under study in zeolites, in nanoporous sol–gel glasses, in mesoporous membranes, in blockcopolymers, etc. With its extraordinary dynamic range (in frequency *and* intensity), broadband dielectric spectroscopy enables one to unravel the subtle interplay between surface and confinement effects and to contribute to the solution of basic questions, such as for instance

- is there an inherent length scale of cooperativity in glass-forming liquids and how does it vary with temperature,

F. Kremer, R. Stannarius, Molecular Dynamics of Liquids in Confinement, Lect. Notes Phys. **634**, 275–300 (2004)
http://www.springerlink.com/

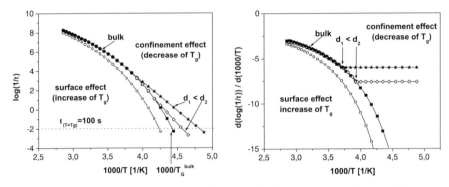

Fig. 1. (a) Schematic description of molecular dynamics in geometric confinement; (b) plot of the temperature dependence of the derivative of the bulk dynamics compared with a surface or a confinement effect

- what is the effect of a lubricant layer in decoupling the dynamics of a liquid from a solid wall, and
- what is the difference between H-bond-forming and van der Waals liquids?

2 H-Bond-Forming Liquids in Zeolitic and Nanoporous Media

Zeolites [25, 26] offer a unique possibility to vary the dimensions and the topology of spatial confinement on a subnanometer scale in controlled manner. Silica sodalite consists of identical so-called β-cages, with an inner diameter of 0.6 nm. Ethylene glycol (EG) is one of the structure-directing agents which control the formation of silica sodalite [27, 28]. Exactly one EG molecule becomes occluded in one sodalite cage during synthesis and cannot escape from it unless the cage is thermally decomposed [28]. Silicalite-I, zeolite beta, and AlPO$_4$-5 have channel-like pore systems (see Fig. 2). In silicalite-I, consisting of pure SiO$_2$, rings of 10 Si and 10 O Atoms form a three-dimensional pore system, with two types of elliptical channels having cross sections of 0.56 nm – 0.53 nm and 0.55 nm – 0.51 nm [25]. In zeolite beta (a 12-ring system) the channels in the [100] and [010] directions have diameters of 0.76 nm × 0.64 nm, whereas the channels in the [001] direction have smaller pores (0.55 nm × 0.55 nm) [29]. The Si:Al ratio of the sample used was 40, to reduce the number of counterions in the channels. AlPO$_4$-5 has a one-dimensional pore system. In this aluminophosphate, the channels, with diameters of 0.73 nm, are arranged in a hexagonal array.

Besides sodalite, which is already loaded with EG after synthesis, all nanoporous hosts were heated to 600 K with a temperature increase of 20 K h^{-1} and evacuated at 10^{-5} mbar for 36 h to remove water and other volatile impurities. Afterwards the zeolitic host systems were filled with EG from the vapor phase in a closed vacuum chamber at 448 K. The samples were cooled down to room temperature and remained in the vacuum chamber for 24 h before the dielectric measurements were carried out.

a) silica-sodalite

b) silicalite

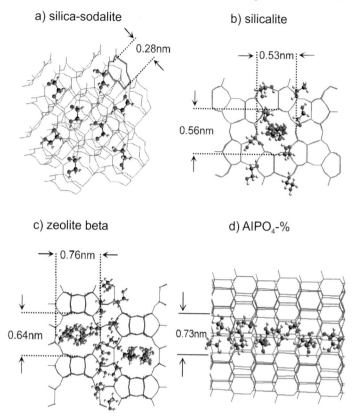

c) zeolite beta

d) AlPO$_4$-%

Fig. 2. Scheme of the zeolitic host systems in which the guest molecule ethylene glycol was confined. (a) Silica sodalite (SiO$_2$) has cubic cages with a lattice constant of 0.89 nm. The cages are connected by channels with a diameter of 0.28 nm. Only *one* molecule is confined to each cage. (b) Silicalite consists of pure SiO$_2$ and has a three-dimensional pore system with two different types of elliptical channels, having cross sections of 0.56 nm × 0.53 nm and 0.55 nm × 0.51 nm. (c) Zeolite beta is an aluminosilicate with a three-dimensional pore system having pore diameters of 0.76 nm × 0.64 nm and 0.55 nm. (d) AlPO$_4$-5 is an aluminophosphate with one dimensional channels (diameter 0.73 nm) arranged in a hexagonal array. Taken from [1]

Isothermal data (Fig. 3) of the dielectric loss ε'' were fitted by a superposition of a relaxation function given by Havriliak and Negami (HN) and a conductivity contribution [30, 31],

$$\varepsilon'' = \frac{\sigma_0}{\varepsilon_0} \frac{a}{\omega^s} + \mathrm{Im}\left[\frac{\Delta\varepsilon}{(1 + (\mathrm{i}\omega\tau)^\alpha)^\gamma}\right]. \tag{1}$$

In this notation, ε_0 is the vacuum permittivity, σ_0 the DC conductivity, and $\Delta\varepsilon$ the dielectric relaxation strength. α and γ describe the symmetric and asymmetric broadening of the relaxation peak. The exponent $s = 1$ holds for pure electronic conduction; deviations ($s < 1$) are caused by electrode polarization or Maxwell–

Wagner polarization effects. The factor a has the dimension s^{1-s}. The uncertainty in the determination of $\log \tau$ is ≤ 0.1 decades and less than 5% for $\Delta\varepsilon$. Owing to the fact that ε' and ε'' are connected by the Kramers–Kronig relations, a fit to ε' does not improve the accuracy. From the fits according to (1), the relaxation rate $1/\tau_{\max}$ corresponding to the frequency of maximum dielectric loss ε'' at a certain temperature can be deduced. A second way to interpret the data is the use of a relaxation time distribution $L(\tau)$ of Debye relaxators with relaxation times τ. The imaginary part of the dielectric function is expressed by

$$\varepsilon'' = (\varepsilon_s - \varepsilon_\infty) \int \frac{L(\tau)}{1 + \omega^2 \tau^2} \, d\tau \qquad (2)$$

where ε_s and ε_∞ denote the low- and high-frequency limits of the permittivity. $L(\tau)$ can be extracted numerically from the data [32] or calculated (analytically) from the fit to HN functions [30, 31]. To characterize the temperature dependence of the relaxation behavior, the averaged logarithmic relaxation time $\log \tau_{\mathrm{med}}$ is calculated:

$$\log \tau_{\mathrm{med}} = \langle \log \tau \rangle = \left(\int_{-\infty}^{+\infty} \log \tau \, L(\tau) \, d\tau \right) / \left(\int_{-\infty}^{+\infty} L(\tau) \, d\tau \right) \qquad (3)$$

$\log \tau_{\mathrm{med}}$ equals $\log \tau_{\max}$ if the peak of a relaxation process is symmetrically broadened. The calculation of $\log \tau_{\mathrm{med}}$ can be done with high accuracy only if the relaxation time distribution function is known over a broad range. For that reason, $\log \tau_{\max}$ was determined for molecules confined to zeolites (where the frequency range is limited), and $\log \tau_{\mathrm{med}}$ was calculated whene nanoporous sol–gel glasses were used as the host.

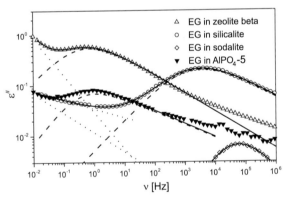

Fig. 3. The dielectric loss ε'' versus frequency for ethylene glycol confined to zeolitic host systems as indicated. The solid lines are a superposition of a Havriliak–Negami relaxation (dashed line) and a conductivity contribution (dotted line). Taken from [1]

Figure 3 shows the dielectric spectra for ethylene glycol (EG) confined to different zeolitic host systems at 160 K. The relaxation rates $\tau_{\max}^{-1}$ for EG in the zeolitic host systems differ by up to six orders of magnitude. In zeolites with smaller pores (silicalite and sodalite), the relaxation rates of EG are significantly higher compared with zeolite beta and AlPO$_4$-5. Especially for EG in sodalite, the relaxation strength is comparatively weak. This is caused by EG molecules which are immobilized owing to the interaction with the zeolitic host matrix. Figure 4 shows the relaxation rate as a function of inverse temperature for EG as a bulk liquid and confined to zeolites. EG in zeolite beta (solid triangles) and in AlPO$_4$-5 (open triangles) has a relaxation rate like that of the bulk liquid (squares) following a temperature dependence according to the empirical Vogel–Fulcher–Tammann (VFT) equation [33–35]:

$$\frac{1}{\tau} = A \exp\left(\frac{-DT_0}{T - T_0}\right), \tag{4}$$

where A is a prefactor, D is the fragility parameter, and T_0 is the Vogel temperature. The relaxation rates of EG in silicalite and sodalite show an Arrhenius-type temperature dependence.

The single-molecule relaxation of EG in sodalite at $T \approx 155$ K is by about six orders of magnitude faster compared with the bulk liquid. Its activation energy

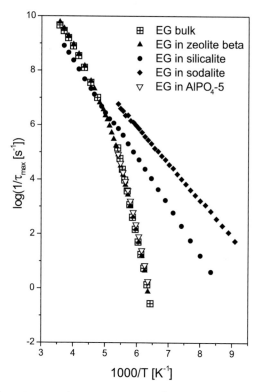

Fig. 4. The relaxation rate versus inverse temperature for ethylene glycol confined to different zeolitic host systems, as indicated. The errors are smaller than the size of the symbols. Taken from [1]

is 26 ± 1 kJ mol^{-1} and corresponds to the value for bulk EG at high relaxation rates (29 ± 2 kJ mol^{-1}) [36]. The relaxation process of EG in silicalite has a larger activation energy (35 ± 2 kJ mol^{-1}) which is still smaller than the apparent activation energy (tangent to the VFT-temperature dependence) of the bulk liquid close to T_g. Its Arrhenius-like temperature dependence resembles the single-molecule relaxation of EG in sodalite.

To study the molecular arrangement of the molecules in confining space, the molecular simulation program Cerius^2 was used on a Silicon Graphics workstation to model a finite zeolite crystal with four unit cells surrounded by vacuum. By "filling" the pores with EG, a completely loaded nanoporous host/guest-system can be simulated and structural parameters such as the distance between molecules, density, and length of H bonds can be determined. The simulations were carried out using three different force fields: the Dreiding force field [36], the Burchart universal force field [38, 39], and the consistent force field [40]. The three force fields provide the same results within the uncertainty for the quantities shown in Table 1.

The computer simulations of EG in zeolitic host systems show that in silicalite the molecules are aligned almost single-file-like along the channels and that in zeolite beta and in AlPO$_4$-5 two EG molecules are located side by side in the channels. But neither for the distance between molecules, for the average length of hydrogen bonds, nor for the density is a significant change found between the bulk liquid and the molecules in the restricting geometry (Table 1). However, for the number of neighboring molecules (coordination number), a pronounced difference is observed (Fig. 5). The coordination number of 11 corresponds to the maximum value in the case of the random close-packing model [40] and is found for the bulk liquid within a radius of $r = 0.66$ nm. Within that radius, EG in zeolite beta and in AlPO$_4$-5 has only five neighboring molecules. As AlPO$_4$-5 has one-dimensional channels and no intersections between them, in contrast to zeolite beta, the dimensionality of the host system seems to play only a minor role in the dynamics of H-bonded guest molecules. Further reduction in the channel size (as in the case of silicalite) decreases the average number of neighboring molecules by about 1. This results in a sharp transition from

Table 1. Distance between molecules, average length of hydrogen bonds (O–H $\cdots$ O bonds with a length up to 0.3 nm), and density, as calculated from the molecular simulations for ethylene glycol confined to zeolite beta and silicalite and for the bulk liquid. For simulation of the bulk liquid a limited volume (6.64 nm^3) was filled with EG molecules until the bulk density of 1.113 g cm^{-3} was reached. In contrast, the densities of EG confined in zeolites are results of the simulation. The error is caused mainly by the uncertainty in calculating the accessible volume of the zeolitic channels.

	Distance between molecules (nm)	Average length of H–bonds (nm)	Density (g cm^{-3})
Bulk liquid	0.42 ± 0.01	0.23 ± 0.02	1.113
Zeolite beta	0.41 ± 0.01	0.25 ± 0.02	1.0 ± 0.1
Silicalite	0.42 ± 0.01	0.24 ± 0.02	1.0 ± 0.1

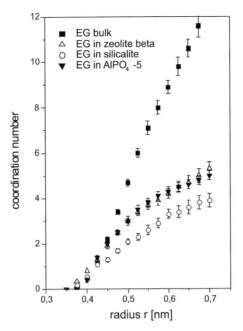

Fig. 5. The average number of neighboring molecules (coordination number) as a function of the radius of a surrounding sphere as calculated from the simulations for EG bulk liquid (squares), EG confined to zeolite beta (triangles), EG confined to silicalite (circles), and EG confined to AlPO$_4$-5 (solid triangles). Taken from [1]

a liquid-like dynamics to that of single molecules. In AlPO$_4$-5, only two molecules are located side by side in the one-dimensional channels; hence the interactions are dominated by the nearest neighboring molecules and an ensemble as small as six EG molecules is sufficient to show a liquid-like dynamics.

Sol–Gel glasses (Geltech Inc., USA) offer another possibility to realize a confining space [3–14]. They are available with pore sizes of 2.5 nm, 5.0 nm, and 7.5 nm and a narrow pore size distribution. The porous glass is disk-shaped (diameter 10 mm, thickness 0.2 mm), so the outer surface is negligible compared with the huge inner surface (520 m^2 g^{-1} – 620 m^2 g^{-1}). After evacuating the porous glasses to 10^{-5} mbar at 570 K for 24 h to remove water and other volatile impurities, the pores were filled by capillary wetting for 48 h at a temperature of about 20 K above the melting point of the liquid. For that purpose, the glass-forming liquid was injected into the (closed) vacuum chamber by use of a syringe. Both sides of the sample disks were covered with aluminum foil (thickness 800 nm) to ensure a homogeneous field distribution and were mounted between gold-plated brass electrodes of a capacitor. As H-bond-forming liquids, the homologous sequence of propylene glycol, butylene glycol, and pentylene glycol (PG, BG, PeG) was used, having averaged molecular radii of $r_m = 0.306$ nm, $r_m = 0.329$ nm, and $r_m = 0.349$ nm, respectively.

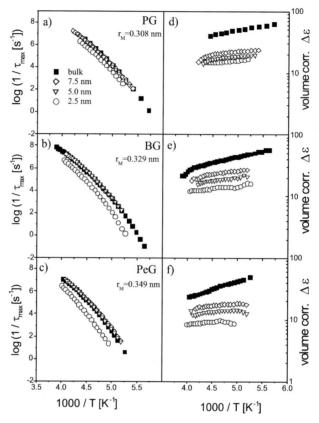

Fig. 6. Relaxation rate $1/\tau_{\max}$ (a–c) and relaxation strength $\Delta\varepsilon$ (d–f) of the α-relaxation of PG, BG, and PeG in pores versus the inverse temperature T. Pore sizes: 2.5 nm (circles), 5.0 nm (triangles), 7.5 nm (diamonds), and bulk (filled squares). The error is smaller than the size of the symbols. r_M is the mean van der Waals radius of the molecules. Taken from [9]

Figures 6a–c show the relaxation rates of the α-relaxation of the bulk and confined liquids. No effect of confinement is observed on the relaxation rates of PG and BG in 7.5 nm and 5.0 nm pores. For PeG the α-relaxation in 7.5 nm and 5.0 nm pores becomes slightly faster compared with the bulk at low temperatures. In 2.5 nm pores the relaxation rate is slowed down compared with the bulk; this effect is more pronounced for the larger molecules. The volume-corrected dielectric relaxation strengths of the α-relaxation of bulk and confined PG, BG, and PeG are shown in Figs. 6d–f. For all liquids the absolute value of the dielectric strength $\Delta\varepsilon$ decreases with decreasing pore size, indicating the existence of a fraction of immobilized molecules which is dielectrically inactive (complete pore filling was checked by weighing). This layer is caused by H bonds formed between the glycol molecules and the inner surface of the sol–gel glass.

For confined PG and BG, the temperature dependence of $\Delta\varepsilon$ is comparable to that of bulk liquids while for PeG, $\Delta\varepsilon$ decreases for low temperatures for all pore sizes. This decrease and increase, respectively, of the relaxation rate may be comprehended by assuming exchange between the molecules bound to the pore surface (solid-like layer) and molecules in the center of pores.

More instructive than the HN parameters τ_{HN}, α, and γ is the relaxation time distribution function $L(\tau)$. It can be deduced from the dielectric spectra, either analytically from the HN fit parameters [30, 31] or by a regularization technique [32]. Figure 7 shows the relaxation time distributions of confined PG at different temperatures. On the long-term side, a broadening is observed with decreasing pore size. For 2.5 nm pores, even the maximum is shifted to lower relaxation times.

The confined molecules can be classified (Fig. 8a) according to their dynamics into three fractions: (i) liquid-like with relaxation rates as in the bulk, (ii) interfacial with reduced mobility, and (iii) immobilized. Assuming that the density of the

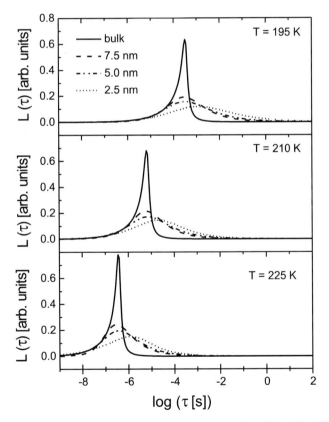

Fig. 7. Relaxation time distribution of the α-relaxation as calculated from Havriliak–Negami fits for bulk and confined PG at 195 K, 210 K, and 225 K (solid line: bulk, dashed line: 7.5 nm, dash-dotted line: 5.0 nm, dotted line: 2.5 nm pore diameter). Taken from [9]

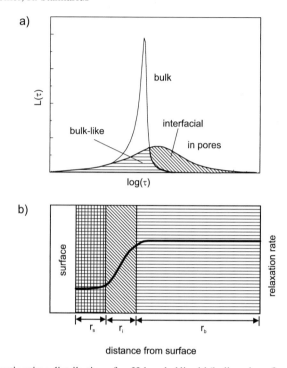

Fig. 8. (a) Relaxation time distribution of an H-bonded liquid (bulk and confined to nanopores). Shaded area with horizontal lines, bulk-like molecules; shaded area with inclined lines, interfacial molecules. (b) Sketch of the presumed spatial relaxation time distribution in the pores. Checked, solid-like molecules (thickness r_s); inclined lines, interfacial layer (thickness r_i); horizontal lines, bulk-like molecules (radius r_b). Taken from [9]

Table 2. Thickness of the solid-like layer r_s and the radius of the bulk-like phase r_b of molecules of PG, BG, and PeG for three different temperatures. The interfacial layer $r_i = R - r_b - r_s$ is ≈ 0.3 nm for all samples. The absolute error of r_s, r_i, and r_b is 0.2 nm

Temperature	Pore size	PG		BG		PeG	
(K)	R (nm)	r_b (nm)	r_s (nm)	r_b (nm)	r_s (nm)	r_b (nm)	r_s (nm)
	7.5	2.5	0.8	2.8	0.5	2.8	0.6
225	5.0	1.5	0.7	1.8	0.5	1.7	0.5
	2.5	0.7	0.3	0.7	0.4	0.6	0.4
	7.5	2.5	0.8	2.8	0.6	2.6	0.8
210	5.0	1.5	0.7	1.6	0.5	1.6	0.7
	2.5	0.7	0.4	0.6	0.4	0.6	0.5
	7.5	2.4	0.9	2.8	0.6		
195	5.0	1.5	0.7	1.6	0.5		
	2.5	0.6	0.4	0.6	0.4		

molecules in confinement is only negligibly influenced and that the dielectric strength is proportional to the number of molecules participating in a relaxation process, a quantitative three layer model (Fig. 8b and Table 2) can be deduced. Here the fraction of immobilized molecules is estimated from the difference in the dielectric strength which is expected from the filling factor of the confined molecules and which is measured as the contribution of the interfacial and bulk-like molecules. Considering a spherical or cylindrical shape for the pores delivers similar results. Within the experimental accuracy, no significant temperature dependence of the layer structure is found. The layer thicknesses r_s and r_i have only a weak pore size dependence, while the radius r_b of the bulk-like phase scales roughly with the pores. From the mean van der Waals radius r_m and r_b the number of bulk-like molecules in the pores can be derived. As in the experiments with zeolitic host systems, it turns out that the number of molecules which is necessary to perform a bulk-like dynamics is small.

Propylene glycol is an H-bonded glass-forming liquid. Owing to the fact that a freshly prepared SiO_2 surface is hydrophilic, one has to expect that the PG molecules form H bonds with the solid surfaces of the nanoporous system. This surface interaction can be hindered by a silanization as shown schematically in Fig. 9.

For PG in uncoated nanopores, compared with the bulk liquid, there is a pronounced broadening of the width of the dielectric loss curve and the relaxation time distribution function (Figs.10a–c). Silanization of the inner surfaces counteracts this effect (Figs. 10c and 11), which is – as expected – strongly temperature-dependent and weakens with increasing thermal activation (Fig. 10): at a temperature of 185 K, the low-frequency broadening of the relaxation time distribution function for PG in uncoated pores is completely removed and the mean relaxation time for PG in coated nanopores becomes even faster than in the bulk. The broadening is interpreted as being caused by interactions of the molecules with the surface [3]. The lubrication

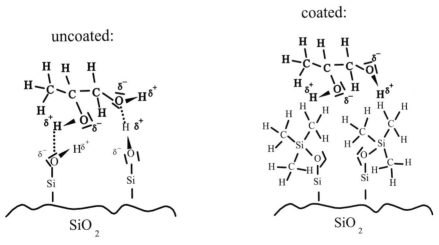

Fig. 9. Scheme of propylene glycol in the neighborhood of an uncoated SiO_2 surface (left) and a silanized SiO_2 surface (right)

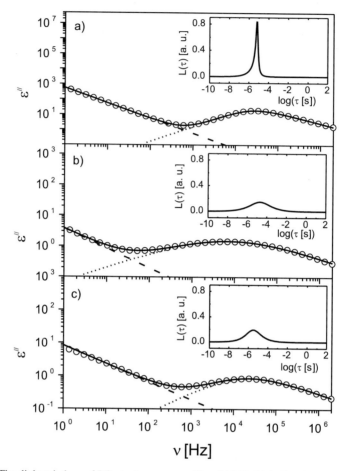

Fig. 10. The dielectric loss of PG at a temperature $T = 210$ K; (a) bulk PG, (b) PG confined to 2.5 nm uncoated pores, (c) PG confined to 2.5 nm coated pores. The error of the measured data is smaller than the size of the symbols. Solid line, superposition of a Havriliak–Negami relaxation function (dotted line) and a conductivity contribution (dashed line). The inset shows the resulting relaxation time distribution $L(\tau)$. Taken from [9]

hinders the formation of H bonds to the solid wall of the nanoporous ambience, hence decoupling its dynamics.

To characterize the temperature dependence of the relaxation rate, the averaged relaxation time $\log \tau_{med}$ was calculated as defined in (3). The temperature dependence follows, for all (uncoated and coated) pores, the well-known VFT law (Fig. 12). The difference in relaxation rate of PG in uncoated and coated pores is maximal for the smallest pore size (2.5 nm) and lowest temperature ($\approx$ 190 K). Despite the fact that, owing to the silane layer having a thickness of about 0.4 nm, the space for the PG molecules becomes even more confining, one finds a dynamics which is – within experimental accuracy – identical to that of the bulk liquid.

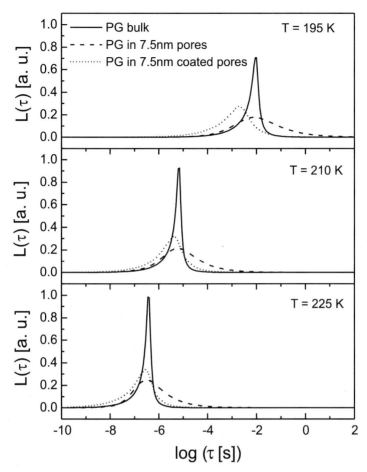

Fig. 11. Relaxation time distribution function versus relaxation time for propylene glycol as a bulk liquid (solid line), in uncoated pores (dashed line) and in coated pores (dotted line) at different temperatures

It is well known that for glass-forming liquids the product of the temperature T and dielectric relaxation strength $\Delta\varepsilon$ increases with decreasing temperature [41] (According to the Langevin function it should be constant, neglecting density effects). This effect vanishes for the dynamic glass transition if it takes place in nanoporous systems (Fig. 13). The fact that the dielectric strength becomes smaller with decreasing pore diameter is attributed to the change in the surface-to-volume ratio of the nanoporous system.

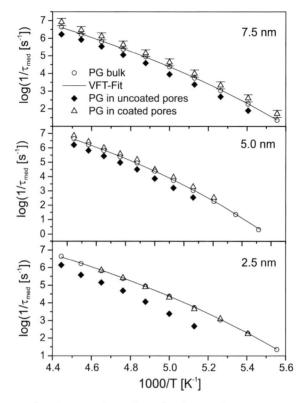

Fig. 12. The mean relaxation rate of propylene glycol versus inverse temperature for the bulk liquid (open circles; solid curve VFT fit) and for propylene glycol confined to uncoated pores (solid diamonds) and coated pores (open triangles) of different pore diameters as indicated

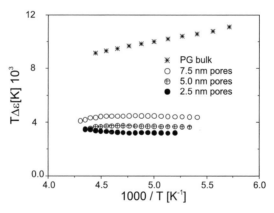

Fig. 13. Dielectric relaxation strength $\Delta\varepsilon$ times temperature T versus inverse temperature for propylene glycol as a bulk liquid (stars) and confined to silanized pores of a sol–gel glass having different diameters of 7.5 nm (open circles), 5.0 nm (cross-centered circles), and 2.5 nm (filled circles)

3 The van der Waals Liquid Salol
in (Lubricated) Nanoporous Sol–gel Glasses

Salol (phenyl salicylate) is one of the most studied organic glass-forming liquids. It is regarded as a "quasi"-van der Waals molecule despite the fact that it can form (mainly intramolecular) H bonds (see scheme in Fig. 14). In nanoporous sol–gel glasses with hydrophilic inner surfaces, an interfacial layer of molecules in the neighborhood of the solid wall is established that has a dynamics which is slowed down compared with the bulk liquid [12].

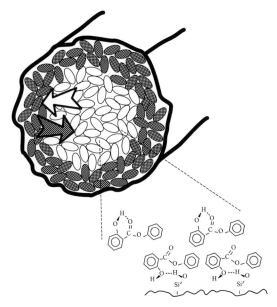

Fig. 14. Schematic view of a pore filled with a glass-forming liquid. The pore walls are covered by a surface-bound layer of molecules; the remaining volume is filled with bulk-like molecules. Owing to H bonding, the salol molecules can bind directly to the SiO_2 surface of the inner walls of the nanoporous system

In the experiment, two molecular relaxation processes (I and II) are observed, which can be separated in frequency and fitted by generalised Havriliak–Negami relaxation functions (inset in Fig. 15). The dielectric spectra can be interpreted in terms of a two-state model with dynamic exchange between a bulk-like phase in the pore volume and an interfacial phase close to the pore wall [10].

The formulas describing the change of the polarization p_i in subsystem i form a set of coupled linear first-order differential equations if $p_i(t)$ are purely exponential and if the jump rate c_{ij} is time-independent:

$$\frac{\mathrm{d}p_i(t)}{\mathrm{d}t} = c_{ij}p_j, \qquad i = 1, 2. \qquad (5)$$

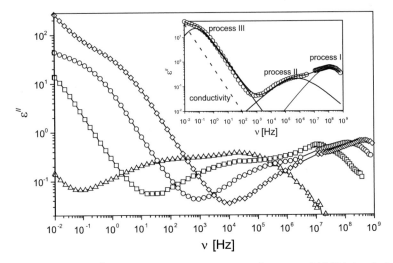

Fig. 15. Dielectric loss ε'' of salol in 7.5 nm pores versus frequency: 245 K (triangles), 265 K (squares), 285 K (circles), 305 K (diamonds). The error of the measured data is smaller than the size of the symbols. The inset illustrates the deconvolution of the data for $T = 285$ K. Relaxation process I is assigned to fluctuations of bulk-like molecules, process II originates from molecules close to the inner walls and process III is caused by Maxwell–Wagner polarization. Taken from [10]

The polarization in subsystem 1 changes by internal relaxation with the function $p_1(t)$ and by transfer to and from state 2 as a consequence of jumps:

$$\frac{dp_1}{dt} = -s_1 p_1 - c_{12} p_1 + c_{21} p_2.$$

the polarization in state 2 is described analogously:

$$\frac{dp_2}{dt} = -s_2 p_2 - c_{21} p_2 + c_{12} p_1.$$

The number of jumps per unit time is $c_{ij} N_i$, where N_i is the number of particles in system i; the transferred polarization for a single particle is p_i / N_i. The constants s_i ($i = 1, 2$) describe the relaxation rates in the uncoupled system. Both equations are collected to construct the relaxation matrix

$$c_{ij} = \begin{pmatrix} -s_1 - c_{12} & c_{21} \\ c_{12} & -s_2 - c_{21} \end{pmatrix}$$

$$= \begin{pmatrix} -s - c - \Delta s - \Delta c & c - \Delta c \\ c + \Delta c & -s - c + \Delta s + \Delta c \end{pmatrix},$$

with rates

$$s = \frac{1}{2}(s_1 + s_2), \qquad \Delta s = \frac{1}{2}(s_1 - s_2), \qquad c = \frac{1}{2}(c_{12} + c_{21}),$$

$$\Delta c = \frac{1}{2}(c_{12} - c_{21})$$

and eigenvalues

$$u_1 = -r + Q \text{ (slow process)}, \quad u_2 = -r - Q \text{ (fast process)},$$

where $Q^2 = c^2 + \Delta s^2 + 2\Delta s \cdot \Delta c$, $r = s + c$. Diagonalization of (5) delivers

$$\begin{pmatrix} \dot{\beta}_1 \\ \dot{\beta}_2 \end{pmatrix} = \begin{pmatrix} u_1 & 0 \\ 0 & u_2 \end{pmatrix} \begin{pmatrix} \beta_1 \\ \beta_2 \end{pmatrix}, \tag{6}$$

where $\beta_i = D_{ij} p_j$ and the diagonalisation matrix

$$D_{ij} = \frac{1}{2Q(c - \Delta c)} \begin{pmatrix} -\Delta s - \Delta c + Q & c - \Delta c \\ \Delta s + \Delta c + Q & -c + \Delta c \end{pmatrix},$$

$$D_{ij}^{-1} = \begin{pmatrix} c - \Delta c & c - \Delta c \\ \Delta s + \Delta c + Q & \Delta s + \Delta c - Q \end{pmatrix}.$$

The solution of (6) is

$$\beta_i = e^{u_i t} \beta_i(0) \quad \text{with} \quad \beta_i(0) = D_{ij} p_j(0).$$

Inserting the relative population numbers $p_{\{1,2\}}(0) = 1/2 \mp \Delta c / c$ one finds

$$\begin{pmatrix} \beta_1 \\ \beta_2 \end{pmatrix} = \frac{1}{4cQ} \begin{pmatrix} (Q + c - \Delta s) e^{-(r-Q)t} \\ (Q - c + \Delta s) e^{-(r+Q)t} \end{pmatrix}, \tag{7}$$

and after transformation back to $p_i = D_{ij}^{-1} \beta_j$ the relaxation functions for states 1 and 2 are obtained [10]:

$$p_1(t) = (c - \Delta c)/(4cQ) \left[(Q + c - \Delta s) e^{-(r-Q)t} + (Q - c + \Delta s) e^{-(r+Q)t} \right],$$

$$p_2(t) = (c + \Delta c)/(4cQ) \left[(Q + c + \Delta s) e^{-(r-Q)t} + (Q - c - \Delta s) e^{-(r+Q)t} \right].$$

Their sum $p(t)$ is the total relaxation function of the system

$$p(t) = \left(\frac{1}{2} + \frac{C}{2Q} + \frac{\Delta s \, \Delta c}{2cQ} \right) e^{-(r-Q)t} + \left(\frac{1}{2} - \frac{C}{2Q} - \frac{\Delta s \, \Delta c}{2cQ} \right) e^{-(r+Q)t}. \tag{8}$$

Identical results are obtained by calculating the probability of a particle remaining in its original state 1 or jumping to state 2.

Typical graphs of these relaxation functions are shown in Fig. 16, where the apparent relaxation rates $1/\tau_i = r \pm Q$ and the corresponding apparent relaxation strengths $1/2 \mp (c + \Delta c \, \Delta s)/(2Qc)$ are depicted in dependence on c/s. The meaning of these apparent rates is connected with the standard interpretation of dielectric relaxation spectra, which are commonly decomposed into model processes (Debye, Havriliak–Negami, etc.)

Keeping the exchange rate constant and lowering the relaxation rates s_1, $s_2 = 100 s_1$ exponentially from left to right (dotted lines) allows us to model the typical slowing down of relaxation rates with temperature. In Fig. 16a, the single-process relaxation rates and, in Fig. 16b, the corresponding relaxation strengths are shown on a logarithmic scale. The apparent shift to higher relaxation rates is of course only an effect of the artificial separation of the relaxation processes. Molecules starting in the fast relaxation state do not relax faster by exchange with a slower state. This is easily verified by calculating the time derivatives of $p_i(t)$.

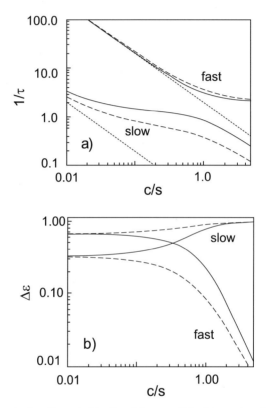

Fig. 16. (a) Apparent relaxation rate $1/\tau$ (in arbitrary units) of two relaxation processes with relaxation time ratio $s_2/s_1 = 100$ at fixed exchange rate c, and s slowing down from left to right. The dotted lines visualize the assumed exponential decrease of the undisturbed rates s_2 and s_1. The relative strengths of the original fast and slow processes are $1:2$ (dashed lines) and $2:1$ (solid lines) respectively. ($\Delta c/c = (n_1 - n_2)/(n_1 + n_2) = \pm 1/2$). (b) The apparent relaxation strengths $\Delta\varepsilon$, notation as above. The relative strength of the slow process approaches 1; the strength of the fast process decays to zero with increasing exchange, irrespective of the strengths of the uncoupled processes. Taken from [11]

These theoretical results have been used to interpret the experimental dielectric data on salol confined in porous sol–gel glasses (Fig. 17). In bulk salol, one observes a single relaxation process. Its relaxation rate decreases with lower temperature (according to the VFT equation). When salol is brought into porous glasses with nanometer pore diameters, new characteristic features are observed. One process, which can obviously be attributed to the free salol in the pores, shows at high temperatures a relaxation rate equal to that of the bulk, but with decreasing temperature it becomes faster than the bulk relaxation at the same temperature. A second process, attributed to a layer of surface-bound salol, appears at relaxation rates which are almost two decades slower. At lower temperatures, it gradually approaches the bulk rate. (A third process at low frequencies, due to Maxwell–Wagner polarisation, will not be considered here and has been omitted in the representation.) As temperature decreases, the fast (volume) process loses its dielectric strength while the second (surface) process gains such that the sum of both processes roughly follows the temperature curve of the bulk value.

Comparison with Fig. 16 suggests the following interpretation. Both the surface and volume relaxation rates in the pores are fast compared with the molecular exchange process between surface-bound and free salol at high temperatures. Their ratio is roughly 1:100. On the basis of the relation between the pore radii and molecular sizes (≈ 0.5 nm), one expects that the ratio of free salol in the pores to the amount of molecules bound in a monomolecular surface layer will be 1:2 or 1:3 for a pore diameter of 2.5 nm or 7.5 nm, respectively. This is roughly equal to the respective experimental high-temperature ratio of the dielectric strengths. As the temperature is lowered towards the calorimetric glass transition, the dielectric relaxation rates reach the order of magnitude of the exchange rate between free and surface-bound salol, which has a weaker temperature dependence. With decreasing temperature, random exchange between surface layer and free molecules leads to an apparently faster relaxation of the volume process compared with the bulk curve, and also to an increased rate of the surface process, which gradually approaches the bulk curve. The measured relaxation strengths show exactly the predicted behavior; the slow process apparently gains intensity from the fast process. In summary, it has been shown that the experimental data obtained for salol in nanoporous sol–gel glasses must be interpreted in terms of a surface volume exchange of salol molecules on the characteristic timescale of the dielectric experiment.

In order to suppress the formation of an interfacial layer of molecules the inner surfaces of the nanoporous sol–gel glasses can be silanized [12]. By that the formation of H bonds between the salol molecules and the solid surfaces is strongly hindered. For the dynamics, this means that the corresponding interfacial relaxation process (process II) is no longer observed (compare Figs. 18b and c). It is remarkable that the relaxation time distribution function of unbound molecules in the silanized pores is only weakly broadened (Fig. 18f) compared with the bulk liquid (Fig. 18d). Figure 19 shows the relaxation time distribution of salol confined to silanized pores of different diameter and at varying temperatures. A broadening is observed at high temperatures owing to various unspecific influences of the random confinement. With decreasing temperature $L(\tau)$ shifts to shorter relaxation times for confined salol. This effect is

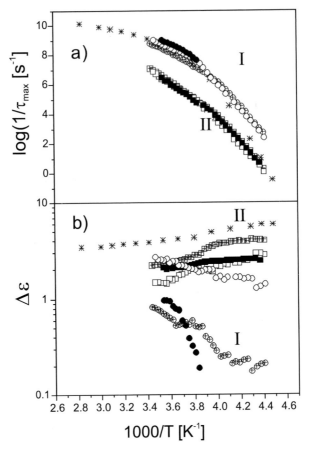

Fig. 17. (a) Relaxation rate $1/\tau_{max}$ and (b) volume-corrected dielectric strength $\Delta\varepsilon$ of salol in pores versus inverse temperature. Pore sizes: 2.5 nm, solid symbols; 5.0 nm, cross-centered symbols; and 7.5 nm, open symbols. Different processes: dynamic glass transition (○) and interfacial relaxation process (boxes). Bulk salol (*). Data from [11]

more pronounced for smaller pores. The analysis of the dielectric strengths shows that the silane layer has a thickness of 0.38 ± 0.02 nm for all pore sizes.

In order to characterize the temperature dependence, the averaged relaxation time $-\log \tau_{med}$ was calculated according to (3). The activation plots of the relaxation rate $-\log \tau_{med}$ are compared for bulk and confined salol in Fig. 20. At high temperatures, the relaxation rates of the confined liquids for all pore sizes are identical to that of the bulk liquid, while with decreasing temperature the molecules in the confining geometry fluctuate faster compared with the bulk liquid, i.e. the glass transition temperature of the confined liquids is shifted to lower temperatures. This shift is more pronounced for smaller pores. A comparable effect is observed by means of DSC [12]. The calorimetric glass transition temperature in the confining geometry is

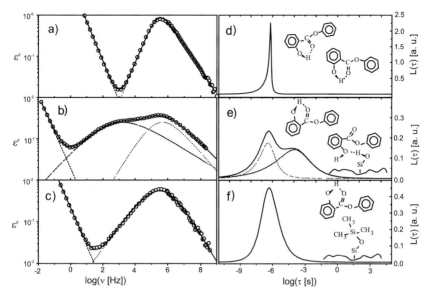

Fig. 18. Dielectric loss ε'' of salol at a temperature $T = 253$ K. The error of the measured data (circles) is smaller than the size of the symbols. The dotted and dash–dotted lines indicate fits to the data according to Havriliak–Negami functions. The dashed lines correspond to conductivity contributions and loss processes caused by polarization effects. The plots on the right-hand side (d, e, f) show the relaxation time distributions $L(\tau)$ that correspond to the relaxational processes. The sketches illustrate possible conformations of the molecules. (a) Bulk salol: the molecules form intramolecular H bonds ($\cdots$); one relaxation process having the characteristic shape of $L(\tau)$ for a bulk glass-forming liquid is observed. (b) Salol confined to 7.5 nm uncoated pores: salol molecules in the direct vicinity of the surface form H bonds ($\cdots$) to the pore surface, resulting in an additional interfacial relaxation process (dashed–dotted lines) next to the relaxation of unbound molecules (dotted lines). The corresponding $L[\log(\tau)]$ has a bimodal shape (solid line). (c) Salol confined to 7.5 nm coated pores: the formation of H-bonds is strongly suppressed. The resulting unimodal $L(\tau)$ is shifted to shorter relaxation times with respect to the bulk phase. Taken from [12]

shifted by 8, 11, and 15 K to lower temperatures for the 7.5, 5.0, and 2.5 nm pores, respectively (indicated by arrows in Fig. 20).

The confinement effect can be unambiguously explained on the basis of the co-operativity of molecular reorientations in glass-forming van der Waals liquids. At high temperatures, the range of cooperativity ξ is smaller than the extension of the nanoporous restrictions. Hence, no difference between molecules in the bulk and in the confining space is to be expected. With decreasing temperature, deviations from the Arrhenius-type temperature dependence occur, indicating the onset of coopera-tivity. The correlation length ξ increases until it becomes limited by the pore diameter ("hindered glass transition") [3,4]. Because of the lubricant coating of the inner sur-faces, the cooperatively rearranging molecules in the pores are decoupled from the solid walls and may reorient within the pore volume. In contrast, the reorientational

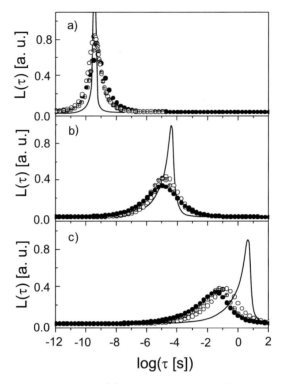

Fig. 19. Relaxation time distribution $L(\tau)$ for salol confined to 7.5 nm (circles), 5.0 nm (cross-centered circles), and 2.5 nm (solid circles) pores with a hydrophobic coating, and bulk salol (solid line) at different temperatures, (a) 305 K, (b) 243 K, (c) 223 K. Taken from [12]

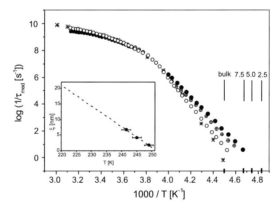

Fig. 20. Activation plots for salol confined to coated pores: 7.5 nm (circles), 5.0 nm (cross-centered circles), 2.5 nm (solid circles), and bulk salol (stars). The error of the data is smaller than the size of the symbols. The arrows indicate the calorimetric glass transition temperatures for salol in pores of different sizes and for bulk salol. Inset: size of cooperatively rearranging domains ξ versus temperature. Taken from [12] and modified

dynamics in the bulk liquid is increasingly retarded owing to the unhindered growth of ξ. This leads to a faster dynamics of the confined molecules compared with the bulk liquid.

Following the simple model described above, the temperature dependence of the length scale of cooperativity can be estimated from the pore size dependence of the shift of the relaxation rate: it starts to increase compared with the bulk rate when the length scale of cooperativity ξ reaches the size of the confining geometry. For the 2.5 nm pores, deviations from the bulk rate are observed at 250 K. At lower temperatures, even for salol in 7.5 nm pores, a significant increase of the relaxation rate compared with bulk salol is observed, so one can estimate the length scale ξ of cooperativity to be greater than 7 nm in the vicinity of the calorimetric glass transition temperature.

The question arises of whether the confinement effect could be caused by a pore-size-dependent decrease of the density of the confined liquid. This conjecture can be ruled out on the basis of the following consideration. Assuming that the density and hence the mean relaxation rate $1/\tau$ obey a VFT law where the Vogel temperature varies with the pore size in a similar way to the measured calorimetric glass transition temperature (Fig. 21a) the derivative $d(\log(1/\tau))/d(1000/T)$ delivers a temperature dependence as shown in Fig. 21b. This can be compared with the difference quotient as determined *experimentally* from the relaxation rate *measured* in temperature steps of 0.5 K (Fig. 21c). It turns out that the difference quotient behaves qualitatively differently. In the temperature interval between 333 K and 260 K, the apparent activation energies for the bulk and the confined (2.5 nm and 7.5 nm) liquids coincide within experimental accuracy. But for lower temperatures, suddenly the charts bend off; this takes place for the 2.5 nm pores at 256 K $\pm$ 3 K and for the 7.5 nm pores at 245 K $\pm$ 3 K. The temperature dependence is in sharp contrast to the results (Fig. 21b) which one would expect from a dependence like that displayed in Fig. 21a, assuming a weakly varying temperature dependence of the density. In reverse, a nonmonotonic change of the density seems to be unreliable. Instead, it is suggested that the measured confinement effects are caused by the cooperative nature of the dynamic glass transition. With decreasing temperature the size of cooperatively rearranging domains grows and the apparent activation energy increases. If, owing to the confinement of the nanoporous system, further growth is prohibited, the VFT dependence turns suddenly into an Arrhenius-like thermal activation.

4 Conclusions

Broadband dielectric spectroscopy enables one to unravel in detail the counterbalance between the surface and confinement effects determining the molecular dynamics in a confining space. This has been exemplified for a variety of different host/guest systems.

Ethylene glycol in zeolitic host systems shows a pronounced confinement effect. Beyond a threshold channel size, the liquid character is lost as indicated by a dramatically increased relaxation rate and an Arrhenius-like temperature dependence.

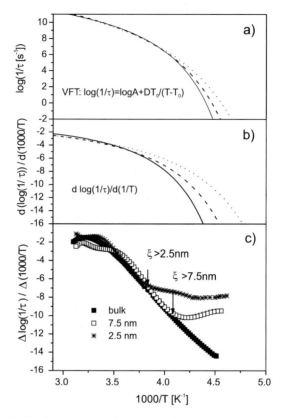

Fig. 21. (a) The calculated dependence of the mean relaxation rate versus inverse temperature assuming a VFT law, where T_0 is shifted according to the calorimetrically measured shift of T_g. Solid line, bulk liquid, $T_g = 222$ K; dashed line, salol confined to a silanized sol–gel glass having a mean diameter of 7.5 nm, $T_g = 214$ K; dotted line, salol confined to a silanized sol–gel glass having a mean diameter of 2.5 nm, $T_g = 207$ K. (b) The calculated derivative $\mathrm{d}(\log(1/\tau))/\mathrm{d}(1000/T)$ assuming the temperature dependence shown in (a). The units are omitted for graphical reasons. They are the same as in part (c) of this figure. (c) The experimentally determined difference quotient $\Delta(\log(1/\tau))/\Delta(1000/T)$ for the data shown in Fig. 20 for the bulk liquid (solid squares), salol confined to silanized nanopores of 7.5 nm (open squares), and salol confined to silanized nanopores of 2.5 nm (stars). Taken from [1]

Computer simulations of the molecular arrangement in a confining space prove that an ensemble as small as six molecules is sufficient to exhibit the dynamics of a bulk liquid.

Propylene glycol, butylene glycol, and pentylene glycol in nanoporous sol–gel glasses show, for (untreated) hydrophilic inner surfaces a surface effect, which can be fully removed by making the boundary layer between the guest molecules and the solid host system hydrophobic. From an analysis of the relaxation time distribution, a quantitative three-layer model can be deduced, with immobilized, interfacial, and bulk-like molecules.

The quasi-van der Waals liquid salol shows, in (untreated) hydrophilic sol–gel glasses, a dynamics which must be interpreted in terms of a two-state model with exchange between a bulk-like phase in the pore volume and an interfacial phase close to the pore wall. This enables us to analyze the interplay between the molecular dynamics in the two subsystems, and hence their growth and decline in dependence on temperature and the strength of the molecular interactions. Analogously to the glycols, the surface-induced relaxation process can be fully removed by silanization of the walls of the solid host system. Under these conditions, a confinement effect for the dynamics of the embedded salol molecules is observed. It is proven by calorimetric studies as well. A refined analysis (Fig. 20) enables us to estimate the size of cooperatively rearranging domains and its temperature dependence.

Acknowledgments

The authors would like to thank Dipl. Phys. L. Hartmann and Dipl. Phys. A. Serghei for critically reading the manuscript. The fruitful collaboration with Prof. Dr. H. Behrens (Hannover) is thankfully acknowledged.

References

1. F. Kremer, A. Huwe, M. Arndt, P. Behrens, W. Schwieger: J. Phys.: Condens. Matter **11**, A175 (1999)
2. G. Adam, J.H. Gibbs: J. Chem. Phys. **43**, 139 (1965)
3. E. Donth: *Glasübergang* (Akademie Verlag, Berlin, 1981)
4. E. Donth: *Relaxation and Thermodynamics in Polymers, Glass Transition* (Akademie Verlag, Berlin, 1992)
5. E.W. Fischer, E. Donth, W. Steffen: Phys. Rev. Lett. **68**, 2344 (1992)
6. E.W. Fischer: Physica A **201**, 183 (1993)
7. D. Sappelt, J. Jäckle: J. Phys. A **26**, 7325 (1993)
8. A. Huwe, F. Kremer, P. Behrens, W. Schwieger: Phys. Rev. Lett. **82**, 2338 (1999)
9. W. Gorbatschow, M. Arndt, R. Stannarius, F. Kremer: Europhys. Lett. **35**, 719 (1996)
10. M. Arndt, R. Stannarius, W. Gorbatschow, F. Kremer: Phys. Rev. E **54**, 5377 (1996)
11. R. Stannarius, F. Kremer, M. Arndt: Phys. Rev. Lett. **75**, 4698 (1995)
12. M. Arndt, R. Stannarius, H. Groothues, E. Hempel, F. Kremer: Phys. Rev. Lett. **79**, 2077 (1997)
13. G. Barut, P. Pissis, R. Pelster, G. Nimtz: Phys. Rev. Lett. **80**, 3543 (1998)
14. P. Pissis, A. Kyritsis, D. Daoukaki, G. Barut, R. Pelster, G. Nimtz: J. Phys.: Condens. Matter **10**, 6205 (1998)
15. D. Daoukaki, G. Barut, R. Pelster, G. Nimtz, A. Kyritsis, P. Pissis: Phys. Rev. B **58**, 5336 (1998)
16. P. Pissis, A. Kyritsis, G. Barut, R. Pelster, G. Nimtz: J. Non-Cryst. Solids **235–237**, 444 (1998)
17. H. Wendt, R. Richert: J. Phys: Condens. Matter **11**, A199 (1999)

18. A. Schönhals, R. Stauga: J. Non-Cryst. Solids **235–237**, 450 (1998)
19. A. Schönhals, R. Stauga: J. Chem. Phys. **108**, 5130 (1998)
20. F. Rittig, A. Huwe, G. Fleischer, J. Kärger, F. Kremer: Phys. Chem. Chem. Phys. **1**, 519 (1999)
21. G. Liu, Y. Li, J. Jonas: J. Chem. Phys. **95**, 6892 (1991)
22. J. Zhang, G. Liu, J. Jonas: J. Chem. Phys. **96**, 3478 (1992)
23. C. Streck, Yu.B. Mel'nichenko, R. Richert: Phys. Rev. B **53**, 5341 (1996)
24. J. Schüller, Yu.B. Mel'nichenko, R. Richert, E.W. Fischer: Phys. Rev. Lett. **73**, 2224 (1994)
25. W.M. Meier, D.H. Olson, C. Baerlocher: *Atlas of Zeolite Structure Types* (Elsevier, Amsterdam, 1996)
26. J. Kärger, D.M. Ruthven: *Diffusion in Zeolites and Other Microporous Solids* (Wiley, New York, 1992)
27. M. Bibby, M.P. Dale: Nature **317**, 157 (1985)
28. C.M. Braunbarth, P. Behrens, J. Felsche, G. van de Goor: Solid State Ionics **101–103**, 1273 (1997)
29. J.M. Newsam, M.M.J. Treacy, W.T. Koetsier, C.B. de Gruyter: Proc. Roy. Soc. (London) **420**, 375 (1988)
30. S. Havriliak, S. Negami: J. Polym. Sci. Part C **14**, 99 (1966)
31. S. Havriliak, S. Negami: Polymer **8**, 161 (1967)
32. H. Schäfer, E. Sternin, R. Stannarius, M. Arndt, F. Kremer: Phys. Rev. Lett. **76**, 2177 (1996)
33. H. Vogel: Phys. Zeit. **22**, 645 (1921)
34. G.S. Fulcher: J. Am. Ceram. Soc. **8**, 339 (1925)
35. G. Tammann, G. Hesse: Anorg. Allgem. Chem. **156**, 245 (1926)
36. B.P. Jordan, R.J. Sheppard, S. Szwarnowski: J. Phys. D **11**, 695 (1978)
37. S.L. Mayo, B.D. Olafson, W.A. Goddard III: J. Phys. Chem. **94**, 8897 (1990)
38. A.K. Rappe, C.J. Casewit, K.S. Colwell, W.A. Goddard III, W.M. Skiff: J. Am. Chem. Soc. **114**, 10024 (1992)
39. E. Burchart: *Ph.D. thesis* (Technische Universiteit Delft, 1992)
40. N.E. Cusack: *The Physics of Structurally Disordered Matter* (Adam Hilger, Bristol, 1987)
41. A. Schönhals, F. Kremer, A. Hofmann, E.W. Fischer, E. Schlosser: Phys. Rev. Lett **70**, 3459 (1993)

Liquid Crystals in Confining Geometries

Ralf Stannarius[1] and Friedrich Kremer[2]

[1] Otto-von-Guericke-Universität Magdeburg, Institut für Experimentalphysik,
   ralf.stannarius@physik.uni-magdeburg.de
[2] Universität Leipzig, Institut für Experimentalphysik I,
   kremer@physik.uni-leipzig.de

Abstract. The properties of liquid-crystalline (LC) phases under submicrometer geometrical confinement can change substantially with respect to bulk samples. We give an overview over studies of the structure and dynamics of confined LC phases. Systems with different adsorber geometries and sizes are compared. Studies of the molecular and collective dynamics of nematic and smectic phases have been performed by means of dielectric and NMR spectroscopy. The experimental results show that the confinement, even in nanometer pores, leaves the molecular dynamics rather uninfluenced. On the other hand, the collective dynamic modes are sensible to pore geometries and sizes already in micrometer pore systems. Moreover, we discuss restricted-volume effects on the phase transitions, and the director configurations and electro-optical properties of nematics in regular and random pore systems.

1 Adsorber Geometries and Confinement Effects

In general, the restricted-volume effects that are observed in *isotropic* liquids confined in porous matrices (see previous chapter) can be found in confined *mesogenic* materials as well. Additionally, the liquid-crystalline (LC) state is characterized by molecular order, anisotropic physical properties, and collective dynamic processes that affect the electric and optical properties of the system. It is natural to expect that in the vicinity of surfaces, the interactions of the mesogens with the boundary will influence the order and orientation of the mesophases to a considerable extent.

One of the prominent effects related to the surface anchoring of the nematic director field, the Fréedericksz transition in nematic sandwich cells, had already been observed "in the dawn" of liquid-crystal research. A uniform ground state of the director field parallel or perpendicular to the confining plates of thin planar cells can be achieved, by proper surface treatment, across cell gaps of a few hundred micrometers. This ground state can be switched by electromagnetic fields, coupling to the respective anisotropic susceptibilities of the material in the nematic mesophase. Elastic forces restore the undeformed ground state after the external field is removed, and the sample returns to the ground state imposed by the boundary conditions. This surface-induced director alignment forms the basis for the most attractive application of liquid-crystal materials today, in display technology.

Although, from an application point of view, surface-induced alignment is the most important aspect, studies of thermodynamic phenomena in spatial confinement, the modification of liquid crystal phases, and the induction or suppression of orientational order in adsorbed mesogenic materials are of equal importance for

R. Stannarius, F. Kremer, Liquid Crystals in Confining Geometries, Lect. Notes Phys. **634**, 301–336 (2004)

fundamental research. A comprehensive introduction to the topic is given in the book *Liquid Crystals in Complex Geometries* [1]. The reader is referred to this book and references collected therein for a comprehensive overview of physical systems, modeling, experimental methods, and fundamental results obtained during the last decades. The present contribution will focus on the description of confinement effects in a few selected systems, starting from a recollection of surface-induced orientation in ordered and disordered systems, via experiments that study the induction or suppression of mesogenic order, to the investigation of dynamic processes in confined liquid crystals. As opposed to the well-understood phenomena observed in thin LC sandwich cells, attention will be focused on systems with complex geometries, such as porous matrices with cylindrical, spherical or irregular cavities, filled with mesogenic materials.

There are a variety of experimental methods that explore directly the surfaces and surface interactions of liquid crystals, such as for example second-harmonic generation, surface plasmon spectroscopy, and atomic force microscopy (AFM). The experiments described in this chapter mainly involve spectroscopic (bulk) methods, polarizing microscopy and electro-optic measurements. Confinement of liquid-crystal phases in porous adsorbers with large inner-surface-to-volume ratios opens up the opportunity to study the influence of boundary and confined-volume effects with such macroscopic detection methods. For example, differential scanning calorimetry (DSC) [2–20], nuclear magnetic resonance (NMR) [21–39], dynamic light scattering [40–42], dielectric and infrared spectroscopy [32, 33, 43–56], and X-ray or neutron scattering [7, 57, 58] have proved very successful in the study of surface and confinement effects in LC mesophases.

Apart from the specific interactions between the adsorbed LC molecules and the matrix material, there are at least two important factors to be considered in a classification of confined liquid crystal systems: the pore size and the geometry of the pores. Liquid crystals contained in volumes larger than $\approx 1\,\mu m$ exhibit bulk behavior, and the macroscopically observed influences of the surfaces reduce to the "classical" alignment of the director field at the container walls, which can be controlled by appropriate surface treatment. These boundary conditions determine to a great extent the corresponding equilibrium solutions of the director field in the volume, e.g. textures, walls, or defects in thin cells or capillaries. Such phenomena have been investigated in great detail, and they will not be considered in this article. In pores with diameters of approximately one nanometer and below, the role of the surfaces dominates, and the characteristics of liquid-crystalline phases are completely lost. In such small pores, the mesogenic properties of the molecules are completely covered by molecular interactions with the adsorber. The effects on mesogen orientation and dynamics reduce basically to those described in the previous chapter. The intermediate, interesting pore size range covers approximately three orders of magnitude, from several nanometers to about one micrometer. From the *size* aspect, the systems can be roughly divided into microporous and macroporous adsorbers, where the term "micropore" comprises pore sizes from a few nanometers to about 100 nm (in at least one dimension). In macroporous adsorbers, all pore dimensions are above 100 nm.

From the viewpoint of pore *geometry*, adsorbers can be divided into regular, ordered pore systems and those containing irregular cavities of more or less random geometry (see Fig. 1).

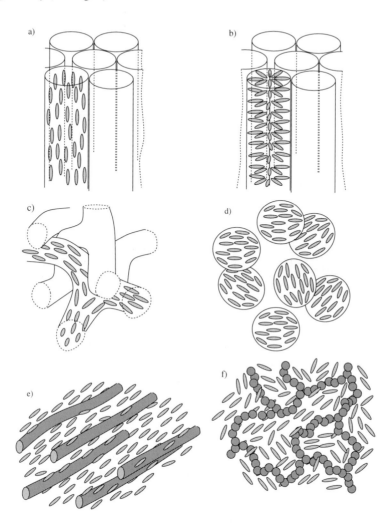

Fig. 1. Simplified classification of confining geometries. (a, b) Regular cylindrical adsorbers (Anopore and Nuclepore filters). Depending on the surface treatment of the channels, axial or radial director configurations can be prepared. (c) Irregular pores (sol–gel glasses and cellulose membranes). For simplicity, the adsorbed LC is depicted in only one pore of Figs. (a)–(c). (d) Spherical droplets (PDLC). (e,f) Dispersions and gels (filled nematics, and inorganic or polymeric networks in the LC phase). In the latter two cases, ordered (e) or disordered (f) director fields can be achieved

Classical representatives of ordered macroporous systems are liquid crystals adsorbed in porous filters such as *Anopore* [27, 47–49, 59] and *Nuclepore* membranes [21–24, 28]. Anopores are amorphous inorganic Al_2O_3 membranes, they are provided commercially as sheets of 60 μm thickness, with sizes of several square centimeters. Internally, they are composed of long, regular channels with polygonal cross sections, and a narrow pore size distribution (Fig. 2). In modeling, they are often approximated as cylindrical tubes. The channel axes are perfectly aligned normal to the filter plane. Figure 2 shows a scanning tunneling microscope image of a broken edge of such a filter sheet. The director field of a nematic adsorbed in an untreated Anopore matrix is depicted schematically in Fig. 1a. Nuclepore membranes consist of a polymer matrix containing cylindrical pores. They are available in a broad range of cylinder sizes. A certain disadvantage (compared with Anopore filters) for spectroscopic experiments is their comparatively poor filling factor, i.e. pore volume per membrane volume.

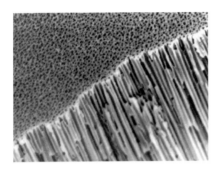

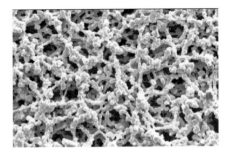

Fig. 2. STM image of an Anopore filter with pore diameters of 200 nm and pore length of 60 μm, (by courtesy of U. Boehnke)

Fig. 3. STM image of an empty cellulose nitrate membrane (Synpore), nominal pore size 0.83 μm (by courtesy of S. Różański)

Polymer Dispersed Liquid Crystals, (PDLC) [60–63] consist of a polymer matrix with roughly spherical inclusions of (mostly nematic or cholesteric) LC material (Fig. 1b). The droplets containing the mesogens are formed by phase separation of the polymer matrix and the LC adsorbate [61]. In thermally induced phase separation (TIPS), the LC is dissolved in the polymer melt and the droplets form during cooling of the mixture when the LC becomes immiscible. A solvent-induced phase separation (SIPS) starts from a solution of both polymer and LC in a solvent and phase separation occurs when the solvent evaporates. A polymerization-induced phase separation (PIPS) uses the effect that the LC and the precursor monomer are miscible, while the LC phase separates during the polymerization reaction. In all cases, droplet sizes are of the order of micrometers or below. Commonly, the LC director aligns parallel to the droplet boundary, forming an axial director field structure (Fig. 1d). Since the refractive indices of the LC and the polymer do not match, and the alignment axes of the individual droplets are oriented at random, the system is optically in a scattering state. Under electric fields, one can reversibly align the orientation axes of the individual droplets and switch the PDLC into a transparent state. There have been promising attempts at technological applications of PDLCs as displays or switchable windows.

An example of disordered macroporous adsorbers that has been in the focus of confined-LC research is macroporous glasses [44, 64]. A crude picture of random cavities and the director field in the adsorbed nematic is given in Fig. 1c. These glases are produced in a controlled sol–gel process. The range of cavity sizes in such a glass matrix is comparatively broad, the pores are interconnected and of random shapes. The director field within the cavities is fixed by the boundary conditions at the irregular pore surfaces. Thus, the ground state in the absence of aligning fields is a highly deformed director configuration with frozen-in defects. Similar macroporous geometries are found in organic membranes, for example in cellulose nitrate filters (Synpore). These *cellulose membranes* consist of randomly arranged fibrils forming irregular cavities of submicrometer sizes (see Fig. 3). When they are filled with nematic liquid-crystalline material, the fibrils induce a state of frozen random disorder of the director field, containing many defects. The optical appearance of such filled cellulose membranes is highly scattering. Experiments with LCs confined in such random-pore organic membranes are discussed below.

Examples of microporous systems with randomly disordered cavities are various types of microporous glases. From the topology of the pores, they are probably comparable to macroporous sol–gel glasses. Controlled *sol–gel glasses* with relatively narrow pore size distributions are available with pore sizes of a few nanometers (see e.g. [13, 32, 33, 44, 65]). In such cavities, dramatic influences of confinement on the formation of mesogenic phases and on transition temperatures as well as the character of phase transitions are found.

Finally, it is possible to disperse solid particles in the liquid crystal in order to influence and to control the orientational state of the mesophase. The alignment of the director at the dispersed-particles surfaces imposes a director configuration on the nematic material (e. g. [66, 67]). Such systems include, for example, "filled" nematics [68], chemically [35–38, 69–74] or physically [75–79] cross-linked gels of organic additives, and silica aerogels [6, 7, 11, 14, 15, 25, 40, 41, 57, 58] The dispersed material can induce, in some cases, random disorder of the director field. This is particularly the case in the silica aerogels and filled nematics. The idea for technological applications of filled nematics is the opportunity to switch the nematic into a transparent state by external electric fields. On the other hand, one can also prepare systems where the dispersed particles align the director uniformly (or induce liquid-crystalline order in an isotropic liquid [35]. Polymerization of photoreactive monomers or the formation of physical networks of hydrogen-bonding additives may lead to structures that can locally be considered as an assembly of rod-like strands stabilizing an aligned nematic director field. Such a situation is schematically depicted in Fig. 1e. Likewise, other director configurations such as twisted nematic director fields can be stabilized by a polymeric network [73].

The following sections are not organized according to this classification of adsorber geometries but consider the physical effects on confined LC mesogens. The next section deals with aspects of the molecular and collective dynamics of mesogens in porous confinement. Then, a few examples of induction or suppression of nematic and smectic ordering will be discussed. Finally, influences of confining geometries on the director field configurations are described.

2 Molecular and Collective Dynamics

A well-suited experimental technique for dynamic studies is dielectric spectroscopy [80]. This method measures the complex dielectric function $\varepsilon = \varepsilon' - i\,\varepsilon''$. An ansatz of superimposed Havriliak–Negami functions [81] and a conductivity contribution is chosen for data deconvolution in the dielectric loss spectra (see (1) on p. 277),

$$\varepsilon''(\omega) = \sum_{k=1}^{n} \mathrm{Im}\left[\frac{\Delta\varepsilon_k}{\left(1 + (i\omega\tau_k)^{\alpha_k}\right)^{\beta_k}}\right] + c(\omega), \tag{1}$$

where $\Delta\varepsilon$ is the dielectric relaxation strength and τ is the mean relaxation time. The index k refers to the n different relaxation processes that contribute to the dielectric response. The parameters α and β describe symmetric and asymmetric broadening of a distribution of Debye ($\alpha = \beta = 1$) relaxation rates for each process. In the case of $\beta = 1$, τ coincides with the position of the loss peak maximum τ_{max}. The conductivity term

$$c(\omega) = \frac{\sigma}{\varepsilon_0}\frac{a}{\omega^s}$$

depends upon the sample conductivity σ and the exponent s of the angular frequency, the latter is 1 for pure ohmic behavior. The constant factor a has the unit s^{-s}.

2.1 Molecular Dynamics of Nematics in Confinement

The molecular dynamics of nematic liquid crystals in *macroporous* systems is basically the same as in the bulk phase. However, even macroporous adsorbers can prove advantageous for the study of dynamics of LCs because of the aligning properties of regular cavities for the director field. For example, Anopore filter channels allows the convenient control of the sample orientation. A uniform alignment with respect to the electric field facilitates the discrimination of individual molecular processes. In the usual capacitor setup, the liquid crystal is sandwiched between metal (e.g. brass) electrodes, and it is not trivial to achieve well-defined sample orientations. The untreated metal electrodes provide a poor, preferentially homeotropic orientation. Additional organic alignment layers may improve sample alignment but contribute additional loss processes and deteriorate the dielectric spectra. Sometimes, magnetic fields can be used to align the samples. When LC-filled Anopore filters are sandwiched between the capacitor electrodes, the field is exactly axial and well defined with respect to the director.

For broadband dielectric studies of the molecular dynamics, a standard nematic liquid-crystalline compound, 5CB (4-n-pentyl-4'-cyanobiphenyl), has been chosen. The bulk material exhibits a nematic phase from 295 to 308 K. The 5CB molecule possesses a strong axial dipole moment (that of the cyano group) nearly parallel to the long molecular axis; its components perpendicular to that axis are negligibly small. As a consequence, molecular reorientations around the long axis are not observed dielectrically. Molecular reorientation around the short molecular axes (in the following referred to as the δ-relaxation process) changes the axial component of the dipole moment; it is most effective in dielectric spectra when the director is in the field direction. Librations of the long axis ("tumbling"), i.e. small angular excursions around the short molecular axis lead to fluctuations of the component of the dipole moment perpendicular to the director axis. They are effective in dielectric relaxation when the electric field is perpendicular to the director. In bulk samples prepared between the metal electrodes of the capacitor in the dielectric spectrometer, the director field is not uniform. Without special treatment, the metal surfaces induce a director alignment that is more homeotropic than planar. Figure 4 shows a dielectric spectrum of 5CB in the high-frequency range 1 MHz to 1 GHz. Two dynamic processes are necessary to fit the observed frequency dependence. The relaxation strengths $\Delta\varepsilon$ of both processes are of comparable orders of magnitude and they suggest a molecular origin. The bulk spectrum is dominated by the slower of the two processes, which is related to the reorientation about the short axis (δ-process). The second relaxation process appears on the high-frequency shoulder; it is assigned to the fast "tumbling" of the molecules around the director axis.

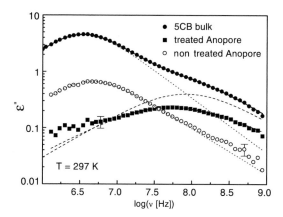

Fig. 4. Dielectric loss ε'' vs. frequency for 5CB in the bulk (●), and in untreated (boxes) and treated (○) Anopore membranes at 297 K. The solid line is a superposition of two Havriliak–Negami fits, assigned to the reorientation about the short axis (δ-relaxation, dotted line) and the tumbling mode (dashed line). In the confined system, the δ-relaxation dominates in untreated pores, where the axial director field is parallel to the electric field, and the tumbling mode dominates in treated pores, where the preferably radial director field is more or less perpendicular to the electric field. Figure taken from [80].

In nematic 5CB confined to nontreated Anopore membranes, the nematic director is aligned perfectly parallel to the pore walls of the membrane (axial configuration). The effectiveness of the fast dynamic process for dielectric relaxation is thus considerably reduced, and the $\varepsilon''(f)$ curve is dominated by the slow δ-relaxation processs.

The alignment of the 5CB director at the Al_2O_3 pore walls can be altered by special surface treatment. In order to modify the interaction between the liquid-crystal molecules and the inner filter surfaces, the latter were treated with a 2 wt.% solution of decanoic acid ($C_9H_{19}COOH$) in methanol. The acid headgroup binds chemically to the Anopore surface and the aliphatic chains form a compact array perpendicular to the inner surfaces. A homeotropic orientation of the director at the wall is induced, leading to a preferably radial nematic director field configuration. In that geometry, the director is dominantly perpendicular to the measuring electric field and one observes the dielectrically effective tumbling mode, while the effective relaxation strength of the δ-process is dramatically decreased (Fig. 4).

Temperature-dependent measurements in Anopore filters of different nominal pore sizes (0.2 μm, 0.1 μm, and 0.02 μm) were performed in the interval between 285 K and 320 K [47,49]. The different nominal pore sizes of the Anopore filters are not relevant here. In the isotropic phase, only one relaxation process is observed, the isotropic reorientation of the molecules. At the clearing point, this process splits into the two relaxation processes mentioned above. The dielectric relaxation strength of the δ-relaxation is comparable to the static dielectric anisotropy of 5CB; the relaxation strength of the second process is one order of magnitude lower.

The activation plot, Fig. 5, shows that in the isotropic phase, the bulk and the confined sample have nearly identical relaxation rates as expected. At the isotropic–nematic phase transition, two well-separated relaxation modes occur in the bulk

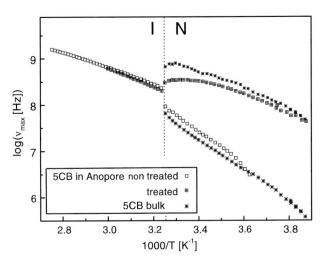

Fig. 5. Activation plot: relaxation rate vs. inverse temperature for 5CB in the bulk and confined to native and decanoic-acid-coated Anopore membranes with nominal pore size 0.2 μm. Data taken from [47]

sample. The different director configurations in treated and untreated Anopores allow a separate observation of both processes. While the phase transition observed in the dielectric spectra of the confined nematic is almost unchanged with respect to the bulk phase, a slight deviation of the relaxation rates is found below the transition temperature T_{NI}. The deviations from bulk behavior are most pronounced in the treated Anopore system. The most probable explanation is that the decanoic acid or other impurities dissolve in small concentrations in the liquid crystal and decrease the order parameter slightly in the vicinity of the clearing temperature.

The orientation of the nematic molecules inside the treated and untreated Anopore membranes allows separate measurements of the dielectric dispersions of the real parts of $\varepsilon_\parallel$ and $\varepsilon_\perp$ of the permittivity tensor (parallel and perpendicular, respectively, to the director) in 5CB. If a simple ansatz of rotational diffusion in a cosine potential $V = V_0 \cos \theta$ is assumed, where θ is the angle between the director and the molecular long axis (roughly the direction of the dipolar moment of 5CB), one can estimate the ratio between the dynamics of the fast process (rotation in the potential minimum) and that of the slow process (jumps across the potential barrier):

$$\frac{\tau_{\text{slow}}}{\tau_{\text{fast}}} = I_0^2 \left(\frac{V_0}{k_B T} \right).$$

This ratio decreases with the square of the modified Bessel function of order zero, I_0. The values for V_0 which correspond to the experimental ratios of relaxation rates are 2.44 $k_B T$ at the high-temperature end of the scanned range ($\tau_{\text{slow}}/\tau_{\text{fast}} = 0.1$) and 4.0 $k_B T$ at the low-temperature end ($\tau_{\text{slow}}/\tau_{\text{fast}} = 0.08$). With these values for the mean-field potentials, one would expect order parameters of 0.638 and 0.793, respectively, in Maier–Saupe theory. Compared with order parameters from other experiments [82], the potential barriers estimated from the relaxation time ratios are slightly too high but, in view of the rough estimates, they are of a reasonable order of magnitude. In the polar liquid crystal 5CB, short-range dipolar antiparallel ordering may be responsible for a deviation from mean-field predictions. For further details, see [47].

In cellulose nitrate membranes (Synpore), the pore sizes are slightly larger than the Anopore channel diameters, and the cavity surfaces are distributed randomly (Fig. 3). Empty Synpore membranes are dielectrically inactive: the real part of the dielectric function is constant in the measured frequency range and no loss peak can be observed. The nematic director aligns at the inner surfaces and hence is oriented isotropically [53]. Synpore filters were cut to small disks (diameter 5 mm) and filled, without special previous surface treatment, with 5CB in its isotropic phase.

Considering the results obtained for bulk and Anopore membranes, two relaxation processes, the δ-relaxation and the tumbling mode, are expected (Fig. 6). Compared with the bulk, the apparent δ-relaxation strength is strongly reduced in the confining system. This is not surprising, because in the bulk phase the liquid-crystalline material is oriented predominantly perpendicular to the electrode planes. Hence the angle between the director and the field axis is close to zero. In that geometry, the reorientation of the 5CB molecules about their short axes, the δ-relaxation, is dielectrically very effective. In the Synpore filter confinement, the director orientation is isotropic,

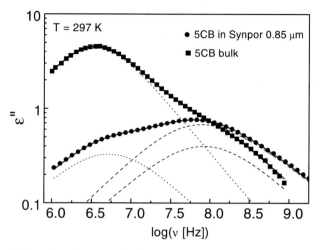

Fig. 6. Dielectric loss ε'' vs. frequency for 5CB in the bulk and in untreated Synpore membranes of 0.85 μm nominal diameter. The solid line is a superposition of two Havriliak–Negami relaxation processes, assigned to the δ-relaxation (dots) and the tumbling mode (dashes). In the confined system both relaxation processes are observable as well, but the δ-relaxation has a decreased dielectric relaxation strength. Data taken from [53]

and thus the tumbling mode has an apparently larger relaxation strength. Figure 7 presents the measured activation plot for the relaxation times, as well as the apparent dielectric relaxation strengths of the bulk and Synpore-filter-confined materials. The relaxation times are unchanged in confinement, except for a slight reduction of the ratio of the δ-relaxation and tumbling-mode frequencies in a small temperature range below the clearing point. This may be a consequence of a slightly reduced nematic order parameter in the confined phase. The activation energy in the isotropic phase is 37 kJ mol^{-1}. In the nematic phase it is 67 kJ mol^{-1} for the δ-relaxation and 23 kJ mol^{-1} for the tumbling mode. The different apparent relaxation strengths (Fig. 7, bottom) in the isotropic phase result from filling-factor differences.

The dielectric loss spectra measured for 5CB in microporous glass are qualitatively different from the bulk spectra, as is seen in Fig. 8. In the isotropic and the nematic phase, two additional processes appear (II, III). The origin of these processes can be explained as follows; Process III arises from a polarization in the pores due to restricted ionic migration in the random pore system. It is not directly connected with the molecular dynamics of the liquid crystal. Process II is related to the retarded dynamics of molecules in partially immobilized layers directly near the random pore walls. It follows the temperature curve of the molecular dynamic processes Ia,b with a retardation of approximately 2 orders of magnitude. From the apparent dielectric strength one can estimate that the thickness of such an immobilized layer is of the order of a few Å, i.e. it corresponds roughly to a monomolecular surface layer. The differences between porous glass matrices of 2.5 nm, 5 nm, and 7.5 nm pore size are not relevant to the dynamics of the relaxation processes observed, but the ra-

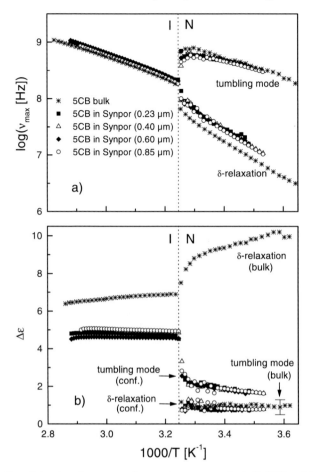

Fig. 7. Activation plot; relaxation rate vs. inverse temperature for bulk 5CB and 5CB confined to Synpore membranes (top) and dielectric relaxation strength vs. inverse temperature for the same processes (bottom). Data taken from [53]

tio of molecules in the free pore volume and in the retarded surface layer depends upon pore size. Consequently, the apparent dielectric relaxation strengths of these processes vary with pore size.

A characteristic feature of dielectric spectra in this system is the considerably reduced splitting of processes Ia,b near the clearing point. The transition from the isotropic relaxation rate to that of the δ-process is smooth. Only several degrees below the transmission, the high-frequency tumbling process is resolvable. This indicates a gradually increasing order parameter below the bulk clearing point, in agreement with findings from NMR spectra (see below).

In Fig. 9, the bulk data are added for comparison. The nematic bulk sample (crystallization at 295 K) can be supercooled by several degrees, to approximately 268

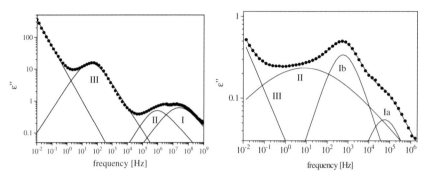

Fig. 8. Dielectric loss ϵ'' of 5CB confined to 5 nm porous glass versus frequency; (a) $T = 310\,\mathrm{K}$ (isotropic phase) and (b) $T = 230$ K. Note the different scales in ϵ'' and frequency. The solid line is a superposition of a conductivity contribution (dashed line) and the contributions of the relaxation processes (dotted lines). Data taken from [32]

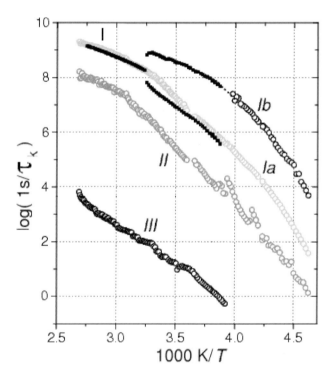

Fig. 9. Activation plot of the temperature dependence of the relaxation rate maximum $1/\tau_{\mathrm{max}}$ for 5CB confined to porous glass (5 nm pore diameter). (Ia) libration of the molecules, (Ib) δ-process, (II) surface-induced retarded relaxation process, (III) Maxwell–Wagner polarization. The bulk data (solid symbols) are given for comparison The weak fast process (Ia) cannot be resolved from the high-frequency wing of process (Ib) at temperatures immediately below the (bulk) clearing point. Data taken from [32]

K. At lower temperatures, the molecular processes are frozen in. It is noticeable that the nematic phase of the porous-glass-confined material can be supercooled considerably more, at least to 220 K. The temperature curves of both the δ-relaxation and the tumbling mode show clear deviations from Arrhenius behavior; they follow VFT (Vogel–Fulcher–Tammann) characteristics. Note that all three molecular processes have qualitatively the same temperature characteristics (except for a constant factor) at low temperatures.

2.2 Molecular and Collective Dynamics of Confined Smectics

The dielectric spectra of ferroelectric smectic C* samples differ in two features from those of the nematic phase. In addition to molecular dynamic processes, one observes collective modes connected with the spontaneous polarization in the smectic C* phase. Figure 10 shows the real and imaginary parts of the complex dielectric function of DOBAMBC (4-decyloxybenzylidene-4-amino-2-methylbutylcinnamate) in the ferroelectric smectic C* phase. Only the low-frequency part of the dielectric spectrum is shown; another, high-frequency molecular process is found above 10 MHz. The strong loss process near the kHz region is connected with collective fluctuations, phasons, of the molecular tilt azimuth (the Goldstone mode, GM). This process affects the direction of the in-plane spontaneous polarization, it is dielectrically effective only when the electric field has a component in the smectic layer plane. A second collective process is connected with fluctuations of the tilt angle (polarons). Its relaxation strength increases at the phase transition SmC*–SmA, and its frequency decreases dramatically towards the transition. In DOBAMBC, the relaxation strength of this soft mode appears to be very small; it is hardly detectable in the dielectric spectra even close to the phase transition.

While the molecular processes are basically unaffected by confinement, Fig. 11 shows the dramatic influences on the collective mode. Graph 11a is a 3D presentation of the frequency and temperature dependence of the dielectric loss in bulk DOBAMBC. In the background, the conductivity contribution ($\propto f^{-1}$) in the low-frequency wing is dominant. The process observed in the high-frequency range (left front side of the 3D image) is the δ-relaxation process. The temperature dependence of its relaxation time τ_{max} is nearly Arrhenius-like, except for those temperature ranges where order parameter changes are superimposed, in particular below the clearing point. The strong relaxation process at intermediate frequencies which sets in at the smectic A–smectic C* transition is the Goldstone mode, with a rather temperature-independent relaxation frequency. The relaxation strength is approximately 3 orders of magnitude larger than that of the molecular process, even though the preferred alignment of the layers in the sample sandwiched between bare brass electrodes is probably not optimal for the observation of this mode. The small superimposed soft mode near the phase transition to SmC* is indicated only by a small low frequency shift of the relaxation peak. It cannot be resolved in the spectra. The sharp drop of the dielectric loss signal at temperatures below 338 K is connected with the phase transition of the sample into the SmI* phase.

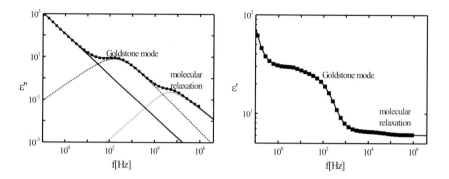

Fig. 10. Complex dielectric function of the mesogen DOBAMBC in the smectic C^* phase at 30 °C. The Goldstone mode is a collective process related to fluctuations of the tilt azimuth, the molecular process is a δ-relaxation. A second, higher-frequency molecular process is found near the GHz region (not shown here). Note the large difference in relaxation strength of the collective and molecular processes

The 3D spectrum of Fig. 11b was obtained with a DOBAMBC-filled Anopore filter. The dielectric signal was measured with the filter sheet sandwiched between the capacitor electrodes (electric field along the pore axes). The dielectric spectra of the confined mesogen show that the GM has completely disappeared, while the molecular process is still present, with unchanged temperature characteristics. Two possible reasons for the disappearance of the GM have to be considered; either the dynamic process is suppressed in the confined system, or it is dielectrically not active. If the smectic layers were arranged exactly in the filter plane (smectic layer normals in the direction of the pore axes), the GM would not be observable when the electric field was along the pore axes. The application of an electric field perpendicular to the pore axes is technically problematic. The spectra shown in Fig. 11c were obtained from an Anopore "powder" sample. After filling with DOBAMBC, the filter material was ground and then placed between the capacitor plates, covered with aluminum foil for better contact. In this experiment, a substantial part of the Anopore material should be in a random orientation. Thus, the dielectric signal comes in part from pores that are aligned at some angle or even perpendicular to the electric field. While a quantitative evaluation of relaxation strengths is not possible in this geometry, it is obvious that a relaxation mode with a nearly temperature-independent relaxation frequency is present in the SmC* temperature range (the frequency range above 1 kHz with the molecular processes has been clipped in the picture for better clarity). The identification of the process as the GM is unambiguous. It sets in at the SmA–SmC* transition. Its dynamics is rather temperature-independent. The apparent relaxation strength is much lower than the bulk signal, primarily because of the random alignment of pores. A detailed analysis shows that the relaxation rate (≈ 200 s^{-1}) is retarded by a factor of approximately 5 with respect to the bulk GM (10^3

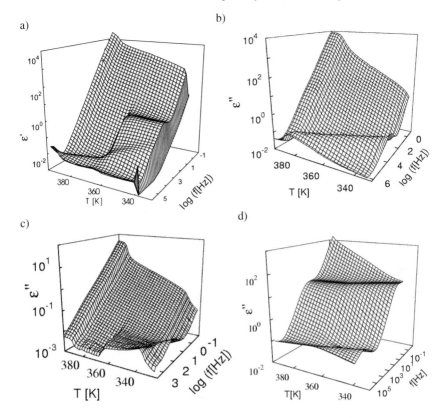

Fig. 11. 3D presentations of the dielectric loss spectra (low-frequency part) of DOBAMBC; (a) Bulk sample with δ-relaxation (flat shoulder in the front of the image) and Goldstone mode (high shoulder in the central part, in the SmC* temperature range). (b) Ordered Anopore sample. The GM is absent because the pores axes are aligned in the electric field direction and the layers are arranged in the filter plane. Only the flat shoulder of the δ-relaxation is seen. (c) Anopore "powder" with random alignment of the filter sheets. The GM is detectable in this geometry as a flat plateau in the SmC* temperature range. The region with the δ-relaxation has been clipped. (d) porous glass sample, 7.5 nm pore size. The GM has completely disappeared. Only the δ-relaxation at high frequencies and a huge Maxwell–Wagner polarization at low frequencies are visible. In all geometries, another high-frequency molecular process exists above 10 MHz. Data taken from [51]

s^{-1}, see Fig. 12). The shift of the GM relaxation frequency is a clear indication that the observed process is not related to some free liquid-crystalline material at the surfaces of filter fragments. A similar slowing down of collective modes has been reported for LC–aerogel systems (e.g. [45]). There is, however, a discrepancy with observations in a high-spontaneous-polarization material (see p. 319) confined in random macropores, so that the observed collective relaxation processes in the

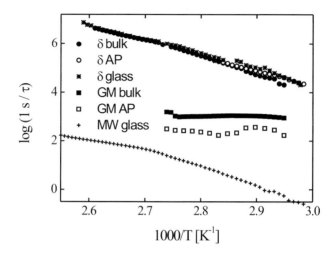

Fig. 12. Dielectric relaxation frequencies vs. inverse temperature for the DOBAMBC data shown in Fig. 11. The notations are "δ" for the δ-process, "GM" for the Goldstone mode and "MW" for Maxwell–Wagner polarization. The symbol "AP" denotes filled Anopore filters, "glass" denotes a porous glass matrix of 7.5 nm cavity size, filled with the LC. Data taken from [51]

confined ferroelectric LC phases may have a more complex relation to the soft and Goldstone mode dynamics than in the bulk material.

Figure 11d shows the dielectric loss in porous glass of 7.5 nm pore size. The relaxation frequencies obtained in 2.5 nm and 5.0 nm pore sizes are equivalent. The first striking observation is the complete disappearance of the GM in the nanometer pores. This time, it is not caused by the orientation of the pores with respect to the field, since in the random porous-glass geometry all directions are equivalent. The GM is suppressed as a consequence of the restricted pore volumes. The most probable interpretation of the dielectric measurements is that the disappearance of the collective dynamic modes indicates the suppression of the smectic phase structure. A proof of the existence of smectic layer ordering by X-ray diffraction has not been achieved. In contrast to the collective dynamic modes, all individual (molecular) modes are unchanged with respect to the bulk dynamics.

The dielectric loss spectra of DOBAMBC confined to an irregular macroporous matrix (Synpore filter membrane) are presented in Fig. 13. In these spectra, the GM is completely absent, too. In the Synpore filters, DSC and X-ray diffraction experiments have successfully proved that the smectic C* phase of DOBAMBC is present in the same temperature range as in the bulk [55]. Consequently, the disappearance of the collective dynamic processes has a different origin. The pore sizes of the Synpore membranes are of the same order of magnitude as the diameter of the Anopore filter channels, where the GM is detectable. Thus, the most probable explanation is that the random surfaces of the Synpore matrix freeze a state of random disorder of the

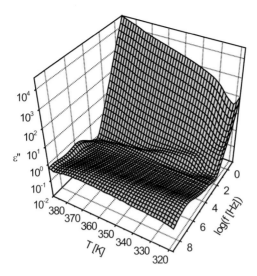

Fig. 13. 3D presentation of the temperature-dependent dielectric loss spectra of the mesogen DOBAMBC. Comparison with Fig. 11a shows that the GM is completely absent in the random macroporous cavities. The weak, temperature dependent loss signal in the central part is not related to the liquid crystal. Figure from [55]

smectic phase, with a high defect density, and the collective reorientation dynamics is suppressed by frustration of the system. Some remaining GM dynamics with a considerably reduced dielectric relaxation strength may still be present, but for the low-spontaneous-polarization material DOBAMBC it is below the detection level for broadband dielectric spectroscopy.

A test of this assumption is the study of a material with a short pitch and a high spontaneous polarization, the compound C7 (4-(3-methyl-2-chloropentanoyloxy)-4′-heptyloxybiphenyl), in bulk and Synpore filters. In the bulk system (Fig. 14 left), a strong collective GM process is seen in the SmC* temperature range. The slight temperature trends of the relaxation time and the dielectric relaxation strength are caused by superimposed soft-mode contributions (see also Fig. 15). The graphs for the dielectric relaxation rates (Fig. 15 left) and relaxation strengths (Fig. 15 right) indicate that the dynamics of molecular processes is, again, unchanged. Since the molecular dipole moment is not strictly axial in this material, two molecular processes are already dielectrically active in the isotropic phase, a high-frequency process which can be assigned to fast rotations of the molecule about its long axis (this process is observed in all phases with little peculiarity at the phase transitions), and the lower-frequency δ-relaxation peak. The latter should split, below the clearing point, into two processes, as observed for 5CB in Fig. 9, p. 312. However, only the reorientation about the short axis, the δ-relaxation process, is resolved in the dielectric spectra. Instead of the GM and soft modes in the SmC* phase, a strong collective process is

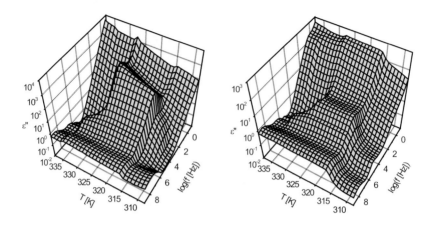

Fig. 14. 3D presentation of the temperature-dependent dielectric spectra of C7 (4-(3-methyl-2-chloropentanoyloxy)-4′-heptyloxybiphenyl) in the bulk and confined to Synpore membranes (850 nm nominal pore size). The number of loss processes in both spectra is the same. Two molecular processes are observed, near the transition to SmC*, the soft mode is superimposed and covers the δ-process. In SmC*, a collective process (the GM relaxation in the bulk material) is dominant. The deconvolution into individual loss processes is shown in Fig. 15. Data taken from [55]

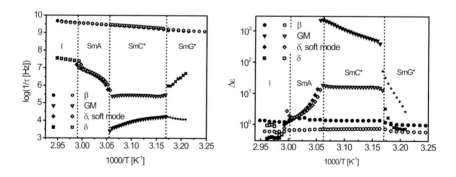

Fig. 15. Relaxation rate (left) and dielectric relaxation strength (right) vs. inverse temperature for C7 in the bulk and confined to Synpore membranes (850 nm nominal pore size). Data taken from [55]

found in Fig. 15 at a frequency about one decade above the GM. Its relaxation strength is two orders of magnitude lower than that of the GM. There are some indications that support its assignment to a residual GM relaxation in the frustrated phase: the relaxation peak is temperature-independent, and it is observed only in the SmC* phase (it covers one of the molecular modes there). If the same reductions apply for the Synpore-confined DOBAMBC, the GM is below the detection level there, in agreement with the argumentation on p. 317. This interpretation supports the model of a frustrated smectic C* phase in random-macropore-confined ferroelectric LC materials. A remaining problem is the frequency shift of the observed GM to higher frequencies in this system. This trend is opposite to the observation in Anopore-confined DOBAMBC (see p. 315).

Summarizing, dielectric spectroscopy experiments have shown that the molecular dynamics of LC mesogens is independent of the confinement. A retarded layer of molecules directly attached to the adsorber (as in isotropic liquids) has been observed for a number of compounds when the surface-to-volume ratio is large enough (nanoporous glass matrices). In some cases (DOBAMBC in nanoporous glass), the loss process connected with such a retarded layer (if present) has not been observed. The collective dynamic processes are present in ordered macroporous materials, they are completely suppressed in nanoporous adsorbers, and they are considerably damped even in macroporous systems (pore sizes up to 1 μm) with irregular cavities, suggesting a frustrated smectic phase in such systems.

3 Surface-Induced and Suppressed Order

In addition to the influence on collective dynamic processes, interfaces may have an impact on orientational order parameters found in confined LCs. Nuclear magnetic resonance is a convenient tool to get direct quantitative access to the order parameter. In systems with ordered pores, it is sufficient to have a thin ordered layer at the interfaces and dynamic exchange within the NMR time (microseconds to milliseconds) between the bulk and the ordered surface layer, in order to detect surface order from dipolar or quadrupolar splitting. In disordered systems, such an induced surface order and even bulk order may be completely motion-averaged in the NMR detection time, and more sophisticated models have to be applied for a quantitative evaluation of the NMR line shapes in the presence of diffusional and exchange averaging.

A model that predicts the induction of nematic order in the isotropic phase near the clearing point can be developed using a Landau–de Gennes expansion. The free-energy density f of a nematogen is composed of the contributions [27, 83–87]

$$f = f_0 + f_v(S) + f_s(S_0),$$

where f_0 is an order-parameter-independent constant, $f_v(S)$ is the free-energy density in the volume, and $f_s(S_0)$ represents the surface part of the free energy. The nematic order parameter S is defined as the average of the second Legendre polynomial, $S = \langle P_2(\cos\theta)\rangle$, of the angles θ between the molecular long axes and the local director. S_0 is the order parameter at the surface. In the vicinity of the clearing point, the bulk part f_v can be described by a Landau–Ginzburg functional

$$f_v(S) = \frac{a}{2}(T - T^*)S^2 - \frac{b}{3}S^3 + \frac{c}{4}S^4 + \frac{1}{2}L(\nabla S)^2 \qquad (a, b, c, L > 0). \quad (2)$$

In the absence of gradients of the order parameter, this functional yields a trivial equilibrium state $(df/dS = 0)$ with $S = 0$. An additional pair of equilibrium solutions,

$$S = \frac{b}{2c} \pm \sqrt{\frac{b^2}{4c^2} - \frac{a(T - T^*)}{c}} \quad (3)$$

is found for values $b^2 > 4ac(T-T^*)$, i.e. for temperatures $T \leq T_1 = T^* + b^2/(4ac)$. At T_1, the metastable branch sets in, with an order parameter of $S(T_1) = b/(2c)$. Below T_1, a metastable nematic state exists. The local free-energy minimum belongs to the positive root of (3). At temperatures $T \leq T_2 = T^* + 2b^2/(9ac)$, the free-energy minimum related to the positive root of (3) becomes lower than that of the isotropic state $S = 0$. The bulk clearing point is reached at T_2 with $S(T_2) = 2b/(3c)$. A nonzero value of L in (2) in combination with a finite value for the order parameter S_0 induced at the surface leads to the formation of a region near the LC surface with a nonzero order parameter, even above the bulk clearing temperature ("paranematic"). If S_0 is sufficiently small, one can neglect the terms with b and c in (2) in one-dimensional treatment (near a planar boundary), an exponentially decaying order parameter

$$S(x) = S_0 \exp\left(-\frac{x}{\xi}\right), \quad \text{where} \quad \xi = \sqrt{\frac{L}{a(T - T^*)}}, \quad (4)$$

is found. The nematic order decays from the value S_0 at the boundary towards $S = 0$ in the volume across a certain temperature-dependent correlation length ξ. The coordinate x is taken along the surface normal. Reasonable values for low-molecular-mass liquid crystals are of the order of $a = 1.3 \times 10^5$ J/m^{-3} K^{-1}, $L = 1.7 \times 10^{-11}$ J m^{-1} [27], yielding a correlation length $\xi = \xi_0 \sqrt{T^*/(T - T^*)}$, where

$$\xi_0 = \sqrt{\frac{L}{aT^*}} = 6.6 \,\text{Å}.$$

In the complete description, the bulk free-energy part in (3) has to be supplemented by the surface parts

$$f_s = -GS_0 + \tilde{G}S_0^2,$$

where S_0 is the order parameter at the surface [88]. The term G has the effect of an increased order parameter at the surface, even at temperatures far above the bulk clearing point. It is related to interactions of the nematic molecules with the surface, which depend upon the orientation of the nematic molecules with respect to the interface. With a nonzero G, the nematic–isotropic transition is smeared and shifted to higher temperatures. The quadratic term is related to reduced interactions with mesogenic neighbors at the surface; $\tilde{G}$ should therefore be positive and tend to reduce the nematic order parameter. Since the term has the same functional dependence on S as the first term in the bulk expression, it can be summarized in a cavity-size-dependent shift of T^* to lower temperatures, without other changes of the temperature characteristics.

In the case of a nonzero prefactor G, which defines the ordering strength of the interface, a nonzero order parameter

$$S_0 = \frac{G}{\sqrt{2a(T - T^*)L}} \tag{5}$$

is observed above the bulk clearing point. In the vicinity of the bulk transition, a sufficiently high value of G can lead to a surface phase transition. In some materials (mixtures), such surface transitions are observed optically [92] over broad temperature ranges. In the case of weak anchoring conditions of the director at the surface, they can obviously influence even the bulk phase transition [92].

Of particular importance are such surface induced ordered layers in the case of restricted-volume systems. The induction of nematic order above the bulk clearing point has been considered in detail for spherical cavities [93]. The formation of nematic "bridges" between dispersed particles may lead to an attractive potential between these particles [94] that can destabilize colloidal suspensions. Experimentally, the existence of nematic bridges (capillary condensation) between a planar surface and a nearby solid sphere has been derived from AFM measurements [95].

For the scenario of a surface-induced order above the bulk clearing point, experimental evidence has been found, e.g. in Anopore-confined material. A suitable method for the observation of nematic order is NMR. In deuterium NMR, the quadrupolar splitting $\delta\nu_q = \delta\nu_{q0}S$ is a direct measure of the order parameter (averaged over the NMR signal acquisition time). The constant ν_{q0} depends upon the electric field gradient tensor and its orientation in the molecule. It can be determined from the doublet splitting and known order parameters in the nematic phase. When nematic material is confined in Anopore filters, the molecules in the ordered surface layers formed above the clearing temperature T_{NI} exchange quickly (fast on the NMR timescale) with the isotropic bulk. In the spectrum, one measures a residual line splitting which is proportional to the averaged order parameter $\langle S \rangle = V^{-1} \int_{(V)} S \, dV$, where V is the pore volume.

The material chosen in Fig. 16 is the nematogen 5CB, selectively deuterated at the second chain carbon position counted from the benzene ring. The splitting in the magnetic-field-aligned nematic phase of this deuterium position is $\delta\nu = 87.5 \, \text{kHz} \times S$. At T_{NI}, a sharp phase transition is observed. In the bulk material, the lines combine to a single Gaussian line above T_{NI}, with a linewidth of $\approx 50 \, \text{Hz}$. In Anopore-confined samples, a small quadrupolar splitting remains even at temperatures of more than 15 K above T_{NI} (Fig. 16). The temperature-dependent splitting $\delta\nu$ is depicted in Fig. 17 for the untreated Anopore surfaces. When a simple exponential decay of the order

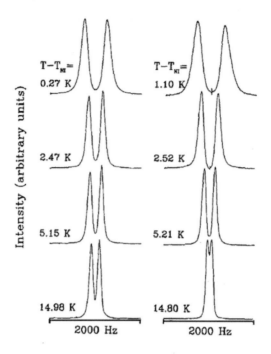

Fig. 16. ^{2}H NMR spectra in the isotropic phase of partially deuterated 5CB-βd$_2$ confined to Anopore channels (axial alignment). The left-hand sequence was measured in untreated channels; the right-hand sequence was measured in channels treated with lecithin (radial alignment). Figure from [27]

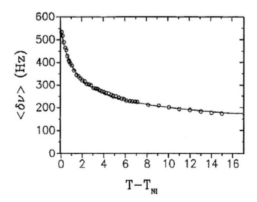

Fig. 17. ^{2}H NMR quadrupolar splitting of partially deuterated 5CB-βd$_2$ confined to untreated Anopore channels. The solid line is a fit with the model discussed in the text. Figure from [27]

parameter according to (4) and a surface induced order according to (5) are assumed, the temperature dependence of the average order parameter $\langle S \rangle = 2\xi/RS_0$ in a cylindrical pore of radius R should follow a $(T - T^*)^{-1}$ dependence ($S_0 \propto \xi \propto 1/\sqrt{T - T^*}$). Actually, the temperature curve of the quadrupolar splitting follows a dependence $\delta\nu_q(T) = \delta\nu_{q0}\langle S \rangle = \delta\nu_{q0}(A'/\sqrt{(T - T^*)} + B')$. If a constant surface order parameter S_0' and, instead of the exponential decay directly from the surface, an additional layer of uniform order parameter with thickness l_0 (of the order of molecular dimensions) next to the walls are assumed, the parameters A' and B' can be interpreted as $A = 2\xi_0 S_0'\sqrt{T^*}R^{-1}$ and $B = 2l_0 S_0' R^{-1}$, and the fit in Fig. 17 yields (using the value $\xi_0 = 0.65$ nm from the literature) $S_0 = 0.021 \pm 0.002$, $l_0 = 1.93 \pm 0.2$ nm, and $T^* = T_{NI} - (0.87 \pm 0.05)$ K [27]. The temperature curve for the lecithin-coated adsorber follows a dependence $\delta\nu_q(T) = \delta\nu_{q0}\langle S \rangle = \delta\nu_{q0}(A/(T - T^*)^{-1} + B/\sqrt{(T - T^*)})$. Again, an additional layer of constant order parameter with thickness l_0 is assumed at the walls, but the surface-induced order is assumed to follow the temperature dependence of (5). The parameters A and B are related to the material constants by $A = 2\xi_0 S_{00} T^* R^{-1}$ and $B = 2l_0 S_{00}\sqrt{T^*}R^{-1}$, where $S_{00} = G/\sqrt{4aLT^*}$. One has to take into account that the splitting is reduced by an additional factor $1/2$, because of the radial alignment of the director, perpendicular to the axial NMR field. The parameters extracted from the fit of the quadrupolar splitting curves are $S_{00} = 0.005 \pm 0.001$, $l_0 = 1.14 \pm 2$ nm, and $T^* = T_{NI} - (1.16 \pm 0.05)$ K for lecithin-coated walls.

Summarizing, NMR can detect and quantitatively evaluate ordered nematic layers in the isotropic bulk phase of a nematogen confined to Anopore filters. Even though the volume share of the ordered surface layers is small, and the induced order parameter is between one and two orders of magnitude smaller than the bulk nematic order parameter, it is clearly measurable. The temperature dependence of the surface-induced order parameter has different characteristics at pure Anopore walls and at lecithin-coated surfaces. It should be noted that in the surface layer model ($l_0 \neq 0$) only the product $l_0 S_0$ enters the coefficients B, B'. While the assignment of the term related to the spatially decaying order parameter with correlation length ξ, which enters the coefficients A, A', is unambiguous, it is possible as well that the first, partially immobilized layer has a higher order parameter but a correspondingly smaller thickness. The exchange of the molecules in the first layer with the unbound phase is expected to occur at the rate of process II in Fig. 9, i.e. fast on the NMR timescale.

An example of suppressed orientational order in confined systems is found in microporous glasses. The detection method is again NMR. In proton-decoupled ^{13}C NMR, the spectra are governed by the chemical-shift anisotropy. In an ordered material, the resonance line of an individual carbon position k shifts according to

$$\omega^{(k)} = \left(\frac{3}{2}\cos^2\phi - \frac{1}{2}\right)\omega_a^{(k)} \cdot S + \omega_{iso}^{(k)}, \tag{6}$$

where $\omega_{iso}^{(k)}$ are the isotropic chemical shifts with respect to a reference substance (here tetramethylsilane, TMS); they can be determined from spectra in the isotropic bulk material (see Fig. 19a). The $w_a^{(k)}$ are related to the anisotropic part of the time-averaged chemical shielding tensor. They can be determined for the individual carbon positions from measurements of ordered samples with known order parameters in the nematic phase (Fig. 19b). The angle ϕ describes the director orientation with respect to the spectrometer magnetic field B_0. A completely disordered sample yields a spectral powder pattern as shown in Fig. 18, bottom, for a single carbon site. The singularity appears at the position corresponding to $\phi = 90°$, $w^{(k)} = -1/2w_a^{(k)}S + w_{iso}^{(k)}$. The complete proton-decoupled ^{13}C NMR spectrum is obtained from the superposition of all lines of individual carbon sites in the molecule (Fig. 19c) Owing to molecular diffusion in the pore matrix, the NMR spectra of the confined material are partially averaged. During the NMR acquisition time, the molecules migrate in the porous channels between domains of different alignment ϕ. A kinetic model that considers the averaging effect can be developed on the basis of the time evolution of the probability distribution of momentary resonance frequencies for

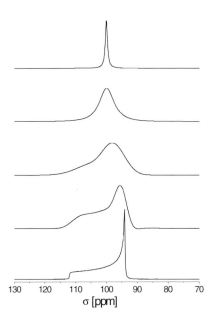

Fig. 18. Calculated ^{13}C NMR powder patterns for a single carbon site with $w_{iso} = 100$ ppm, $w_a = 20$ ppm (the characteristic time τ_D is given in units of $10^6/2\pi f_0$, with f_0 being the ^{13}C resonance frequency). The order parameter is set to $S = 0.6$. Changes in the line shape are due to motional averaging. From top to bottom: $\tau_D = 0.01$, complete averaging, the line is a narrow Gaussian around the isotropic position w_{iso}; $\tau_D = 0.1$, broad Gaussian near w_{iso}; $\tau_D = 1.0$; $\tau_D = 10$, weak motional averaging of the powder pattern; and $\tau_D = 100$, quasi-static line shape. Figure taken from [32]

carbon spins in a diffusing molecule [32]: when the line shape of the rigid lattice is given (the powder pattern $f_0(\omega)$ of a single carbon line) and a Poisson jump process between all possible orientations is assumed, one can write down the conditional probability density $p(\omega_2|\omega_1, t_2 - t_1)$ for the momentary frequency ω_2 of a nuclear spin at time t_2 if the spin started with ω_1 at time t_1. The momentary phase of an individual spin is calculated from the time integral of the random frequencies, and the nuclear magnetization is found by summing over an ensemble of spins. Under the assumption of a random jump process (molecules exchange stochastically between domains of different orientations ϕ) and a Gaussian distribution of phases, the probability density of the phases can be expanded in terms of its moments, and with a suitable truncation of this expansion (after the second moment), an analytical expression for the free induction decay, and consequently for the line shape, can be derived. The only parameters necessary to describe the line shape are the nematic order parameter S, which determines the width of the line splitting, and a correlation time τ_D that represents an average rate of orientational jumps in the angle ϕ, i.e. the ratio of the square of an average straight channel length and the diffusion coefficient of the mesogen.

The fit of the experimental spectra with a superposition of partially motion-averaged powder spectra yields the temperature-dependent order parameter. The jump rate τ_D provides additional information about the diffusion dynamics of 5CB in the glass pores [32]. A test of the measured order parameters can be performed with proton NMR. The dipolar broadening of the ^{1}H NMR signal can be calculated along similar lines, as discussed above for the ^{13}C chemical shift spectra. The advantage of the proton NMR is higher signal sensitivity because of the rich natural abundance of ^{1}H nuclei. Furthermore, the NMR powder pattern has a line shape that is much closer to a Gaussian line than the ^{13}C powder pattern of Fig. 18, bottom. In the diffusional averaging model, a Gaussian shape is preserved during motional averaging, and the linewidth and the second moment are both basically determined by the order parameter. In Fig. 20, order parameters of 5CB in nanoporous glass determined from ^{1}H NMR spectra [33] are presented by open circles. They agree fairly well with the ^{13}C data. In contrast to the theoretical ansatz on p. 321, the experiment yields both a reduced order parameter and a smeared-out transition. Qualitatively, the order parameter characteristics below T_{NI} are comparable to what is observed in the dielectric spectra of 5CB confined to microporous glass (the two processes IIa,b in Fig. 9). Obviously, the quadratic coefficient $\tilde{G}$ plays the dominant role here, but owing to the random pore geometry and the distribution of pore sizes, $\tilde{G}$ is not uniform in the sample volume. This may explain the order parameter curve observed in the microporous glasses at least qualitatively. It seems that at temperatures sufficiently below T_{NI}, the bulk order parameter is gradually approached, even in the microconfined material.

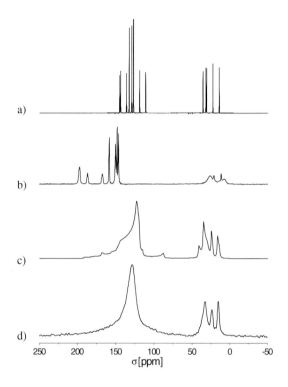

Fig. 19. Proton-decoupled ^{13}C NMR spectra of 5CB: (a) isotropic bulk spectrum ($T = 313$ K), all lines at $\omega_{\text{iso}}^{(k)}$; (b) oriented bulk nematic phase ($T = 297$ K), $\phi = 0$, all lines shifted to $\omega_{\text{iso}}^{(k)} + S\omega_{\text{a}}^{(k)}$; (c) calculated powder spectrum for order parameter $S = 0.6$; (d) spectrum of 5CB confined to 5 nm porous glass ($T = 275$ K), partially motion-averaged powder spectrum with $S = 0.4$. Figure taken from [32]

4 Director Configurations in Confined Phases

For the discussion of director alignment, only macroporous systems are of interest. In nanopores, one obtains a highly complex nematic configuration, which will basically be imposed by the local pore walls. The density of defects will be correspondingly large and the director fields cannot be influenced by external fields, because the electric or magnetic correlation length is large compared with the characteristic cavity dimensions.

The director field of confined nematic phases is found from minimization of the free energy, which contains elastic contributions, and in the case of applied external electromagnetic fields, additional electric and magnetic terms. In cylindrical channels, the uniform axial director field for planar anchoring is the simplest solution. Radial configurations in the case of homeotropic boundary conditions are more complex. In that case, a defect line in the core of the cylinder results, and depending

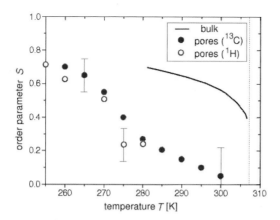

Fig. 20. Comparison of the order parameter measured in bulk 5CB (^{1}H NMR) and in nanoporous glass (^{13}C and ^{1}H NMR). Data taken from [32, 33]

on the pore sizes, the director field chooses "escaped" radial configurations with point defects on the cylinder axis [96]. The situation is more complex for cholesteric liquid crystals when the formation of the helix is in conflict with the boundary conditions. For planar axial surface alignment, a radial helical structure with defects forms [89, 90]. In cholesterics confined to untreated Anopore filters, it has been found by NMR measurements that the director alignment is not perfectly axial as for nematics. In addition, a large optical rotatory power has been measured [91]. This observation suggests that a helical structure along the pore axis is formed in the channels. Such a "conic" helical structure has not been found in bulk phases; in cholesterics, the director is perpendicular to the helix axis. One of the interesting properties of such a helical tilted phase is that the local symmetry allows a spontaneous polarization, which has not been confirmed experimentally so far. In the case of regular geometries, e.g. thin cells, cylindrical pores and droplets, the calculation and experimental determination of director fields have been studied in great detail in the literature [28, 96].

Here, we shall focus on the case of random geometry. A qualitative discussion of the director field is performed on the basis of the free-energy analysis of a nematic with dispersed solid particles. Under the assumption that the director field is anchored at the particle surfaces, the influence of external electromagnetic fields on the sample alignment is calculated. For simplicity, only the *magnetic* case will be considered, because the distortion of the inner field can be neglected there. The free energy per volume of the nematic is given by

$$F = \frac{1}{2}K_{11}(\operatorname{div} \boldsymbol{n})^2 + \frac{1}{2}K_{33}(\boldsymbol{n} \times \operatorname{rot} \boldsymbol{n})^2 + \frac{1}{2}K_{22}(\boldsymbol{n} \operatorname{rot} \boldsymbol{n})^2 - \frac{1}{2}\frac{\chi_a}{\mu_0}(\boldsymbol{B} \cdot \boldsymbol{n})^2, \tag{7}$$

where χ_a is the anisotropy of the magnetic susceptibility, and K_{ii} ($i = 1, 2, 3$) are splay, twist, and bend elastic constants, respectively, of the nematic. Equations for the

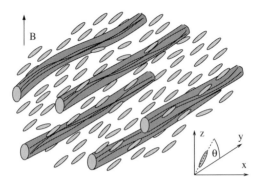

Fig. 21. Geometry of a nematic with embedded parallel fibrils and definition of the coordinates and angles

electric case can be derived analogously when the last term is replaced by the electric torque. In general, the director configuration is nonplanar and three-dimensional. In the following, the calculations are simplified to a two-dimensional treatment. A system that can be considered as 2D in a first approximation is represented, for example, by long parallel fibrils embedded in the nematic (Fig. 21). However, even if the boundary conditions and the electric/magnetic field are uniform in one spatial coordinate, the equilibrium director distortions need not strictly be uniform along that coordinate [97], particularly when the elastic anisotropy is large ($K_{11}, K_{33} > K_{22}$). In addition, even when the director ground state n_0 is spatially uniform, a homogeneous external field that is not parallel to n_0 may in general induce nonplanar deformations where the director escapes the plane formed by n_0 and the field direction [98,99]. Such effects will be neglected in the following.

A planar director configuration can be expressed by the tilt angle profile $\theta(x, z)$, with $n = (0, \cos\theta, \sin\theta)$. In a two-constant approximation, $K = K_{11} = K_{33}$, (7) takes the form

$$F(x, z) = \frac{1}{2}K\theta_z^2 + \frac{1}{2}K_{22}\theta_x^2 - \frac{1}{2}\frac{\chi_a}{\mu_0}B^2 \sin^2(\theta + \theta_H).$$ (8)

where θ_x, θ_z are the spatial derivatives of θ, and θ_H is the angle of the magnetic field with the y axis. Minimization of the free energy with appropriate boundary conditions yields the stationary director field. In a random geometry, where an analytical solution is not available, this can be done numerically by means of a relaxation technique.

We introduce the magnetic correlation length $\xi_m = \sqrt{K\mu_0/(\chi_a B^2)}$, the relaxation time $\tau_0 = \xi_m^2\gamma/K$ (where γ has the meaning of a rotational viscosity of the nematic) and the elastic ratio $\kappa = \sqrt{K/K_{22}}$. After transformation of the torque balance equation to dimensionless coordinates $\eta = \kappa x/\xi_m$, $\zeta = z/\xi_m$, $\tau = t/\tau_0$, one obtains

$$\dot{\theta} = \theta_{\zeta\zeta} + \theta_{\eta\eta} + \frac{\sin 2(\theta + \theta_H)}{2}.$$ (9)

With this equation, the minimum magnetic field necessary to distort a nematic director field anchored at randomly dispersed parallel fibrils can be calculated. It is assumed that at the surfaces of the fibrils, n is fixed (rigid anchoring, $\theta = 0$). Any influence of the director field on the dispersed fibrils is neglected. Since reorientation of an inhomogeneous director field in general involves macroscopic flow which couples to the director rotation, (9) is not an adequate description of the reorientation *dynamics*, but only a crude approximation in the case of weakly distorted director fields and weak magnetic torques. However, any arbitrarily chosen, defect-free initial nematic director field relaxes into the stationary equilibrium state ($\dot{\theta} = 0$).

The number of fibrils per area in the (x, z) cross section is C, and a characteristic length that can be related to this parameter is $\xi_0 = C^{-1/2}$ as a measure of the average distance between neighboring fibrils. The equilibrium director field for given field strengths and fibril concentrations C was calculated numerically on a square lattice with periodic boundary conditions. In order to minimize statistical errors, several ensembles of random fibril distributions were evaluated for a given field strength. Fibril diameters were assumed to be small compared with their spacing ξ_0. In this case, (9) describes a universal behavior of the director field for a given fibril arrangement. Material parameters influence the scaling of the coordinates η and ζ, which affects only the concentration $c = \xi_m^2/\kappa C$ of fibrils in the (η, ζ) plane and not the statistics of a random distribution in that plane [100]. The only parameter that influences the statistical properties (e.g. maximum and average deflection angles) of the distributions $\theta(x, z)$ for a given concentration of fibrils is the reduced magnetic field $B/B_0 = \sqrt{\kappa}\xi_0/\xi_m$, where we have defined B_0 as

$$B_0 = \frac{1}{\xi_0}\sqrt{\frac{K\mu_0}{\chi_a\kappa}} \propto \sqrt{C}.$$

Tilt angle profiles $\theta(\eta, \zeta)$ calculated for the same reduced magnetic fields are independent of K, κ and C [100]. 2D relief plots of calculated equilibrium tilt profiles for an exemplary realization of the random fiber distribution and two field strengths are shown in Fig.22. The average tilt deformation $\langle\theta\rangle$ as a function of the reduced field for a given fibril arrangement is shown in Fig. 23 left. It is close to zero at fields below $B_{th} \approx 1.09\,B_0$, whereas both the maximum and the average tilt angles increase continuously with increasing magnetic field above that value. At first sight, it may be surprising that in the case of a random distribution, a threshold-like behavior is observed. Between different random arrangements of fibers with the same concentration C (approximately 100 fibers were used in the numerical calculations), the numerical prefactor varies by approximately $\pm 5\%$. The conclusion drawn from the numerical calculations is that a critical concentration

$$C_c = \frac{1}{\xi_{0c}^2} \approx \frac{\kappa}{(1.18 \pm 0.1)\,\xi_m^2(B)} \tag{10}$$

of fibrils with strong director anchoring is sufficient to stabilize the uniform-director ground state in an external magnetic field B perpendicular to n_0.

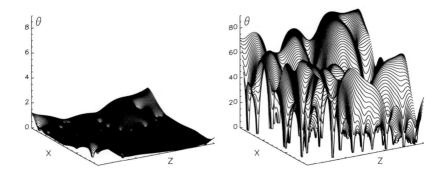

Fig. 22. Example of a typical profile of the director deflection in a random dispersion of fibrils; the fibrils extend in parallel order along y. The director is aligned at their surface along y, θ is the local director angle with respect to y in the (x, z) plane, and the magnetic field is along z ($\theta_H = 90°$). The inductions are $B = 1.13B_0$ (left) and $B = 2.26B_0$ (right)

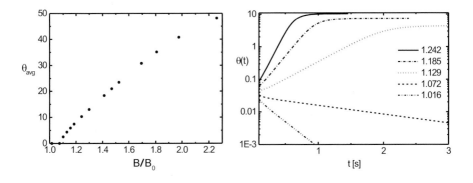

Fig. 23. Left: average tilt angle $\langle\theta\rangle = \int_V \theta \, dV/V$ of the director as a function of the external magnetic field for a given random fiber distribution. In the calculation, a cross section containing 100 fibers was selected. Because of the relatively poor statistics, the threshold can vary by $\approx 5\%$ between different random distributions. Right: time dependence of the average tilt $\langle\theta\rangle$, starting from a very small initial deformation. The numbers denote the parameter B/B_0 of the individual curves

Table 1. Magnetic threshold fields for different geometries (one-constant approximation, $\kappa = 1$). The electric thresholds E_{th} are found from the corresponding magnetic fields B_{th} when $\sqrt{K\mu_0/\chi_{\mathrm{a}}}$ is replaced by $\sqrt{K/(\varepsilon_{\mathrm{a}}\varepsilon_0)}$

Geometry	Director profile	Threshold
Planar Fréedericksz cell, thickness d	$\theta(z) \propto \cos(\pi z/d)$	$\xi_{\mathrm{m}} = d/3.1416$
Circular pore, radius R	Bessel function $\theta(r) \propto J_0(2.4045r/R)$	$\xi_{\mathrm{m}} = R/2.4045$
Square fiber lattice, separation D	See Fig. 24	$\xi_{\mathrm{m}} \approx D/1.2$ (numerical result)
Random fiber lattice structure factor ξ_0	See Fig. 22 left	$\xi_{\mathrm{m}} \approx \xi_0/1.09$ (numerical result)

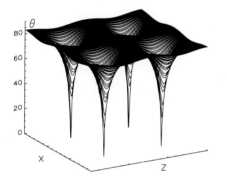

Fig. 24. Profile of the director deflection in a regular grid of fibrils with periodicity D, at $B = 2.82/D\sqrt{K\mu_0/\chi_{\mathrm{a}}}$. The fibrils are aligned parallel to y. The director is fixed at their surfaces along y, θ is the local director angle with respect to y in the (x, z) plane, and the magnetic field is along z ($\theta_H = 90°$)

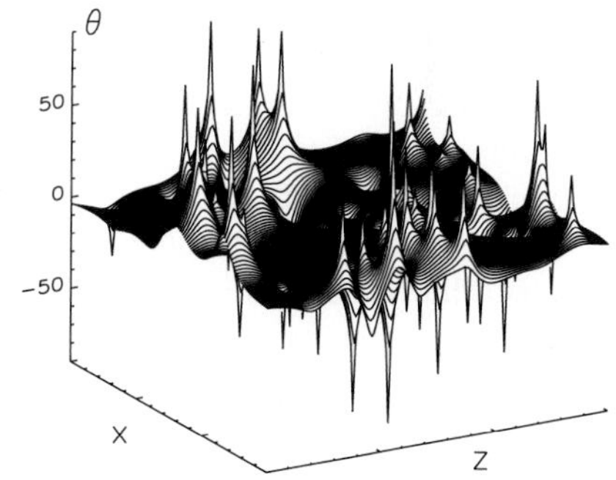

Fig. 25. Profile of the director deflection with random alignment at irregularly dispersed fibrils with concentration $C = 0.25\ \mu\mathrm{m}^{-2}$. The field is $B = 2.26B_0$, it is directed along the fibril axes y ($\theta_H = 0°$)

The threshold field can be compared with the critical fields in a number of regular geometries (Table 1). In a planar sandwich cell (electrode gap d), the ground mode is a cosine deformation of θ, and the critical field is inversely proportional to the cell gap, $B_{\mathrm{th}} = \pi/d\sqrt{K\mu_0/\chi_{\mathrm{a}}}$, i.e. the deformation sets in at $\xi_{\mathrm{m}} = d/\pi$. In a cylindrical cavity, the ground deformation is a Bessel function of order zero, and the threshold field is inversely proportional to the pore radius. For a regular grid (Fig. 24) of thin fibrils, the threshold has been determined numerically (see Table 1).

The characteristic relaxation time of the director deformation (if flow coupling is neglected) is

$$|\tau_{\text{off}}| = \frac{\gamma\mu_0}{\chi_a B_{\text{th}}^2}.$$

The relaxation time for a nematic with dispersed fibrils is related to that of a sandwich cell by

$$\frac{\tau_{\text{disp}}}{\tau_{\text{fred}}} \approx \frac{8.3\xi_0^2}{d^2}.$$

Because ξ_0 can be in the submicrometer range, orders of magnitude below d, the embedded fibrils provide an opportunity to decrease the response times of the nematic, which is an important goal in display applications. Figure 23 right shows the numerical results for the response of $\langle\theta\rangle$ to changes of the reduced field on a logarithmic scale. The field dependence of the exponential response rates is $\tau^{-1}(B) = \tau_{\text{off}}^{-1}((B/B_{\text{th}})^2 - 1)$.

An opportunity to measure the director distribution in the sample is NMR [37]. Measurements have been performed, for example, with chemical gels prepared from nematic mesogens with dispersed diacrylates that are photo-crosslinked in the oriented nematic state. The diacrylate network consists of fibrils which fix the director configuration present during the crosslinking reaction. It has been shown that a concentration of gel formers of $\approx 2\%$ is sufficient to stabilize the director field in a magnetic field of 4.7 T ($\xi_m \approx 0.8$ μm for 5CB).

In the same way as the network can stabilize a uniform nematic ground state, one can achieve the stabilization of a disordered nematic director field by means of particle dispersions (see Sect. 1) or confinement in irregular cavities. In that case, quantitatively correct numerical simulations require a three-dimensional treatment, which has not been attempted here. The qualitative effect is demonstrated in Fig. 25. When the director distribution is calculated in a 2D geometry, and a *random* distribution of anchoring angles at the individual fibrils is assumed, the director distribution is a powder pattern in the field-free state, whereas a sufficiently strong external field aligns the nematic director partially. Since the field is not perpendicular to the initial n_0 directions, the deformations set in much earlier than in the case of a uniform ground state perpendicular to the magnetic field; there is no well-defined threshold anymore. Experimentally, the gradual alignment of the director field with increasing external field can be detected from the optical scattering or transmission curves of a sufficiently thin sample, when the average fibril distances are in the optical wavelength range. In that case, the sample is strongly scattering at zero field, while the scattering efficiency decreases with increasing alignment of the director.

As an alternative to such a dispersion of solid particles in the nematic, a random initial orientation can be achieved in an irregular matrix such as a cellulose nitrate filter. Such filters are available with pore sizes of ≈ 230 to 850 nm. In a study of nematic 5CB confined to Synpore filter material, the optical transmission has been determined as a function of the electric field applied across the filter sheet [53]. The filter was sandwiched between glass plates with transparent ITO electrodes. In Fig. 26, the optical transmission is depicted as a function of the electric correlation length. At low electric field strengths (large ξ), the scattering of the sample is constant, but when the electric correlation length becomes comparable to the nominal pore size

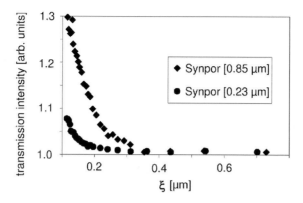

Fig. 26. Relative optical transmission (scaled to the value in the field-free state) of liquid-crystal (5CB)-filled Synpore filter sheets as a function of the applied electric field. The electric correlation length ξ is related to the electric field strength by $\xi = \sqrt{K/(\varepsilon_a \varepsilon_0 E^2)}$ where $\varepsilon_a = \varepsilon_\| - \varepsilon_\perp$ is the anisotropy of relative electric constant of 5CB. Data taken from [53, 101]

of the filters, the scattering decreases continuously, while the transmission coefficient increases with increasing field strength. As discussed above, a well-defined threshold behavior is not found in this system. Rather, the transmission intensity increases smoothly. The pore size dependence is also evident, but the dependence upon the nominal pore size is much weaker than expected. Since the pore geometry is irregular, the onset of director alignment should be expected somewhere between the threshold for cylindrical pores and that for the randomly-dispersed-fiber geometry, which are in the range $\approx 0.2\xi_{pore}$ to ξ_{pore} for the 2D system (ξ_{pore} representing an average pore size). Whereas in the small-pore system (0.23 μm), the slope of the transmission curve sets in approximately when ξ reaches the nominal pore size, the factor for the 0.85 μm filters is only ≈ 2. This indicates that the simplified model provides only a rough qualitative picture of the experimental situation, and a detailed 3D analysis taking into account the exact internal pore structure has to be performed to obtain quantitative agreement between experiment and model.

In the same way as chemically crosslinked networks [69–73] hydrogen-bonded gels [75–78] may stabilize the director structure of nematics or smectic mesophases. A study of the electro-optic properties of free-standing films of a ferroelectric hydrogen-bonded LC gel has been published in [78]. In that study, it was shown that the hydrogen-bonding gel former organizes in a practically 2D network of fibrils in the smectic film plane and stabilizes the c-director in the film. The network structure can be detected optically by means of polarizing microscopy in this case. The switching behavior in electric fields can be treated in an analogous way, as described above, when the dielectric/diamagnetic torque is replaced by the torque on the spontaneous polarization in the SmC* phase, and the nematic director is replaced by the c-director. The thin-film geometry allows one to measure the dynamics of the c-director reorientation in the electric field and allows one to retrieve semiquantitative information about the strength of the interactions between the network and the smectic c-director.

Acknowledgments

The authors acknowledge fruitful collaboration with Stanis law Rózański (dielectric spectroscopy of confined liquid crystals) and contributions from Lama Naji and Christiane Cramer (NMR and dielectric spectroscopy) and Jianjun Li (electro-optics of free-standing films).

References

1. G.P. Crawford and S.Žumer (Eds.): *Liquid Crystals in Complex Geometries* (Taylor & Francis, London, 1996)
2. M.D. Dadmun, M. Muthukumar: J. Chem. Phys. **98**, 4850 (1992)
3. G.S. Iannacchione, D. Finotello: Phys. Rev. Lett. **69**, 2094 (1992)
4. G.S. Iannacchione, G.P. Crawford, J.W. Doane, D. Finotello: Mol. Cryst. Liq. Cryst. **222**, 205 (1992)
5. D. Finotello, G.S. Iannacchione, S. Qian: "Phase transitions in restricted geometries", in [1], pp. 325–343
6. T. Bellini, N.A. Clark, C.D. Muzny, L. Wu, C.W. Garland, D.W. Schaefer, B.J. Olivier: Phys. Rev. Lett. **69**, 788 (1992)
7. N.A. Clark, T. Bellini, R.M. Malzbender, B.N. Thomas, A.G. Rappaport, C.D. Muzny, D.W. Schaefer, L. Hrubesh: Phys. Rev. Lett. **71**, 3505 (1993)
8. G.S. Iannacchione, D. Finotello: Liq. Cryst. **14**, 1135 (1993)
9. G.S. Iannacchione and D. Finotello: Phys. Rev. E **50**, 4780 (1994)
10. G.S. Iannacchione, J.T. Mang, S. Kumar, D. Finotello: Phys. Rev. Lett. **73**, 2708 (1994)
11. L. Wu, B. Zhou, C.W. Garland, T. Bellini, D.W. Schaefer: Phys. Rev. E **51**, 2157 (1995)
12. K.M. Unruh: Nanostruct. Mater. **9**, 709 (1997)
13. G.S. Iannacchione, Sihai Qian, D. Finotello, F.M. Aliev: Phys. Rev. E **56**, 554 (1997)
14. B. Zhou, G.S. Iannacchione, C.W. Garland, T. Bellini: Phys. Rev. E **55**, 2962 (1997)
15. T. Bellini, A.G. Rappaport, N.A. Clark, B.N. Thomas: Phys. Rev. Lett. **77**, 2507 (1996)
16. T. Bellini, C. Chiccoli, P. Pasini, C. Zannoni: Phys. Rev. E **54**, 2647 (1996)
17. G.S. Iannacchione, S. Qian, G.P. Crawford, S.S. Keast, M.E. Neubert, J.W. Doane, D. Finotello: Mol. Cryst. Liq. Cryst. **262**, 1301 (1995)
18. S. Qian, G.S. Iannacchione, D. Finotello, L.M. Steele, P.E. Sokol: Mol. Cryst. Liq. Cryst. **265**, 2961 (1995)
19. S. Qian, G.S. Iannacchione, D. Finotello, Phys. Rev. E **53**, R4291 (1996)
20. D. Finotello, G.S. Iannacchione: Int. J. Mod. Phys. B **9**, 2247 (1995)
21. G.P. Crawford, L.M. Steele, R. Ondris-Crawford, G.S. Iannacchione, C.J. Yeager, J.W. Doane, D. Finotello, J. Chem. Phys. **96**, 7788 (1992)
22. G.P. Crawford, D.K. Yang, S. Žumer, J.W. Doane: Phys. Rev. Lett. **66**, 723 (1991)
23. G.P. Crawford, M. Vilfan, J.W. Doane, I. Vilfan: Phys. Rev. A **43**, 835 (1991)
24. G.P. Crawford, D.W. Allender, M. Vilfan, I. Vilfan, J.W. Doane: Phys. Rev. A **44**, 2570 (1991)
25. A. Zidansek, S. Kralj, G. Lahajnar, R. Blinc: Phys. Rev. E **51**, 3332 (1995)
26. S. Žumer, S. Kralj, M. Vilfan: J. Chem. Phys. **91**, 6411 (1989)

27. G.P. Crawford, R. Stannarius, J.W. Doane: Phys. Rev. A **44**, 2558 (1991)
28. G.P. Crawford, D.W. Allender, J.W. Doane: Phys. Rev. A **45**, 8693 (1992)
29. S. Kralj *et al.*: Phys. Rev. E. **48**, 340 (1993)
30. J.H. Erdmann, S. Žumer, J.W. Doane: Phys. Rev. Lett. **64**, 1907 (1990)
31. P. Ziherl, M. Vilfan, S. Žumer: Phys. Rev. E **52**, 690 (1995)
32. C. Cramer, T. Cramer, F. Kremer, R. Stannarius: J. Chem. Phys. **106**, 3730 (1997)
33. C. Cramer, T. Cramer, M. Arndt, F. Kremer, L. Naji, R. Stannarius: Mol. Cryst. Liq. Cryst. **303**, 209 (1997)
34. O. Jarh, M. Vilfan: Liq. Cryst. **22**, 61 (1997)
35. S. Žumer, G.P. Crawford: "Polymer network assemblies in nematic liquid crystals", in [1], pp 83–101
36. M. Vilfan, N. Vrbančič-Kopač: "NMR of liquid crystals with an embedded polymer network", in [1], pp. 159–186
37. R. Stannarius, G.P. Crawford, L.C. Chien, J.W. Doane: J. Appl. Phys. **70**, 135 (1991)
38. M. Vilfan et al.: J. Chem. Phys. **103**, 8726 (1995)
39. S. Žumer, P. Ziherl, M. Vilfan: Mol. Cryst. Liq. Cryst. **292**, 39 (1997)
40. T. Bellini, N.A. Clark: "Light scattering as a probe of liquid crystal ordering in silica aerogels", in [1], pp. 38–409, and refs. therein
41. T. Bellini, N.A. Clark, D.W. Schaefer: Phys. Rev. Lett. **74**, 2740 (1995)
42. F.M. Aliev, V.V. Nadtotchi: Mater. Res. Soc. Symp. Proc. **407**, 125 (1996)
43. F.M. Aliev, J. Kelly: Ferroelectrics **151**, 263 (1994)
44. F.M. Aliev: "Liquid crystals and Polymers in pores: the influence of confinement on dynamic and interfacial properties", in [1], p. 345–370
45. H. Xu, J.K. Vij, A. Rappaport, N.A. Clark: Phys. Rev. Lett. **79**, 249 (1997)
46. H. Miyata, M. Maeda, I. Suzuki: Liq. Cryst. **20**, 303 (1996)
47. S.A. Rózański, R. Stannarius, H. Groothues, F. Kremer: Liq. Cryst. **21**, 59 (1996)
48. S.A. Rózański, F. Kremer, H. Groothues, R. Stannarius: Mol. Cryst. Liq. Cryst. **303**, 319 (1997)
49. G.P. Sinha, F.M. Aliev: Phys Rev E **58**, 2001 (1998) F.M. Aliev, G.P. Sinha: Mol. Cryst. Liq. Cryst. **364**, 435 (2001)
50. Yu.P. Panarin, C. Rosenblatt, F.M. Aliev: Phys. Rev. Lett. **81**, 2699 (1998)
51. L. Naji, F. Kremer, R. Stannarius: Liq. Cryst. **25**, 363, (1998)
52. S.A. Rózański, L. Naji, F. Kremer, R. Stannarius: Mol. Cryst. Liq. Cryst. **329**, 483 (1999)
53. S.A. Rózański, R. Stannarius, F. Kremer: Z. Phys. Chem. **211**, 147 (1999)
54. I. Rychetsky, N. Novotna, M. Glogarova: J. Physique IV **10**, Pr7-119-122 (2000)
55. S.A. Rózański, R. Stannarius, F. Kremer, S. Diele: Liq. Cryst. **28**, 1071 (2001)
56. S.A. Rózański, R. Stannarius, F.Kremer: IEEE Trans. Dielectr. Electr. Insulat. **8**, 488 (2001)
57. A.G. Rappaport, N.A. Clark, B.N. Thomas, T. Bellini: "X-ray scattering as a probe of smectic A liquid crystal ordering in silica aerogels", in [1], pp. 411–466, and refs. therein
58. A. Jakli, G. Kali, L. Rosta: Physica B **234**, 297 (1997)
59. J. Hoffmann: Am. Lab. **21**, 70 (1989)
60. J.W. Doane, N.A. Vaz, B.G. Wu, S. Žumer: Appl. Phys. Lett. **48** 269 (1986)
61. J.W. Doane: "Polymer dispersed liquid crystals", in *Liquid Crystals: Applications and Uses*, ed. by B. Bahadur vol. 1 (World Scientific, Singapore, 1990)
62. A. Goleme, S. Žumer, D.W. Allender, J.W. Doane: Phys. Rev. Lett. **61**, 1937 (1988)
63. H.S. Kitzerow, H. Molsen, G. Heppke: Appl. Phys. Lett. **60**, 3039 (1992)
64. F.M. Aliev: "Dynamics of liquid crystals confined in random porous matrices", in *Access in porous Materials*, ed. by T.J. Pinnavaia, M.F. Thorpe (Plenum Press, New York, 1995)
65. G.S. Iannacchione, G.P. Crawford, S. Žumer, J.W. Doane, D. Finotello: Phys. Rev. Lett. **71**, 2595 (1993)

66. A. Borštnik, H. Stark, S. Žumer: Phys. Rev. E **60**, 4210 (1999); A. Borštnik, H. Stark, S. Žumer: Phys. Rev. E **61**, 2831 (2000)
67. H. Stark: Phys. Rev. E **66**, 041705 (2002)
68. M. Kreuzer, R. Eidenschink: "Filled nematics", in [1], pp. 307–324
69. R.A.M. Hikmet: J. Appl. Phys. **68**, 4406 (1990)
70. A. Jakli, L. Bata, K. Fodor-Csorba, L. Rosta, L. Noirez: Liq. Cryst. **17**, 227 (1994)
71. A. Jakli, K. Fodor-Csorba, A. Vajda: "Liquid crystals gel dispersions prepared in the isotropic phase", in [1], pp. 143–157
72. R.A.M. Hikmet: "Anisotropic gels obtained by photopolymerization in the liquid crystal state", in [1], pp. 53–82
73. D.K. Yang, L.C. Chien, Y.K. Fung: "Polymer stabilized cholesteric textures: materials and application", in [1], pp. 103–142
74. D.J. Broer: "Liquid crystalline networks formed by photoinitiated cross-linking", in [1], pp. 239–254
75. T. Kato, T. Kutsuna, K. Hanabusa, M. Ukon: Adv. Mater. **10**, 606 (1998) .
76. C. Tolksdorf, R. Zentel: Adv. Mater. **13**, 1307 (2001)
77. J. Prigann, C. Tolksdorf, H. Skupin, R. Zentel, F. Kremer: Macromolecules **35**, 4150 (2002)
78. J. Li, R. Stannarius, C. Tolksdorf, R. Zentel: Phys. Chem. Chem. Phys. **5**, 916 (2003)
79. J. Li, D. Geschke, R. Stannarius: Liq. Cryst. **31**, 21 (2004)
80. F. Kremer, A. Huwe, A. Schönhals, S.A. Rózański: "Molecular dynamics in confined space", in *Broadband Dielectric Spectroscopy*, ed. by F. Kremer, A. Schönhals (Springer, Berlin, Heidelberg, 2003), pp. 210ff
81. S. Havriliak, C. Negami: J. Polym. Sci. C **14**, 99 (1966)
82. A.P.Y. Won, S.B. Kim, W.I. Goldburg, W.H.M. Chan: Phys. Rev. Lett. **70**, 954 (1993)
83. P. Sheng: Phys. Rev. Lett. **37**, 1059 (1976); Phys. Rev. A **26**, 1610 (1982)
84. K. Miyano: Phys. Rev. Lett. **43**, 51 (1979)
85. J.C. Tarczon, K. Miyano: J. Chem. Phys. **73**, 1994 (1980)
86. A. Mauger, G. Zribi, D.L. Mills, J. Toner: Phys. Rev. Lett. **53**, 2485 (1984)
87. P.G. de Gennes, J. Prost: *The Physics of Liquid Crystals* (Oxford University Press, Oxford, 1993)
88. A. Poniewierski, T.J. Sluckin: Liq. Cryst. **2**, 281 (1987); G. Barbero, E. Miraldi, A. Stepanescu: J. Appl. Phys. **68**, 2063 (1990)
89. R.J. Ondris-Crawford, G.P. Crawford, S. Žumer, J.W. Doane: Phys. Rev. Lett. **70**, 194 (1994)
90. R.J. Ondris-Crawford, M. Ambrošič, J.W. Doane and S. Žumer: Phys. Rev. E **50**, 4773 (1994).
91. H. Schmiedel, R. Stannarius, G. Feller, C.Cramer: Liq. Cryst. **17**, 323 (1994); H. Schmiedel, R. Stannarius, G. Feller, C. Cramer: Z. Phys. Chem. **190**, 135 (1995); H. Schmiedel, R. Stannarius, C. Cramer, G. Feller, H.E. Müller: Mol. Cryst. Liq. Cryst. **262**, 167 (1995)
92. M.I. Boamfa, M.W. Kim, J.C. Maan, T. Rasing: Nature **421**, 149 (2003)
93. S. Kralj, S. Žumer, D.W. Allender: Phys. Rev. A **43**, 2943 (1991)
94. P. Galatola, J.B. Fournier: Phys. Rev. Lett. **87**, 3915 (2001)
95. K. Kočevar, A. Borštnik, I. Muševič, S. Žumer: Phys. Rev. E **64**, 051711 (2001)
96. D.W. Allender, G.P. Crawford, J.W. Doane: Phys. Rev. Lett. **67**, 1442 (1991)
97. F. Lonberg, R.B. Meyer: Phys. Rev. Lett. **55**, 718 (1985)
98. G. Haas: *Thesis* Universität Karlsruhe (1991)
99. C. Cramer, U. Kühnau, H. Schmiedel, R. Stannarius: Mol. Cryst. Liq. Cryst. **257**, 99 (1995)
100. R. Stannarius, M. Grigutsch: Mol. Cryst. Liq. Cryst. **262**, 67 (1995)
101. S. A. Rózański: Synth. Met. **109**, 245 (2000)

Surfaces and Interfaces
of Free-Standing Smectic Films

Heidrun Schüring[1] and Ralf Stannarius[2]

[1] Universität Leipzig, Institut für Experimentalphysik I,
  pge91dsf@studserv.uni-leipzig.de
[2] Otto-von-Guericke-Universität Magdeburg, Institut für Experimentalphysik,
  ralf.stannarius@physik.uni-magdeburg.de

Abstract. Free-standing smectic films are unique fluid objects with an exceptionally large surface-to-volume ratio. They are excellently suited for studies of surface and interface properties of liquids. We describe surface tension measurements in the smectic and isotropic phases, with particular emphasis on anomalies near the phase transitions. A model that considers the excess surface entropy of ordered surface layers is applied to describe the experimental observations qualitatively and quantitatively. From the geometrical properties of isotropic droplets in free-standing films and from the forces acting on these droplets, interface tensions between the smectic and isotropic phases are derived. Finally, we report measurements of the gas permeation through smectic films and develop a model for the description of the film-thickness dependence of the permeation coefficient.

1 Introduction

Smectic liquid crystals of calamitic (rod-like) mesogens are characterized by an orientational order of the molecules. They lack the true three-dimensional long-range lattice characteristic of crystalline solids. However, in addition to the orientational order that is present also in the nematic and cholesteric phases, a positional order of the molecules exists in at least one spatial direction. The simplest case of a smectic phase is smectic A (SmA), with the mean orientation of the long axes of the prolate molecules directed normal to the smectic layers. Within these layers, the molecules are arranged liquid-like without any positional long-range order (Fig. 1).

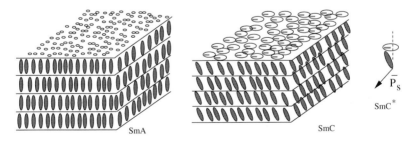

Fig. 1. Molecular arrangements in the smectic A and C phases (see text).

H. Schüring, R. Stannarius, Surfaces and Interfaces of Free-Standing Smectic Films, Lect. Notes Phys. **634**, 337–381 (2004)
http://www.springerlink.com/   © Springer-Verlag Berlin Heidelberg 2004

The smectic A phase is nonpolar, its symmetry is $D_{\infty h}$, and it is optically uni-axial with the optic axis along the layer normal. In the smectic C phase (SmC), the preferential axis of the molecules is tilted with respect to the layer normal, and the symmetry reduces to C_{2h}. The projection of the tilt direction on the layer plane, called the c-director, defines the orientation of the tilt azimuth in the plane. This c-director has the freedom to reorient in the layer plane. Optically, the smectic C phase is weakly biaxial. In a crude approximation, it may often be treated as nearly uniaxial with the optic axis along the tilt direction. As in the smectic A phase, the molecules in SmC can move freely within the liquid-like smectic layer planes, with-out in-plane positional order, and like the smectic A phase it is nonpolar, too. More highly ordered smectic phases possess, in addition, a positional order within the layer planes (smectic B, F, I, ...). Smectic B is a nontilted hexagonal phase, smectic F and I are tilted and they are distinguished by the tilt azimuth with respect to the next neighboring molecule in the layer.

Of particular interest from the technical points of view are smectic phases formed by chiral molecules, in the first place smectic C*. The SmC* phase has the symme-try C_2, and it forms a spontaneous polarization in the layer plane, perpendicular to the tilt plane [1] (as indicated in Fig. 1, right). The variety of smectic phases be-comes much larger if one considers arrangements of molecules that deviate from a simple rod-like shape. So called "banana-shaped" mesogens [2,3] can form phases with spontaneous polar in-plane ordering. In addition, double-layer or multiple-layer phases can be distinguished, such as the bilayer smectic A phase. The different arrangement of azimuthal orientations of subsequent layers in tilted smectics can produce antiferroelectric or ferrielectric bulk phases.

Because of their layered structure, smectics can form (meta)stable free-standing films (FSFs), sometimes also called freely suspended films, very similar to the well-known soap films. Ordered molecular structures with lateral extensions of up to several square centimeters and with uniform thickness can be formed by only a few (even two) molecular layers. The smectic layers in free-standing films are arranged parallel to the film surfaces and stabilize the films against thickness fluctuations. These films form Plateau surfaces between the edges of their supports (surfaces of mean curvature zero), which are perfectly flat when the support edges lie in a plane. Other typical shapes are catenoids between two circular rings on the same axis [4,5]. If the two surfaces of the film are at different pressures, the equilibrium shape of the film is spherical, with the Laplace pressure of the curved surface compensating the pressure difference [6–8]. Such structures may be considered as the equivalents of soap bubbles.

Free-standing smectic films have been in the focus of scientific interest since their first description more than two decades ago [9–12]. An introduction to their physical properties is given, for example, in [13–15]. These films are physically unique in many aspects. Their large surface-to-volume ratio is exceptional. The attractiveness of these systems for scientific research lies in the access to surface-induced molecular order effects and properties of thin liquid layers. Phase transitions in thin smectic FSFs can be significantly shifted with respect to the bulk, the very character of phase transitions can change, new phases may be induced, or bulk phases can be suppressed [16–48]. In particular, melting of smectic films into the isotropic phase (or transitions between smectic modifications) has been observed as a process of layer-by-layer melting of the core of the film (e.g. [31,32,49–59]. Often, the surface of the smectic film is covered with one or more molecular layers of a higher order phase [18,26,35,40,48,60–71]. Above the bulk clearing point, when the core of the smectic material starts to melt, smectic regions with thicknesses of a temperature-dependent smectic coherence length remain at the free surfaces. On approaching the isotropic–smectic bulk transition from above, this coherence length diverges, while even several K above the melting point, it may still be larger than the thickness of a single smectic layer. When a thick freely suspended film is heated above the bulk clearing temperature, the inner part of the film melts and the isotropic material usually flows out to the meniscus. A layer-by-layer thinning of the smectic film with increasing temperature is the consequence [31,49]; the equilibrium thickness at a given temperature is a measure of the smectic layer coherence length.

Another well-known example of surface-induced order is found in the smectic A phase. SmA films are covered by at least one tilted smectic layer at their surfaces (e.g. [18,35,64]). Surface-enhanced ordering is a phenomenon quite often encountered for mesogenic materials. It is, of course, not restricted to freely suspended films. "Swimming" smectic surface layers on isotropic liquid droplets have been observed at temperatures above the clearing point (see e.g. [72–74]). Likewise, solid substrates may induce orientational ordering at temperatures above the bulk clearing point (see e.g. [75,76] and references therein). The observation of such ordering phenomena in thin layers and detailed structure assessment are conveniently performed in freely suspended films, which provide large, molecularly flat surfaces. For example, ellipsometry (e.g. [36,64,66,67,77–80]) and X-ray investigations [17,19–21,27,45,54,61,62,71,80–86] have been successfully applied.

Directly related to the structure of the films is their ferroelectric, antiferroelectric and ferrielectric behavior [23,36,66,77–80,87–94]. In films of a few molecular layers, these properties can change dramatically with respect to the bulk samples. For example, thin films of antiferroelectric bulk phases are characterized by an odd–even alternation of the polar electric properties if antiferroelectricity is the consequence of a compensation of the spontaneous polarization of adjacent layers. While even-numbered films behave qualitatively similarly to the bulk material, films with an odd number of layers possess a macroscopic spontaneous polarization of the uncompensated layer.

Other important research topics in freely suspended films include the study of dynamic processes in quasi-two-dimensional systems induced by external fields. In SmC or SmC* films, the local orientation of the c-director can be influenced by external electric fields. Among the effects observed in electrically driven FSFs are orientational dynamics of the c-director, macroscopic flow, and dynamic pattern formation [3,11,88–91,93,95–112]. From the study of uniform reorientation dynamics and the structure of domain walls (kink solutions of the azimuthal orientation angle), viscoelastic material parameters are obtained. Electrically driven convection has been studied in SmA [113–115], and SmC and SmC* [116,117] phases. Vortex flow patterns are generated in smectic A and C films by the interaction of free charges on the film surface and in the film with laterally applied electric fields. In tilted phases, coupling of this flow field to the c-director creates spiral or target textures which are observable with polarizing microscopy. When free-standing smectic films are exposed to temperature gradients in the film plane, thermally driven two-dimensional convection patterns can be formed [118–121]. As in electrically driven patterns, the flow field is in the smectic layer plane.

Because of the coupling of the c-director to shear flow in tilted smectic phases, dynamic structures can be generated by mechanical excitation as well [122–124]. Even the first-order phase transition between SmA- and SmC*-like structures in freely suspended films can be induced by transient changes of the film tension [125]. The vibrations of smectic films have been studied to demonstrate, and to investigate experimentally, the problem of isospectral drums [126].

A potential application of these films is the preparation of macroscopically ordered ferroelectric elastomers (e.g. [94,127,128]). Smectic films can be transferred onto solid substrates [129,130] as an alternative strategy to create ordered organic layers without involving molecular-beam deposition. Among the topics listed above, this chapter focuses on two aspects of the investigation of thin smectic films connected with surface and interface properties of the mesogenic material. The first part describes the measurement of surface and interface tensions in the FSF geometry, the second part introduces experiments for the study of gas permeation through the liquid–gas interface.

2 Surface and Interface Tensions of Thin Smectic Films

Liquids minimize their surfaces because of the attractive intermolecular forces on molecules at the liquid–gas boundary directed into the inner part of the material. The physical quantity surface tension, σ, describes the amount of energy per area necessary to increase the surface.

The most important features of surface tensions are described by two basic principles: according to Eötvös's rule, the surface tension of common liquids decreases linearly with increasing temperature [131]. If surface energies are compared between different molecular liquids, it may be more convenient to refer the surface energy to the area occupied by a single molecule at the surface, or to a surface formed by a fixed number of molecules, say $N_A^{2/3}$ (N_A = Avogadro's number). Inserting the molar volume V_M, composed of N_A molecules, one obtains the molar free surface enthalpy

$$\sigma_M = \sigma V_M^{2/3}. \tag{1}$$

From the equivalence of the liquid and gaseous states at the critical temperature T_c, it follows that the surface tension between the liquid and gas phases disappears at T_c. The molar free surface enthalpy is proportional to $T_c - T$, with

$$\sigma_{\mathrm{M}} = \kappa_{\mathrm{M}}(T_{\mathrm{c}} - T). \tag{2}$$

For nonassociated, nonmetallic liquids, the constant κ_{M} is determined mainly by dispersive intermolecular forces; therefore this factor is rather independent of the specific liquid, $\kappa_{\mathrm{M}} \approx 2.1 \times 10^{-7}$ J K^{-1}mol$^{-2/3}$.

Langmuir's principle of independent surface action assigns localized contributions to the surface tension to each part of a molecule [132]. In isotropic liquids, these local contributions are very difficult to discriminate because of the random orientational disorder of the molecules. To a certain extent, molecular order is induced by the surfaces in isotropic liquids, but a well-defined orientation of molecules at a liquid surface is found in mesogenic substances. In particular, nematic and smectic fluids should be well suited to investigate the specific contributions of different submolecular moieties to the surface energy.

Thin free-standing smectic films with their robust and stable layer structure, the exceptionally large surface-to-volume ratio, and their macroscopically ordered molecular arrangement, allow one to measure the surface tension of these anisotropic fluids with a variety of methods. Especially, Langmuir's principle can be tested and contributions to the surface tension can be attributed to individual parts of the molecules [52, 133–137]. Another aspect is the investigation of the surface tension in the vicinity of liquid–liquid phase transitions. Deviations from the normal temperature dependence, (2), connected with entropic contributions to the surface tension can be observed [70].

The part describing surface and interface tension measurements is organized as follows. Section 2.1 is focused on the influence of the chemical composition of smectogens on the surface tension. An overview of different methods used for surface tension measurements in liquid crystals is given in Sect. 2.2. Results on the temperature dependence of surface tensions are presented in Sect. 2.3. Interface tensions between different fluid phases are derived from the study of isotropic inclusions in thin smectic films in the final section.

2.1 Surface Tension of Anisotropic Fluids

In contrast to isotropic fluids, the surface tension of smectic and nematic materials is a tensor and depends on the degree of orientational order as well as the orientation of the mesogens with respect to the liquid–gas interface [138]. Against this background, freely suspended smectic films are ideal for surface tension measurements. They are characterized by a perfect layer order, parallel to the free surfaces. Within these layers, the molecular long axes of the rod-shaped molecules in all smectic phases have a preferred orientation, whereas an additional positional order of the molecules exists only in the low temperature smectic phases (smectic B, F, I, G, H, ...). Owing to this well-defined orientational order of the molecules with respect to the smectic layer normal, and the smectic layer arrangement parallel to the film surfaces, most experiments performed with smectic FSFs determine the same component of the surface tension. In Sect. 2.4, however, it will be shown that in principle information on other components of σ can be obtained, if inhomogeneities, layer steps, or inclusions in the smectic films are studied.

In extensive and systematic studies, Mach et al. [137] have found that the surface tensions of smectic materials are mainly determined by specific terminal groups or molecular substructures. According to the authors, surface tensions of liquid crystals can be classified in the following way:

- materials with close-packed, fluid-like, CH_3-containing surfaces have surface tensions of about 21×10^{-3} N/m,
- an increase of CH_2- or polar terminal groups leads to surface tensions between 24×10^{-3} N/m and 27×10^{-3} N/m,
- the surface tension decreases to 13×10^{-3} N/m for compounds with partly fluorinated alkyl chains, and
- complete fluorination results in surface tensions of $\approx 11 \times 10^{-3}$ N/m.

In [136], it has been demonstrated that the replacement of a single atom in groups located at the free surface can have a dramatic effect on the surface tension. By those results it was proved that Langmuir's principle can be successfully applied in the mesogenic materials investigated.

2.2 Methods for Surface Tension Measurements

In this section, a review of experimental methods for the assessment of surface tension is given. The pendant drop method (Fig. 2a), used for example in [139–141], is a well-known standard method for isotropic liquids. The shape of a pendant drop is given by the equilibrium of forces related to gravitation and surface energy. The surface tension can be calculated from the droplet profile by an equation of the form

$$\sigma = g \rho d^2 / H \tag{3}$$

with the acceleration of gravity g, the density ρ, the maximum equatorial diameter d, and a correction factor H which depends on the droplet shape. This correction factor is a complex function of the droplet shape and size. However, one can refer to tabulated values from the literature [142]. A better accuracy of the measurements can be achieved with some computational effort when the complete droplet shape is fitted. This method enables qualitatively correct measurements of surface tensions of isotropic, nematic and smectic liquid crystals. However, the exact determination of droplet profiles is very complex and a couple of factors influence the reliability of data. First, the establishment of the equilibrium droplet shape and size seems to be a general problem for mesogenic materials. In particular, the smectic materials are characterized by a large viscosity, and one has to be very careful to ensure that the equilibrium droplet shape is reached. But even in the nematic phase, the macroscopic equilibration of the macroscopic shape and director field may be time-consuming and may involve unpredictably long timescales. "Time-dependent" surface tensions reported for nematic materials investigated with the pendant drop method indicate that systematic errors may be involved. Besides this, the orientation of the molecules at the surface is not known in these measurements.

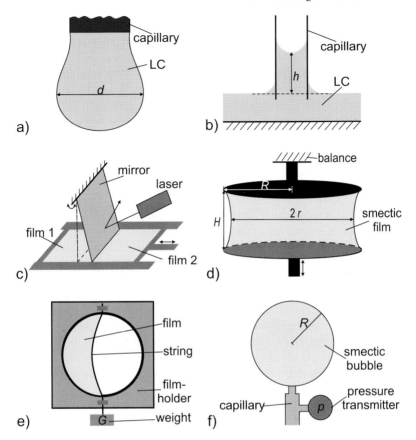

Fig. 2. Methods for surface tension measurements in liquids and liquid crystals: (a) pendant drop method, (b) method of capillary elevation, (c) (modified) Wilhelmy balance, (d) catenoid film balance (see text for details), (e) string tensiometer, (f) bubble method

The well-known capillary elevation method (Fig. 2b) uses the curvature pressure at the meniscus of liquids confined in thin capillaries. In a gap between two parallel, vertical walls wetted by the liquid, or in a thin capillary tube, this pressure counteracts the gravitational pressure and elevates the liquid surface. For measurements of the surface tension anisotropy in nematics, a modification of this method has been proposed in [143]. Special treatment of the inner surface of a capillary with a rectangular, elongated section preserves a homeotropic orientation ($\perp$, director normal to the surface) in one part and a planar orientation ($\parallel$) of the director in another part. From the different elevations h_i ($i = \parallel, \perp$), the surface tensions σ_i are calculated according to

$$\sigma_i = \frac{\rho g d h_i}{2 \cos \theta}, \tag{4}$$

where $d \approx 200 \, \mu m$ is the distance between the opposite walls and $\cos \theta$ is the contact angle, determined from a drop of liquid crystal on an appropriately treated glass plate.

Capillary waves generated by ac electric fields and detected via laser reflection provide another method to access information on surface tensions [144, 145].

Whereas the previous methods use bulk liquid or liquid-crystalline material, methods based on freely suspended smectic films are presented in the following. These methods work in lyotropic systems such as soap films as well as in thermotropic smectic phases. One approach is in many respects comparable to the previous method. Instead of capillary waves on bulk samples, vibrations of smectic films are generated by an ac electric field [146, 147]. The surface tension is deduced from the resonance frequency $\omega_r \propto \sqrt{2\sigma}/R$, where R is the radius of the film. A certain problem of this method is the accurate determination of the film radius R. Any freely suspended film is surrounded by a meniscus, a rim of disordered bulk material, at its support. This meniscus is partially involved in the vibrational motions. Therefore, measurements of absolute values of surface tensions are difficult, while relative measurements such as a qualitative assessment of the temperature dependence can be obtained with high precision.

The Wilhelmy balance is a simple device for surface tension measurements that measures directly the force necessary to increase the surface of a liquid film. The film is spanned by a rectangular film holder consisting of a U-shaped frame and a movable wire. An application of this method to the investigation of smectic films has been proposed in [148] and [149]. The smectic film is prepared on a rectangular frame with a movable side attached to a pendulum. The tension of the film causes a deflection of the pendulum, which is either measured directly or compensated by an electromagnet. In both cases, the setup must be calibrated with a material of known surface tension. In a modified setup [148] films can be drawn on each side of the pendulum, as sketched in Fig. 2c. This allows, in principle, measurements of the film thickness dependence of the surface tension. Technical problems arise, for example from the excess material at the frame that shifts the zero position of the pendulum. This might explain why experimental results obtained with this method are rather sparse.

A solution of the problem of the lateral meniscus in the modified Wilhelmy balance is the choice of a cylindrical geometry. The principle of the setup is sketched in Fig. 2d. A high-resolution balance measures the force resulting from the film tension [5]. The smectic film is drawn between two circular rings 10 cm in diameter. The film deforms to a catenoid shape where the local mean curvature is zero everywhere (Plateau surface). The upper ring is attached to a balance that measures the weight of the ring and the excess material at the ring, reduced by the forces generated by the film tension. When the film is destroyed, the balance measures the pure weight alone. The force difference ΔF between the two measurements is related to the surface tension by $\Delta F = 4\pi\sigma r$, where r is the minimum film radius (in the middle between the rings, where the local film plane is vertical). The minimum film radius r can be deduced from the radius R of the rings and the distance H between them.

The basic principle of a string tensiometer, as used in [133], is shown in Fig. 2e. The opening of the film holder is divided into two parts by a thin flexible string, spanned by a weight G. The preparation of a smectic film on one side of the string leads to deformation of the originally straight suture into an arc. The surface tension can be calculated, using the relation

$$\sigma = \frac{G}{2R},$$ (5)

from the radius of curvature R of the string. An exact determination of the radius of curvature is difficult. Therefore, the resolution of this method is not high enough to resolve the temperature dependence of the surface tension. Despite this fact, it is up to now the most successful method for absolute surface tension measurements [52, 133–137].

All measurements presented in Sect. 2.3 were obtained with the bubble method [7]. For this reason, a more detailed description of this method is given here. The bubble method is based on the idea that planar smectic films can be inflated into spherical bubbles, as shown in Fig. 2f. Smectic films are drawn onto the open end of a glass capillary. The other end is airtight-connected to a microliter syringe, which is used for injection of air into the capillary. Smectic bubbles have many features in common with soap bubbles. They are stabilized by the internal layer structure of the liquid film. However, in contrast to soap films that consist of mixtures of water and amphiphilic surfactants, they are homogeneous and consist of a single component; they can be prepared with a well-defined uniform film thickness and they are stable in the long term. According to the Laplace–Young equation

$$\sigma = \frac{pR}{4}$$ (6)

the surface tension can be calculated directly from the inner excess pressure p (the difference between the pressure of the inner bubble volume and the environmental pressure) and the radius of curvature R of the bubbles. In the experiment, bubbles are illuminated with parallel light and observed in transmission by means of a camera with a macro objective. Typical bubble radii for surface tension measurements are between 1 mm and 3 mm. This is a compromise between convenient observability of bubble radii and sufficient difference pressure according to (6). The bubble radii can be determined from digital images with an accuracy of 0.2 to 0.5%. The pressure difference is measured by means of a highly sensitive pressure transmitter. For temperature control, the bubbles are enclosed in a thermobox. The accuracy of this method allows not only a reliable absolute measurement of surface tensions but also the analysis of the temperature dependence of σ. The thickness of the smectic film forming the bubble turns out to be practically irrelevant for the determination of its surface tension. It has been confirmed by measurements of bubbles in a broad thickness range (a few smectic layers to thousands of layers) that the pressure vs. radius of curvature dependence is film-thickness-independent within experimental resolution. The determination of the film thickness of smectic bubbles is described in [150], and a short summary is given below on p. 363.

For the comparison of surface tensions of different samples, a slightly modified setup can be used. Instead of a single bubble, spherical caps are prepared on two communicating capillaries, as shown in Fig. 3. One of the caps is made from a reference substance, while the second cap consists of the material under investigation. The surface tension ratio of the two substances can be determined with high accuracy from the ratio of the two radii of curvature.

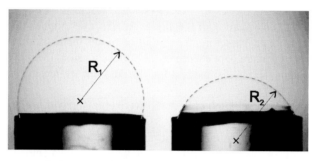

Fig. 3. Reference method: communicating capillaries of the same radius, with two films of the material DOBAMBC inflated on the open ends. Both spherical caps have the same radius of curvature; they complement one another to form a complete sphere. Measurements are restricted to $R_{1,2}$ in the vicinity of the capillary radius

All methods presented above are useful for an examination of the surface tension with respect to the gas phase. Differences in surface tensions of the smectic and isotropic phases of the same material and interface tensions between both phases can be accessed by the analysis of the shape of isotropic inclusions in the smectic film plane in a biphasic range. Such inclusions are observable in the vicinity of the smectic-to-isotropic phase transition. This approach is introduced in detail in Sect. 2.4.

2.3 Temperature Dependence and the Role of Phase Transitions

The results presented in this section were obtained by the bubble method and (6). Before details of the experiments are discussed, it is necessary to mention the influence of the film thickness on the surface tension of free-standing films. Being a genuine surface property, it should in principle be film-thickness-independent. Experimental observations over a broad film thickness range are in accordance with this expectation [7, 137]. It is not excluded, however, that in extremely thin films (of only a few layers) systematic deviations from the surface tension measured in bulk samples or thick films could be found [151].

The temperature dependence of the surface tension of DOBAMBC (4-*decyloxy-benzylidene-4-amino-2-methylbutylcinnamate*; the mesomorphism and the phase transition temperatures of the liquid crystals investigated are listed in Table 1 on p. 375) is presented in Fig. 4a. DOBAMBC has been chosen as a well-characterized standard substance [152] with a broad smectic range and rich polymorphism. The first set of data (open circles) corresponds to a bubble prepared at 80 °C and heated stepwise, and the second set (filled symbols) represents a bubble prepared at 90 °C and cooled down successively. Below 48 °C, the viscosity of the material is so high that the bubbles become very sensitive to external disturbances, they burst quickly. Above 108 °C, bubbles are very susceptible to destruction by an increasing convective flow in the film. The measured surface tension values are between 19.7×10^{-3} N/m and 20.8×10^{-3} N/m. The surface tension decreases with increasing temperature as in ordinary isotropic liquids. A slight change of the slope of the temperature curve is observed at the bulk phase transition temperatures (SmI*–SmC*, SmC*–SmA). DOBAMBC consists of molecules with two alkyl chains attached to the rigid aromatic core. The mean value of 20.3×10^{-3} N/m is in accordance with the classification presented in Sect. 2.1.

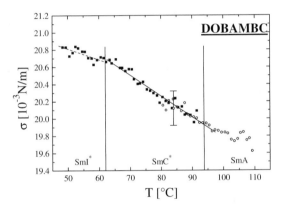

Fig. 4. Temperature dependence of the surface tension of DOBAMBC [70]. Open and solid symbols are data from two different samples

A similar monotonic decrease of surface tension with temperature is characteristic, for the standard material 8CB (4-N-octyl-4′-cyanobiphenyl) in the SmA phase, for example. The absolute value of σ is considerably higher than for DOBAMBC [70]. The mean value of $\sigma \approx 27.4 \times 10^{-3}$ N/m again fits into the values of the classification given by Mach et al. Molecules of the material 8CB, possess only one alkyl chain; a polar terminal cyano group is substituted instead of the second alkyl chain. At least for the first material, DOBAMBC, the observable temperature range of the smectic phases is large enough to compare the temperature dependence with the predictions of Eötvös's rule, (2). The coefficient κ_M is calculated from the molar mass $M = 0.492$ kg/mol, the density $\rho \approx 1000$ kg/m^3, and $\Delta\sigma/\Delta T = 2.4 \times 10^{-5}$ N m^{-1} K^{-1} as $1.5 \cdot 10^{-7}$ J K^{-1} mol$^{-2/3}$. Thus, for a highly anisotropic material the value of κ_M is in rather good correspondence with the predictions for isotropic liquids. Actually, for an anisotropic material the surface per N_A molecules should be significantly smaller than for an isotropic material with the same molar volume, since the average space occupied by the molecule is prolate normal to the free surface. The deviation of the experimental value from the literature value κ_M for nonassociative liquids, however, is in just the opposite direction. A slight change of the temperature slope of σ at the phase transition from SmC* to SmI*, and probably also at the SmC*–SmA transition, is indicated, but is within the experimental uncertainty. For 8CB, with $M = 0.291$ kg/mol, one finds $k_M \approx 2 \times 10^{-7}$ JK^{-1}mol$^{-2/3}$, of the same order of magnitude, but because of the limited temperature range of the smectic A phase (≈ 12 K), the uncertainty of this value is rather large.

In contrast to the previous examples, the temperature dependence observed for the surface tension of HOPDOB (4-hexyloxyphenyl-4-decyloxybenzoate) differs significantly from the normal linear slope. In Fig. 5 it can be seen that the surface tension *increases* at about 11×10^{-5} N m^{-1} K^{-1} with increasing temperature in the range between 45 °C and 50 °C. At higher temperatures, the slope changes its sign and the surface tension decreases at 1×10^{-5} N m^{-1} K^{-1} [70]. This anomalous behaviour has been confirmed by Veum et al. [147] from the frequency analysis of vibrating films.

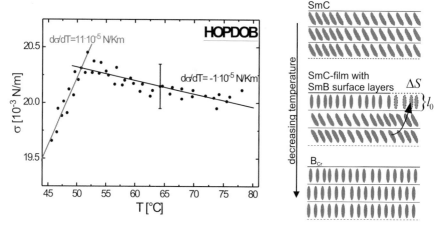

Fig. 5. Temperature dependence of the surface tension of HOPDOB (left) and model for surface phase transition (right)

According to the phase sequence of the bulk material, there is no *bulk* phase transition at 50 °C. At 44.5 °C the transition to the monotropic B_{Cr} phase, a nontilted, crystalline phase with hexagonal in-plane order, occurs. It is impossible to cool HOPDOB bubbles below 45 °C, because the B_{Cr} material is brittle, more solid-like than liquid, bubbles cannot be inflated or deflated, and the radius/pressure relation no longer reflects σ. Thus, surface tension data are available only in the temperature range of the SmC bulk phase.

Following the Maxwell relation

$$-\left(\frac{\partial \sigma}{\partial T}\right)_A = \left(\frac{\partial S}{\partial A}\right)_T, \tag{7}$$

a negative slope $-\partial\sigma/\partial T > 0$ means that the entropy S is lowered when the surface area A is increased at constant temperature T. Thus, an explanation of the measured temperature dependence of σ can be given under the assumption of an increased order at the surfaces with respect to the interior of the smectic film. As discussed in the introduction, such more highly ordered surface layers are quite common for smectogenic materials. A surface phase transition at 50 °C is supposed. Above 50 °C the entire film is in the SmC phase; at lower temperatures the surface layers show a (probably hexagonal SmB-like) intralayer order. The situation at one of the film surfaces is sketched at the right-hand side of Fig. 5. An increase of the surface area of such a film requires a transport of molecules from the interior of the film to the surface layers. These molecules undergo a phase transition into a more highly ordered state connected with an excess entropy. The consequences of this hypothesis can be tested with a simple calculation. The surface excess entropy ΔS_S, given by [153]

$$\Delta S_S = -\frac{\Delta\sigma}{\Delta T}, \tag{8}$$

can be compared with the molar SmC–B_{Cr} transition enthalpy ΔH_M known from the literature. Values of ≈ 5 kJ/mol have been reported [154–156]. The specific transition entropy ΔS_V can be calculated from

$$\Delta S_V = \frac{1}{T_{BC}} \frac{\rho}{M} \Delta H_M, \tag{9}$$

when the appropriate values for the phase transition temperature $T_{BC} = 323$ K, the sample density $\rho \approx 10^3$ kg/m^3, and the molar mass $M = 0.454$ kg/mol are inserted. Under the assumption that the specific transition enthalpy for the surface layer is comparable to that of a bulk sample, a layer thickness $l_0 = \Delta S_S/\Delta S_V$ of 3.2 nm is determined. This value is in very good agreement with the thickness of a single smectic layer of 3.3 nm for HOPDOB in the B_{Cr} phase determined from X-ray measurements [156, 157]. This quantitative result should not be overestimated, since the above calculation involves some rough approximations. It presupposes that the thickness of the surface layer is temperature-independent, that the specific entropy gain can be equated to the specific SmC–B_{Cr} transition enthalpy, and that the contributions of the "normal" temperature dependence, (2), may be neglected. Qualitatively, the model of an induced higher-order "interphase" with molecular thickness at the film surfaces explains the experimental data reasonably. We note that the assumption of such a more highly ordered surface layer is not arbitrary but is backed by electron diffraction, calorimetry, and optical reflectivity measurements obtained on another smectic A material (e.g. [37, 48]). It has been found in the substance 4O.8 (N-(4-n-butoxybenzylidene)-4-n-octylaniline) that upon cooling from the SmA phase, the outer surface layers display a distinct hexatic SmB order and subsequently transform into the B_{Cr} phase.

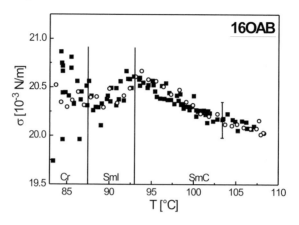

Fig. 6. Temperature dependence of the surface tension of 16OAB. Open and solid symbols are data from two different samples

A variety of materials have been investigated in order to find further examples of an anomalous temperature dependence of the surface tension. 16OAB (di-hexadecyl-oxyazoxybenzole) shows a comparable phase sequence, with a transition from the SmC to the SmI phase. SmI is a tilted hexagonal phase. The temperature dependence of σ for this sample is shown in Fig. 6. In the SmC phase, the surface tension decreases with increasing temperature by about $-3.5 \times 10^{-5}\,\mathrm{N\,m^{-1}\,K^{-1}}$. In the vicinity of the bulk phase transition temperature to SmI, the slope changes its sign. Over the tempera-ture range of the SmI phase, the surface tension increases with increasing temperature by $\approx 10^{-4}\,\mathrm{N\,m^{-1}\,K^{-1}}$. At temperatures below 88 °C, the bubble method no longer yields reliable data because the material becomes highly viscous. Tiny temperature fluctuations in the bubble volume are no longer compensated by radius changes of the bubble but lead to large pressure fluctuations around the equilibrium value given by (6). Figure 7 shows experimental data for N-(4-n-pentoxybenzylidene)-4-n-hexylaniline (5O.6) and N-(4-n-heptoxybenzylidene)-4-n-heptylaniline (7O.7), both homologues of the above-mentioned compound 4O.8, with a rich polymorphism of smectic phases. In the temperature range where reliable surface tension data are available (solid squares in the figure), one finds the slopes $7.9 \times 10^{-5}\,\mathrm{N\,m^{-1}\,K^{-1}}$ and $3.3 \times 10^{-5}\,\mathrm{N\,m^{-1}\,K^{-1}}$, respectively.

Reference measurements provide a cross-check of the individual surface tension data. The reference method (Fig. 3) can determine surface tension differences of a few % of the absolute σ values. FELIX 16/100, a commercial smectic mixture (Hoechst) with a broad SmC* phase range, served as the reference substance. The surface tension of this material is shown in Fig. 8, it decreases monotonically at $\Delta\sigma/\Delta T \approx -5.0 \times 10^{-5}\,\mathrm{N\,m^{-1}\,K^{-1}}$, from 22.8×10^{-3} N/m at 24 °C to 20.5×10^{-3} N/m at 70 °C. As in the DOBAMBC sample, a flattening slope in the smectic A phase seems to be indicated. A possible reason for this could be the existence of tilted surface layers in SmA.

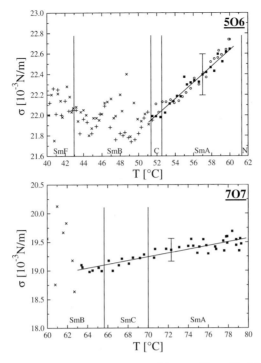

Fig. 7. Surface tensions of the disubstituted benzylidene-anilines 5O.6 (top) [70] and 7O.7 (bottom). Full boxes represent the reliable surface tension data, crosses indicate the data for $pR/4$ in the temperature regions where because of the large viscosity, this value fluctuates around the equilibrium value of σ.

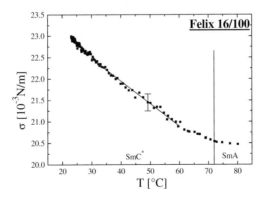

Fig. 8. Temperature dependence of the surface tension of FELIX 16/100

The comparison of 8CB (Fig. 9 left) with the reference yields a relatively temperature-independent ratio of 1.22. Several measurements of the curvature ratios were performed at each temperature (open squares). The mean values for these

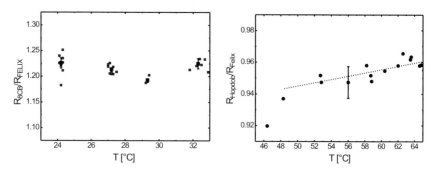

Fig. 9. Ratios of surface tensions determined from reference measurements with communicating tubes. Left: ratio of surface tensions of 8CB vs. FELIX 16/100. Right: HOPDPB vs. FELIX 16/100

ratios, corresponding to the ratio of surface tensions, are represented by the solid squares. The right-hand graph of Fig. 9 shows the surface tension ratio of HOPDOB vs. FELIX 16/100. Their surface tensions differ only slightly at high temperatures (above 50 °C). In that case, the method has the highest sensitivity to small surface tension differences. In the low-temperature part of the graph, the anomalous temperature dependence of HOPDOB can be clearly recognized in the systematic decay of the curvature ratio. The calculated σ for HOPDOB at 46 °C, 19.96×10^{-3} N/m, is considerably lower than the values at 48 °C and 52 °C, 20.30×10^{-3} N/m and 20.33×10^{-3} N/m, respectively. In accordance with the temperature behavior shown in Fig. 5, a monotonic decrease of the surface tension is observed at $T > 50$ °C.

2.4 The Smectic–Isotropic Interface

Isotropic Droplets in Free-Standing Smectic Films

As mentioned on p. 339, the transition of a smectic FSF into the isotropic phase may occur via successive layer-by-layer melting of the central parts of the film. If this process is sufficiently slow, the isotropic material is expelled from the film plane and enters the meniscus, while the film thickness shrinks to a new equilibrium value. Another possible scenario is the trapping of isotropic islands in the smectic film plane [158]. If isotropic material is completely enclosed by the smectic film, it minimizes its surface energy by taking the shape of a thin lens. Since both the isotropic bulk material covered by smectic layers with the smectic coherence length and completely smectic FSFs with thicknesses of less than twice this coherence length are thermodynamically stable, such droplets may represent equilibrium structures above the bulk clearing transition point. This mechanism is certainly not the only possible way leading to droplet formation in FSFs. Similar isotropic, nematic, or cholesteric droplets in smectic FSFs have been observed in a number of situations [158–161]. In particular, they exist in films that are thicker than the smectic coherence length.

In some materials that are characterized by a transition interval SmA–isotropic (polymers, samples containing impurities, mixtures, etc.), the interval where isotropic droplets are found may be rather large. The thermodynamic stability of these objects is not considered here. For the purpose of surface tension studies, it is sufficient that the droplets reach a stationary shape within a fraction of a second, and that they can be observed in that stationary state for hours as long as the film temperature is stabilized. If the film is heated further, one observes the growth of existing droplets as well as the creation of new droplets in the film. On cooling, the droplets shrink and disappear as the film becomes completely smectic again. The three-dimensional shape of such droplets is easily accessible with polarizing microscopy. The horizontal cross section

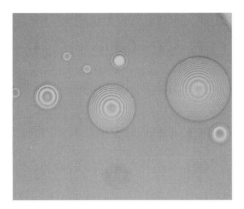

Fig. 10. Isotropic droplets in a region of uniform thickness of a free standing smectic film, observed under monochromatic illumination, $\lambda = 590$ nm. The material is PBOT (3-n-Octyl-5-[4-(4-n-pentyloxybenzoyloxy)benzylidene]-4-oxo- thiazolidin-2-thione), Image size $100\,\mu m \times 80\,\mu m$

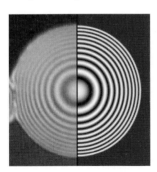

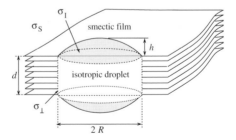

Fig. 11. Optical image of a single droplet (left) and corresponding calculated image (right). Parameters are: wavelength $\lambda = 551$ nm, droplet radius $R = 35\,\mu m$, film thickness $d = 200$ nm and droplet elevation $h = 945$ nm [158]

Fig. 12. Schematic model of an isotropic droplet in a free-standing smectic film and definition of the parameters R, h, and d. The droplet surfaces are spherical caps (possible smectic surface layers on the isotropic droplet are not shown). Note that the vertical scale is unrealistically enhanced

of a droplet (in the film plane) in a region of uniform film thickness is circular. The film thickness is determined from the wavelength-dependent reflectivity

$$R(\lambda) = \frac{4\rho^2/(1-\rho^2)^2 \sin^2 \varphi}{1 + 4\rho^2/(1-\rho^2)^2 \sin^2 \varphi}, \quad \varphi = 2\pi n_0 \frac{d}{\lambda}, \quad \rho = \frac{n_0 - 1}{n_0 + 1} \quad (10)$$

with the refractive index n_0 being equal to the ordinary index of refraction n_o in the smectic A phase; the vacuum wavelength of the normally incident light is λ and the film thickness is d. Since the droplets are sufficiently flat, one can neglect refraction of light at the upper and lower droplet surfaces, and (10) can be employed to determine the thickness profile of the individual droplets. In monochromatic light, one observes regions of equal droplet elevation as concentric dark or bright rings (see Figs. 10 and 11).

Since the isotropic liquid/air interface forms a minimal surface, one can pre-suppose that the droplet surfaces on both sides of the film are spherical. Figure 11 shows one half of the experimental image of an isotropic droplet, observed in 551 nm monochromatic light (left), and the corresponding calculated image (right half). In all optical calculations, it has been assumed that the effective refractive index is $n_0 = 1.5$ in both the isotropic and the smectic phases. Further it is assumed that the droplet shape is that of a biconvex spherical lens (Fig. 12), the droplet radius in the image is $R = 35$ μm. The uniform film thickness around the droplet is $d = 200$ nm and there remains only one fit parameter to adjust the positions of all fringes, the droplet height (elevation above the film surface), $h = 945$ nm. These values yield a radius of curvature of the droplet surface of $R_c = 0.65$ mm, i.e. the use of plane wave optics is justified. Since droplets are created in a broad size range, it is possible

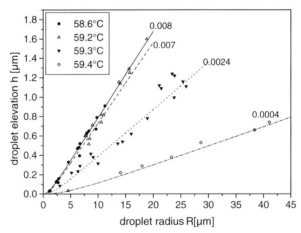

Fig. 13. Droplet elevation h vs. radius R for a smectic CM11B film of 230 nm thickness. The fit curves correspond to the surface tension difference model (see text below), with ratios $\Delta\sigma/(\sigma_S + \sigma_I)$ as indicated. The film thickness is independent of temperature and much larger than the smectic coherence length here (picture taken from [158])

to determine the droplet elevation vs. radius curves at different temperatures. The geometrical droplet shape is controlled by the minimization of the surface and interface energies, and all other influences are negligible. Thus, with a suitable model, information on surface and interface tensions can be obtained from the shape vs. size curves of the droplets. Such a graph of the droplet elevation h vs. radius R for the material CM11B ((2S, 3S)-2-chloro-3-methylpentanoic acid-[4'-(undec-10-enyloxy)-biphenyl]-ester) is shown in Fig. 13. It is obvious that for droplets with lateral extensions large compared with the film thickness, the slope h/R, as the only shape parameter, remains rather constant. In a certain temperature range, the droplet shape parameter is constant, while at high temperatures, towards the clearing temperature of the film, droplets become flatter. Similar observations have been made with a smectogen polymer sample, which was characterized by a clearing interval of about 5 K [158].

Droplet Shape Model

The only parameters that have to be considered here for a description of the droplet shapes are surface and interface tensions. As mentioned above, the curvature elasticity (of smectic boundary layers at the droplet surface) plays no role since the radius of curvature is at least one order of magnitude larger than the droplet size. Unlike the situation in liquid droplets on isotropic liquid films, one cannot use Young's equation to describe the contact angle of the droplet at the film surface. The smectic film in the vicinity of the isotropic droplet remains perfectly flat, all forces connected with surface tensions that act normal to the film plane are compensated by the smectic layers. The free energies connected with the surfaces of the droplet are $E_{\mathrm{I}} = 2\sigma_{\mathrm{I}} \cdot A_{\mathrm{I}}$ for the two spherical caps (shaded areas in Fig. 12), $A_{\mathrm{I}} = \pi(R^2 + h^2)$, and $E_{\perp} = \sigma_{\perp} \cdot 2\pi R d'$ at the peripheral interface between the isotropic and smectic materials. The thickness d' is the actual film thickness, reduced by the thickness of the smectic boundary layers at the droplet surfaces. When the droplet covers the film area $A_{\mathrm{S}} = \pi R^2$, an energy term $E_{\mathrm{S}} = 2\sigma S A_{\mathrm{S}}$ (smectic surface area replaced by droplet) has to be subtracted from the total film area. The surface tensions are introduced as shown in Fig. 12: σ_{I} for the boundary between the isotropic material and air, σ_{S} for the smectic material with respect to air at a surface parallel to the layer plane, $\Delta\sigma = \sigma_{\mathrm{I}} - \sigma_{\mathrm{S}}$, and $\sigma_{\perp}$ for the boundary between the isotropic and smectic material at an interface perpendicular to the smectic layers. Under the assumption of a constant droplet volume $V_{\mathrm{droplet}} = \pi h(3R^2 + h^2)/3 + 2\pi R^2 d$, the total free surface energy is minimized. An analytical solution for the droplet height h vs. radius is

$$h(R) = -\frac{\sigma_{\mathrm{I}}}{\Sigma} d' + \sqrt{\left(\frac{\sigma_{\mathrm{I}}}{\Sigma} d'\right)^2 + \frac{\sigma_{\mathrm{I}} - \sigma_{\mathrm{S}}}{\Sigma} R^2 + \frac{\sigma_0^{\perp}}{2\Sigma} R d'}, \tag{11}$$

where

$$\Sigma = \sigma_{\mathrm{I}} + \sigma_{\mathrm{S}} - \sigma_0^{\perp} \frac{d'}{2R}.$$

The droplet shape is stabilized by the interplay of forces quenching the droplet (the "line tension" around its circumference, connected with $E_\perp$, in a two-dimensional treatment, or differences in the surface tensions σ_I and σ_S) and effects that favor a spreading of the droplet in the film plane (the additional surface of thick droplets with respect to the plane film). For a qualitative discussion of (11), two limiting cases are considered. If the smectic–isotropic interface ($E_\perp$) provides the dominant contribution to droplet confinement, (11) reduces for large droplets ($R \gg d, d'$) to

$$h(R) = \sqrt{\frac{\sigma_0^\perp}{2\Sigma}d'} \cdot R^{1/2}. \tag{12}$$

Since the droplet diameter grows linearly with R while the droplet surface grows with R^2, the effectiveness of the confinement decreases, and large droplets tend to become much more oblate than small droplets, in contradiction to the experimental observations. The second limiting case neglects the smectic–isotropic interface and $\sigma_\perp$, but allows for $\Delta\sigma \neq 0$. In that case, $\Delta\sigma < 0$ (isotropic surface tension larger than that of the smectic material) would lead not to stable droplet shapes but to the spreading of the droplet to $h = 0$. If one assumes $\Delta\sigma > 0$, the limit $\sigma_\perp = 0$ yields

$$h(R) = -\frac{\sigma_I d'}{\sigma_I + \sigma_S} + \sqrt{\frac{\sigma_I^2 d'^2}{(\sigma_I + \sigma_S)^2} + \frac{\sigma_I - \sigma_S}{\sigma_I + \sigma_S}R^2} \tag{13}$$

or, for large droplets ($R \gg d, d'$),

$$h(R) = \sqrt{\frac{\Delta\sigma}{2\sigma_S}}R, \tag{14}$$

where in the last equation, $\sigma_I + \sigma_S$ has been approximated by $2\sigma_S$ (i.e. $\Delta\sigma \ll \sigma_S$). For sufficiently large droplets, one obtains a size-independent shape as found in the experiment, i.e. a linear slope of the $h(R)$ graph, with an offset $-d'/2$ with respect to the origin.

Fit curves of the experimental data are shown as solid lines in Fig. 13. More examples of other smectogens can be found in [158, 162]. One can conclude that all experimental data can be fitted satisfactorily under the assumption that the influences of the smectic–isotropic interface tension at the droplet circumference are negligible with respect to the effects of the surface tension difference $\Delta\sigma$. The value for $\Delta\sigma$ can be extracted from the fit curves, for the material CM11B (Fig. 13), it decays from 1.6% at the lowest temperatures where droplets were found towards zero with increasing temperature. For all materials studied so far [158, 162], values of the same order of magnitude (2–4 %) have been found. Such small differences are far below the resolution of the classical surface tension measurements. From the order of magnitude of $\sigma \approx 2 \times 10^{-2}$ N/m, one can conclude that $\Delta\sigma$ is well below 10^{-3} N/m.

For very large droplets, where the curvature of the smectic–isotropic interface can be neglected, an alternative, somewhat simpler access to the surface tension ratios

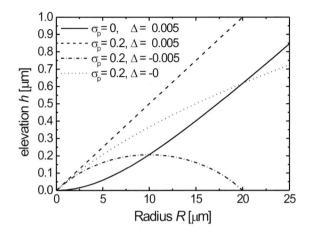

Fig. 14. Droplet shape vs. size calculated from (11) for different combinations of $\Delta\sigma$ and $\sigma_\perp$: solid line, $\Delta\sigma > 0$, $\sigma_\perp = 0$; dashed line, $\Delta\sigma > 0$ plus peripheral interface tension; dotted line, $\Delta\sigma = 0$; and dash–dotted line, $\Delta\sigma < 0$ in combination with positive $\sigma_\perp$. Abbreviations $\sigma_p = \sigma_\perp/\sigma_S$ and $\Delta = \Delta\sigma/\sigma_S$ have been used, and a film thickness of $d' = 1$ μm has been assumed

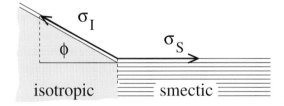

Fig. 15. "Contact angle" of large droplets and equilibrium of in-plane tensions. The force equilibrium implies $\sigma_I \cos\phi = \sigma_S$

of the isotropic and smectic phases can be found. One can employ the balance of in-plane forces to derive the ratio $\Delta\sigma/\sigma_S$. Figure 15 shows a side view of the droplet boundary, only one surface of the film is shown. The in-plane forces are balanced when $\sigma_S = \sigma_I \cos\phi$, where ϕ plays the role of a "contact" angle. Forces normal to the layers are mediated through the film by the smectic structure and balanced by the opposite component on the bottom film surface. The above equation can be written as

$$\sigma_S(1 - \cos\phi) = \Delta\sigma \cos\phi.$$

The term $(1 - \cos\phi)$ is identical to the ratio h/R_c of the droplet. If one substitutes the radius of curvature of the droplet surface, $R_c = (h^2 + R^2)/(2h) \approx R^2/(2h)$, by the measurable droplet radius R, one obtains

$$\frac{\Delta\sigma}{\sigma_S} = \frac{h}{R_c \cos\phi} \approx 2\left(\frac{h}{R}\right)^2, \qquad (15)$$

where the approximation holds for small ϕ (flat droplets). This result is identical to (14).

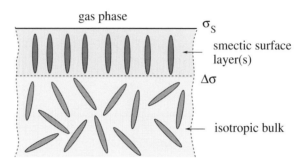

Fig. 16. Model of the isotropic-liquid/gas interface with smectic surface layers

At first sight, a positive $\Delta\sigma$ seems counterintuitive. The density in the isotropic liquid phase is lower than in the smectic phase, and intermolecular forces should be weaker. On the other hand, if one considers that the surface of the isotropic droplet near the smectic temperature range consists in fact of two interfaces, as sketched in Fig. 16, then $\Delta\sigma > 0$ can be explained naturally. The surface tension contribution related to the liquid/gas boundary is identical to σ_S, and the additional interface between the smectic boundary layers and the isotropic core contributes $\Delta\sigma$ to the total surface tension σ_I. From this conclusion, one may speculate that $\sigma_\perp$, the interface tension between the smectic and the isotropic liquid at the phase boundary perpendicular to the layer planes is of the same order of magnitude as $\Delta\sigma$, the tension of the interface parallel to the layers. If $\sigma_\perp$ is of the order of a few percent of $\sigma_{I,S}$, the corresponding terms in (11) can indeed be neglected for droplets with $R \gg d'$; it leads to a slight vertical shift of the $h(R)$ curve upwards but leaves the linear slope unchanged (cf. Fig. 14). The role of $\sigma_\perp$ is to maintain the circular shape of the droplet cross section in the film plane when the droplet is completely embedded in a plateau of constant film thickness.

Droplets in Films of Inhomogeneous Thickness

So far, droplets enclosed by a uniformly thick film area have been considered. Such droplets not only have a stationary shape and size, but moreover are at rest (except for a random, "Brownian" motion) with respect to the surrounding smectic film. In regions of inhomogeneous thickness, droplets experience a directed force towards the thicker film region. This effect is comparable to the observations made by Sur and Pak in isotropic films [163], where solid inclusions are driven by capillary forces in the direction of the local film thickness gradient. In films with large thickness gradients (the average distance of adjacent thickness steps is small compared with the lateral dimensions of the droplets) this motion is quasi-continuous; when the layer steps are separated by approximately the droplet size, the droplets move in discrete jumps towards the thicker film regions until they reach a plateau in the film or the meniscus. The physical origin is the wetting of the droplet by the surrounding smectic film material. A qualitative evaluation of the capillary forces driving this motion can be made from a close inspection of films at discrete film thickness steps. The boundary of two uniformly thick regions forms a trap, since droplets formed at such steps remain pinned there, and droplets that have formed in thinner film regions and that get in contact with a film thickness step are first drawn into the thicker plateau and then remain pinned at the step (Fig. 17). As a consequence, even monolayer thickness steps in the films are decorated by lines of droplets. At small thickness steps, the droplets are almost completely embedded in the higher film plateau next to the step.

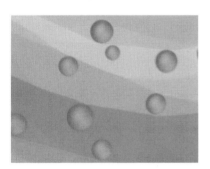
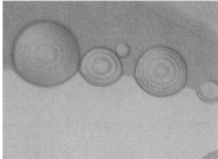

Fig. 17. Left: isotropic droplets in an inhomogeneously thick film (white light image, CM11B). Regions of different film thicknesses are distinguished by interference colors (grayscales in the image). All droplets line up at small thickness steps (one or a few smectic layers high), on the upper plateau. Image size 77 μm $\times$ 60 μm. Right: droplets trapped at a high step ($\approx$ 90 nm), image taken in monochromatic light (PBOT). Image size 160 μm $\times$ 115 μm, $\lambda = 550$ nm. Droplet cross sections deviate considerably from circular shape

Figure 18 shows a droplet at the edge of a 92 nm step in the film height; the droplet radius is 5 μm, and its height h above the upper plateau is 0.5 μm. Note that in the image, the z-axis normal to the film surface is drastically enhanced. The forces pinning the droplet can be qualitatively estimated in a two-dimensional approximation of the geometry. In the model, we consider that the surface tension difference $\Delta\sigma$ which is responsible for the confinement of the droplet can be neglected when only the in-plane geometry is considered. We assume that the droplet consists of a disk (or is composed of sections of disks) in the film plane, and consider the forces on the outer boundaries of the 2D structure (Fig. 19).

In the 2D model, the interface tensions along the smectic–isotropic boundary reduce to "line tensions" γ_i. We shall assign the line tension γ_1 to the droplet interface in the thick film region (thickness d_1), and γ_2 to the interface in the thin film region

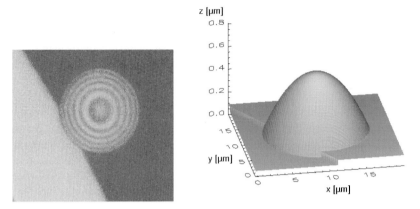

Fig. 18. Single isotropic droplet at the edge (92 nm high) of a plateau in an FSF: optical image in reflection, $\lambda = 550$ nm (left), and height profile of the top film surface corresponding to the image (right). Note that in the calculation, symmetry between the top and bottom of the film was assumed

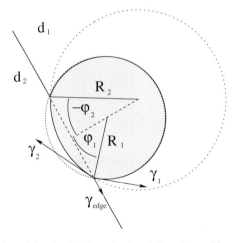

Fig. 19. 2D presentation of the droplet (top view) and directions of forces connected with the line tensions γ_i

(thickness d_2); γ_{edge} attributes a tension to the film thickness step. Experimentally, one finds that the droplet boundaries within the uniform regions are to a good approximation circular, with radii of curvature R_1, R_2. In all cases where these two radii are resolvable in the experiment, one finds $R_2 > R_1$. The equilibrium of forces necessary for the stationary droplet position and shape requires that the components of the forces acting on the edge and on the interfaces vanish. From Fig. 19, one finds

$$\gamma_1 \sin \varphi_1 = \gamma_2 \sin \varphi_2, \quad \text{(forces normal to the edge),} \quad (16)$$

$$\gamma_{\text{edge}} + \gamma_1 \cos \varphi_1 = \gamma_2 \cos \varphi_2, \quad \text{(forces along the edge),} \quad (17)$$

$$R_1 \sin \varphi_1 = R_2 \sin \varphi_2, \quad \text{(trigonometry).} \quad (18)$$

A combination of the first and the last equation provides the relation between the radii $R_{1,2}$ and the line tensions $\gamma_{1,2}$:

$$\frac{\gamma_2}{\gamma_1} = \frac{R_2}{R_1},$$

and the second equation transforms to

$$\frac{\gamma_{\text{edge}}}{\gamma_1} = \sqrt{(R_2/R_1)^2 - \sin^2 \varphi_1} - \cos \varphi_1.$$

The geometrical quantities $R_{1,2}$ and φ_1 are accessible in the experiment, at least at high thickness steps, and allow one to estimate ratios of the γ_i. In the limit of $\Delta R = R_2 - R_1 \ll R_1$ (small step height) and small $\varphi_{1,2}$, a good approximation of the latter equation is $\gamma_{\text{edge}}/\gamma_1 \approx \Delta R/(R_1 \cos \varphi_1)$. From the droplets shown in Fig. 17 (right), a value of $\gamma_{\text{edge}} \approx 0.3\gamma_1$ was obtained for all droplets. The smectic film is ≈ 1.65 μm thick, and the step height is 5.5% of the total film thickness. The radius R_2 is about 10% larger than R_1 (rather independent of the droplet size); this difference provides the capillary force that moves the droplet into the upper plateau. If one equates γ_1 naively with the film height d_1 multiplied by the interface tension $\sigma_\perp$, and γ_{edge} with the height of the thickness step $d_1 - d_2$ multiplied by a surface tension of the smectic material with respect to air at a boundary perpendicular to the layers, $\sigma_{S\perp} \approx \sigma_S$, the ratio is $\sigma_S \approx 5.6\,\sigma_\perp$. However, this oversimplified assignment certainly does not describe the physical situation correctly. A systematic investigation of droplets at steps of different heights is necessary to understand the physical origin of the tensions related to layer steps and isotropic–smectic interfaces in FSFs.

3 Gas Permeation Through Thin Smectic Films

A second aspect of free-standing smectic films where surfaces are involved in a specific way is their gas permeability. The penetration of gas molecules through thin liquid films is essential for many biological, technological and physico-chemical processes (see, e.g. [164–166] and references therein). As an example, gas exchange in lung tissue is based on the permeation of oxygen and carbon dioxide through biological membranes, in foam-like structures [167]. The stability of foams is determined to a large extent by the permeability of the thin liquid films encapsulating the foam blisters. "Microbubbles" have gained relevance in medical applications as potential drug and gene delivery containers, ultrasound contrast agents, and blood substitutes. They are composed of gaseous cores enclosed in organic membranes. The gas permeation resistance of the membranes is one of the important factors for their dissolution behavior [168]. Finally, (polymer) membranes used in technological gas separation have an important application potential [169].

The permeation of gas molecules through a liquid membrane is a combined sorption–diffusion process that acts to equilibrate pressure gradients (or differences in partial pressures in gas mixtures) across the membrane. Since thermally activated transport processes in the membrane are essential, the permeability of liquid membranes usually increases with temperature. Measurements of gas permeation through liquid membranes have been reported for a variety of systems (e.g. [170–175]). So far, most experimental investigations have been performed with polymeric membranes (e.g. [176–180]). For quantitative measurements, special membrane techniques have been developed [181–183]. In many of the experiments, the basic interest lies in the selectivity of the membranes for gas permeation [184–191] in view of their application for gas separation. The following section will primarily deal with the influences of surfaces on membrane permeability, and thus we shall focus on measurements in thin (submicrometer) films.

For the stabilization of thin liquid films, two strategies can be applied. Either the films are deposited on a porous support or the liquid material is confined in a porous matrix, or the films are stabilized by an internal layer structure or electric double-layer forces between the surfaces. A classical example is Newton black films (NBFs), which consist of two monolayers of amphiphilic surfactants adsorbed on each other without a water layer in between. Common black films contain an additional layer of water, and the surfaces are stabilized by electrical forces. The thickness of the liquid layer can be controlled by the ion concentration in the core electrolyte [192]. A number of studies of gas permeation through monolayers of surfactants and Langmuir–Blodgett (LB) films have been published in recent years (see [166, 192–195] and references therein). LB films may be particularly useful for gas separation because of their extremely low thickness, i.e. their high permeation rate, and the well-ordered structure, which may favor a highly selective transport of gas particles through the membrane. For better stability, such films can be deposited on porous support material [166]. It has been demonstrated that the permeation across very thin surfactant films (NBFs) shows an unexpected thickness dependence. The permeability of thicker films can be larger than that of a thinner NBF, because of an increased adsorption density of the surfactants in the latter systems [192–195].

In this section, the first experiments measuring gas permeation through smectic free standing films are described. The latter represent well-defined, simple model membranes, stable and robust enough for mechanical experiments. Their thickness can be controlled to a certain degree by the preparation conditions, and, unlike LB films, the internal structure is rather homogeneous (although more highly ordered boundary layers may exist in some phases).

The gas flow through smectic films is measured as a function of film thickness and temperature, and a molecular model is presented that reproduces standard permeation theory for thick films but explains, in addition, deviations from the classical sorption–diffusion model that are observed in thin films. While gas permeation into the bulk liquid or through thick liquid layers is basically determined by the pressure-dependent solubility and diffusivity, the permeation through ultrathin films behaves qualitatively differently. Deviations from the static equilibrium concentrations (Henry's law) of gas molecules near the film surfaces are observed. A dynamic equilibrium is established between the diffusion stream through the film and the intrusion of gas particles from the gas phase into the film.

3.1 Measurements of Gas Permeation Through Smectic Films

Preparation

Smectic bubbles represent ideal systems for gas permeation measurements, their surface and excess pressure can be easily controlled and the gas volume flow through the film can be determined straightforwardly from the bubble volume. Smectic bubbles can be prepared within a broad range of uniform film thicknesses. As the gas contained in the bubble penetrates the smectic film, the radius is recorded during its shrinking as a function of time. If the surface tension of the smectic material is known, it is sufficient to record only one parameter (the bubble radius R or the excess pressure Δp) to obtain the permeation characteristics. In order to avoid confusion of absolute pressures and pressure differences, the Laplace pressure between both sides of the curved film (p in Sect. 2) will be denoted in the following by Δp.

Smectic bubbles were inflated as described in the previous Sect. on a thin glass capillary. The opposite end of the capillary was then closed so as to be airtight, so that the enclosed gas particles from inside the bubble could leave the bubble volume only by migrating through the smectic film. The test of the device and estimation of the gas loss through other parts of the apparatus could be performed with a thick bubble (film thickness in the micrometer range). The interesting range for measurements of the film thickness dependence of gas permeation is from two smectic layers to about 100 nm. The bubble radius was determined by digital processing of transmission images, taken with a conventional digital camera with a macro objective, and the excess pressure was monitored in addition. Since the excess pressure in the bubble was rather small and its measurement represented the largest statistical error source, it was convenient and more accurate, however, to calculate the excess pressure from the bubble radius by means of (6), and to use the pressure gauge only for an independent confirmation of the data. The surface tension was determined in independent experiments (see previous sections) from the average of multiple individual measurements of the radius/pressure relation.

Optics

The thickness of the smectic films was determined from transmission images in monochromatic light. If one considers the transmission and reflection coefficients of monochromatic light at the film surfaces and interference with multiply reflected light at the film surfaces, one finds the transmission profile [150, 196] of a thin optically uniaxial film (optical axis normal to the film surface) in an arbitrary orientation θ to the transmitted light beam, in an analogous way to (10). Since for an oblique film the

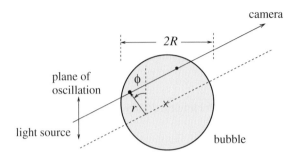

Fig. 20. Observation of bubbles in transmission: r is the distance between the observation axis through the bubble center and a selected beam; ϕ is the angle between the polarization of the beam (oscillation plane) and the projection of $\boldsymbol{r}$ on a plane normal to the propagation direction

polarization direction of the incident light matters, two intensities $I_{\parallel}$ and $I_{\perp}$ describe the transmission for light polarized in the plane of incidence and perpendicular to it, respectively. In the bubble geometry, the image is constructed from the local transmission intensities (analogously to the construction of Newton fringes). For a bubble with radius R and a beam at a distance r from the axis through the bubble center (Fig. 20), the incident angle on the film (with respect to the layer normal) is $\theta_0 = \arcsin(r/R)$, and the polar angle of the beam propagation in the film (after refraction at the film surface), θ_1, can be determined from θ_0 by Snell's law [150]. The normalized transmission intensity (incident intensity I_0 through one hemisphere is

$$T_{\parallel,\perp}(r) = I_{\parallel,\perp}(r)/I_0 = \frac{1}{1 + 4\rho_{\parallel,\perp}^2/(1 - \rho_{\parallel,\perp}^2)^2 \sin^2(2\pi n_o d \cos\theta_1/\lambda)}, \quad (19)$$

where $\rho_{\parallel}(r) = \tan(\theta_0 - \theta_1)/\tan(\theta_0 + \theta_1)$ and $\rho_{\perp}(r) = \sin(\theta_0 - \theta_1)/\sin(\theta_0 + \theta_1)$. The beam passes through the bubble twice, so that thickness differences between the rear and front hemispheres of the bubble cannot be resolved. In most experimental situations, however (see Fig. 21), the thickness of the rear and front parts are the same when the bubble is observed in a horizontal direction. Discrete steps, or gradients, in the film thickness are sometimes formed in the vertical direction, owing to gravity. In most bubbles, the film thickness is uniform over the complete sphere, and only such bubbles have been considered for the permeation experiments. Then, one can treat the transmittivities of the rear and front hemispheres as independent of each other, and one obtains the intensities $T_{\parallel}^2(r)I_0$ and $T_{\perp}^2(r)I_0$ for the local transmission through the bubble. The bubble image (under polarized light), as a two-dimensional projection, can be constructed from the local intensities, $I(r, \phi) = [T_{\parallel}^2(r)\cos^2\phi + T_{\perp}^2(r)\sin^2\phi]I_0$, where ϕ is the angle in the projection plane with respect to the oscillation plane of the beam. When a uniformly thick bubble is observed in nonpolarized light, all radial directions are equivalent and the image $I(r) = [T_{\parallel}^2(r) + T_{\perp}^2(r)]/2\,I_0$ has circular symmetry. Figure 21 shows two

Fig. 21. Experimental (left) vs. simulated (right) images of smectic bubbles. Top: 1.45 μm thick homogeneous bubble, with circular rings reminding one of Newton fringes. Bottom: bubble with three regions of different thicknesses, 24 nm, 70 nm, and 460 nm from top to bottom ([150])

examples of measured and calculated bubble images. The bubble diameters are of the order of 1 cm.

Determination of the Gas Loss

When the bubbles are larger than hemispheres, which is the usual experimental situation, an interesting peculiarity of the system is that during gas loss, the excess pressure inside the bubble *increases*. Gas permeation per film area therefore becomes more efficient with decreasing bubble volume, while the bubble surface decreases. The timescale for the shrinking process depends crucially on temperature and film thickness. Thin (two-layer) bubbles can shrink completely within a few minutes, while bubbles of micrometer-thick films at low temperatures persist for several days in the laboratory. Before a relation to the usual permeation coefficients is established, we develop a phenomenological description of the bubble shrinkage due to gas loss [197]. For a complete sphere, one can derive a simple analytical formula for the radius shrinkage dynamics. The loss of gas volume contained in the bubble and the capillary, $\dot{V} = 4\pi R^2 \dot{R}$, is proportional to the smectic-film area $S \approx 4\pi R^2$ and to the pressure difference $\Delta p = 4\sigma/R$ (Laplace pressure, (6); see remark on p. 363) between both sides of the smectic film. A proportionality factor $\gamma(d, T)$ is introduced as a phenomenological parameter, which will later be related to molecular quantities:

$$\dot{V} = -\gamma S\,\Delta p \tag{20}$$

Substituting the volume, surface area, and pressure difference, one finds the bubble radius dynamics to be

$$4\pi R^2 \dot{R} \approx -\gamma \cdot 4\pi R^2 \cdot \frac{4\sigma}{R}. \tag{21}$$

The solution of this differential equation is

$$R^2(t) \approx R_0^2 - 8\sigma\gamma t = R_0^2 - gt, \quad \text{where} \quad g(d,T) = 8\sigma\gamma \tag{22}$$

and the initial radius is $R_0 = R(0)$. This square dependence holds as long as the bubble is large compared with the radius of the capillary opening r_c. A first-order correction takes into account that the area of the capillary opening must be subtracted from the bubble surface. An approximation by a disk of radius r_c yields the corrected surface $S = 4\pi R^2 - \pi r_c^2$ and the analytical solution

$$R^2 + \frac{r_c^2}{4}\ln\left(R^2 - \frac{r_c^2}{4}\right) = R_0^2 + \frac{r_c^2}{4}\ln\left(R_0^2 - \frac{r_c^2}{4}\right) - gt. \tag{23}$$

Figure 22 shows the two curves for the shrinkage of the bubble radius. In the experiments, the initial radius of the bubbles was chosen large enough, $R_0 > 3r_c$, and (23) was used to obtain the only fit parameter g from the fit of the $R(t)$ curve. Gas permeation through the smectic film can be unambiguously distinguished from gas loss through a potential leakage in the apparatus, since the radius change in the latter case is $R^4 - R_0^4 \propto t$.

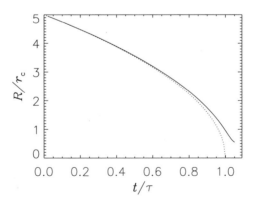

Fig. 22. Bubble shrinkage by gas loss: approximate (22) (dotted line) and corrected (23). Except for $R \approx r_c$ (bubble almost shrunk to a hemisphere), the graph of (23) is practically identical to the exact (numerical) solution. $1/\tau = g = 8\sigma\gamma$ is the gas permeation rate

3.2 Experimental Results

The results of permeation measurements with the method described above [197] are shown in Fig. 23 as a function of film thickness for two standard materials, HOPDOB and DOBAMBC (see pp. 346 and 348 and Tab. 1 on p. 375). Each point is obtained from an individual bubble with the fit of the bubble radius change to (23). Films have been measured in a thickness range up to $\approx 1\ \mu m$, but only results for thin films are shown; the thick film data follow a $1/d$ dependence as expected from the sorption–diffusion model of (27). The dashed lines in the figures symbolize

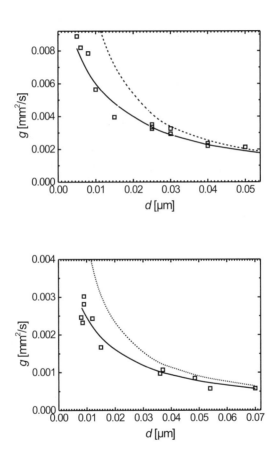

Fig. 23. Gas loss (air) through thin films of HOPDOB at 82 °C (top) and DOBAMBC at 85 °C (bottom). The parameter g is defined in (22). The solid curves are fits to $g = G_0/(d - d_0)$; the phenomenological fit parameters are $d_0 = 8.0\ nm$, $G_0 = 1.1 \times 10^{-16}\ m^3/s$ (HOPDOB), and $d_0 = 8.4\ nm$, $G_0 = 4.5 \times 10^{-17}\ m^3/s$ (DOBAMBC). Dashed curves mark a $1/d$ characteristic as expected from (27)

this inverse proportionality to the film thickness d. As can be seen in both graphs, the experimental data deviate considerably from the model curve in the thin-film range (< 10 smectic layers) to systematically lower values. At first sight, this is in qualitative agreement with measurements on Langmuir films (e.g. [195]). However, another explanation has to be found for the smectic material than the increased adsorption density of surfactants at the film surfaces [192]. Although the induction of orientational or positional order at the smectic/air interface is a well-established observation (see p. 339), there is no indication that the permeability of the outer layers of smectic films differs from the interior. If, ad hoc, an extrapolation length d_0 is introduced, and the film thickness dependence of g is described by an equation of the form

$$g(d) = \frac{G_0}{d + d_0},\qquad(24)$$

the experimental data are consistently represented by the fit curves (solid lines in Fig. 23). The interpretation of this model in terms of molecular parameters is attempted in the next section. At film thicknesses $d \gg d_0$, the sorption–diffusion model describes the experimental data adequately and G_0 can be related directly to the parameters H and D of the sorption–diffusion model (see (27) on p. 370).

The temperature dependence of the gas permeation rate is shown in Fig. 24 for both materials, HOPDOB and DOBABMC. Each individual data point is a fit of the shrinkage of R to (23). In order to avoid experimental uncertainties of the film thickness measurements when data at different temperatures are compared, all data on HOPDOB were obtained from the same bubble, which was repeatedly reinflated after a sufficiently long shrinking process for one temperature had been recorded. In the DOBAMBC graph, data obtained from two bubbles have been combined. In both cases, the permeability shows an Arrhenius-like temperature dependence over a large temperature range. This is not astonishing, since it involves diffusion as a thermally activated process. Typical activation energies of the diffusion constants in smectics for diffusion of small tracer molecules normal to the smectic layers are of the order of 100 kJ/mol [198]. In the plots of Fig. 24, Arrhenius curves $g(T) \propto \exp(-E_a/(N_A k_B T))$ with $E_a = 70$ kJ/mol (HOPDOB) and $E_a = 59$ kJ/mol (DOBAMBC) are included for comparison (k_B is the Boltzmann constant). It seems that far from the transitions into the low-temperature smectic phases, within the temperature ranges of the in-plane liquid-like SmC (SmC*) and SmA phases, these curves describe the thermal behavior quite accurately. At low temperatures, there is a systematic deviation of the permeability to lower values. The reason for this decreasing permeability has to be sought by systematic measurements of the temperature-dependent permeability of films with different thicknesses. Possible structural changes that may influence the temperature curve of the permeability could be the formation of more highly ordered surface layers, with a hexagonal packing of the molecules. This would have an impact on the solubility of gas molecules in the smectic material, or some pretransitional intralayer ordering of the mesogens. In order to distinguish between the effects of such surface layers and mechanisms that may hinder the gas transport in the bulk (i.e.

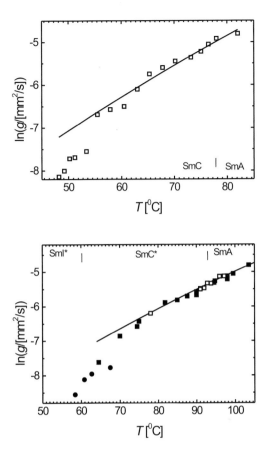

Fig. 24. Temperature dependence of the gas loss (air) through thin films of HOPDOB (top) and DOBAMBC (bottom). The parameter g is defined in (22). The solid squares and open squares were obtained during stepwise cooling of two different bubbles; the circles are data measured in an upward temperature scan, where the bulk material (and possibly the film as well) was in the SmI* phase. In both data sets, bubbles with film thicknesses of 11 ± 1 nm have been chosen

hexagonal packing in the SmI* phase of DOBAMBC), a comparison with thicker films is needed.

The most interesting question for technological applications, selectivity for differ-ent gases, has not been investigated within the experiments with smectics performed so far. All quantitative experiments have been performed with air. For a qualitative comparison, a few measurements of helium-filled bubbles were set up. The perme-ation rate determined from the shrinkage of helium-filled DOBAMBC bubbles was found to be faster by a factor of approximately four compared with air permeation. It is not surprising that the much smaller helium atoms penetrate the liquid film much faster than air molecules. In the experiments with helium, the bubbles were inflated in air, so that actually one has to take into account the time-dependent partial pressures of air and helium in the bubble. An exact quantitative study would require that the bubbles were inflated in a helium atmosphere. At the present state, it seems that the low excess pressure that can be achieved limits any prospects for technological applications. It seems that a possible way to circumvent this problem might be the preparation of smectic foams instead of single membranes, but such systems have not been investigated so far, and no studies of the stability of smectic foams have been reported.

3.3 Permeation Through Thick Liquid Layers

In the limit of thick layers, the gas concentrations on both surfaces of the film are practically equal to the pressure-dependent static equilibrium concentrations given by Henry's law. The concentrations ξ_i ($i = 1, 2$) of gas particles in the film near surfaces 1 and 2 are given by

$$\xi_{1,2}(p) = \frac{1}{H} p_{1,2} \tag{25}$$

where the ξ_i are related to the density ρ_L and molar mass M_L of the liquid and the gas particle density per volume, n_i, by

$$\xi_i = \frac{M_L}{\rho_L N_A} n_i, \tag{26}$$

H is Henry's constant, and p is the partial pressure of the gas near the liquid surface. The concentration gradient $(\xi_1 - \xi_2)/d$ across a liquid film of thickness d causes a diffusive flow of gas particles through the film. Since the exchange of gas particles between the liquid film and the air is fast compared with the diffusion through thick films, the concentrations $\xi_{1,2}$ are not influenced by the diffusion process. The permeation flow $j = D(n_1 - n_2)/d$ of gas particles per film area

$$j = \frac{D}{H} \frac{\rho_L N_A}{M_L} \cdot \frac{\Delta p}{d} \quad \text{where} \quad \Delta p = p_1 - p_2, \tag{27}$$

at a given pressure gradient and film thickness is thus determined by both the sorption of gas particles and their diffusion in the liquid. From (27), it follows that the permeation of gases through thick liquid films is inversely proportional to the film thickness. (The sorption and diffusion processes in polymer films, for example, may involve different competing mechanisms (Langmuir sorption and percolation [199, 200]), and the permeability may be more complex, for example, it may depend explicitly upon pressure).

3.4 Molecular Model for Thin Films

The standard permeation model considers sorption of gas in the liquid and the diffusion of gas molecules in the film, and an inverse proportionality between gas permeation and film thickness is the consequence. It has been mentioned in the previous section that the predictions of the simple sorption/diffusion model are qualitatively correct for thick films but that there are characteristic deviations for films of a few nanometers thickness. The fit curves for the permeation parameter $g(d)$ reach a saturation value G_0/d_0 of the order of 10^{-8} mm^3/s for $d \to 0$. This value represents some permeation limit for the infinitely thin film. The limiting factor can be neglected in micrometer-thick films, where diffusion in the liquid is the bottleneck for the gas particle transport, and the gas concentrations on the two surfaces of the penetrated membrane are given by Henry's law.

If one looks for an adequate model that reproduces (24), one has to include two dynamic effects. First, the molecular diffusion of gas particles in the liquid film leads to a flow of the penetrant between the two membrane surfaces. Second, the penetrant exchanges with the surrounding gas phase through the surface of the liquid. In the limit of infinitely thin films, the second process will always take the dominating role, since the dynamic exchange of gas particles between the liquid and gas phases in nonequilibrium is not arbitrarily fast.

The model introduced here describes the permeation through a plane membrane [197] (the curvature of the smectic film can be neglected on a molecular scale), it can therefore be treated one-dimensionally. Outside the film, the concentration of gas particles per volume on each side is homogeneous (uniform pressures inside and outside the bubble). We consider only one species of penetrant. Thus, the model will not exactly correspond to the experimental situation where air permeation was measured. In principle, if several different penetrants are involved, all pressures have to be replaced by the partial pressures and a set of dynamic equations for each species of penetrant has to be considered. The dynamic equations are derived in terms of the particle densities n_i ($i = $ u, 1, 2, d), giving the number of gas molecules per volume in four relevant locations in the system. The particle density *outside* the liquid film on the upstream side (inside the bubble) is given by n_u, and the gas pressure is p_1. Likewise, the particle density *outside* the liquid film on the downstream side of the layer is given by n_d, and the gas pressure is $p_2 = p_1 - \Delta p < p_1$. The relations between $n_{u,d}$ and $p_{1,2}$ are

$$n_{u,d} = \frac{p_{1,2}}{k_B T}. \tag{28}$$

Inside the film, n_1 and n_2 describe the number of adsorbed gas particles per volume near the film surfaces, and a linear gradient from n_1 to n_2 is assumed (Fig. 25). For the particle exchange rates per film area between these concentrations n_i, we make the following ansatz:

$$j_{\text{diff}} = j_{1 \to 2} = D(n_1 - n_2)/d$$

(diffusion in a linear concentration gradient in the liquid). The coefficient D is the diffusion constant of the penetrant, in the direction perpendicular to the smectic layers. We have

$$j_{\text{in}} = j_{u \to 1} = \alpha(n_u - hn_1)$$

(solvation dynamics at the upstream surface) from the higher pressure gas reservoir into the film, and

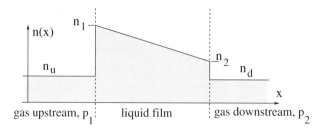

Fig. 25. Geometry of the model and definition of the concentrations n_i.

$$j_{\text{out}} = j_{2\to d} = \alpha(hn_2 - n_d)$$

from the film into the downstream gas reservoir. The coefficient α determines the exchange rate between the dissolved and free gas concentrations. The constant h determines the ratio of the gas particle concentrations in the gas phase and in the liquid near the surface. In equilibrium, $p_1 = p_2$, $n_u = n_d$, $j_{\text{in}} = j_{\text{out}} = 0$, and Henry's law must be recovered, hence the constant h must be related to H by

$$h = \frac{M_L}{N_A k_B T \rho_L} \cdot H. \tag{29}$$

Since in a stationary flow n_1 and n_2 are constant, $j_{\text{in}} = j_{\text{diff}} = j_{\text{out}} = j$ is required, and one obtains

$$\alpha(n_u - hn_1) = \frac{D}{d}(n_1 - n_2) = -\alpha(n_d - hn_2) \tag{30}$$

and, when the first and the second equation are added,

$$\alpha(n_u - n_d) = \left(h\alpha + \frac{2D}{d}\right)(n_1 - n_2). \tag{31}$$

In the right-hand side of this equation, $n_1 - n_2$ can be replaced by $j_{\text{diff}}/\gamma = j/\gamma$, and the permeation flow through the film is given by

$$j = \frac{\alpha D/d(n_u - n_d)}{h\alpha + 2D/d}. \tag{32}$$

Substitution of (28) and (29) yields

$$j = \frac{D}{(dh + 2D/\alpha)} \cdot \frac{\Delta p}{k_B T} = \frac{D}{H} \frac{N_A \rho_L}{M_L} \frac{1}{(d + 2D N_A \rho_L k_B T(\alpha H M_L)^{-1})} \Delta p \tag{33}$$

for the number of gas particles permeating the film per unit time and area. First, one acknowledges that in the limit of thick films (large d, i.e. $\alpha H M_L d \gg 2N_A \rho_L k_B T$), the second term in the denominator of (33) can be neglected and (27) is recovered. In that limit, the term $(2D/d)$ in (31) can be dropped and $n_{1,2}$ are identical to the equilibrium values given by (25) and (26).

With the rate $S \cdot j$ being the number of gas particles passing a liquid film of surface area S per time, the volume loss in the upstream reservoir is given by

$$\dot{V}_\mathrm{u} = -S j \frac{k_\mathrm{B}T}{p_1},$$

where we have tacitly assumed that the pressure dynamics (due to the coupling of R and Δp in (6) is slow and can be neglected in the determination of the momentary flow rates. A comparison with the definitions of γ (20) and g (22),

$$\dot{V}_\mathrm{u} = -\frac{S g}{8 \sigma} \Delta p,$$

leads to the assignment

$$g = \frac{8\sigma k_\mathrm{B}T}{p_1} \frac{D}{H} \frac{N_\mathrm{A}\rho_\mathrm{L}}{M_\mathrm{L}} \frac{1}{(d + 2DN_\mathrm{A}\rho_\mathrm{L}k_\mathrm{B}T(\alpha H M_\mathrm{L})^{-1})}, \qquad (34)$$

and it is obvious that in this model, g has the same structure as in (24) with the assignments

$$G_0 = \frac{8\sigma k_\mathrm{B}T}{p_1} \frac{D}{H} \frac{N_\mathrm{A}\rho_\mathrm{L}}{M_\mathrm{L}} \qquad (35)$$

and

$$d_0 = \frac{2DN_\mathrm{A}\rho_\mathrm{L}k_\mathrm{B}T}{\alpha H M_\mathrm{L}}. \qquad (36)$$

3.5 Discussion and Access to Material Parameters

As can be seen from the last two equations, the model describes the experimental observations qualitatively correctly. Material parameters can be extracted from the measured $g(d)$ characteristics (Fig. 23). The two equations (35) and (36) relate the experimental data to three unknown quantities, the diffusion constant D, Henry's constant H, and the rate constant α. First, it is reasonable to evaluate the ratio G_0/d_0 of the fit parameters. Experimentally, this value corresponds to the extrapolation of the $g(d)$ curve to $d \to 0$. Both of the unknown parameters D and H are eliminated in this ratio, and one can access α directly,

$$\alpha = \frac{G_0}{d_0} \cdot \frac{p_1}{4\sigma}. \qquad (37)$$

For the two mesogenic materials investigated in the experimental Sect. 3.2, α is evaluated as 17×10^{-3} m/s (HOPDOB) and 6.6×10^{-3} m/s (DOBAMBC), respectively. The interpretation of α has to be sought in a molecular theory describing the gas exchange dynamics between the liquid and gas phase. Here, we shall use the simple model that gas particles in the gas phase follow a Maxwell–Boltzmann velocity distribution, and they hit the liquid surface S at a rate $S\alpha'n$ (where $n =$

$p/k_{\mathrm{B}}T \approx n_{\mathrm{u}}, n_{\mathrm{d}}$). If v_x is the component of the thermal velocity of a gas particle normal to the surface and μ is the mass of the gas molecule, the Maxwell–Boltzmann velocity distribution integrated over the two velocity components v_y, v_z gives

$$\Pi(v_x) = \sqrt{\frac{\mu}{2\pi k_{\mathrm{B}}T}} \cdot \exp\left(-\frac{\mu v_x^2}{2k_{\mathrm{B}}T}\right),$$

and $\alpha' = \int_0^\infty v_x \Pi(v_x)\mathrm{d}v_x$ is evaluated as

$$\alpha' = \sqrt{\frac{k_{\mathrm{B}}T}{2\pi\mu}}.$$

This value is equivalent to 1/4 of the average thermal velocity $\bar{v}$. The discrepancy between $\alpha' = 128$ m/s (nitrogen, 350 K) and the experimental α is a factor of the order of about 13×10^{-5} (HOPDOB) and $\approx 5 \times 10^{-5}$ (DOBAMBC). If one excludes the possibility that the limiting factor for gas permeation of thin smectic films is in the surface layer, and there is no indication of a different packing density or other factors that might hinder the gas transport in smectic surface layers in excess to the bulk material, then one may attribute this factor to an energy barrier at the smectic/gas interface. Barrier heights of 26.4 kJ/mol and 29 kJ/mol, respectively, for the two materials would be sufficient to explain the ratio of the order of 10^4 between the number of gas particles reaching the surface and those entering the liquid film per time interval. A possible way to obtain more reliable information about this limiting factor α/α' could be obtained by measurements of the pressure dependence of $g(d)$, particularly of G_0/d_0, and of the sorption dynamics of gases in smectic phases.

Since α does not enter the coefficient G_0, it is reasonable to substitute all known parameters in G_0 to access the ratio D/H, independent of the above discussion of the rate α. This value can be determined with great precision, since the $1/d$ characteristics of the $g(d)$ curve can be measured over a large range of film thicknesses in the limit $d \gg d_0$. For the material HOPDOB, after substitution of all known material parameters, a value $D/H = 10 \times 10^{-18}$ m^2 s^{-1} Pa^{-1} is found at the temperature 82 °C. The corresponding ratio for DOBAMBC at 85 °C is 4.4×10^{-18} m^2 s^{-1} Pa^{-1}.

It is desirable to compare these findings with known experimental data from the literature. However, only a few experimental data are available for the solubility and diffusion of gases in liquid-crystalline materials. One cannot expect an accurate estimate for the diffusion coefficients of the penetrants N$_2$ and O$_2$, since their solubilities may differ substantially. Both of the parameters D and H have been determined individually for nitrogen and the nematogenic material MBBA (4-methyloxy-benzylidene-4'-N-butylaniline) [201]. From the high pressure data reported in [201], one can estimate a value $H \approx 7.2 \cdot 10^7$ Pa for N$_2$ in the nematic phase of MBBA at room temperature. The solubility of N$_2$ increases strongly with temperature. The diffusion coefficient of N$_2$ in MBBA at room temperature is about 1.7×10^{-11} m^{2}s^{-1} [201]. The combination of both parameters yields the ratio $D/H = 2.4 \times 10^{-19}$ m^{2}s^{-1}Pa^{-1}. This is more than one order of magnitude lower than the ratios given above for the two smectogens at ≈ 85 °C. However, both the solubility and the diffusion coefficient are expected to increase strongly with temperature, and one has to extrapolate the MBBA ratio to 85 °C. At 50 °C, where MBBA is already in the isotropic state, the diffusion coefficient D is approximately $2.5 \times 10^{-11} \ \mathrm{m^2 \, s^{-1}}$ and H (at atmospheric pressure) $\approx 5.2 \times 10^6$ Pa, thus D/H has increased to $\approx 4.8 \times 10^{-18} \ \mathrm{m^2 \, s^{-1} \, Pa^{-1}}$. This value is already in the same range as the ratios for the smectic materials. Of course, one cannot compare the MBBA data directly with DOBAMBC and HOPDOB, since the data on the latter materials have been obtained in mesophases with higher order. In particular, the diffusion in smectics normal to the molecular layers follows different mechanisms and has a considerably higher activation energy than diffusion in the nematic phase. No solubility data are available to our knowledge for gases in smectic materials so far.

Summarizing, the study of the shrinkage dynamics of smectic bubbles provides a valuable and direct tool to access information about gas transport in liquid films. While the molecular model presented gives a reasonable description of the observed qualitative and quantitative permeability characteristics, and both the film thickness and the temperature dependence can be explained consistently, many questions remain open, and more systematic data on the permeability of smectic membranes, in particular in relation to the selectivity with respect to different gases, are desirable.

Table 1. Polymorphism of the mesogenic compounds mentioned in the text

Material	Phase transitions
16OAB	Cr 75.5 °C SmA 113.0 °C N 121 °C I
5O.6	Cr 34.7 °C SmG 37.4 °C SmF 42.9 °C SmB 51.4 °C SmC 52.6 °C SmA 61.3 °C N 73.0 °C I
7O.7	Cr 33.2 °C SmG 48.0 °C SmB 65.9 °C SmC 70.0 °C SmA 83.6 °C N 84.4 °C I
8CB	Cr 21.5 °C SmA 33.5 °C N 40.5 °C I
CM11B	Cr 50.0 °C Sm A* 57.0 °C I
DOBAMBC	Cr 74.6 °C (SmI* 62.0 °C) SmC* 94.0 °C SmA 117.0 °C I
FELIX 16/100	X −20 °C SmC* 72 °C SmA 85 °C N 94–90 °C I
HOPDOB	(35.0 °C) Cr 62.5 °C SmE 38.0 °C (SmB 44.5 °C) SmC 77.5 °C SmA 83.3 °C N 88.9 °C I
PBOT	Cr 106 °C SmA 119 °C I

References

1. R.B. Meyer, L. Liebert, L. Strzelecki, P. Keller: J. Physique Lett. **36**, L69 (1975); R.B. Meyer: Ferroelectrics **28**, 319 (1980)
2. T. Niori, F. Sekine, J. Watanabe, T. Furukawa, H. Takezoe: J. Mater. Chem. **6**, 1231 (1996); T. Sekine, Y. Takanashi, J. Watanabe, H. Takezoe: Jpn. J. Appl. Phys. **36**, L1201 (1997)
3. D.R. Link, G. Natale, R. Shao, J.E. Maclennan, N.A. Clark, E. Körblova, D.M. Walba: Science **278**, 1924 (1997)
4. M. Ben Amar, P.P. Da Silva, N. Limodin, A. Langlois, M. Brazovskaia, C. Even, I.V. Chikina, P. Pieranski: Eur. Phys. J. **B 3**, 197 (1998)
5. F. Schneider: Rev. Sci. Instrum. **73**, 114 (2002)
6. P. Oswald: J. Physique **48**, 897 (1987)
7. R. Stannarius, C. Cramer: Liq. Cryst. **23**, 371 (1997)
8. R. Stannarius, C. Cramer: Europhys. Lett. **42**, 43 (1998)
9. C.Y. Young, R. Pindak, N.A. Clark, R.B. Meyer: Phys. Rev. Lett. **40**, 773 (1978)
10. C. Rosenblatt, R. Pindak, N.A. Clark, R.B. Meyer: Phys. Rev. Lett. **42**, 1220 (1979); Phys. Rev. A **21**, 140 (1980)
11. R. Pindak, C.Y. Young, R.B. Meyer, N.A. Clark: Phys. Rev. Lett. **45**, 1193 (1980)
12. R. Pindak, D.J. Bishop, W.O. Sprenger: Phys. Rev. Lett. **44**, 1461 (1980)
13. T. Stoebe, C.C. Huang: Int. J. Mod. Phys. B **9**, 2285 (1995)
14. C. Bahr: Int. J. Mod. Phys. B **8**, 3051 (1994)
15. A.A. Sonin: *Freely Suspended Liquid Crystalline Films*, 1st edn (Wiley, New York, 1998)
16. W.H. de Jeu, E.A.L. Mol, G.C.L. Wong: Liq. Cryst. Today **8**, 1 (1998)
17. S.C. Davey, J. Budai, J.W. Goodby, R. Pindak, D.E. Moncton: Phys. Rev. Lett. **53**, 2129 (1984)
18. S. Heinekamp, R.A. Pelcovits, E. Fontes, E.Y. Chen, R. Pindak, R.B. Meyer: Phys. Rev. Lett. **52**, 1017 (1984)
19. E.B. Sirota, P.S. Pershan, S. Amador, L.B. Sorensen: Phys. Rev A **35**, 2283 (1987)
20. E.B. Sirota, P.S. Pershan, L.B. Sorensen, J. Collett: Phys. Rev. A **36**, 2890 (1987)
21. E.B. Sirota, P.S. Pershan, M. Deutsch: Phys. Rev. A **36**, 2902 (1987)
22. T. Stoebe, R. Geer, C.C. Huang, J.W. Goodby: Phys. Rev. Lett. **69**, 2090 (1992)
23. C. Bahr, D. Fliegner: Phys.Rev. Lett. **70**, 1842 (1993)
24. A. Poniewierski, R. Holyst: Phys. Rev. B **47**, 9840 (1993)
25. I. Kraus, P. Pieranski, E. Demikhov, H. Stegemeyer, J. Goodby: Phys. Rev. E **48**, 1916 (1993)
26. E. Demikhov, U. Hoffmann, H. Stegemeyer: J. Physique II **4**, 1865 (1994)
27. E. Gorecka, Li Chen, W. Pyzuk, A. Krowczynski, S. Kumar: Phys. Rev. E **50**, 2863 (1994)
28. L. Reed, T. Stoebe, C.C. Huang: Phys. Rev. E **52**, R2157 (1995)
29. A.A. Sonin, A. Yethiraj, J. Bechhoefer, B.J. Frisken: Phys. Rev. E **52**, 6260 (1995)
30. M.S. Spector, J.D. Litster: Phys. Rev. E **51**, 4698 (1995)
31. E.I. Demikhov, V.K. Dolganov, K.P. Meletov: Phys. Rev. E **52**, R1285 (1995)
32. E. Demikhov: Mol. Cryst. Liq. Cryst. **265**, 2969 (1995)
33. T. Stoebe, L. Reed, L.M. Veum, C.C. Huang: Phys. Rev. E **54**, 1584 (1996)
34. C. Biensan, B. Desbat, J.-M. Turlet: Thin Solid Films **284**, 293 (1996)
35. C. Bahr, C.J. Booth, D. Fliegner, J.W. Goodby: Phys. Rev. Lett. **77**, 1083 (1996)

36. H. Moritake, M. Terayama, S. Uto, Y. Fuwa, M. Ozaki, K. Yoshino: Trans. IEEE Jpn., Pt. A **116A**, 503 (1996)
37. C.F. Chou, A.J. Jin, C.Y. Chao, S.W. Hui, C.C. Huang, J.T. Ho: Phys. Rev. E **55**, R6337 (1997)
38. C.Y. Chao, J.E. Maclennan, J.Z. Pang, S.W. Hui, J.T. Ho: Phys. Rev. E **57**, 6757 (1998)
39. C.Y. Chao, S.W. Hui, J.E. MacLennan, J.Z. Pang, J.T. Ho: Mol. Cryst. Liq. Cryst. **330**, 1495 (1999)
40. A. Fera, B.I. Ostrovskii, D. Sentenac, I. Samoilenko, W.H. de Jeu: Phys. Rev. E **60**, R5033 (1999)
41. E.E. Gorodetskii, E.S. Pikina, V.E. Podnek: JETP **88**, 35 (1999)
42. N. Sakamoto, K. Sakai, K. Takagi: J. Chem. Phys. **112**, 946 (2000)
43. J.A. Collett, P.T. Kondratko, M.E. Neubert: Phys. Rev. E **62**, 6760 (2000)
44. N.J. Mottram, T.J. Sluckin, S.J. Elston, M.J. Towleri: Phys. Rev. E **62**, 5064 (2000)
45. A. Fera, I.P. Dolbnya, R. Opitz, B.I. Ostrovskii, W.H. de Jeu: Phys. Rev. E **63**, 020601/1 (2001)
46. P. Cluzeau, G. Joly, N.T. Nguyen, C. Gors, V.K. Dolganov: Liq. Cryst. **29**, 505 (2002)
47. A.J. Jin, T. Stoebe, C.C. Huang: Phys. Rev. E **49**, R4791 (1994)
48. A.J. Jin, C.C. Huang, C.F. Chou, C.Y. Chao, J.T. Ho: Mod. Phys. Lett. B **10**, 765 (1996)
49. T. Stoebe, P. Mach, C.C. Huang: Phys. Rev. Lett. **73**, 1384 (1994)
50. F. Picano, P. Oswald, E. Kats: Phys. Rev. E **63**, 021705/1 (2001)
51. E.I. Demikhov, V.K. Dolganov: Ferroelectrics **181**, 179 (1996)
52. P.M. Johnson, P. Mach, E.D. Wedell, F. Lintgen, M. Neubert, C.C. Huang: Phys. Rev. E **55**, 4386 (1997)
53. Y. Martínez-Ratón, A.M. Somoza, L. Mederos, D.E. Sullivan: Rhys. Rev. E **55**, 2030 (1997)
54. E.A.L. Mol, G.C.L. Wong, J.M. Petit, F. Rieutord, W.H. de Jeu: Physica B **248**, 191 (1998)
55. S. Pankratz, P.M. Johnson, H.T. Nguyen, C.C. Huang: Phys. Rev. E **58**, R2721 (1998)
56. S. Pankratz, P.M. Johnson, R. Holyst, C.C. Huang: Phys. Rev. E **60**, R2456 (1999)
57. S. Pankratz, P.M. Johnson, A. Paulson, C.C. Huang: Phys. Rev. E **61**, 6689 (2000)
58. D.L. Lin, J.T. Ou, Shi Longpei, X.R. Wang, A.J. Jin: Europhys. Lett. **50**, 615 (2000)
59. P. Cluzeau, G. Joly, H.T. Nguyen, C. Gors, V.K. Dolganov: Phys. Rev. E **62**, R5899 (2000)
60. B.D. Swason, H. Stragier, D.J. Tweet, L.B. Sorensen: Phys. Rev. Lett. **62**, 909 (1989)
61. J. Collett, L.B. Sorensen, P.S. Pershan, J. Als-Nielsen: Phys. Rev. A **32**, 1036 (1985)
62. B.M. Ocko, A. Braslau, P.S. Pershan, J. Als-Nielsen, M. Deutsch: Phys. Rev. Lett. **57**, 94 (1986)
63. A. Bottger, D. Frenkel, J.G.H. Joosten, G. Krooshof: Phys. Rev. A **38**, 6316 (1988)
64. C. Bahr, C.J. Booth, D. Fliegner, J.W. Goodby: Phys. Rev. E **52**, R4612 (1995)
65. V.K. Dolganov, E.I. Demikhov, R. Fouret, C. Gors: JETP **84**, 522 (1997)
66. E.G. Bortchagovsky, A. Deineka, M. Glogarova, V. Hamplova, L. Jastrabik, M. Kaspar: Ferroelectrics **244**, 611 (2000)
67. X.F. Han, D.A. Olson, A. Cady, D.R. Link, N.A. Clark, C.C. Huang: Phys. Rev. E **66**, 40701 (2002)
68. V.K. Dolganov, E.I. Kats, S.V. Malinin: JETP **93**, 533 (2001)
69. A.N. Shalaginov, D.E. Sullivan: Phys. Rev. E **63**, 031704/1 (2001)
70. H. Schüring, C. Thieme, R. Stannarius: Liq. Cryst. **28**, 241 (2001)
71. L.V. Mirantsev: Liq. Cryst. **10**, 425 (1991)
72. R. Najjar, Y. Galerne: Mol. Cryst. Liq. Cryst. **328**, 489 (1999)

73. Y. Galerne, L. Liebert: Phys. Rev. Lett. **64**, 906 (1990)
74. V. Candel, Y. Galerne: Liq. Cryst. **15**, 541 (1993)
75. G.P. Crawford, R. Stannarius, J.W. Doane: Phys. Rev. A **44**, 2558 (1991)
76. P. Ziherl, M. Vilfan, N. Vrbancic-Kopac, S. Zumer: Phys. Rev. E **61**, 2792 (2000)
77. C. Bahr, D. Fliegner, C.J. Booth, J.W. Goodby: Phys. Rev. E **51**, R3823 (1995)
78. H. Kobayashi, N. Matsuhashi, Y. Okumoto, T. Akahane: Jpn. J. Appl. Phys. Pt. 1 **38**, 6428 (1999)
79. S. Shibahara, Y. Takanishi, K. Ishikawa, H. Takezoe: Ferroelectrics **244**, 595 (2000)
80. A. Fera, R. Opitz, W.H. de Jeu, B.I. Ostrovskii, D. Schlauf, C. Bahr: Phys. Rev. E **64**, 021702/1 (2001)
81. R. Holyst: Phys. Rev. A **42**, 7511 (1990)
82. S. Gierlotka, P. Lambooy, W.H. de Jeu: Europhys. Lett. **12**, 341 (1990)
83. R. Holyst: Phys. Rev. A **44**, 3692 (1991)
84. Q.J. Harris, D.Y. Noh, D.A. Turnbull, R.J. Birgeneau: Phys. Rev. E **51**, 5797 (1995)
85. A. Fera, I.P. Dolbnya, G. Grubel, H.G. Muller, B.I. Ostrovskii, A.N. Shalaginov, W.H. de Jeu: Phys. Rev. Lett. **85**, 2316 (2000)
86. D. Sentenac, A. Fera, R. Opitz, B.I. Ostrovskii, O. Bunk, W.H. de Jeu: Physica B **283**, 232 (2000)
87. B. Rovsek, M. Cepic, B. Zeks: Phys. Rev. E **54**, R3113 (1996)
88. A. Becker, H. Stegemeyer: Ber. Bunsenges. **101**, 1957 (1997)
89. A. Becker, H. Stegemeyer: Proc. SPIE **3318**, 49 (1998)
90. P.O. Andreeva, V.K. Dolganov, K.P. Meletov: JETP Lett. **66**, 442 (1997)
91. P.O. Andreeva, V.K. Dolganov, C. Gors, R. Fouret, E.I. Kats: Phys. Rev. E **59**, 4143 (1999)
92. B. Rovsek, M. Cepic, B. Zeks: Phys. Rev. E **62**, 3758 (2000)
93. D.R. Link, J.E. Maclennan, N.A. Clark: Phys. Rev. Lett. **77**, 2237 (1996); H.R. Brand, P.E. Cladis, H. Pleiner: Phys. Rev. Lett. **86**, 4974 (2001); D.R. Link, J.E. Maclennan, N.A. Clark: Phys. Rev. Lett. **86**, 4975 (2001)
94. W. Lehmann, H. Skupin, F. Kremer, E. Gebhard, R. Zentel: Ferroelectrics **243**, 107 (2000); W. Lehmann, H. Skupin, C. Tolksdorf, E. Gebhard, R. Zentel, P. Krüger, M. Lösche, F. Kremer: Nature **410**, 447 (2001)
95. J. Xue, M.A. Handschy, N.A. Clark: Ferroelectrics **73**, 305 (1987)
96. J. Maclennan: Europhys. Lett. **13**, 1435 (1990)
97. W.S. Lo, R.A. Pelcovits, R. Pindak, G. Srajer: Phys. Rev. A. **42**, 3630 (1990)
98. P.E. Cladis, H.R. Brand: Liq. Cryst. **14**, 1327 (1993)
99. P.E. Cladis, P.L. Finn, H.R. Brand: Phys. Rev. Lett. **75**, 1518 (1995)
100. C. Dascalu, G. Hauck, H.D. Koswig, U. Labes: Liq. Cryst. **21**, 733 (1996)
101. S. Uto, H. Ohtsuki, M. Terayama, M. Ozaki, K. Yoshino: Jpn. J. Appl. Phys. **35**, L158 (1996)
102. G. Hauck, H.D. Koswig: Ferroelectrics **122**, 253 (1995)
103. G. Hauck, H.D. Koswig, C. Selbmann: Liq. Cryst. **21**, 847 (1996)
104. E.I. Demikhov, S.A. Pikin, E.S. Pikina: Phys. Rev. E **52**, 6250 (1995); R. Stannarius, N. Klöpper, T.M. Fischer, F. Kremer: Phys. Rev. E **58**, 6884, (1998)
105. N. Klöpper, F. Kremer, T.M. Fischer: J. Physique II **7**, 57 (1997)
106. E. Hoffmann, H. Stegemeyer: Ber. Bunsenges. **100**, 1250 (1996)
107. C. Langer, R. Stannarius: Ferroelectrics **244**, 347 (2000); R. Stannarius, C. Langer: Mol. Cryst. Liq. Cryst. **358**, 109 (2001)
108. D.R. Link, L. Radzihovsky, G. Natale, J.E. Maclennan, N.A. Clark, M. Walsh, S.S. Keast, M.E. Neubert: Phys. Rev. Lett. **84**, 5772 (2000), and refs. therein

109. R. Stannarius, C. Langer, W. Weißflog: Phys. Rev. E **66**, 031709 (2002)

110. R. Stannarius, C. Langer, W. Weißflog: Ferroelectrics **277**, 177 (2002)

111. K. Nakano, M. Ozaki, K. Yoshino: J. Appl. Phys. **92**, 6384 (2002)

112. R. Stannarius, C. Langer, W. Weißflog: Phys. Rev. Lett. **90**, 025502 (2003)

113. S.W. Morris, J.R. de Bruyn, A.D. May: Phys. Rev. Lett. **65**, 2378 (1990)

114. S.W. Morris: Phys. Rev. Lett. **65**, 2778 (1990)

115. S.W. Morris, J.R. de Bruyn, A.D. May: Phys. Rev. A **44**, 8146 (1991)

116. A. Becker, H. Stegemeyer, R. Stannarius, S. Ried: Europhys. Lett. **39**, 257, (1997)

117. C. Langer, R. Stannarius: Phys. Rev. E **58**, 650 (1998)

118. Y. Marinov, P. Simova: Liq. Cryst. **12**, 657 (1992)

119. M.I. Godfrey, D.H. van Winkle: Phys. Rev. E **54**, 3752 (1996)

120. J. Birnstock, R. Stannarius: Mol. Cryst. Liq. Cryst. **366**, 815 (2001)

121. J. Birnstock, R. Stannarius: Phys. Rev. Lett. **86**, 4187 (2001)

122. P.E. Cladis, Y. Couder, H.R. Brand: Phys. Rev. Lett. **55**, 2945 (1985)

123. I. Mutabazi, P.L. Finn, J.T. Gleeson, J.W. Goodby, C.D. Andereck, P.E. Cladis: Europhys. Lett. **19**, 391 (1992)

124. D. Dash, X.L. Wu: Phys. Rev. Lett. **79**, 1483 (1997)

125. I. Kraus, P. Pieranski, E. Demikhov: J. Phys. Condens. Matter **23A**, suppl. A415 (1994)

126. C. Even, P. Pieranski: Europhys. Lett. **47**, 531 (1999); C. Even, S. Russ, V. Repain, P. Pieranski, B. Sapoval: Phys. Rev. Lett. **83**, 726 (1999); C. Even, P. Pieranski: Proc. SPIE **3318**, 386 (1998)

127. H.M. Brodowsky, U.C. Boehnke, F. Kremer, E. Gebhard, R. Zentel: Langmuir **15**, 274 (1999)

128. H. Schüring, R. Stannarius, C. Tolksdorf, R. Zentel: Macromolecules **34**, 3962 (2001) R. Stannarius, R. Köhler, U. Dietrich, M. Lösche, C. Tolksdorf, and R. Zentel: Phys. Rev. E **65**, 041707 (2002)

129. J. Maclennan, G. Decher, U. Sohling: Appl. Phys. Lett. **59**, 917 (1991)

130. Y.C. Chen, T. Geue, U. Pietsch, S. Manukow, E. Schmeer: Mol. Cryst. Liq. Cryst. **329**, 1013 (1999)

131. C. Czeslik, H. Seemann, R. Winter: *Basiswissen Physikalische Chemie* (BG Teubner, Stuttgart, 2001)

132. I. Langmuir: J. Am. Chem. Soc. **38**, 2221 (1916)

133. T. Stoebe, P. Mach, C.C. Huang: Phys. Rev. E **49**, R3587 (1994)

134. T. Stoebe, P. Mach, S. Grantz, C.C. Huang: Phys. Rev. E **53**, 1662 (1996)

135. P. Mach, C.C. Huang, H.T. Nguyen: Langmuir **13**, 6357 (1997)

136. P. Mach, C.C. Huang, H.T. Nguyen: Phys. Rev. Lett. **80**, 732 (1998)

137. P. Mach, C.C. Huang, T. Stoebe, E.D. Wedell, H.T. Nguyen, W.H. de Jeu, F. Guittard, J. Naciri, R. Shashidhar, N.A. Clark, I.M. Jiang, F.J. Kao, H. Liu, H. Nohira: Langmuir **14**, 4330 (1998)

138. C. Papenfuss: *Contribution to a Continuum Theory of Two Dimensional Liquid Crystals*, Ph.D. Thesis (Wissenschaft und Technik Verlag, Berlin, 1995).

139. S. Krishnaswamy, R. Shashidhar: Mol. Cryst. Liq. Cryst. **35**, 253 (1976)

140. S. Krishnaswamy, R. Shashidhar: Mol. Cryst. Liq. Cryst. **38**, 353 (1977)

141. B. Song: *Untersuchung der Oberflächen- und Grenzflächenspannung flüssigkristalliner Verbindungen mittels der computergestützten Pendant-drop-Methode*, Ph.D. Thesis (TU Berlin, 1994)

142. R.-J. Roe, V.L. Bacchetta, P.M.G. Wong: J. Phys. Chem. **71**, 4190 (1967)

143. B. Stry la, W. Kuczyński, J. Ma lecki: Mol. Cryst. Liq. Cryst. **1**, 33 (1985)

144. C.H. Sohl, K. Miyano, J.B. Ketterson: Rev. Sci. Instrum. **49**, 1464 (1978)

145. C.H. Sohl, K. Miyano, J.B. Ketterson, G. Wong, Phys. Rev. A **22**, 1256 (1980)
146. K. Miyano: Phys. Rev. A **26**, 1820 (1982)
147. M. Veum, C. Pettersen, P. Mach, P.A. Crowell, C.C. Huang: Phys. Rev. E **61**, R2192 (2000)
148. P. Pieranski, L. Beliard, J.-P. Tournellec, X. Leoncini, C. Furtlehner, H. Dumoulin, E. Riou, B. Jouvin, J.-P. Fénerol, P. Palaric, J. Heuving, B. Cartier, I. Kraus: Physica A **194**, 364 (1993)
149. M. Eberhardt, R.B. Meyer: Rev. Sci. Instrum. **67**, 2846 (1996)
150. R. Stannarius, C. Cramer, H. Schüring: Mol. Cryst. Liq. Cryst. **329**, 1035 (1999); Mol. Cryst. Liq. Cryst. **350**, 297 (2001)
151. Y. Martinez, A.M. Somoza, L. Mederos, D.E. Sullivan: Phys. Rev. E **53**, 2466 (1996)
152. Liq. Cryst. database (http://licryst.chemie.uni-hamburg.de)
153. C.C. Huang, in: *Handbook of Liquid Crystals* (Wiley-VCH, Weinheim, 1998) pp. 441ff
154. R. Shashidhar: Mol. Cryst. Liq. Cryst. **64**, 217 (1981)
155. D. Demus, M. Pohl, S. Schönberg, L. Weber, A. Wiegeleben, W. Weissflog: Wissenschaftl. Beiträge d. Univ. Halle **41**, 18 (1983)
156. P. Keller, P.E. Cladis, P.L. Finn, H.R. Brand: J. Physique **46**, 18 (1985)
157. V.K. Dolgnov, E.I. Demikhov, R. Fouret, C. Gors: Phys. Lett. A **220**, 242 (1996)
158. H. Schüring, R. Stannarius: Langmuir **18**, 9735 (2002)
159. P. Cluzeau, V. Dolganov, P. Poulin, G. Joly, H.T. Nguyen: Mol. Cryst. Liq. Cryst. **364**, 381 (2001)
160. R. Najjar, Y. Galerne: Mol. Cryst. Liq. Cryst. **367**, 3263 (2001)
161. E.I. Demikhov, M. John, K. Krohn: Liq. Cryst. **23**, 443 (1997)
162. H. Schüring, R. Stannarius: Mol. Cryst. Liq. Cryst., in press (2004)
163. J. Sur, H.K. Pak: Phys. Rev. Lett. **86**, 4326 (2001)
164. J.H. Fendler: J. Mater. Sci. **30**, 323 (1987)
165. T. Moriizumi: Thin Solid Films **160**, 413 (1988)
166. B. Tieke: Adv. Mater. **3**, 532 (1991)
167. D. Exerowa, Z. Lalchev: Langmuir **2**, 668 (1986)
168. M.A. Borden, M.A. Longo: Langmuir **18**, 9225 (2002) and refs. therein
169. R.W. Baker: Int. Eng. Chem. Res. **41**, 1393 (2002) and refs. therein
170. G.D. Rose, J.A. Quinn: J. Colloid Interface Sci. **27**, 193 (1968); Science **159**, 636 (1968)
171. N.N. Li: Chem. Eng. J. **17**, 459 (1971)
172. N. Ramachandran, A.K. Didwania, K.K. Sirkar: J. Colloid Interface Sci. **83**, 94 (1981)
173. O. Albrecht, A. Laschewsky, H. Ringsdorf: Macromolecules **17**, 937 (1984); J. Membr. Sci. **22** 187 (1985)
174. K. Sujatha, T.R. Das, R. Kumar, K.S. Gandhi: Chem. Eng. Sci. **43**, 1261 (1988)
175. R.L. Cook, R.W. Tock: Sep. Sci. **9**, 185 (1974)
176. J.-J. Shieh, T.S. Chung: J. Polym. Sci. Pt. B **37**, 2851 (1999)
177. T.C. Merkel, V.I. Bondar, K. Nagai, B.D. Freeman, I. Pinnau: Macromolecules **32**, 8427 (1999); J. Polym. Sci. Part B **38**, 415 (2000)
178. Z. Zhang: J. Polymer Sci. Part B **38**, 1833 (2000)
179. C.K. Yeom, J.M. Lee, Y.T. Hong, K.Y. Choi, S.C. Kim: J. Membr. Sci. **166**, 71 (2000)
180. P. Molyneux: J. Appl. Polym. Sci. **79**, 981 (2001)
181. G. Obuskovic, T.K. Poddar, K.K. Sirkar: Ind. Eng. Chem. Res. **37**, 212 (1998)
182. Z. Mogri, D.R. Paul: J. Membr. Sci. **175**, 253 (2000)
183. P.S. Rallabandi, D.M. Ford: J. Membr. Sci. **171**, 239 (2000)
184. M. Teramoto, K. Nakai, N. Ohnishi, Q. Huang, T. Watari, H. Matsuyama: Ind. Eng. Chem. Res. **35**, 528 (1996)

185. W.J. Ward, W.L. Robb: Science **156**, 1481 (1967)
186. J.H. Meldon, P. Stroeve, C.E. Gregoire: Chem. Eng. Commun. **16**, 263 (1982)
187. R.R. Bhave, K.K. Sirkar: J. Membr. Sci. **27**, 41 (1986)
188. H. Chen, A.S. Kovvali, S. Majumdar, K.K. Sirkar: Ind. Eng. Chem. Res. **38**, 3489 (1999); H. Chen, A.S. Kovvali, K.K. Sirkar: Ind. Eng. Chem. Res. **39**, 2447 (2000)
189. A.S. Kovvali, H. Chen, K.K. Sirkar: J. Am. Chem. Soc. **122**, 7594 (2000)
190. Y. Hirayama, Y. Kase, N. Tanihara, Y. Sumiyama, Y. Kusuki, K. Haraya: J. Membr. Sci. **160**, 87 (1999)
191. C. Nagel, K. Günther-Schade, D. Fritsch, T. Strunskus, F. Faupel: Macromolecules **35**, 2071 (2002)
192. R. Krustev, H.J. Müller: Langmuir **15**, 2134 (1999)
193. R. Krustev, D. Platikanov, M. Nedyalkov: Colloids Surf. **79**, 129 (1993)
194. R. Krustev, D. Platikanov, M. Nedyalkov: Colloids Surf. **123–134**, 383 (1997)
195. R. Krustev, H.J. Müller: Rev. Sci. Instrum. **73**, 398 (2002)
196. M. Born: *Optik* (Springer, Berlin, 1985)
197. J.-J. Li, H. Schüring, R. Stannarius: Langmuir **18**, 112 (2002)
198. G. Krüger: Phys. Rep. **82**, 231 (1982); M. Hara et al.: Jpn. J. Appl. Phys. **24**, L777 (1985); W. Urbach, H. Hervet, F. Rondelez: J. Chem. Phys. **83**, 4 (1985)
199. M.A. Coleman, W.J. Koros: Macromolecules **32**, 3106 (1999)
200. S.Y. Ha, H.B. Park, Y.M. Lee: Macromolecules **32**, 2394 (1999)
201. G.-H. Chen, J. Springer: Mol. Cryst. Liq. Cryst. Sci. Technol. A **339**, 31 (2000)

Pattern Formation in Langmuir Monolayers Due to Long-Range Electrostatic Interactions

Thomas M. Fischer[1] and Mathias Lösche[2,3]

[1] Florida State University, Department of Chemistry and Biochemistry, Tallahassee, FL,
`tfischer@chem.fsu.edu`
[2] Universität Leipzig, Institut für Experimentelle Physik I, Leipzig, Germany
[3] Johns Hopkins University, Department of Biophysics, Baltimore, MD

Abstract. A distinctive characteristic of Langmuir monolayers that bears important consequences for the physics of structure formation within membranes is the uniaxial orientation of
the constituent dipolar molecules, brought about by the symmetry break which is induced by
the surface of the aqueous substrate. The association of oriented molecular dipoles with the
interface leads to the formation of image dipoles within the polarizeable medium – the subphase – such that the effective dipole orientation of every of the individual molecules is strictly
normal to the surface, even within molecularly disordered phases. As a result, dipole-dipole
repulsions play an eminently important role for the molecular interactions within the system
– independent of the state of phase (while the dipole area density does of course depend on
the state of phase) – and control the morphogenesis of the phase boundaries in their interplay
with the one-dimensional (1D) line tension between coexisting phases. The physics of these
phenomena is only now being explored and is particularly exciting for systems within a three-
phase coexistence region where complete or partial wetting, as well as dewetting between the
coexisting phases may be experimentally observed by applying fluorescence microscopy to
the monolayer films. It is revealed that the wetting behavior depends sensitively on the details
of the electrostatic interactions, in that the apparent contact angles observed at three-phase
contact points depends on the sizes of the coexisting phases. This is in sharp contrast to the
physics of wetting in conventional 3D systems where the contact angle is a materials property,
independent of the local details. In 3D systems, this leads to Young's equation – which has
been established more than two centuries ago.

We report recent progress in the understanding of this unusual and rather unexpected behavior of a quasi-2D system by reviewing recent experimental results from optical microscopy
on equilibrium phase shapes, non-equilibrium phenomena – such as relaxation of the shapes
after distortions inferred by Laser tweezers or local impulse heating – and rheological properties of the system. The theoretical analysis of the underlying molecular interactions leads to a
comprehension of the observed phenomena and reveals microscopic properties of the system
in quantitative terms. In view of the recently proposed *"lipid raft"* hypothesis, a particularly
fascinating implication of our results is the possibility that biochemical reactions which depend on complex interactions between membrane-bound proteins might be controlled by the
non-conventional physics of the 2D system: As an electrogenic event – such as ion transfer
across the membrane – changes the electrostatic properties of the membrane surface it might
concurrently infer wetting between 2D phases and thus lead to the conjunction of membrane
areas that were originally separated within the plane. If two reactants (e.g., membrane-bound
enzymes) are dissolved in distinct phases, such a colloidal reorganization might rearrange the
micro-evironment to bring them into close vicinity – and thus trigger the biochemical reaction.

T.M. Fischer, M. Lösche, Pattern Formation in Langmuir Monolayers Due to Long-Range Electrostatic Interactions,
Lect. Notes Phys. **634**, 383–394 (2004)
`http://www.springerlink.com/`

1 Introduction

Langmuir monolayers are monomolecular layers of insoluble amphiphiles at the air/water interface. Molecular self-organization has these amphiphiles with their hydrophilic headgroup immersed in the water and their hydrophobic tail dangling into the air. As such, Langmuir monolayers are interesting because on the one hand they enable us to study peculiarities of quasi two-dimensional (2D) systems. On the other hand, Langmuir monolayers represent half of a lipid bilayer, which in turn is prevalent in biology and forms the local environment where, for example signal cascades and signal transduction reactions occur.

Both Langmuir monolayers and biological membranes are asymmetric. In a Langmuir monolayer, the asymmetry arises from the juxtaposition of two different bulk phases – air and water – that are separated by the monolayer. In the biological membrane, the composition of the constituent lipids in the two leaflets of the bilayer is different, and there may be a gradient of salt concentration across the membrane. As a result of this asymmetry, electric dipole moments persist and are the origin of interface potentials, which in Langmuir monolayers give rise to long-range electrostatic interactions; in bilayers, these electrostatic interactions are screened and thus confined to the nanometer range. Interestingly, both Langmuir monolayers and biological membranes are also *laterally* heterogeneous on length scales comparable to the range of the electrostatic interactions. In Langmuir monolayers, micron-sized phase regions of liquid condensed domains – which are hexatic in their nature – are embedded in a fluid – i.e. molecularly more disordered – environment. In membranes, so-called lipid rafts – detergent resistant domains rich in sphingomyelin – of nanometer size coexist with their in-plane environment rich in phospholipids. It is well established that the phase patterns in Langmuir monolayers arise from an interplay of long-range electrostatic interactions and the line tension between the phases; it is then an interesting question to ask whether similar reasoning might be true for the nanoheterogeneities in membranous systems.

2 Wetting in Langmuir Monolayers

It has been demonstrated that the electrostatic free energy is not an extensive quantity for monolayers [1]. As a consequence, the domains that coexist with the matrix phase in the course of a first-order phase transition may undergo shape instabilities [2] as their size, A, increases. The range of possibilities for instabilities to occur is greatly enlarged in a three-phase coexistence regime, where not only the domain shape but also the phase topology might change with changes in the relative proportions of phase areas. Different topologies may result from differences in the wetting behavior of phase 1 at phase 2 depending on whether or not phase 3 intervenes between the two other phases. This behavior is characterized by the contact angle. In a finite-size system, where the free energy is not an extensive quantity, all sizes have to be measured with respect to a characteristic length scale. For monolayer systems, it has been assumed that this role is taken by a dipolar in-plane correlation length, Δ, which is interpreted as the distance at which the dipole density changes from the value for one phase to that for the other as one crosses a phase boundary.

Within the framework of such a theory, a contact angle of the order $\alpha \approx \Delta^2/A$ characterizes complete wetting. A value of $\alpha \approx \pi - \Delta^{1/2}A^{-1/4}$ is characteristic of complete dewetting, and partial wetting/dewetting corresponds to $\Delta^2/A < \alpha < \pi - \Delta^{1/2}A^{-1/4}$. In the thermodynamic limit, $A \to \infty$, these results coincide with the usual definitions: complete wetting, $\alpha = 0$; complete dewetting, $\alpha = \pi$; and partial wetting/dewetting, $0 < \alpha < \pi$. In order to fully characterize wetting in the three-phase coexistence region and to assess the influence of the electrostatic interaction, one needs to determine (a) the line tensions, $\lambda_{i,j}$, between the phases (see Sect. 2.1), (b) the surface potential density contrast, $\Delta\mu_{i,j}$, between the phases (see Sect. 2.2), (c) the characteristic length scale Δ (see Sect. 2.3), (d) the nature of the coexisting phases (see Sect. 2.4), and (e) the topology arising from the interplay between the phases (see Sect. 2.5). Section 3 concludes this short review by summarizing the pecularities of the physics of wetting in 2D that are due to the impact of long-range electrostatic interactions.

2.1 Line Tension

There are different definitions of the line tension λ between two coexisting monolayer phases possible, which are used to describe the energy per unit length associated with the phase boundaries in different contexts. One may define the line tension as a response function, i.e. as the change of free energy upon an infinitesimal elongation of the phase boundary. In the absence of long-range electrostatic interactions, this free-energy change does not depend on how the elongation is brought about. Hence, the line tension is a quantity entirely defined by equilibrium properties. However, in the presence of electrostatic interactions, the free-energy increase becomes dependent on the deformation mode and is thus no longer an equilibrium property.

Another possibility for characterizing the line tension is by using the Young–Laplace equation for an assessment of the equilibrium shape, e.g., of a phase droplet within its surrounding matrix. Here, the curvature at one point of the droplet's periphery results from a compensation between the Laplace pressure across the phase boundary and the line tension. Again, without long-range electrostatic interactions, the definitions of the Laplace pressure and the line tension are unique. The electrostatic interaction, however, leads to a renormalization of both the Laplace pressure and the line tension. This renormalization involves a new length scale, which can be arbitrarily chosen such that there are many ways of defining an equilibrium line tension in the presence of long-range electrostatic interactions.

Finall,y one may define the "bare" line tension as a fictive quantitity that would be in place if one were able to switch off all electrostatic interactions.

There have been numerous attempts to measure each of the line tensions defined above. Using shear flow, Benvegnu and McConnell distorted domains into "bola" shapes, whose relaxation to their original circular shapes was observed upon switching off the shear field [3]. From this relaxation, the "response-function type" line tension was determined by calculating the hydrodynamic forces active during the relaxation. Experiments based on this technique have also been performed by Mann et al. [4] and by Läuger et al. [5]. Instead of using shear flow, Wurlitzer et al. deformed domains using optical tweezers, which provides a more localized possibility for manipulating the phase boundaries. Thus, the "response-function type" line tensions between the liquid expanded (LE) and gaseous phase (Fig. 1) as well as between the liquid condensed (LC) and the LE phases, (Fig. 2), of methyloctadecanoate have been measured [6, 7]. Determination of the Young–Laplace line tension has been performed by Heinig et al. [8] (Fig. 3).

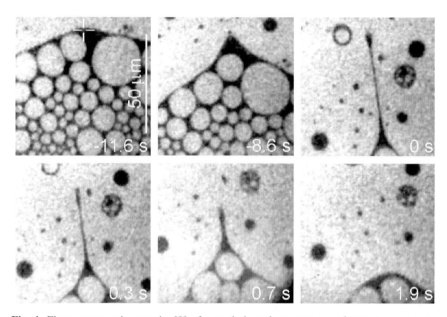

Fig. 1. Fluorescence micrographs [9] of a methyloctadecanoate monolayer on pure water, visualized using ≈ 0.5 mol% nitrobenzoxadiazol-hexadecylamine (NBD-HDA). The phase boundary between the gas phase (dark) and the LE phase (bright) is manipulated with an optical tweezer that holds a nonfluorescent silica bead in place (crosshair) which is immersed in the monolayer. Owing to a slow (a few 10 μm/s) convective flow of the monolayer on top of its aqueous subphase, an elongated strip of the gas phase develops ($t = -11.6$ s ... 0 s). The tweezer is subsequently switched off ($t = 0$) and a shortening of the stripe is observed, which is due to a balance between the "response-function type" line tension and hydrodynamic forces. Reproduced with permission from [6]. A movie that visualizes the dynamics of the process is provided as supplementary material [15]

2.2 Surface Potential

For the analysis of the electrostatic effects on wetting in Langmuir monolayers, one needs to assess their interaction strength. The sources of the electrostatic interaction are projections of the permanent electric dipoles on the direction normal to the

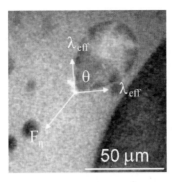

Fig. 2. Fluorescence micrograph of a methyloctadecanoate monolayer (≈ 0.5 mol% NBD-HDA) on pure water showing the stationary shape of an LC droplet within its LE surroundings. An optical tweezer holds the LC droplet in place via a silica bead against the convective flow of the monolayer. Owing to the larger line tension between the coexisting phases than that observed in Fig. 1, the force balance leads merely to the formation of a cusp at the droplet perimeter. From the angle Θ at the cusp, the line tension between the LE and LC phases is determined. Reproduced with permission from [7]. A movie that visualizes the dynamics of the process is provided as supplementary material [15]

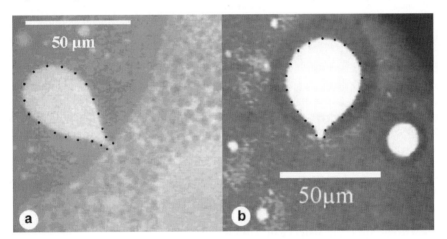

Fig. 3. Fluorescence micrographs of methyloctadecanoate (≈ 0.5 mol% NBD-HDA) on water in the three-phase coexistence region showing LE droplets (bright) immersed in gas (dark areas) that wet LC phase (gray)/gas boundaries partially. Fitting the observed shape with an electrostatic Young–Laplace equation (dotted contours) yields the "Young–Laplace" line tension. Reproduced with permission from [8]

air/water interface, whose lateral densities, μ_i, may differ for the various phases. Macroscopic determination of monolayer surface potentials and the assessment and interpretation of their changes due to molecular reorientation within the different monolayer phases have been pioneered by Alexander and Schulman using the Kelvin probe technique [10]. Since then, the method has been extensively used and refined to become a standard characterization technique for the electrostatics of surfaces [11].

The surface potentials of the relevant homogeneous phases, investigated extensively in the wetting studies described below, have been determined by Wurlitzer and coworkers [12].

2.3 Characteristic Length Scale

As briefly described in Sect. 2.1, the equilibrium line tension is renormalized by the long-range electrostatic interactions. On shorter length scales (nm), other interactions dominate, and in this regime the line tension corresponds to the "bare" line tension, i.e. to the 2D line tension of that definition which resembles most closely its 3D equivalent. This leads to the conclusion that very close to a three-phase intersection point, the angle α between the incoming boundary lines is determined by the "bare" line tension values, while macroscopic shapes of the phase boundaries are determined by the renormalized line tensions. Thus, three-phase intersection points follow Antonow's rule for the "bare" line tension values. As shown above (Fig. 3), the renormalized line tension can be determined from droplet shapes. The renormalization of the line tension involves the ratio between a macroscopic length, e.g. the square root of the droplet area, $\sqrt{A}$, and the dipolar in-plane correlation length, Δ. If this ratio becomes larger than a certain threshold value, either the "bare" line tension is no longer compatible with Antonow's rule or the renormalized line tension is no longer compatible with the shape of the droplet. Observations of pendant droplets, such as those displayed in Fig. 3 therefore set a *lower* limit on the dipolar correlation length Δ. Such measurements on pendant LE droplets partially wetting LC/gas phase boundaries have been used to demonstrate that the characteristic length is on the order of 100 nm in methyloctadecanoate [13].

2.4 Three-Phase Coexistence or Modulated Phase?

Equilibrium thermodynamics distinguishes between phase coexistence in a first-order phase transition and modulated phases. In a phase coexistence regime, two or three phases coexist with each other, and by changing extensive external control parameters – such as the average area per molecule in the film – the experimenter may convert one of the phases into the other. During this conversion process, every intensive property of each individual phase remains the same. In distinction, modulated phases are characterized by spatial variations of the intensive properties of the phase such as the density. These properties usually change upon changing an extensive external field.

In Langmuir monolayers, the simultaneous occurrence of areas with different physical properties – such as local in-plane molecular density differences and solubility differences of a fluorescent dye – is frequently observed [9]. Such observations are usally attributed to the coexistence of gaseous, LE, and/or LC phases [14]. In order to understand the wetting behavior of those coexisting areas, a correct assignment as a "true" phase coexistence or a modulated phase is necessary. In methyloctadecanoate, this issue has been clarified by examining the behavior of "LE" droplet shapes immersed in "gaseous" cavities (the 2D equivalent of bulk "gas bubbles") within an "LC" matrix (Fig. 4) upon partial interconversion of the phases into each other. One observes a destabilization of the shape of the original, circular "LE" droplet upon compressing the surface film, and hence reducing the area of the cavity.

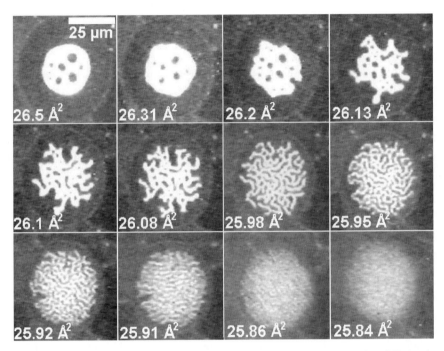

Fig. 4. Fluorescence micrographs (methyloctadecanoate on pure water) of an LE droplet (bright) immersed in a gas bubble (dark) within an LC matrix (gray). Upon compression, gas phase is converted into the LE and/or LC phase. The bubble radius shrinks and the circular shape of the liquid droplet is destabilized, indicating that the simultaneous occurrence of areas with (at least) three different sets of physical properties is due to a modulated phase rather than a three-phase coexistence in the proper thermodynamic sense. A much better appreciation of the dynamical changes associated with the transition is obtained from a film clip provided as supplementary material to this paper [15]. Reproduced from [16] with permission

A theoretical stability analysis [16] of the shapes assuming thermodynamic *phase* coexistence (which assumption inherently implies that no changes in either the line tension or the dipole density upon compression of the monolayer should occur) predicts that a decrease in cavity size should *stabilize* the circular droplet shape, rather than destabilize it as observed in the actual experiment shown in Fig. 4. The disagreement between experiment and theory then argues against the hypothesis that there are no changes in the properties of the *phases* upon interconversion; one hence has to identify the simultaneous occurrence of areas with different physical properties in the monolayer as the signature of a *modulated phase*.

In view of the rather subtle distinction, and the conventional designation of the co-existing areas as coexisting *phases*, we adhere to this designation and the established nomenclature of the phases [14] throughout this paper.

2.5 Signature of Electrostatic Impact on Wetting

The electrostatic repulsion among molecular dipoles with parallel components along the surface normal in a two-phase coexistence region leads to phase boundary in-stabilities. Droplets tend to elongate as they grow larger than a certain threshhold size [17]. If such a droplet is located close to a third phase, there are two possible growth directions – parallel or perpendicular to the boundary – along which elon-gation can occur. In addition, an elongation might be augmented or suppressed by the electrostatic interaction with the third phase. A theoretical analysis of the pos-sibilities for such a process close to an interface has been performed by Khattari et al. [13]. In a circular segment approximation (which is the exact solution for the shape in the absence of electrostatic interaction), the apparent contact angle of such droplets has been predicted as a function of the "bare" (nonelectrostatic) spreading coefficient $S = \cos\alpha - 1 = (\lambda_{2,3} - \lambda_{1,3} - \lambda_{1,2})/\lambda_{1,2}$, the $\Delta\mu_{i,j}$ and the ratio, $\sqrt{A}/\Delta$ (where $i = 1$ signifies the droplet phase, 2 is the low-λ phase surrounding the droplet, and 3 is the high-λ phase forming the supporting base). For the case, $\Delta\mu_{1,2}, \Delta\mu_{1,3} > \Delta\mu_{2,3}$, the droplet is predicted to spread at the interface between phases 2 and 3 as its size increases, e.g. upon compression of the monolayer film. This spreading occurs at a critical value of $\Delta\mu^2/\lambda$, which is, however, lower than the shape instability threshold in the bulk.

Figure 5 shows experimental results where such a behavior is observed. The experiment starts at large areas (panel a). The fluorescence micrograph shows two circular disks of the LE phase (brightest areas) with gas bubbles (dark) enclosed. The largest proportion of the image area is occupied by the LC phase (dark gray). Thin layers of a gas "atmosphere" surround the LE droplets (white arrow), separating them entirely from the LC matrix. Owing to low contrast, gas and LC phase areas are difficult to discriminate; the gas rings around LE are, however, much better recognized in the live movie (see supplementary material to this paper [15]), from which the micrographs shown in Fig. 5 have been clipped. In the situation shown in Fig. 5a, gas areas thus wet the LE/LC boundary line completely ($\alpha = 0$). In addition, elongated gas bubbles are observed in the interior of the brighter and smaller of the two LE disks (right-hand side of panel a).

Upon decreasing the average area A of the monolayer (Fig. 5 b–i) and thus reducing the proportion of gas phase by compression of the surface film, a shape instability and subsequent continuous change in the wetting behavior from complete wetting ($\alpha = 0$), via partial wetting/dewetting to complete dewetting ($\alpha = 180°$) is observed. Monolayer compression thus leads first to a destabilization of the elongated gas bubbles into a chain of smaller circular bubbles (panels b, c). Upon further compression, resulting in a further decrease of the gaseous areas below the bulk shape instability threshold, the gas film that surrounds the larger of the two LE droplets initially dewets the LE/LC phase boundary partially (panel d, where arrows indicate the resulting three-phase contact points). In subsequent panels, arrows indicate that the gas phase gradually dewets the LE/LC boundary in the course of the experiment until eventually (frame i) dewetting is complete. Again, the dynamic change of this reversal from complete wetting to complete dewetting is much better appreciated in the live video provided as supplementary material [15].

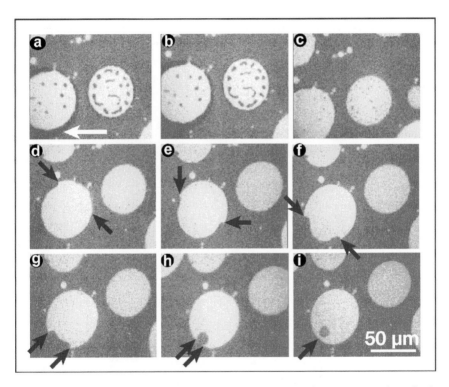

Fig. 5. Sequence of fluorescence micrographs of a methyloctadecanoate monolayer in the three-phase coexistence gas/LE/LC during compression. The reduction of the total amount of gas phase leads to a transition from elongated toward circular gas bubbles in the bulk. A reversal of the wetting characteristics is observed upon further compression (black arrows). Such a behavior is predicted to occur if $\Delta\mu_{1,2}$, $\Delta\mu_{1,3} > \Delta\mu_{2,3}$. Reproduced from [13] with permission. See also supplementary material [15]

2.6 Extrapolation to Biomembranes?

Wetting transitions such as those displayed in Fig. 5 are induced by compression or expansion of the monolayer with a barrier. In biological systems, changes within the membrane structure or composition are triggered by enzymes rather than changes in lipid density. Figure 6 shows that wetting transitions in Langmuir monolayers may equally well be induced by enzymatic reactions as by molecular density changes. The micrographs show a binary monolayer comprising 58 weight% DMPC and 40 weight% cholesterol (plus 2 weight% of fluorescent dye). Morphological changes within the monolayer upon enzymatic cleavage of the phospholipid into a fatty acid and a lysophospholipid after injection of phospholipase A_2 into the aqueous (buffer) subphase are documented. Prior to enzyme injection, the LE phase wets the gas/LC interface partially (panel a). After injection, a complete dewetting of the LE phase is observed in panel (b).

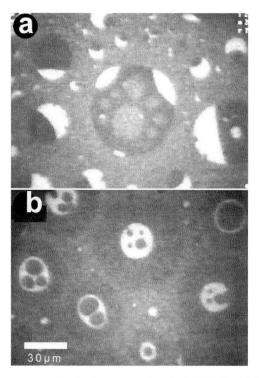

Fig. 6. Fluorescence micrographs of a dimyristoylphosphatidylcholine/cholesterol monolayer (DMPC:Chol = 58:40 with 2% NBD-HDA) on physiological buffer (10 mM Tris, 5 mM CaCl$_2$, pH 9, $T = 25\ °C$, $\pi \approx 0$) in the three-phase coexistence gas/LE/LC: (a) prior to and (b) after injection of phospholipase A_2 (final concentration 30 ng/l) into the subphase. The phospholipase causes a change in lipid composition, which causes a wetting transition in a similar way to the situation depicted in Fig. 5

Hence, if a similar control mechanism for wetting via electrostatic interactions to that in Langmuir monolayers were also applicable to biological membranes, such a mechanism would at the same time control the mutual reactivity of proteins via the phase behavior of the constituent lipids. An important requirement for wetting instabilities to occur is the long-range nature of the electrostatic interactions (where "long-range" implies that the interaction distance exceeds the in-plane dimension of the wetting layer). In membranes, the length scale of the electrostatic interactions is set by the Debye length, which is typically in the nanometer range at the high salt concentrations in a physiologically relevant environment. Lipid rafts are believed to be on the order of 50 nanometers in diameter [18], and a thin film of a *third* phase – if it exists – would presumably be thinner than that. Although the physical criteria for a hierarchy of length scales similar to that observed in Langmuir monolayers are thus definitely at the edge of being relevant, it is nevertheless exciting to speculate on the ramifications of their existence. Only future experiments on membranous nanosystems will show whether or not similar wetting phenomena to the ones documented here play an important role in controlling the reactivity among molecular species within biomembranes.

3 Conclusions

Long-range electrostatic interactions in Langmuir monolayers cause the development of mesoscopic patterns. Shape transitions and topological transformations triggered by changes of the area per molecule, electrostatic contrast, or line tensions between coexisting phases affect the morphology of the monolayer critically: a wealth of nontrivial dynamic reorganization events are observed upon manipulation of these quantities. It is thus clear that Young's equation – established over two centuries ago for 3D systems – does *not* hold in a straightforward extrapolation to 2D arrays of molecules at interfaces: The physics of 2D molecular materials is distinctly different from that applicable to the 3D world!

Owing to the range of electrostatic interactions operative in Langmuir monolayers, the size of the resulting patterns is such that they are trivially visualized using optical microscopy [9]. On the other hand, electrostatic interactions in biological membranes – while operating in a similar way to those in Langmuir monolayers which are composed of the same molecular classes – are strongly shielded owing to the fact that physiological bilayer leaflets are immersed in aqueous environments on *both* sides of their interfaces. It is thus tempting to speculate that the same wealth of dynamic morphogenesis and the same range of morphological changes should exist in biomembranes, albeit on a much shorter length scale, reduced by ≈ 3 orders from its magnitude in the monolayer. *If* wetting transitions such as the examples shown in this paper for Langmuir monolayers do exist in biological membranes, they will play an important role in controlling the cross-reactivity of proteins that reside within the membrane or at its periphery. In that sense, the lipid raft hypothesis should be extended to account for dynamic morphology changes that might be instrumental in providing a physics-based control mechanism of biochemical reactions in the membrane.

Acknowledgments

We thank H. Möhwald for generous support and valuable discussions. This work has been supported by the DFG within the priority program *Wetting and Structure Formation at Interfaces* and with a Heisenberg fellowship to TMF. ML enjoyed support from the Fonds der Chemischen Industrie, Frankfurt am Main.

References

1. H.M. McConnell: Annu. Rev. Phys. Chem. **42**, 171 (1991)
2. K.Y.C. Lee, H.M. McConnell: J. Phys. Chem. **97**, 9532 (1993)
3. D.J. Benvegnu, H.M. McConnell: J. Phys. Chem. **96**, 6820 (1992)
4. E.K. Mann, S. Henon, D. Langevin, J. Meunier: Phys. Rev. E **51**, 5708 (1995)
5. J. Läuger, C.R. Robertson, C.W. Frank, G.G. Fuller: Langmuir **12**, 5630 (1996)
6. S. Wurlitzer, P. Steffen, T. M. Fischer: J. Chem. Phys. **112**, 5915 (2000)
7. S. Wurlitzer, P. Steffen, M. Wurlitzer, Z. Khattari, T. M. Fischer: J. Chem. Phys. **113**, 3822 (2000)
8. P. Heinig, P. Steffen, S. Wurlitzer, T.M. Fischer: Langmuir **17**, 6633 (2001)
9. M. Lösche, E. Sackmann, H. Möhwald: Ber. Bunsenges. Phys. Chem. **87**, 848 (1983)
10. A. Alexander, J. Schulman: Proc. Roy. Soc. Ser. A **161**, 115 (1937)
11. D.M. Taylor: J. Colloid Interface Sci. **87**, 183 (2000)
12. S. Wurlitzer, H. Schmiedel, T.M. Fischer: Langmuir **18**, 4393 (2002)
13. Z. Khattari, P. Heinig, S. Wurlitzer, P. Steffen, M. Lösche, T.M. Fischer: Langmuir **18**, 2273 (2002)
14. D.A. Cadenhead, F. Müller-Landau, B.M.J. Kellner: *Ordering in Two Dimensions*, ed. by S.K. Sinha (Elsevier North Holland, Amsterdam, 1980), pp 73–81.
15. T.M. Fischer: A video can be viewed in the online version of this book under http://www.springerlink.com
16. Z. Khattari, T.M. Fischer: J. Phys. Chem. B **106**, 1677 (2002)
17. R. de Koker, H.M. McConnell: J. Phys. Chem. **97**, 13419 (1993)
18. A. Pralle, P. Keller, E.L. Florin, K. Simons, J.K.H. Horber: J. Cell Biol. **148**, 997 (2000)

Characterization of Floating Surface Layers of Lipids and Lipopolymers by Surface-Sensitive Scattering

Peter Krüger[1] and Mathias Lösche[1,2]

[1] Universität Leipzig, Institut für Experimentelle Physik I, Leipzig, Germany
[2] Johns Hopkins University, Department of Biophysics, Baltimore, MD,
   loesche@physik.uni-leipzig.de

Abstract. Nanotechnology and molecular (bio-)engineering are making ever deepening in-roads into everybody's daily life. Physicochemical and biotechnological achievements in the design of physiologically active supramolecular assemblies have brought about the quest for their submolecular-level characterization. We employ surface-sensitive scattering techniques for the investigation of planar lipid membranes – floating monolayers on aqueous surfaces – to correlate structural, functional and dynamic aspects of biomembrane models. This chapter surveys recent work on the submolecular structure of floating phospholipid monolayers – where the advent of third-generation synchrotron X-ray sources has driven the development of realistic, submolecular-scale quasi-chemical models – as well as of more complex systems: cation binding to anionic lipid surfaces; conformational changes of lipopolymers undergoing phase transitions; the conformational organization of phosphatidylinositol and phosphatidyli-nositides, as examples of physiologically important lipids; and the adsorption of peptides (neuropeptide Y, NPY) or solvents (dimethylsulfoxide, DMSO) onto phospholipid surface layers.

1 Introduction

Investigations of monolayer structures on aqueous subphases using X-ray reflection have in the past been limited by a relatively restricted accessible momentum transfer range, and thus, spatial resolution. Recent developments in synchrotron radiation technology – almost doubling this range – have considerably improved the capabilities of the technique. Before that, data interpretation has relied entirely on "box models" which describe the structures as molecularly homogeneous slabs – one hydrophobic and one hydrophilic. We have recently shown that box models of phospholipid monolayers are rather inadequate to model data at the high momentum transfer nowadays available. As an alternative, a hybrid data inversion strategy has been developed that treats the hydrophobic alkane phase as a homogeneous slab and describes the position of submolecular fragments of the lipid headgroups by means of distribution functions along the interface. Thus, for the first time in surface-sensitive scattering, lipid headgroups have been resolved to such detail that their internal structure becomes available. In fact, we observe that these headgroups *must* be modeled as inhomogeneous moieties, because of the increase in available momentum transfer, to describe the data adequately. Within this approach, composition-space refinement – enabling the coupling of data sets from various X-ray and neutron contrasts – in

P. Krüger, M. Lösche, Characterization of Floating Surface Layers of Lipids and Lipopolymers by Surface-Sensitive Scattering, Lect. Notes Phys. **634**, 395–438 (2004)
http://www.springerlink.com/

connection with volumetric constraints enables structural characterization of lipid monolayers in unprecedented detail.

In this contribution, we concentrate largely on describing work in progress, rather than recalling older results obtained in our work within the *Sonderforschungsbereich* that have already been extensively covered in the original publications (although some overlap does occur). Specifically, we shall not cover here extensive work on the adsorption and recrystallization of large proteins at lipid-functionalized aqueous interfaces [1–4], and the morphogenesis of lipid phase domains [5,6] and its use in the nanopatterning of organic rare earth complexes [7], nor studies on the interaction of pharmaceutical compounds [8] or of lung surfactant peptides [9,10] with surface monolayers.

This contribution is organized as follows. We start by recollecting the basics of surface-sensitive scattering with an emphasis on applications to *soft condensed matter* (section 2) and in particular pay tribute to recent developments in data modeling (sections 2.2 to 2.4). We then proceed to discuss recent insights into the structure of floating phospholipid monolayers (3) and attempts at a quantitative assessment of ion binding to charged monolayers (section 4). As an example of a phospholipid monolayer structure of particular interest in cell signaling, the structure of inositol and inositide monolayers is discussed in a section of its own (section 5). We finish this report with a discussion of the adsorption of very small molecules on lipidic interfaces (section 6), as well as the conformational organization of large water-soluble polymers (section 7) grafted to surfaces by means of lipid anchors to round up the field.

2 Surface-Sensitive Scattering and the Structure of Floating Surface Monolayers

2.1 X-Ray and Neutron Optics at Interfaces

The optics of X-rays and neutron beams at surfaces is well established [11–13]. Briefly, the real part of the refraction index n for X-rays is only slightly different from unity – and usually lower – for the relevant frequencies, and the imaginary part of n, i.e. the absorption β, is also small:

$$n(z) = 1 - (\delta + i\beta) = 1 - r_0\rho_{el}(z)\lambda^2/(2\pi) - i\mu(z)\lambda/(4\pi), \qquad (1)$$

where z is the direction of the surface normal, r_0 is the classical Thomson electron radius, ρ_{el} is the local electron density, λ is the wavelength and μ is the linear absorption. (In neutron scattering, the neutron scattering length density ρ_n replaces $r_0\rho_{el}$.) Except in the case of resonant X-ray scattering, the absorption may be neglected for all practical purposes: $\beta \approx 0$. For a locally homogeneous material, the scattering length density (SLD) ρ is directly determined by the scattering-length contributions of its constituents, $\rho = (1/V)\sum_i \nu_i b_i$, where V represents a molecular-size volume containing ν_i atoms of the species i with the scattering length b_i. In the context of nonresonant X-ray scattering, b_i is simply the number of electrons of i.

If a planar wave with wave vector $k = 2\pi/\lambda$ impinges on a planar interface, external total reflection is observed below a critical angle of incidence[1]

$$\alpha_c \approx \sqrt{2\delta} = \frac{\sqrt{4\pi\rho}}{k}. \tag{2}$$

For specular reflectivity, the momentum transfer

$$\boldsymbol{Q} = \boldsymbol{k}_{\text{out}} - \boldsymbol{k}_{\text{in}} \tag{3}$$

is strictly normal to the interface:

$$Q_z = 2k \sin \alpha = \frac{4\pi}{\lambda} \sin \alpha. \tag{4}$$

Hence, the critical momentum transfer for total reflection

$$Q_c = \sqrt{16\pi\rho} \tag{5}$$

is a characteristic property of the medium (e.g., $Q_c^{\text{X-ray}} \approx 0.0217$ Å^{-1} for water). While for X-rays Q_c is always real, it may be imaginary in neutron scattering, for example when a beam hits H_2O, for which $\rho_n < 0$. In such a case, there exists no regime of total external reflection. In cases where it does exist, α_c is in the 100 millidegree range, since δ is of the order of 10^{-5}.

The reflection amplitude for a beam at an ideal interface between two media $j = 1, 2$ is derived from Fresnel's law,

$$
\begin{aligned}
r_{1,2} &\approx \frac{\alpha - \sqrt{\alpha^2 - 2(\delta_1 - \delta_2)}}{\alpha + \sqrt{\alpha^2 - 2(\delta_1 - \delta_2)}} \\
&= \frac{Q_z - \sqrt{Q_z^2 - (Q_{c,1}^2 - Q_{c,2}^2)}}{Q_z + \sqrt{Q_z^2 - (Q_{c,1}^2 - Q_{c,2}^2)}} \\
&\approx \frac{Q_{c,1}^2 - Q_{c,2}^2}{4Q_z^2} = \frac{4\pi(\rho_1 - \rho_2)}{Q_z^2}
\end{aligned}
\tag{6}
$$

and yields the reflectivity, $R = |r|^2$. Above the critical angle, the Fresnel reflectivity of an ideal surface in vacuum (or, for practical purposes, adjacent to a gas volume) decays thus with Q_z as

$$R_F \propto \left(\frac{Q_c}{2Q_z}\right)^4. \tag{7}$$

In real life, even ideal interfaces between two media are not mathematically sharp but are graded on the Ångstrom length scale, owing to atomic roughness in the case

---

[1] As the index of refraction is only slightly smaller than unity, all accessible information is collected at grazing angles. By convention, angles are thus measured *from* the interface in surface-sensitive scattering.

of planar solid state surfaces or thermally excited capillary waves in the case of fluid surfaces. This is usually taken into account by convolution of the step function $\Theta(z - z_0)$, which characterizes an ideal interface, with a Gaussian, yielding the error function (erf) as the relevant profile describing the interface [14]:

$$\Theta(z - z_0) \rightarrow \frac{1}{2}\mathrm{erf}\left(\frac{z - z_0}{\sqrt{2}\sigma}\right) + \frac{1}{2}, \tag{8}$$

where $\mathrm{erf}(z) = (2/\sqrt{\pi}) \int_0^z e^{-t^2} dt$ and σ is a parameter characterizing the amplitude of the roughness. The surface roughness (s.r.) leads to a Debye–Waller-like damping of the reflection amplitude at an ideally sharp interface,

$$r(\text{with s.r.}) = r(\text{w/o s.r.}) \cdot e^{-Q_z^2 \sigma^2/2}. \tag{9}$$

If a molecularly thin homogeneous film (index n_2, thickness d_2) is located on a semi-infinite substrate (index n_3), reflection of a wave impinging on the film-covered surface occurs according to (7) at both the front and the back faces, and gives rise to interference with an intensity pattern in the far field characteristic of the refractive index and thickness of the film. The interference originates from a phase factor that takes into account the propagation of the wave in the medium and leads to the amplitude

$$r_{1,2} = \frac{r_2 + r_3 \exp(2ik_{z,n_2} d_2)}{1 + r_2 r_3 \exp(2ik_{z,n_2} d_2)}. \tag{10}$$

Stratified surface films can obviously be accounted for by recursive application of (10), taking into account a global surface roughness by means of (9), as first suggested by Parratt [15].

The recursive Parratt algorithm treats the film and its substrate as continua. Alternatively, describing the interaction of the beam with each scattering center and summing the amplitudes leads to the kinematical approximation[2] to the reflectivity [11],

$$\frac{R(Q_z)}{R_F(Q_z)} \approx \frac{1}{\rho_{\text{substrate}}^2} \left| \int \frac{d\rho(z)}{dz} e^{iQ_z z} dz \right|^2. \tag{11}$$

Even if not precise, it is analytical and generally more intuitive than the recursion algorithm.

Arbitrary SLD profiles across the interface may be treated either by slicing the profile into a sequence of thin layers and determining the reflectivity by the Parratt recursion algorithm or by the kinematic approximation, (11).

As in any scattering experiment, the experimental data cannot be directly translated into the underlying structure, because of the "phase problem", i.e. the loss of phase information upon obtaining the scattered intensities from the amplitudes by use of either (10) or (11).

---

[2] (11) is an approximation, because multiple scattering in the medium is neglected. It is thus only correct for $Q_z > 5Q_c$, where multiple scattering plays no role.

2.2 Approaches to Reflectivity Data Inversion and Chemical Models

Procedures used for the inversion of reflectivity data fall into two categories: model-free and model-inspired approaches. Model-free inversion algorithms have been intensively investigated in the past decade [16–21]. One of the most general approaches is a constrained least-squares method using B-splines for the parameterization of the SLD $\rho(z)$, developed by Skov Pedersen and Hamley [19],

$$\rho(z) = \sum_{i=1}^{N} a_i B_i(z), \tag{12}$$

where B_i are spline functions and N is in practical terms determined by Q_z^{max}, the momentum transfer available in the data, via the sampling theorem. Results are steered toward physically "plausibl" solutions by constraints that suppress large fluctuations in ρ and define an "expected" average value of the SLD [19]. All model-free approaches have in common that they yield an SLD profile without an intrinsic molecular interpretation of the underlying structure.

In distinction, model-inspired data inversion procedures utilize the fact that molecular layer structures are often amenable to a parameterization that reflects the chemistry of the constituents. In the simplest case – i.e. floating surface monolayers of simple amphiphiles – "chemical intuition" suggests that the surface film comprises at least two distinct layers, one hydrophobic and adjacent to the gas phase and one hydrophilic and adjacent to the aqueous subphase. In such a case, "box models" [11, 13, 14] – or layer models – provide a simple, intuitive approach to data interpretation in submolecular terms using Parratt's algorithm [15]. This is why they have been extensively used in the past for the inversion of X-ray and neutron reflectivity data in studies of Langmuir monolayers at aqueous surfaces [22–24] and water/oil interfaces [25], as well as protein or polymer interactions with lipid surface layers [26–33]. Analogous models for data inversion have also been used to describe molecular layer systems at solid/fluid interfaces [34–39]. However – as shown recently in an analysis of high-resolution X-ray reflectivity data [40] – even in the simple case of monomolecular interface layers they are generally deficient at a realistic description of the chemically heterogeneous (phospho)lipid headgroups. Nevertheless, they have been frequently used – often *without* a molecular interpretation – since they are so simple to implement and lead often to a plausible description of the experimental data, as long as those are confined to low resolution.

More recently, improvements in synchrotron technology upon the commissioning of undulator beamlines [41] – which yielded a boost of the brilliance of the available X-ray intensity by three orders of magnitude – were initially utilized to perform more challenging diffraction experiments [4, 42]. However, for X-ray reflection experiments on molecular layers at aqueous surfaces, these technical improvements also led to a dramatic development: an increase of the available Q_z range by almost a factor of two, pushing the limits in resolution and enabling for the first time structural characterization of planar lipid model systems in submolecular detail [40].

2.3 VRDF Model

A realistic scenario describing the structure of phospholipid monolayers must account for the significant conformational flexibility of the lipid headgroups. The headgroup orientation might change from preferentially in-plane to preferentially out-of-plane, e.g. upon going through the monolayer "main" transition (i.e. the transition from the "liquid expanded" phase, LE, to the hexatic "tilted condensed" phase, TC [43, 44]). This structural flexibility is not possible to implement within the framework of a box model. Consequently, we developed a new approach to the modeling of molecular amphiphilic interface layers [45] which is based on a parameterization of the interface structure in terms of volume-restricted distribution functions (VRDF). In a similar approach to fluid bilayer structure determination, Wiener and White parsed the lipid molecules into different Sects. [46] and modeled the bilayer structure in terms of Gaussian distributions of these fragments [47]. In doing so, they coined the term "*composition-space refinement*" for a whole family of structure-based modeling procedures. Inspired by their work, we introduced a hybrid model in which the phospholipid backbone and lower headgroup are decomposed into fragments whose locations are described by distribution functions along the interface normal [45]. The aliphatic chains are treated just as in the box model, see Fig. 1. In distinction from the box model, volumetric information on the submolecular constituents is used as a constraint in the VRDF approach. Thus, space-filling is explicitly taken into account:

$$\sum_{\xi} dn_{\xi}\left(z\right) V_{\xi} = dV = A_{\text{lipid}}\, dz, \qquad (13)$$

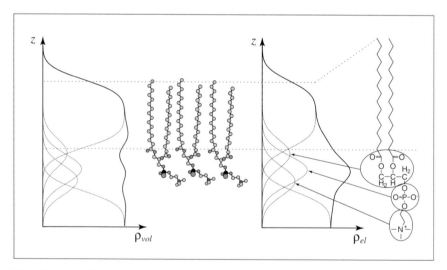

Fig. 1. Parsing of phospholipid molecules – DPPC is shown as an example – in the VRDF model. The volume distribution (left) is compared with the resulting electron density distribution (right). Reproduced from [45]

where $dn_\xi(z)$ is the amount of a fragment ξ in the plane dz located at a distance z from the interface and V_ξ is its partial volume. Handling of this additional constraint requires obviously an *a priori* knowledge of V_ξ.

Thermal broadening of the interface structure in the distribution function model derives from two contributions that need to be well distinguished: a broadening by capillary waves, σ_{cw}, and an intrinsic broadening of fragment positions, σ_{int}, which is the value one would expect to observe *without* capillary waves, i.e. within an interface film at an ideally flat surface. We assume that typical wavelengths for the two processes are well separated in real space. Hence the upper and lower interfaces of the alkane slab – which we consider atomically flat, at least for the ordered phases TC and UC ("untilted condensed") – are affected *only* by σ_{cw}, whereas the distributions of the headgroup fragments are affected by both contributions:

$$\sigma_{\text{total},\xi} = \sqrt{\sigma_{cw}^2 + \sigma_{\text{int},\xi}^2}. \tag{14}$$

This separation of the contributions to interfacial broadening permits a characterization of the distribution width of molecular subfragments within the surface layer and its development along the isotherm. Furthermore, one may check whether the capillary waves' amplitudes depend on the surface pressure π and temperature T as predicted by theory [48–50].

In order to estimate confidence limits on the parameters determined, one may map the deviations χ^2 – defined in the usual way – between the experimental data and the modeled reflectivity while detuning one selected parameter deliberately from its optimum value and readjusting all others to a new local minimum of χ^2. This has been described in detail in [40] and has been systematically used in the evaluation of the experiments described below. Typically we find that one may determine the position of, for example, a phosphate fragment at the interface with a precision of $\pm$ 0.5 Å [40]. The confidence limits on derived values for surface roughness parameters are typically lower than $\pm$ 0.1 Å [40].

2.4 *StringFit*: A Novel Data Evaluation Technique for Polymers at Interfaces

Polymers near interfaces are disordered systems by their very nature. As discussed above, reflectometry is very well suited to characterizing molecular interface layers with intrinsic disorder. However, the application of this technique to molecular polymer layers is complicated by the fact that these are not susceptible to modeling in a simple, intuitive way that reflects the molecular interactions as in the case of amphiphilic surface layers. Thus, data interpretation in slab models contributes at best little to a molecular-scale comprehension of such systems [32, 33], although some progress has been recently made [31, 39].

We have devised a novel data inversion technique that is particularly suited to thread-like (linear) polymer samples [51]. In this approach, an ensemble of individual polymer molecules (e.g. 40 in our published application to mid-sized polyoxazoline lipopolymers [52]) is explicitly modeled in a quasi-molecular description, thereby deliberately overparameterizing the system. We then utilize an evolution strategy

(ES) algorithm [53] to identify *typical* ensemble conformations that are relevant to the problem. In this context, one may use the terms "genotype" (the genetic information encoded in a chromosome) and "phenotype" (the appearance of an individual, here a molecular ensemble or its reflectivity under well-defined constraints) in analogy to their definitions in biological evolution. A chromosome consists of a sequence of information units ("genes") that are represented by an n-dimensional vector. In the implementation of such an approach in the current work, a chromosome corresponds to the genotype of a certain molecular ensemble. The "fitness" of an individual that is used as a selection criterion in the ES is a numerical value and is derived from the correspondence between the reflectivity computed for the molecular density distribution of the molecular ensemble across the interface and the experimentally observed data. To mimic biological evolution processes, individuals may be grouped. Such groups are then called "populations".

In a typical implementation (discussed in detail below, see p. 432), the chemical structure of a lipopolymer is treated as a sequence of cylindrical segments, representing the hydrophobic alkyl anchor and hydrogenated or deuterated polymer chains. Each of the segments is subdivided into subsegments of equal length that are characterized by their effective SLDs per unit length, derived from the scattering lengths and the molecular cross sections. As the chemical composition in the various segments of a lipopolymer changes along its contour line, so does the representation of the molecule in terms of its SLD per unit length and volumetric cross section. Segments and subsegments are connected via flexible joints. One angle between each pair of neighboring (sub)segments is sufficient to determine the density distribution of a molecule along z. Thus, a sequence of orientation angles ψ_i is used to describe the molecular configuration of an individual molecule. In addition, other molecular properties, such as the penetration depth of an individual molecule within the ensemble, may be included in the "genome" description and refined in the fitting. As the algorithm parameterizes the molecular configuration, it enables composition-space refinement [24, 46] and may thus be used for simultaneous fitting to various data sets from isotopically distinct samples or to neutron and X-ray data obtained under similar experimental conditions [52].

2.5 GIXD on Hexatic Alkane Phases in Surface Monolayers

As extensively discussed by Fischer and Lösche in this volume (Chap. 10), phospholipid monolayers at the air–water interface undergo various phase transitions as a function of, for example, temperature or surface pressure. The respective phases bear a resemblance to three-dimensional gaseous, liquid, hexatic, and solid phases. The lateral pressure, temperature, and chemistry of the liquid subphase are thus convenient control parameters to induce and study such transitions. The availability of intense synchrotron X-ray radiation enables one to observe diffraction from (partially) ordered alkane systems even from monolayers in dedicated experiments. The setup and experimental parameters used in this work to characterize monolayers in grazing-incidence X-ray diffraction (GIXD) were similar to those of the X-ray reflectometry experiments already discussed above. The technique was developed in the second half of the 1980s and is described in numerous publications, e.g. [22, 54–56].

For a GIXD measurement, the incident angle α_i is kept constant at a value below the critical angle α_c. An evanescent wave is thus generated which propagates parallel to the surface. The intensity of this evanescent wave decays exponentially toward the depth of the subphase, z. In the experimental setup used at BW1 ($\alpha_i \approx 0.85\alpha_c$), a layer with a thickness of ≈ 100 Å is illuminated by the incident X-ray beam. Such an experimental situation is obviously perfectly tailored to investigating molecular lipid [54] and protein [4, 57] monolayers.

A hexatic alkyl chain lattice within a surface monolayer leads to Bragg diffraction, typically picked up with a position-sensitive detector that records the scattered intensity as a function of the angles 2Θ (within the sample plane) and α_f (perpendicular to the sample plane). Owing to the quasi-two-dimensional (2D) organization of the system, the Bragg spots that are expected for diffraction from a 3D system are smeared into *lines* of intensity, the so-called Bragg rods. The diffracted intensity along a particular Bragg rod – designated by the Miller indices h, k of the 2D lattice – is given by

$$I(h, k, Q_z) = |F(h, k, Q_z)|^2 \, e^{(-Q_{hk}^2 u_{\mathrm{hor}} + Q_z^2 u_{\mathrm{hor}})} \, |T(\alpha_f)|^2 \,. \tag{15}$$

The most prominent feature in this relation is the structure factor $F(h, k, Q_z)$, which represents the Fourier transform of the electron density of a molecule within the unit cell. The exponential term accounts for the Debye–Waller factor which quantifies the horizontal and vertical mean square displacements, u_{hor} and u_z. The factor $|T(\alpha_f)|^2$, finally, represents the so-called Vineyard–Yoneda peak [58, 59] due to interference observed at $\alpha_f = \alpha_c$.

Reciprocal-lattice vectors can be directly obtained from the position of the peaks at $Q_{xy} = Q_{xy}(h, k)$. From the real-space lattice parameters, the minimal molecular area within 2D patches – or domains – in the monolayer is calculated. Analysis of the shape of the positions of the Bragg rods – which may be degenerate – and their transverse widths yields information about the chain tilt angle and azimuth (within the unit cell) on the one hand, and the lattice correlation length on the other.[3] The details of such an analysis have been described in detail [11, 44, 60, 61].

2.6 Experimental Details

Instrumentation for Scattering Experiments

The general principle of an X-ray reflectometer is quite simple.[4] A well-monochromated X-ray beam is guided through a pair of slits toward the surface that bears the molecular layer system under investigation. This beam can be tilted with respect to

---

[3] In practical terms, the information obtained from such a shape analysis is often limited by the resolution of the experimental setup.

[4] Here we refer to the so-called angle-dispersive instruments, while energy-dispersive instruments will not be discussed.

the surface by a monochromator in the case of synchrotron facilities or by moving the X-ray tube itself in laboratory devices. Before the beam hits the sample surface, it passes through a calibrated absorber to increase the dynamic range of the detector. The reflected beam is collected by another pair of slits and guided to the detector in a flight tube. Naturally, home-built and synchotron-associated X-ray reflectometers differ quite a bit in their details. Thus, synchrotron experiments usually require an extra detector that measures a tiny, but constant, fraction of the incoming beam to monitor the source intensity. Home-built devices do not require such a beam monitor as the stability of the X-ray tube is sufficient. The latter ones, however, may incorporate a sophisticated X-ray optics setup (i.e. a multilayer mirror) to increase the efficiency with which photons are directed from the anode to the sample. Since even at synchrotrons, with their high flux, these measurements may take of the order of hours, the experiments are becoming more and more automated to allow continuous operation. Some synchrotron beamlines (e.g. the MUCAT beamline 6-ID-B at Argonne National Laboratory) are equipped with upstream double monochromators, allowing a very narrow selection of the photon energy. This enables anomalous or resonant scattering on absorption edges of heavy elements as described later, see Sect. 4.3.

Sample preparation and precharacterization were routinely performed at an in-house X-ray reflectometer (built at Leipzig in collaboration with JJ X-ray, Roskilde, Denmark). It allows reflectivity measurements down to $R \approx 5 \cdot 10^{-9}$ and has been described in detail [62]. The X-ray instrument located at the undulator beamline BW1 [41] of the DORIS III bypass at HASYLAB, DESY, Hamburg is routinely used for both reflection and GIXD measurements and has also been extensively described [63]. It allows reflectivity measurements down to $R \approx 5 \cdot 10^{-10}$. A station similar in characteristics has been established at the Midwestern Universities Collaborative Access Team (MUCAT) sector of the Advanced Photon Source (APS) of Argonne National Laboratory (ANL) in Argonne, IL. The double monochromator of this beamline selects the X-ray energy better to than 2 eV FWHM. This superior energy resolution can be utilized, for example, in anomalous-scattering experiments using heavy metals [64].

We found that beam damage of the surface monolayer is an important issue in the full beam at high-flux synchrotron beamlines, particularly for phospholipid monolayers in the fluid phase. The monolayer was thus continuously shifted underneath the beam footprint to ensure that data were collected from undamaged areas within the surface film. Reproducibility of the measured data was checked routinely along various fragments of the reflectivity curve.

For routine measurements and to establish experimental procedures at the diffractometer at low resolution, an all-purpose diffractometer [62] in our labs at Leipzig University has been extensively used. This instrument has also been used to perform wide-angle X-ray scattering (WAXS) measurements from bulk lipid dispersions, as shown in Sect. 6.2.

Auxiliary Monolayer Characterization Experiments

Synchrotron X-ray experiments were carried out only after extensive sample characterization with more conventional techniques. These are briefly described in the following few paragraphs.

Surface Isotherms. Isotherms were measured on Langmuir film balances made from Teflon (PTFE) that were of local design. Typically, about $20 - 100\ \mu l$ of a solution of lipids, or of a premixed peptide/lipid cosolution where applicable, in chloroform or chloroform/methanol at a concentration in the range of $0.5 - 1$ g/l was deposited on a clean water or buffer surface (maximum area typically 300 cm^2) using a microsyringe. The volatile solvent was allowed to evaporate for at least 10 minutes. Subsequently, a computer-controlled PTFE barrier decreased the film area continuously at a rate of (again, typically) $1\ \text{Å}^2/(\text{molecule} \times \text{min})$ while the surface tension γ was measured with a Wilhelmy plate made of ash-free filter paper and monitored as a function of the surface area A (measured in Å^2/molecule). The surface pressure π is defined as the difference between the tension of the clean substrate surface and the tension of the film-covered surface at a given lateral molecular density, i.e. $\pi(A) = \gamma_0 - \gamma(A)$.

Fluorescence Microscopy. Since data inversion of reflectivity experiments is model-based (see Sect. 2.1), it is crucial to ascertain that molecular surface films investigated by X-ray scattering are homogeneous at length scales exceeding the in-plane coherence length of the experiment, approx. $10\ \mu\text{m}$ [24]. One characteristic situation in which this condition is *not* met occurs obviously if the surface films are phase separated in the course of a first-order phase transition (see Chap. 10). The homogeneity of surface films, i.e. the absence of lateral phase separation, is typically verified by fluorescence microscopy [65, 66] or Brewster angle microscopy [67, 68]. Care was taken in all surface scattering experiments described here to establish sample preparations in which phase separation was strictly avoided.

Fluorescence microscopy was carried out using an epifluorescence microscope (Carl Zeiss, Jena, Axiotech Vario) attached to a home-built Langmuir film balance as decribed in [69]. In brief, the Langmuir monolayer contains a small amount of fluorescently labeled lipids (typically about 0.5 mol%). As the majority of lipids segregate into coexisting phases, the label dissolves preferentially into the phase of higher local disorder, yielding optical contrast. Since the length scale of the phase separation is on the order of tens of micrometers (Chap. 10), this phase separation is well amenable to optical observation. In the Leipzig setup, fluorescence is excited by illumination of the molecular film on the aqueous surface though the objective lens (typically, a Zeiss Epiplan $50\times$ LD), using a 50 W mercury arc lamp and an appropriate filter/beam splitter combination to select the excitation band of the dye and separate it from its emission wavelength band. The fluorescence is collected through the same lens, redirected toward the observation channel by a dichroic beam splitter, and filtered with an emission band filter. It is then detected using a low-light-level camera with a silicon-intensified target (SIT) cathode (Hamamatsu C2400-08).

The microscopic images are digitally sampled. Where postprocessing was necessary, this was performed using NIH Image v1.63. With the $50\times$ objective lens, the images have a typical width of 150 μm. The entire setup is placed on an vibration isolation table (JRS Systems). To compensate for convective motion of the surface film, an x–y stage allows shifting the position of the surface laterally under the objective lens.

ATR-IR Spectroscopy. Attenuated total-reflection infrared spectroscopy (ATR-IR) measurements of oriented multilayers were carried out with a Bio-Rad (Digilab) FTS 60A spectrometer. The sample compartment was purged with dry N_2. Experiments on pure DPPC and neuropeptide Y, as well as mixed DPPC/NPY multilayers at 8 mol% of the peptide, were performed by spreading 100 μl of the respective solution on a $ZnSe_2$ crystal. The solvent was then allowed to evaporate for 10 min. Measurements were performed at $T = 20 \pm 0.5°C$ by averaging 256 scans with a resolution of $4\,\mathrm{cm}^{-1}$.

Materials

Water was filtered in a Millipore Milli-Q system, yielding a residual specific resistance of 18.2 MΩ cm. Dimyristoylphosphatidic acid (DMPA), dipalmitoyl-phosphatidylcholine (DPPC), DMPC, DM-phosphatidylethanolamine (DMPE), DM-phosphatidylserine (DMPS), and DP-phosphatidylglycerol (DPPG) from Sigma (Munich) were used as supplied. Peptide/lipid monolayers were prepared by co-spreading typically 50 μl of a premixed DPPC/NPY solution in chloroform/methanol (ratio 3:1, Merck, HPLC grade) on pure water subphases. π–A isotherms ($T = 20\pm0.5°C$) were registered while compressing the monolayer with a constant barrier speed of 0.1 mm/s, corresponding to typically 1.8 $\mathrm{\mathring{A}}^2/(\mathrm{molecule} \times \mathrm{min})$. Dipalmi-toylphosphatidylinositol (DPPI) and its monophosphorylated derivatives DPPI-3P, DPPI-4P, and DPPI-5P, from Echelon Inc., Salt Lake City, UT were used without further purification. Lipid solutions ($CHCl_3$/MeOH, 3:1, plus traces of water) were spread to form monolayers at $T = 20\ °C$ on pure water or physiological buffer subphases (100 mM NaCl, 10 mM HEPES, 0.1 mM EDTA).

For fluorescence microscopy investigations, DPPC was doped with 0.4 mol% β-Bodipy-C_2-HPC or, for lipid/peptide dual-label experiments, with 0.2 mol% rho-damine B-DHPE (both labels from Molecular Probes, Leiden, The Netherlands).

3 Resolved Structure of Phospholipid Monolayers on H_2O

Manfred Schalke[1], Markus Weygand[1], Peter Krüger[1], and Mathias Lösche[1,2]

[1]   Universität Leipzig, Institut für Experimentelle Physik I, Leipzig, Germany
[2]   Johns Hopkins University, Department of Biophysics, Baltimore, MD

Figure 2 (upper left) shows the DMPA isotherm ($T = 26.8 \pm 0.5°C$) on pure water and indicates where five reflectivity measurements (lower left) were taken to evaluate the structure of the molecules within the surface monolayer. The isotherm shows the well-known [70] first-order phase transition between LE and TC near $\pi \approx 8$ mN/m. At $\pi \approx 40$ mN/m, a second-order phase transition to the UC phase is observed [54].

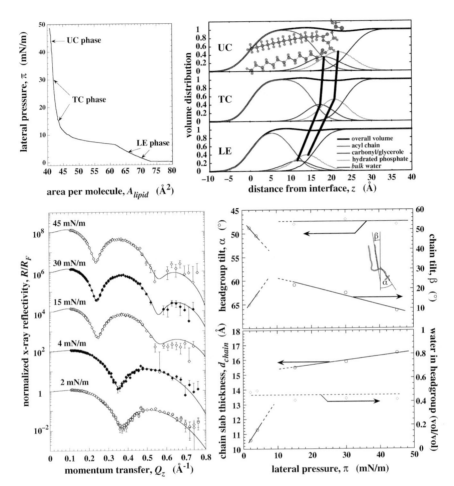

Fig. 2. Upper left: DMPA isotherm on H_2O at $T \approx 27°C$. Lower left: normalized X-ray reflection of a DMPA monolayer on H_2O ($T \approx 27°C$) at various lateral pressures π, as indicated. Best fits to the data (continuous lines) are derived from VRDF models. Data at higher π are offset by factors of 100. Upper right: development of the VRDF-derived volume distributions of DMPA fragments in the various distinct phases along the isotherm. A molecular model of DMPA with highly ordered acyl chains at approximately the correct scale is indicated. Fat black lines indicate the change of the distance between the centers of gravity of the lipid's backbone and its phosphate across the phase transitions. Lower right: quantification of the derived fragment distributions (upper right) in terms of molecular parameters. Reproduced in part from [45]

Fitting of the data with a molecular model was achieved better with the quasi-molecular VRDF model than with slab models [45], although in the LE phase ($\pi \approx 2$ or 4 mN/m), the quality of the fits was comparable. Values of the area per molecule, A_{lipid}, in the surface films for fitting were taken from the isotherm. The component volumes for the submolecular fragments of DMPA (AC, acyl chains; GC, glycerol–carbonyl backbone; and P, phosphate) used in the VRDF fits were taken from the literature [71]. The best-fit VRDF volume distributions are shown in the upper right of Fig. 2. Qualitatively, it is observed (fat lines) that the distance between the centers of the GC and P distributions increases as the monolayer is compressed. This may be interpreted as a swiveling motion of the headgroup toward the subphase upon compression. This is shown quantitatively in the lower right of Fig. 2. One significant result that is in contradiction to earlier interpretations of monolayer structure using slab models [72, 73] is that the headgroup hydration remains constant along the isotherm (lower panel in the lower right of Fig. 2). The VRDF interpretation suggests that on average around 6 water molecules associated with the phosphate are located at the same z distance from the interface. This number, however, does not fully reflect the hydration shell of the phosphate, as there are more water molecules in the subphase associated with P: recent hydration-dependent scattering experiments on DOPC multilayers showed a crossover from partial to full hydration at about 12 water molecules per lipid [74]. This would suggest that about 6 more water molecules are associated with the phosphate in DMPA on top of the semi-infinite water reservoir of the subphase. A full assessment of the DMPA monolayer structure, a determination of the confidence limits on the parameters determined and a more extensive discussion of the implications in terms of membrane structure are given in [45].

Other phospholipid species that have been recently studied in surface monolayers on H_2O include DPPC, DMPC, DMPE, and DPPG [75]. In this work, similar trends as for DMPA were observed. However, owing to the lack of electron density contrast between the terminal fragments of the lipid headgroups (choline, ethanolamine, and glycerol) and water, it is at present not possible to determine these structures conclusively from the available X-ray data alone without making assumptions.

4 Divalent Cations at Charged Surface Monolayers

Jens Pittler[1], Manfred Schalke[1], Peter Krüger[1], David Vaknin[2], and Mathias Lösche[1,3]

[1]   Universität Leipzig, Institut für Experimentelle Physik I, Leipzig, Germany
[2]   Ames Laboratory and Department of Physics and Astronomy, Iowa State University, Ames, IA
[3]   Johns Hopkins University, Department of Biophysics, Baltimore, MD

4.1 High Ba^{2+} Concentrations at $DMPA^-$ and $DPPG^-$

Divalent cations play an extremely important role in cell physiology [76]. On a molecular scale, the binding of proteins, such as annexins, to membranes is controlled by calcium [77]. Thus, the role of such ions has been extensively studied in biomimetic model systems [78]. We have investigated lipid monolayers on ionic subphases as model systems to characterize the interaction of biomembranes with their ionic environment by a host of different techniques. We have particularly focused on the interactions of divalent cations – in the form of metal salts – with the charged headgroups of DMPA molecules at the air–water interface by using combined techniques of X-ray reflectometry and GIXD. It is well established that monolayers become much more ordered in the presence of cations in the subphase. In fact, cation superlattices have been observed in GIXD at ordered lipid monolayers [79, 80]. It is thus clear that the properties of charged organic molecules within monolayers at the air–water interface are strongly affected by such metal ions.

In a recent study [40], X-ray reflectivity measurements of DMPA on $BaCl_2$ subphases indicated that one cannot simply assume a stoichiometric ratio of 2 : 1 ($DMPA^-$: Ba^{2+}) in describing the surface structure. Instead, the electron density observed in the headgroup region was found to be disproportionately large at low lateral pressure and approached the value expected for the stoichiometric (2 : 1) ratio upon compression of the monolayer only at very high pressures [40]. On the other hand, it has been reported that the position of the bound cations along the surface normal coincides with the position of the phosphate [40], such that one may model the phosphate fragment and the cations using one distribution function along the z-axis in a VRDF approach. As a practical consequence, one may directly determine the overall number of water molecules in the headgroup without any assumption on a specific water : cation association ratio. Similarly to the situation observed with DMPA on pure water [45] – and in distinction from results on phospholipids derived from reflectivity measurements via box models [73] that indicated large variations of the headgroup hydration along the isotherm – the VRDF evaluation of high-resolution data suggests that the hydration is essentially constant, irrespective of the surface pressure, phase state, and thus available area per lipid molecule [40]. Similar studies of Ba^{2+} binding to $DPPG^-$ [62] have subsequently confirmed such a picture, so that it appears that both an overstoichiometric binding of the divalent cations to anionic lipid interfaces and the (approximately) constant hydration of phospholipids along the isotherm may be general features of phospholipid structure at aqueous surfaces.

4.2 Concentration Dependence of Ba^{2+} Binding to $DMPA^-$

The high affinity between divalent cations ions and anionic lipids is easily demonstrated in simple surface pressure measurements. Figure 3 shows DMPA isotherms for different Ba^{2+} concentrations. Deviations from the isotherm on pure water appear at $c_{Ba^{2+}}$ as small as 10 nM. At higher Ba^{2+} concentrations, monolayers show significant condensation, particularly obvious at the first-order (LE–TC) phase transition. While this transition is fully developed on pure water at temperatures slightly above room temperature, e.g. at 24 °C, it is largely suppressed at 100 nM Ba^{2+} and entirely abolished at 1 μM and above. Also observed in the isotherm is the second-order TC/UC transition at $\pi \approx 40$ mN/m, which is much less affected by the changes of the Ba^{2+} concentration.

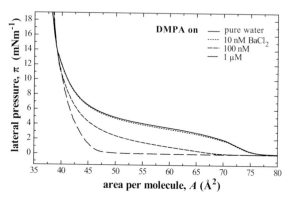

Fig. 3. DMPA isotherms on water at different Ba^{2+} concentrations ($T = 22\,°C$)

To avoid problems with the data interpretation, all DMPA X-ray measurements were obtained in homogeneous phase regions. Figure 4 shows three X-ray scattering scans of DMPA on water at $\pi = 30$ mN/m, normalized to the Fresnel reflectivity R_F, in the absence and presence of Ba^{2+} together with the best-fit VRDF models (see p. 400). The DMPA molecule has been parsed into just three components: the alkane region and the GC and P fragments, as described earlier (see p. 406). The values of A used in the fitting of the different data sets have been measured separately using GIXD: at $\pi = 30$ mN/m, we determined $A = 41.6$ Å^2 on pure water and $A = 41.3$ Å^2 on 1 mM Ba^{2+} solution. The submolecular fragment volumes $V_{GC} = 146.8$ Å^3, $V_P = 53.7$ Å^3, and $V_{Ba^{2+}} = 10$ Å^3 were taken from the literature [71] or calculated from the ion radius. Within the VRDF modeling, we fitted the chain slab thickness d_{chain}, the distance of the phosphate position from the alkane/air interface z_P, the global surface roughness due to capillary waves σ_{cw}, the intrinsic distribution width of the headgroup components along the z-axis σ_{int}, the number of water molecules n_W, and the number of Ba^{2+} ions per headgroup $n_{Ba^{2+}}$. The results as a function of the Ba^{2+} concentration are visualized in Fig. 6. While some of the model parameters do not change significantly with $c_{Ba^{2+}}$, ion binding – and thus the electron density located within the headgroup – increases systematically, as expected. This has been independently confirmed using a model-free data description based on the approach of Skov Peterson and Hamley [16]. Confidence limits on the derived parameters were determined as described earlier (see p. 401). Despite the fact that the headgroup electron density increases systematically, the detailed molecular picture of the lipid's headgroup region is not yet entirely clear. Figure 5 shows the pressure and Ba^{2+} concentration dependence of two selected model parameters in monolayers, d_{chain} and z_P. Their values increase as the monolayers are compressed. On the other hand, however, they show an entirely different Ba^{2+} concentration dependence: while the chain slab thicknesses – and hence, more generally, the chain packing – are very

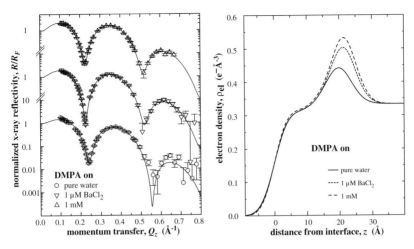

Fig. 4. Normalized X-ray reflectivity R/R_F, of DMPA at $\pi = 30$ mN/m, $T = 22$ °C, and different Ba^{2+} concentrations as indicated (left), and derived electron density distributions (right)

similar at all but the highest pressures,[5] the distance of the phosphate from the interface depends systematically on both π *and* $c_{Ba^{2+}}$. The increase of z_P with $c_{Ba^{2+}}$ indicates progressively larger amounts of cations and/or hydration water incorporated in the headgroup. In most cases – and particularly at any $c_{Ba^{2+}} > 100$ μM – we observe that more cations are bound to the surface than expected for a stoichiometric ($DMPA^- : Ba^{2+} = 2 : 1$) ratio. However, *exactly* how much Ba is bound remains unclear.

The reason for this uncertainty resides in the model dependence of the data inversion. While the electron density profiles shown in the right panel of Fig. 4 are unambiguous – and have been reconfirmed using model-independent fits – their interpretation in terms of a submolecular fine-structure *is not*. The reason is an underdetermination of the model in the case when only *one* reflectivity data set for a monolayer in a particular preparation state is available. This is exemplified in Fig. 6, which shows the ion binding to DMPA monolayers at a specific reference pressure, π = 30 mN/m, as a function of $c_{Ba^{2+}}$, derived from VRDF models *under slightly different assumptions*: while in one set of models, electrostatic neutrality of the surface has been enforced (for details, see p. 416), this boundary condition has been relaxed in the other set of models. As revealed from Fig. 4, the two models deviate significantly, in particular at low Ba^{2+} concentrations. As is indicated with two broken lines, intended as guides for the eye, these two models – which correspond to extreme cases – would have to be interpreted in terms of two very different binding constants and binding mechanisms (which, in view of our incapability to discriminate the validity of the competitive models, are irrelevant to the discussion at this point). Obviously,

---

[5] We found that fits of the data for monolayers above 40 mN/m were consistently poorer than for those at lower surface pressures.

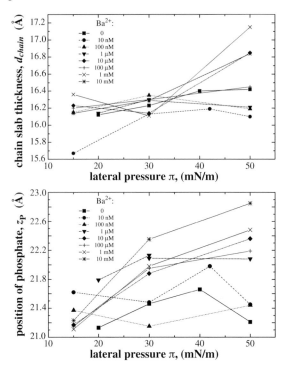

Fig. 5. Dependence of two key parameters describing the structure of DMPA monolayers on water as a function of the lateral pressure and the Ba^{2+} concentration

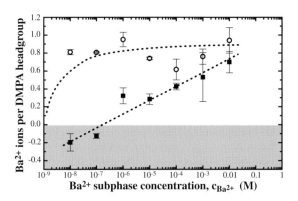

Fig. 6. Number of Ba^{2+} ions per DMPA$^-$ headgroup as a function of the bulk Ba^{2+} concentration, derived from VRDF models under two different assumptions

both assumptions – that of a neutral interface and that of an uncompensated charged one – represent two extreme cases. In that sense, Fig. 6 depicts a worst-case scenario. It must, however, be emphasized that the *electron density distributions* in the monolayers under the different models look alike. A resolution of this matter could thus only be afforded by "labeling" of the molecular species of interest – i.e. the barium ions. One way to achieve this could be by replacing them with a species of similar chemical properties but different electron number. Thus, an exchange of the barium with strontium, calcium, and eventually, magnesium would be a salient strategy to resolve this puzzle. However, different chemical interaction of the phosphate with the divalent ions – specifically, with Mg^{2+} – and different water association in the vicinity of the distinct ions – again, Mg^{2+} is probably the most critical in this group – might conceivably lead again to distinct monolayer constitutions. Another possibility – and this is the one we have chosen, see next section – is to investigate the system of interest under *identical* chemical conditions while varying the effective number of Ba^{2+} ions in anomalous-scattering experiments, i.e. by using photon energies at an absorption edge of barium and comparing the scattering with that at energies far away from any absorption.

4.3 Anomalous X-Ray Reflection of Ba^{2+} near DMPA$^-$ Monolayers

Anomalous X-ray scattering uses the fact that both the effective number of electrons and the absorption of an atomic species contributing to the scattering depend on the photon energy of the scattered radiation. As pointed out earlier, X-rays generally interact weakly with matter. The interaction becomes more pronounced when the energy of the radiation is resonant with the shell energies of atoms in the beam. In this case, absorption becomes significant, changing rapidly across the so-called absorption edges. This feature is utilized for variation of the scattering associated with that particular atomic species, i.e. as a means of contrast variation, comparable to the deuterium labeling used in neutron scattering or IR spectroscopy, with the benefit that samples need not be changed to probe a particular structure at different contrasts. One drawback, however, is that this elegant method can only be used efficiently with heavy atoms. As a rule of thumb, anomalous scattering works well for atoms located in the fifth row of the periodic table and up.

In order to resolve ambiguities associated with the evaluation of the 8.0 keV X-ray reflectivity measurements on divalent-cation binding to anionic monolayers reported above (see Sect. 4.2), we utilized the anomalous scattering of Ba^{2+} at the barium L$_{\text{III}}$ edge [64], which is centered at $E = 5.247$ keV. As revealed from Fig. 7 (left panel), the f_1 value varies by as much as 30% across the edge. In an exploratory experiment, we have observed the X-ray reflectivity from a DMPA monolayer on a subphase that contained a large concentration of $BaCl_2$, $c_{Ba^{2+}} = 10$ mM, at 8.0 keV (far above the edge), 5.247 keV (at the edge) and 5.1 keV (slightly below the edge), see left panel of Fig. 7. A closer inspection of Fig. 7 shows subtle differences between the scattering curves at the various X-ray energies, particularly around the interference minima near 0.22 Å^{-1} and 0.52 Å^{-1}. For a first, semiquantitative evaluation it is also noted that the reflection intensities of the maxima near 0.1 Å^{-1} and 0.33 Å^{-1} differ from

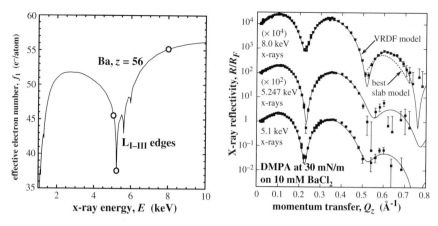

Fig. 7. Left: anomalous dispersion of Ba and X-ray energies where measurements have been taken in this work. Right: X-ray reflectivities of a DMPA monolayer on 10 mM BaCl at different photon energies are shown together with the best VRDF fit using one global parameter set (solid lines). The dotted line shown in the 8.0 keV data set indicates the best-fit reflectivity model using a two-slab approach. The corresponding electron densities of the slab model and the VRDF model are compared in the left panel of Fig. 8. In parts, adapted from [64]

each other in the sequence $R(8.0 \text{ keV}) > R(5.10 \text{ keV}) > R(5.247 \text{ keV})$, indicating that the total amounts of scattering length comprised within the surface layer vary in the same sequence. More quantitative evaluations based on either slab or VRDF models reveal further details of the corresponding differences in the scattering-length density profiles as a function of E [64]. These SLD profiles are shown in Fig. 8.

Fitting the data with a two-slab approach by using a consistent set of energy-independent parameters (i.e. surface roughness, thicknesses, and chain densities in the slabs) for all three data sets, while allowing distinct headgroup SLD and absorption values in a description of the distinct data sets, already yields a fair agreement between the data and the model. However, subtle deviations are noticed at high Q_z, as exemplified in the right panel of Fig. 7, where a dotted line shows the *best* fit to the 8 keV data. Under the constraint that the energy-independent parameters are identical for all three measurements at distinct X-ray energies, these deficiencies cannot be relieved. Moreover, even if each data set is analyzed without cross-reference to the others, significant discrepancies between the data and the models persist. SLD profiles retrieved from such slab fits are depicted in the left panel of Fig. 8 (and compared with the best-fit VRDF model, see below). The slab fitting does not allow us to locate any bound Ba^{2+} within the headgroup slab with submolecular resolution. However, the correlation of the total scattering length located within the headgroup slab, $n_{e,eff}^{headgr}$, with the variation of the effective electron number on the metal cation, $z_{Ba^{2+},eff}$, as a function of E (inset in the left-hand panel of Fig. 8) shows that the variations in the reflectivity with E are consistent with the assumption that these are entirely due to the anomalous scattering of Ba^{2+}. Moreover, the slope of the linear correlation and its extrapolated intercept at $z_{Ba^{2+},eff} = 0$ are indicative of the number of Ba^{2+} bound in the headgroup and the scattering length as it would appear *without* any Ba^{2+} in the headgroup, respectively. From the latter quantity, the number of water molecules that hydrate the phosphate may be determined. These numbers compare well with the corresponding quantities derived from the VRDF approach (see below). Beyond these quantities, the VRDF, however, reveals information about the organization of the bound metal cations *within* the headgroup. For data evalua-

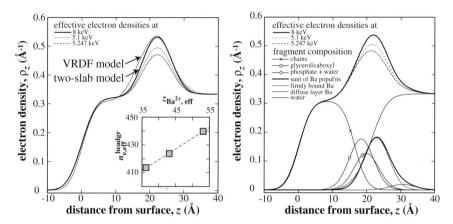

Fig. 8. Electron density profiles describing the reflectivity data shown in Fig. 7. Left: comparison between the two-slab and the VRDF model. Inset: apparent sum of f_1 (per DMPA molecule) of the headgroup constituents – including bound Ba^{2+} and hydration water – as a function of the effective electron number of Ba^{2+} at the relevant E values. The correlation demonstrates that the differences in the scattering curves are *entirely* accounted for by the anomalous dispersion of Ba. Right: decomposition of the VRDF-derived SLD profile into components due to the submolecular fragments. The chemical heterogeneity of the headgroup region is clearly revealed. The data are best described if one allows for an asymmetric Ba distribution, modeled as *two* Gaussian distributions. One may speculate that these two distributions relate to a *Stern* layer closely bound to the interface and a diffuse Debye layer penetrating into the bulk subphase. Adapted from [64]

tion, the VRDF approach is beneficial in two ways. As discussed earlier, the model enables inherently a chemical identification of components within the layer system. Moreover, since molecular moieties are identified within the sample structure, VRDF allows a proper assignment of scattering lengths to the various subfragments as they appear at different X-ray energies. This implies that the complete data set which comprises all reflectivities at the various energies can be modeled *simultaneously* and *consistently*. While this approach naturally highlights the location and quantity of the heavy metal ion, a higher sensitivity of the model to all other parameters than for one single measurement at a fixed X-ray energy is also achieved, as quantified in a χ^2 mapping, enabling the determination of more parameters in the model without over-parameterizing the problem.

Although the number of free parameters fitted in the VRDF approach equals the number of parameters in the slab fitting (the latter is less economical in its use of parameters since there is no possibility to cross-correlate the SLDs and absorptions of the headgroup slabs between data sets taken at distinct X-ray energies), the χ^2 value of the best VRDF model is about 40% below that of the best slab model. As indicated in Fig. 7, the slab model performs particularly poorly at high Q_z, suggesting that the corresponding structural differences between the two models are close to the limit of resolution. Figure 8 (left) shows that these subtle differences are located specifically where the lipid headgroups are bordering the hydrophobic chains on one side and the aqueous subphase on the other. In particular, at the acyl-chain–headgroup interface the electron density profile is rather asymmetric, a feature that cannot easily be reproduced in a slab model without introducing even more parameters.

The right-hand side of Fig. 8 shows the electron density profiles generated from the VRDF modeling of the reflectivity data at the three photon energies. Furthermore, a breakdown of the overall electron density at 8 keV into contributions from the individual molecular subfragments is also shown. The data are significantly better accounted for if one assumes *two* distinct contributions to the Ba^{2+} surface excess [64]. The Ba^{2+} distribution thus obtains a pronouncedly asymmetric envelope. A possible interpretation of these two Ba^{2+} population is as follows. One population – located within the lipid headgroup and overlapping strongly with the phosphate distribution – represents a tightly bound *Stern* layer, while the second population may represent the tail of the diffuse *Debye-Hückel* layer, which is expected to form a cloud of counterions that decays exponentially as one moves away from the interface. Since the charged interface is not a mathematically precise plane, but rather roughened by thermally excited capillary waves, and since coarse-grain effects play a dominant role at the length scales reported here, a quantitative comparison with the Debye–Hückel continuum theory is not straightforward.

In any case – and independent of any theoretical description – both the box and the VRGD models show unequivocally that the number of cations bound per lipid is considerably larger than the stoichiometric ratio, 1 : 2. Since it is certainly unreasonable to assume an *over*compensation of the anionic surface charge, one has to invoke other anionic species in the chemical description of the interface. These extra charges may derive either from the Cl^- reservoir in the bulk subphase or – more likely – from hydroxyl ions in the water, i.e. Ba^{2+} ions within the *Stern* layer may be bound in the form of a $Ba(OH)^-$ species. A similar interpretation of Cd^{2+} binding to fatty acid monolayers has been put forward earlier by Leveiller and coworkers [81].

5 Phosphoinositol and Phosphoinositide Structure

Carsten Selle[1], Undine Dietrich[1], Peter Krüger[1], and Mathias Lösche[1,2]

[1] Universität Leipzig, Institut für Experimentelle Physik I, Leipzig, Germany
[2] Johns Hopkins University, Department of Biophysics, Baltimore, MD

Inositol is a carbohydrate that was isolated for the first time in about 1850 from cardiac muscle. Its systematic name is 1,2,3,4,5,6-cyclohexahexol, for which the term "cyclitol" is also used. Nine stereoisomers can be formed with this constitution. Seven of those are so-called *meso* compounds; they contain a symmetry element and are thus not chiral. The most abundant cyclitol in nature is *myo*-inositol, where only one of the hydroxyl moieties is oriented axially in the most stable (chair) conformation (see Fig. 9). The stereochemistry of *myo*-inositol is rather complex, since chirality can be easily achieved by substitution of hydroxyl groups, and thus special rules and recommendations apply for the nomenclature of *myo*-inositol derivatives [82]. Inositols linked to one or more phosphate groups are commonly denoted phosphoinositols, and any phosholipid including an inositol ring is usually designated a phosphoinositide.

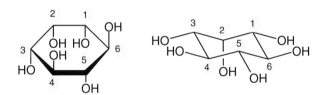

Fig. 9. Structure of *myo*-inositol

5.1 Physiological Significance of PIs and PIPs in Membranes

Biological membranes are characterized by a great variety of the phospholipids that they comprise, i.e. they constitute a complex molecular alloy. This diversity may be interpreted as providing the specific physical environment for the multifarious membrane-associated proteins and as facilitating fulfilling structural physicochemical prerequisites for biological processes such as fusion, exo/endocytosis, or apoptosis. Since the 1980s it has become progressively clearer that some minority-component classes of membrane phospholipids have high relevance for essential cellular functions. Glycerophospholipids with an inositol group represent only 5 – 10 percent of the total lipid in eucaryotic cells. Nevertheless, they have been shown to operate as second messengers in cellular signalling [83]. 1-(3-*sn*-phosphatidyl)-D-*myo*-inositol (or, for short, 1,2-diacyl-phosphatidylinositol, see Fig. 9) serves as the basis for at least seven distinct compounds that are either singly or multiply phosphorylated at the inositol headgroup. These phosphorylated derivatives are synthesized by a variety of lipid kinases, which are in turn activated by interaction with different types of receptors [84]. In distinction from physiological studies of phosphoinositides, which are numerous, investigations of their biophysical properties [85–89] are still rare. This may partially be related to their relatively complicated chemical synthesis. To the best of our knowledge, studies reporting detailed insight into the structural organization of phosphoinositides within monolayers at the air–water interface have not yet been published. Here we report specular X-ray reflectivity (XR) and grazing-incidence X-ray diffraction (GIXD) measurements from stereochemically pure dipalmitoyl-phosphatidylinositol (DPPI) and its respective 3-, 4- and 5-monophosphorylated derivatives (DPPI-3P, DPPI-4P, and DPPI-5P) within monolayers on water subphases and on physiological buffers at room temprature.

5.2 Phosphoinositol Conformation
in DPPI Surface Monolayers on Water and Buffer

Pressure–area (π–A) isotherms (Fig. 10) reveal that DPPI forms condensed phases with surprisingly small molecular areas both on water and on physiological buffer – where the monolayers are slightly more expanded than on pure water. This difference in A, however, decreases with increasing π. Furthermore, slight changes in the slope of the isotherm on buffer between 20 and 30 mN/m may possibly indicate a phase transition.

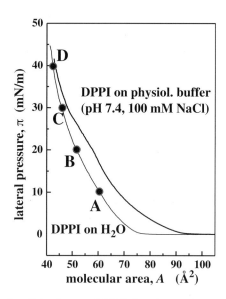

Fig. 10. Pressure–area (π–A) isotherms of DPPI (inset) monolayers recorded on water and physiological buffer. Circles on the isotherm on water denote where scattering measurements have been taken

In order to elucidate the structural implications of the phase behavior reflected by these isotherms, we have undertaken structural investigations of phosphoinositols (PIs) and some of their monophosphorylated derivatives (PIPs) in surface monolayers using X-ray scattering and diffraction. These measurements were carried out at BW1 (HASYLAB/DESY, Hamburg) at a wavelength of $\lambda = 1.304$ Å. For the interpretation of the X-ray reflection data within the VRDF model, independent information about the partial volumes of molecular subfragments within these complex headgroups is required. As a first step toward this end, the volume of inositol in water has been determined by means of density measurements to be $V_{inositol} = 198$ Å^3. For the actual PI and PIP subfragments, the values $V_{InsPI} = 170$ Å^3 and $V_{InsPIP} = 142$ Å^3 have been used. These values are smaller than $V_{inositol}$ owing to the loss of one (PI) or two (PIP) hydroxyl groups due to substitution by phosphate(s). Molecular-fragment volumes for other components were taken from Armen et al. [71].

Initial attempts to derive structural models single-handedly from XR data failed as it was impossible to derive unambiguous χ^2 minima, since some of the structural parameters were strongly coupled. A decrease of the number of free VRDF parameters was, however, achieved by invoking information obtained from GIXD (see e.g. Fig. 11). Results such as the one shown in the figure reveal a *minimum* value of the mean molecular area[6] as well as the tilt angle of the acyl chains within the monolayer. With these results taken as fixed parameters, it was then possible to elucidate by reflectivity measurements the molecular organization of the PI headgroups in the monolayer on pure water, while it became evident that on buffer, the molecules form coexisting hexatic domains, a situation which is not suited for structural investigations with reflectometry. On the contrary, DPPI monolayers on pure water appeared to form homogeneous phases at all lateral pressures and were investigated in X-ray reflectivity measurements (Fig. 12).

Already the model-free electron-density profiles of DPPI on pure water (Fig. 12, right), derived with the Skov Pedersen-Hamley approach [18,19], exhibit pronounced differences between samples measured at distinct lateral pressures. These differences become even more evident in an interpretation of the data within the VRDF approach. Not surprisingly, the chain slab thickness increases with lateral pressure, indicative of a decrease in the chain tilt angles. In fact, the chain tilt vanishes at $\pi \approx 40$ mN/m, consistent with the GIXD results. In a similar vein, the distances of the phosphate and the inositol subfragments[7] from the surface increase with increasing lateral pressure while the mutual distance between the centers of the two distribution functions themselves remains approximately constant, at about 5 Å, over the range of lateral pressures investigated. The model predicts that the hydration of the phosphate group decreases from 9 to 4, and that of the inositol moiety from 6 to 1 water molecules per headgroup, as π increases from 10 to 40 mN/m.

---

[6] Diffraction is only observed from ordered domains within the film; disordered film areas between such domains increase the *mean* molecular area above the value derived from the hexatic chain lattice.

[7] "Distance of a molecular subfragment" denotes here the distance of its center of gravity from the hydrophobic/hydrophilic interface.

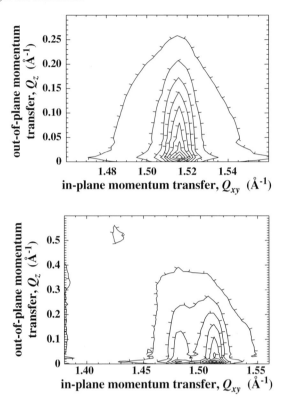

Fig. 11. GIXD from the acyl chains of DPPI in monolayers on water (top) and physiological buffer (bottom). Both data sets were recorded at $\pi = 40$ mN/m. Note that the data set measured for the system on buffer is fundamentally different from that for the system on pure water which indicates the coexistence of two distinct chain lattices.

These structural changes are more clearly visualized in Fig. 13, where the DPPI fragment distributions (i.e. those for the DPPI acyl chains, and the glycerol–carbonyl, phosphate, and inositol subfragments) are depicted for $\pi = 10, 20, 30,$ and 40 mN/m in a monolayer on pure water at 20 °C. The detailed arrangement of these distributions suggests that the inter- and intramolecular inositol–phosphate interactions increase with decreasing hydration at higher lateral pressures. This appears to lead to a tight stacking of the inositol rings and explains why the monolayer is so remarkably condensed – despite the bulky inositol headgroups.

5.3 Structure of DPPI-3P, DPPI-4P, and DPPI-5P in Surface Monolayers

PIPs are distinguished from PIs by virtue of their one extra phosphate group. Because of the resulting high electron density contained within the headgroup, one would expect that these compounds should be easily characterized in XR measurements. It turns out, unfortunately, that the complication inferred by the required addition of

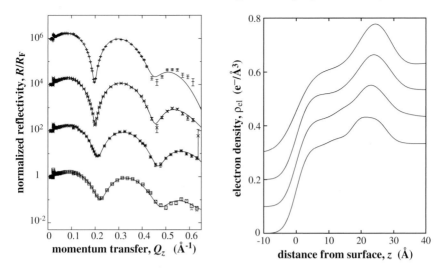

Fig. 12. Left: Fresnel-normalized XR data of DPPI monolayers on pure water at different lateral pressures (bottom to top: $\pi =$ 10, 20, 30, and 40 mN/m). Best fits from the VRDF model are superimposed. Right: corresponding overall electron density profiles along the surface normal

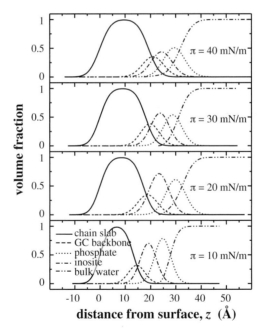

Fig. 13. Volume distribution (VRDF model) for DPPI monolayers at $\pi =$ 10, 20, 30, and 40 mN/m on water

structural parameters overcompensates the advantage of a high electron density, and thus contrast against the subphase. Furthermore, most probably owing to the extraordinarily high polarity of the PIP headgroup, the preparation of stable monolayers is relatively difficult to achieve. As a result, π–A isotherms are not always reproducible. On the other hand, a simplification of the parameterization, as well as independent results from GIXD measurements, helped again to obtain significant and novel structural information. We have approximated the molecular substructure of the headgroup by treating the inositol and the second phosphate moiety as one molecular fragment (with a common center of gravity). As an exemplary result, fitted XR data and the derived electron density are presented for DPPI-4P spread on water at 30 mN/m (lower traces) and 40 mN/m (upper traces) in Fig. 14. Figure 15 shows the exemplary results of Fig. 14 broken down into individual molecular-subfragment distributions.

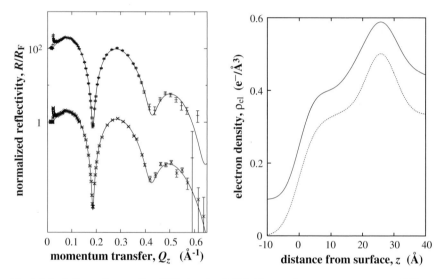

Fig. 14. Left: Fresnel-normalized XR data for DPPI-4P monolayers on pure water at different lateral pressures ($\pi = 30$, lower trace, and $\pi = 40$ mN/m, upper trace). Best fits from a reduced VRDF model, as described in the text, are superimposed as lines. Right: corresponding electron density profiles along the surface normal for the data shown on the left

As in the case of DPPI monolayers, the model-free data inversion approach by Skov Pedersen and Hamley [19] provides a cross-check on the results derived from the VRDF model (see Fig. 16). Moreover, the model-free approach is easier to implement. The results indicate that the center of gravity of the PIP headgroup electron density is slightly shifted from the secondary phosphate position. This analysis reveals that the differences in the electron densities of DPPI-4P monolayers on water and on buffer vanish at higher lateral pressures. Conceivably, this indicates that counterions are squeezed out of the headgroup as the surface films are compressed. A detailed assessment of the PIP structures as a function of lipid headgroup chemistry, subphase composition, and lateral density will be given elsewhere.

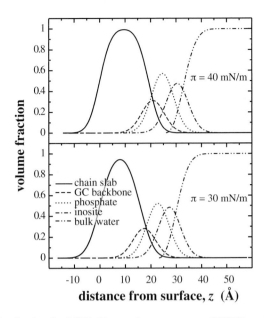

Fig. 15. Volume distribution for DPPI-4P monolayers on water (VRDF model) at $\pi = 30$ and 40 mN/m

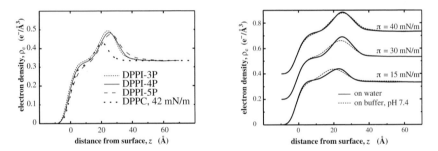

Fig. 16. Model-free electron density profiles. Left: DPPI-3P, DPPI-4P, and DPPI-5P at $\pi =$ 40 mN/m on water. Right: DPPI-4P at different lateral pressures on pure water (solid line) and on buffer at pH 7.4 (broken lines)

6 Adsorption of Small Molecules on Lipid Surface Monolayers

Martina Dyck[1], Peter Krüger[1], Andrea Bettio[2], Annette G. Beck-Sickinger[2], Mikhael A. Kiselev[3], and Mathias Lösche[1,4]

[1] Universität Leipzig, Institut für Experimentelle Physik I, Leipzig, Germany
[2] Universität Leipzig, Institut für Biochemie, Leipzig, Germany
[3] Frank Laboratory of Neutron Physics, Joint Institute for Nuclear Research, Dubna, Russia
[4] Johns Hopkins University, Department of Biophysics, Baltimore, MD

Not only high-molecular-weight biopolymers, such as DNA or proteins, but also small molecules, such as hormones, signal peptides, or pharmaceuticals, play important roles in the living cell. For example, hormones may act as triggers in a chemical signal transduction chain by initiating secondary processes as they affect, for example the conformation of their receptors. Many of these processes occur at biomembranes, where the receptor may reside within the bilayer leaflet as a membrane protein. A characteristic property of such smaller determinants of membrane structure and function is that their physiologically active concentration is *exceedingly* low. For the physico-chemical characterization of their interaction with membranes of receptors, this poses great challenges because the effector may not be observable even if present in large excess – compared with its physiological abundance – at a model surface. One active area of research in our group within the *Sonderforschungsbereich* has been the structural and functional characterization of several classes of small effector molecules. While some work along these lines has already been published in detail, such as a characterization of the lung surfactant peptides SP-B and SP-C in lipid surface monolayers [9, 10] and an assessment of the interaction of the piperidinopyrimidine dipyridamol (DIP) with membrane surfaces [90], we concentrate here on previously unpublished work in progress.

In this section, we shall thus discuss the influence of the tyrosin-rich signal peptide neuropeptide Y (NPY) with membrane surfaces and present preliminary results on the incorporation of dimethylsulfoxide into lipid monolayers. While the first one acts as a neurotransmitter in higher animals and humans, the latter one is used for transdermal transport of pharmaceutical agents. This functionality is widely used in pharamaceutical unguents, but poorly understood, one of the unresolved issues being its equilibrium location within the (lipid) membrane.

6.1 Effect of Neuropeptide Y on Surface Monolayers

NPY is a 36 amino acid neurotransmitter peptide, first isolated from porcine brain [91], but also found throughout the central and peripheral nervous systems of many mammalian species, including humans [92, 93]. It has been one of the peptides most intensively focused on in pharmacological and structural studies in the last years. Its amino acid sequence is YPSKP DNPGE DAPAE DMARY YSALR HYINL ITRQR Y-NH$_2$. Studies on the structure and dynamics of micelle-bound NPY revealed that the peptide is located at the lipid–water interface, with a C-terminal helix parallel to the membrane surface (for a ribbon representation of the NPY solution structure [94], see inset in Fig. 17); it penetrates the hydrophobic membrane interior only *via* insertions of a few long aliphatic or aromatic side chains [95].

Peptide Synthesis

Neuropeptide Y (NPY) was prepared by solid-phase synthesis following the *Fmoc* strategy, as described in [96] using an automated peptide synthesizer (MultisynTech Syro II). NPY was cleaved by treating the resin with 1 M trimethylbromosilane in trifluoroacetic acid/anisol/cresol (90 : 5 : 5) for 3 hours. The resin was removed by filtration and the peptide precipitated from diethyl ether at 0 °C, washed several times with cold diethyl ether, and dried. The crude product was dissolved in *tert*-butanol : water (1 : 4) and lyophilized. The peptide was purified by reverse-phase semipreparative HPLC. The elution system was 0.08% trifluoroacetic acid in acetonitrile (A) and 0.1% trifluoracetic acid in water (B), applying a linear gradient from 20 to 70% (A) over 30 mins at a flow rate of 2.5 ml/min. A more detailed description of the synthesis procedure can be found in [97].

Carboxyfluorescein labeled NPY (CF-NPY) was prepared by coupling the dye (Molecular Probes) to the resin-bound peptide as described in [98]. In the dual-label fluorescence microscopy and peptide-label experiments, CF-NPY was typically used in mixtures with unlabeled NPY in various ratios between 1 mol% and 10 mol%.

Surface Isotherms

Pure DPPC monolayers and NPY/DPPC mixtures at different NPY concentrations were investigated at the air–water interface, see Fig. 17. Upon increasing the NPY concentration, the molecular area per DPPC within the monolayer increases and a new phase transition is observed at surface pressures between 16 mN/m and 18 mN/m. Above 8 mol% NPY, the monolayer expansion levels off, presumably owing to losses of monolayer material to the subphase and/or aggregation of NPY within the surface film.

It is also possible to spread pure NPY at the aqueous surface and record isotherms that are largely reproducible as far as their shape is concerned (data not shown). The phase transition at 16 mN/m is clearly observed in such π–A curves. It appears thus that this feature is related to a property of the peptide. Below 16 mN/m, the pressure is observed to rise very slowly. This rise extends over a very large range of area, such that such isotherms cannot be recorded in a single measurements owing to the insufficient compression ratio (largest surface area : smallest surface area) available in standard Langmuir film balances. We found that patches of the π–A curves that were recorded by spreading successively more material at the water surface overlapped well in their general shape; however, the molecular-area values computed from the amount of spread peptide material varied depending on the actual amount applied to the surface. It is thus clear that these π–A curves do not represent true equilibrium isotherms in the thermodynamic sense. Consequently, we have not been able to assign

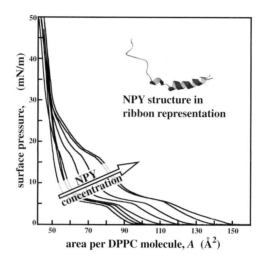

Fig. 17. Isotherms of DPPC and DPPC/NPY monolayers and peptide structure [94]

an area value to the peptide within the surface film which might have permitted an educated guess about the peptide's conformation at the interface.

Fluorescence Microscopy

Dual-label FM of the molecular surface layers reports the in-plane morphologies of coexisting monolayer phases (Chap. 10) – since a phase of lower intrinsic order dissolves a label better than a phase of higher intrinsic order – and reveals preferences of the peptide for associating with a specific lipid phase. Using β-Bodipy-C_2-HPC and unlabeled NPY or CF-NPY and pure DPPC (without lipid label), we observed a marked influence of the neuropeptide on the domain shapes of DPPC (Fig. 18). Dual-label experiments showed further that the peptide dissolves preferentially in the fluid phase. At NPY concentrations above about 4 mol%, corresponding to about 18 μmol/l of NPY in a bulk concentration, we observed the formation of a new feature in the fluorescence micrographs: within the brightly fluorescent fluid lipid phase, dark grainy structures appeared that became more numerous as the surface monolayers were further compressed. Also, these dark areas became more abundant at larger concentrations of the peptide with respect to total lipid. This observation is tentatively assigned to the formation of NPY dimers that are incorporated into bigger aggregates which are no longer fluorescent. This new feature is not obviously related to the phase transition observed in the isotherm at 16 mN/m and higher.

ATR-IR Measurements

ATR-IR measurements were performed to investigate which molecular conformation NPY assumes at the surface, whether this structure depends on the concentration

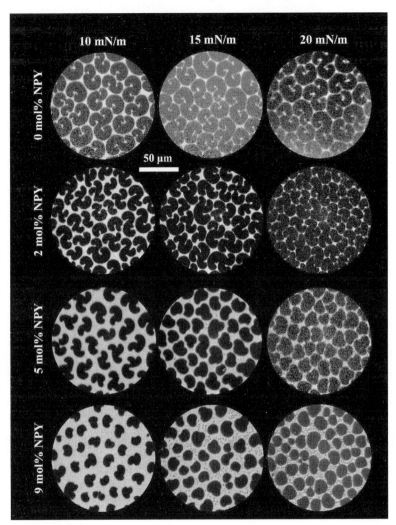

Fig. 18. Fluorescence micrographs of pure DPPC and mixed DPPC/NPY films containing different ratios of NPY at different surface pressures. 10% of the total peptide was labeled in these experiments

of the pure NPY solutions, and whether there are differences between pure NPY solutions and the DPPC/NPY mixture. For both pure peptide films and mixed peptide/lipid films, the intensities, shapes, and positions of the amide I and II bands indicated that the peptide adopts an α-helical structure (Fig. 19). Moreover, no significant differences in the spectra were observed, implying that the conformation of NPY in pure peptide films and at lipid surfaces is comparable.

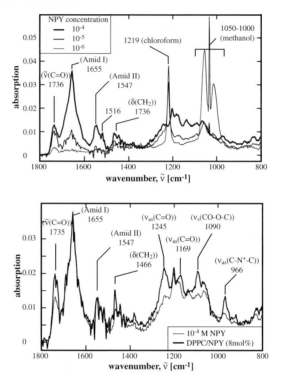

Fig. 19. Top: ATR-IR spectra of pure NPY and mixed DPPC/NPY multilayer films containing 8 mol% NPY. The peptide or lipid/peptide solution was prepared at the solid/air interface, and the measurements were performed on a ZnSe$_2$ crystal. Bottom: ATR-IR spectra of pure NPY films in different concentrations performed on a ZnSe$_2$ crystal

6.2 Localization of DMSO in DPPC Monolayers

Phospholipid monolayers on aqueous subphases containing high concentrations of DMSO were recently investigated by Brezesinski and coworkers using GIXD [99]. We have started a more systematic synchrotron X-ray study, in which we have measured reflection, GIXD, and wide-angle scattering (WAXS) to reinvestigate the question: what is the submolecular-scale architecture of DMSO/phospholipid interface layers in a liquid surface geometry? In addition, WAXS experiments were performed at the home-built multipurpose X-ray diffractometer at Leipzig [62]. Vesicle solutions were prepared with excess solvent, typically some 100 solvent molecules per lipid molecule.

Surface Isotherms

DPPC isotherms on mixed subphases (see Fig. 20) reveal that the main influence of DMSO is a chain condensation, as observed earlier. A transition between a state where the DPPC isotherms are similar to those on pure water and a state that is dominated by DMSO is observed at around 1% DMSO content of the subphase; all isotherms measured on subphases with DMSO concentrations above this value are virtually indistiguishable. In our scattering measurements, we have consequently focused on the lower concentration range of DMSO in order to explore the conformational changes with increasing DMSO influence.

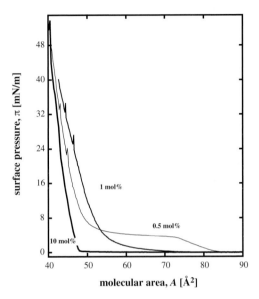

Fig. 20. DPPC monolayer isotherms at 20 °C on aqueous subphases with different DMSO concentrations

Supporting Measurements

To estimate the volume of DMSO in water (needed for the VRDF refinement of the scattering data), we measured the densities of a water/DMSO mixture series at large water excess and determined a linear extrapolation to zero water content. In this procedure, a *water-solubilized volume* of DMSO of $V_{DMSO,aq} \approx 87 \text{ Å}^3$ was obtained. This varies greatly from the volume calculated from the bulk density of DMSO, $V_{DMSO,bulk} = 157 \text{ Å}^3$. We rationalize this apparent discrepancy by taking into account that the volume of a water molecule ($V_{water} = 30 \text{ Å}^3$) is much smaller than that of a DMSO molecule. Hydrogen bridging between the two molecular species then brings about a strong contraction of the total volume as DMSO is transferred into water.

X-Ray Measurements

Reflectivity curves were recorded as a function of surface pressure and DMSO content in the subphase. To obtain a more complete overall picture, WAXS data were also recorded for some points in composition space. At first sight, the reflectivity scans of DPPC monolayers on DMSO-containing subphases appear similar to those from DPPC on pure water. As examples, Fig. 21, panel (A), shows such reflectivities from DPPC obtained with various DMSO concentrations at 30 mN/m, together with the best fits from VRDF data inversion runs. The corresponding electron density profiles are depicted in panel (B) of Fig. 21.

In addition, WAXS reveals information about the chain organization in vesicles, thus complementing GIXD measurements on planar surfaces. The chain ordering peaks disappear above the chain melting temperature, e.g. at $T_{chain_{DPPC}} = 42.7\ °C$ for DPPC, indicating the local melting of the lipid's acyl chains. From the position of the Bragg peaks below T_{chain}, the area per chain and their tilt from the local surface normal can be calculated similarly to the methods described in Sect. 2.5.

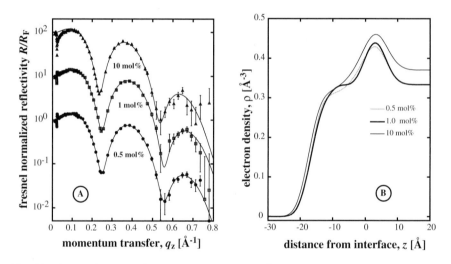

Fig. 21. X-ray reflectivity of DPPC monolayers at 20 °C on subphases containing different DMSO concentrations and their best VRDF model fits (A), and the derived electron density profiles (B)

It is commonly believed that the DMSO is incorporated only into the headgroup region of phospholipids when bilayers are in contact with aqueous media containing DMSO [100, 101]. A critical examination of our recent scattering data, however, suggests a different picture. The VRDF data inversion of reflectivity data such as those shown in Fig. 21 indicates that DMSO, rather than being associated exclusively with the lipid headgroups, is preferentially incorporated into the hydrophobic acyl chain region, close to the lipid headgroup. This alternate localization is also supported

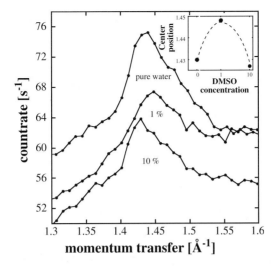

Fig. 22. Wide-angle diffraction patterns of DPPC vesicles in fluid phases containing various DMSO concentrations at $T \approx 35$ °C. The inset quantifies the center positions of the main peak

by the WAXS results (see Fig. 22). In contrast to the sample preparations that led to the established view of DMSO organization *near* lipid interfaces [100, 101], the samples investigated in our lab were prepared with a large excess of solvent – so that the situation is closer to that realized in the monolayer experiments. In conclusion, our surface-sensitive scattering data for DPPC at DMSO-rich aqueous surfaces show clearly – and in deviation from established views – that DMSO inserts in the chain region. This is confirmed by the observation of a widening of the DPPC acyl chain lattice in WAXS experiments.

7 Lipopolymer Phase Behavior

Angelika Wurlitzer[1], Peter Krüger[1], Erich Politsch[2], G. Cevc[2], and Mathias Lösche[1,3]

[1] Universität Leipzig, Institut für Experimentelle Physik I, Leipzig, Germany
[2] Technische Universität München, Institut für Medizinische Biophysik, München, Germany
[3] Johns Hopkins University, Department of Biophysics, Baltimore, MD

Lipopolymers consist of hydrophilic polymers covalently attached to the polar head-groups of amphiphilic molecules. Through their hydrophobic tails, such amphiphiles may be anchored to interfaces that divide a hydrophilic from a hydrophobic compartment, such as an aqueous surface or the interface of a lipid membrane with water. Equilibrium conformations and forces in the resulting "tethered" polymer layers have received considerable attention in recent years owing to a wide range of potential applications, such as the stabilization of colloidal particles, control of adhesion, and the biocompatibilization of interfaces. In particular, attention has been paid to the employment of lipopolymers for the purpose of drug carriers [102]: sterically stabilized vesicles may be prepared by an admixture of small amounts of lipopolymers to physiological lipids in membranes. Vesicles masked in such a way circulate considerably longer in the bloodstream [103, 104]. Lipopolymers also serve as model systems to test theories describing grafted polymers at interfaces [105–107]. The most thoroughly investigated lipopolymers to date are polyethyleneglycol (PEG) lipids [108–110]. However, several more lipopolymers have been described that vary in size and functionality [109, 111].

We investigated the molecular conformations of a lipopolymer with a polyoxazoline headgroup at air–water interfaces as a function of lateral area per molecule with X-ray and neutron reflectometry. The polymer 1,2-dioctadecanyl-sn-glycero-3-poly(2-methyl-2-oxazoline), PMO-$(C_{18})_2$, forms stable surface monolayers. Pressure/area isotherms around room temperature show a plateau region, indicative of a phase transition, whose origin was examined. For data evaluation, a novel approach – described in Sect. 2.4 – was used that acts on explicit quasi-molecular ensemble conformations of the polymer [52].

7.1 Phase Transition of Poly(methyloxazolines) on Water

To obtain insight into the conformational response of PMO-$(C_{18})_2$ to lateral compression that leads to a phase transition as evidenced from the isotherm, Fig. 23, we have conducted X-ray and neutron reflectivity measurements in which the area A per molecule in the floating surface monolayer was well controlled. Investigations of isotopically distinct samples, in which PMO species with different hydrogenation/deuteration patterns on their hydrophilic polymer headgroups were investigated on H_2O or D_2O subphases [112], suggested that the major conformational changes in PMO-$(C_{18})_2$ occur in their hydrophobic anchors rather than in the hydrophilic headgroups. Figure 24 depicts a quasi-molecular rendering of ensembles of 40 PMO-$(C_{18})_2$ molecules used to describe the available data simultaneously for two positions on the isotherm ($\pi = 15$ mN/m, below the onset of the phase transition, and 30 mN/m, above the phase transition at $T = 15$ °C). Note that in spite of the cylindrical symmetry of the problem, horizontal displacements of chain fragments are consistently drawn to occur to the right-hand side, in order to achieve a better visualization of the ensemble conformations.

The overall PMO density distribution in the two configurations is compared in Fig. 25 with the one predicted for molecular brushes of the appropriate chain density [106, 107]. Except near the hydrophobic/hydrophilic interface at $z = 0$, where particularly below the phase transition a significant adsorption of the slightly hydrophobic poly(methylazoline) chains to the alkane phase is observed, the experimental results follow the theoretical predictions remarkably well. It may be concluded that the PMO chains form in both cases a dense "polymer brush" and that the phase transition is thus *not* associated with a reorganization of the hydrophilic polymer

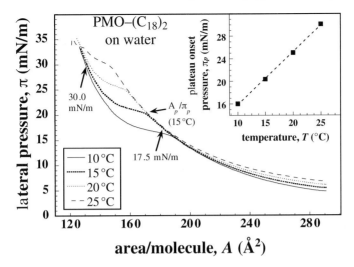

Fig. 23. Dioctadecanyl-poly(oxazoline) isotherms on water (H_2O) at different temperatures as indicated. The inset shows the dependence of the phase transition onset pressure π_p on T. Arrows indicate the positions on the isotherm at $T = 15\ ^\circ C$ where reflectivity measurements have been undertaken to reveal conformational changes across the phase transition. Reproduced from [112]

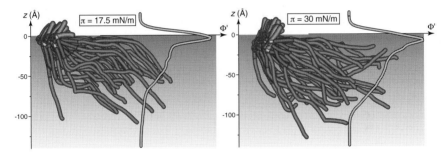

Fig. 24. Dioctadecanyl-poly(methyloxazoline) conformations at the air–water interface at different lateral pressures that correspond to pure phases, below the onset of the phase transition shown in Fig. 23 and above. The illustrations each show 40 molecules that form ensembles, which were used to model X-ray and neutron reflection data simultaneously in the *StringFit* approach. Color code: brightest gray, ether linkage; attached to it, medium gray, alkyl chains; dark gray, deuterated PMO chain fragment; below, in medium bright gray, hydrogenated PMO chain fragment. Envelope volume density distributions of the overall PMO-$(C_{18})_2$ are indicated on the right. Reproduced from [112]

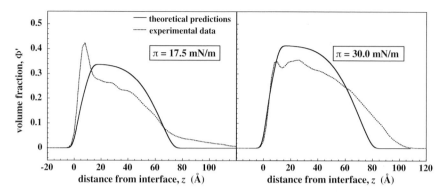

Fig. 25. Comparison of the segment density distributions as a function of distance from the interface derived from the predictions for a polymer brush according to the Alexander/de Gennes theory [106,107] and from experimental results. Reproduced from [112]

headgroup. Rather, as Fig. 24 demonstrates visually, a major difference resides in the hydrophobic anchor and the ether linker: while the anchors are well confined to the interface below the phase transition, they are partially immersed in the aqueous subphase after going through the transition. This may be interpreted as the result of strain exerted on the hydrophobic anchors by confinement of the hydrophilic PMO strands [112,113]: grafting of the PMO strands to the interface reduces their configurational entropy; this effect is augmented as the strands are further confined by compressing the monolayer. At the phase transition point, the energy associated with this elasticity effect has reached such a level that it compensates the energetic penalty associated with a partial immersion of the hydrophobic alkyl anchors in the aqueous medium. Upon immersion, the alkyl chains may aggregate and thus condense to reduce their exposure to water molecules. This may explain the shift of the methylene vibration frequency to lower wavenumbers observed in FTIR spectroscopy on PMO monolayers at the phase transition [114,115] and also at the corresponding phase transition in PEO–lipid monolayers [116]. Figure 26 summarizes our conclusions for the phase transition in PMO monolayers in a pictorial way.

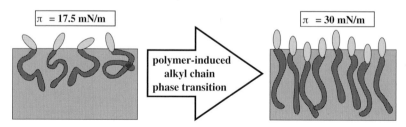

Fig. 26. Changes in molecular organization of dioctadecanyl-poly(methyloxazoline) monolayers in the course of the phase transition at $A \approx 140$ Å^2 as revealed from combined X-ray and neutron reflectometry

Acknowledgments

We are indebted to our numerous collaboration partners, named in connection with work described here or unnamed (as many aspects of our work have not been covered in this chapter), within the *Sonderforschungsbereich* 294, on the national as well as on the international scale. Especially warm thanks are due to Kristian Kjaer of Risø National Laboratory in Roskilde, Denmark, who ran the TAS9 reflectometer at the DR3 reactor at Risø and also constructed the liquid-surfaces beamline station at BW1 of HASYLAB at DESY. He has also been instrumental in his contributions to this research program in its early stages.

Beyond financial funding through the *Sonderforschungsbereich* 294, the research described here was partially supported by grants from the BMBF, the DFG via the HBFG program and the priority program "Wetting and structure formation at interfaces", the DAAD, the Volkswagen Foundation, the Fonds der Chemischen Industrie, the US NIH (1 RO1 RR14812), and The Regents of the University of California.

Last but not least, we are grateful for beam time at the HASYLAB (contracts II-99-078 and II-01-042), Argonne National Laboratory (under GUP-237) and Risø National Laboratory via the EU-TMP programme.

References

1. M. Weygand, B. Wetzer, D. Pum, U.B. Sleytr, K. Kjaer, P.B. Howes, M. Lösche: Biophys. J. **76**, 458–468 (1999)
2. M. Weygand, M. Schalke, P.B. Howes, K. Kjaer, J. Friedmann, B. Wetzer, D. Pum, U.B. Sleytr, M. Lösche: J. Mater. Chem. **10**, 1456–1463 (2000)
3. M. Weygand, K. Kjaer, P.B. Howes, B. Wetzer, D. Pum, U.B. Sleytr, M. Lösche: J. Phys. Chem. B **106**, 5793–5799 (2002)
4. S.A.W. Verclas, P.B. Howes, K. Kjaer, A. Wurlitzer, M. Weygand, G. Büldt, N.A. Dencher, M. Lösche: J. Mol. Biol. **287**, 837–843 (1999)
5. P. Krüger, M. Lösche: Phys. Rev. E **62**, 7031–7048 (2000)
6. M. Lösche, P. Krüger: *Morphology of Langmuir Monolayer Phases, Lecture Notes in Physics*, vol. 600 (Springer, New York, 2002), pp. 159–178
7. R.J. Zhang, P. Krüger, B. Kohlstrunk, M. Lösche: ChemPhysChem **2**, 452–457 (2001)
8. W. Caetano, M. Ferreira, M. Tabak, M. Mosquera Sanchez, O.N. Oliveira Jr., P. Krüger, M. Schalke, M. Lösche: Biophys. Chem. **91**, 21–35 (2001)
9. P. Krüger, M. Schalke, Z. Wang, R.H. Notter, R.A. Dluhy, M. Lösche: Biophys. J. **77**, 903–914 (1999)
10. P. Krüger, J.E. Baatz, R.A. Dluhy, M. Lösche: Biophys. Chem. **99**, 209–228 (2002)
11. J. Als-Nielsen, K. Kjaer: "X-ray reflectivity and diffraction studies of liquid surfaces and surfactant monolayers", *Phase Transitions in Soft Condensed Matter*, ed. by T. Riste and D. Sherrington (Plenum Press, New York, 1989), pp. 113–138
12. J. Penfold, R.K. Thomas: J. Phys.: Condens. Matter **2**, 1369–1412 (1990)
13. J. Als-Nielsen, D. Jacquemain, K. Kjaer, M. Lahav, F. Leveiller, L. Leiserowitz: Phys. Rep. **246**, 251–313 (1994)
14. J. Als-Nielsen, H. Möhwald: "Synchrotron x-ray scattering studies of Langmuir-films", *Handbook on Synchrotron Radiation*, ed. by S. Ebashi, M. Koch and E. Rubinstein (Elsevier North-Holland, Amsterdam, 1991), pp. 1–53

15. L.G. Parratt: Phys. Rev. **95**, 359–369 (1954)
16. J. Skov Pedersen: J. Appl. Cryst. **25**, 129–145 (1992)
17. X.L. Zhou, S.H. Chen: Phys. Rev. E **47**, 3174–3190 (1993)
18. I.W. Hamley, J. Skov Pedersen: J. Appl. Cryst. **27**, 29–35 (1994)
19. J. Skov Pedersen, I.W. Hamley: J. Appl. Cryst. **27**, 36–49 (1994)
20. N.F. Berk, C.F. Majkrzak: Phys. Rev. B **51**, 11 296–11 309 (1994)
21. C.H. Chou, M.J. Regan, P.S. Pershan, X.L. Zhou: Phys. Rev. E **55**, 7212–7216 (1997)
22. C.A. Helm, H. Möhwald, K. Kjaer, J. Als-Nielsen: Europhys. Lett. **4**, 697–703 (1987)
23. M.J. Grundy, R.M. Richardson, S.J. Roser, J. Penfold, R.C. Ward: Thin Solid Films **159**, 43–52 (1988)
24. D. Vaknin, K. Kjaer, J. Als-Nielsen, M. Lösche: Biophys. J. **59**, 1325–1332 (1991)
25. M. Thoma, M. Schwendler, H. Baltes, C.A. Helm, T. Pfohl, H. Riegler, H. Möhwald: Langmuir **12**, 1722–1728 (1996)
26. D. Vaknin, J. Als-Nielsen, M. Piepenstock, M. Lösche: Biophys. J. **60**, 1545–1552 (1991)
27. M. Lösche, M. Piepenstock, A. Diederich, T. Grünewald, K. Kjaer, D. Vaknin: Biophys. J. **65**, 2160–2177 (1993)
28. J. Gallant, B. Desbat, D. Vaknin, C. Salesse: Biophys. J. **75**, 2888–2899 (1998)
29. D. Gidalevitz, Z. Huang, S.A. Rice: Proc. Natl. Acad. Sci. U.S.A. **96**, 2608–2611 (1999)
30. M. Fukuto, K. Penanen, R.K. Heilmann, P.S. Pershan, D. Vaknin: J. Chem. Phys. **107**, 5531–5546 (1997)
31. T.L. Kuhl, J. Majewski, P.B. Howes, K. Kjaer, A. von Nahmen, K.Y.C. Lee, B. Ocko, J.N. Israelacili, G.S. Smith: J. Am. Chem. Soc. **121**, 7682–7688 (1999)
32. A. Wurlitzer, E. Politsch, G. Cevc, T. Gutberlet, K. Kjaer, M. Lösche: Physica B **276–278**, 343–344 (2000)
33. T. Gutberlet, A. Wurlitzer, U. Dietrich, E. Politsch, G. Cevc, R. Steitz, M. Lösche: Physica B **283**, 37–39 (2000)
34. A. Schmidt, J. Spinke, T. Bayerl, E. Sackmann, W. Knoll: Biophys. J. **63**, 1185–1192 (1992)
35. G. Fragneto, R.K. Thomas, A.R. Rennie, J. Penfold: Science **267**, 657–660 (1995)
36. S. Krueger, J.F. Ankner, S.K. Satija, C.F. Majkrzak, D. Gurley, M. Colombini: Langmuir **11**, 3218–3222 (1995)
37. A. Diederich, M. Lösche: "Novel biosensoric devices based on molecular protein hetero-multilayer films", *Protein Array: an Alternate Biomolecular System*, ed. by K. Nagayama (Japan Scientific Societies Press/Elsevier, Tokyo/Limerick, 1997), pp. 205–230
38. A. Malik, W. Lin, M.K. Durbin, T.J. Marks, P. Dutta: J. Chem. Phys. **107**, 645–652 (1997)
39. T.L. Kuhl, J. Majewski, J.Y. Wong, S. Steinberg, D.E. Leckband, J.N. Israelachvili, G.S. Smith: Biophys. J. **75**, 2352–2362 (1998)
40. M. Schalke, M. Lösche: Adv. Colloid Interface Sci. **88**, 243–274 (2000)
41. R. Frahm, J. Weigelt, G. Meyer, G. Materlik: Rev. Sci. Instrum. **66**, 1677–1680 (1995)
42. H. Haas, G. Brezesinski, H. Möhwald: Biophys. J. **68**, 312–314 (1995)
43. D.A. Cadenhead, F. Müller-Landau, B.M.J. Kellner: "Phase transitions in insoluble one and two-component films at the air/water interface", *Ordering in Two Dimensions*, ed. by S.K. Sinha (Elsevier North Holland, Amsterdam, 1980), pp. 73–81
44. V.M. Kaganer, H. Möhwald, P. Dutta: Rev. Mod. Phys. **71**, 779–819 (1999)
45. M. Schalke, P. Krüger, M. Weygand, M. Lösche: Biochim. Biophys. Acta **1464**, 113–126 (2000)
46. M.C. Wiener, S.H. White: Biophys. J. **59**, 174–185 (1991)
47. M.C. Wiener, S.H. White: Biophys. J. **61**, 434–447 (1992)
48. P.S. Pershan: Faraday Discuss. Chem. Soc. **89**, 231–245 (1990)

49. P.S. Pershan: Colloids Surf. A: Physicochem. Engineer. Aspects **171**, 149–157 (2000)
50. A. Plech, T. Salditt, C. Münster, J. Peisl: J. Colloid Interf. Sci. **223**, 74–82 (2000)
51. E. Politsch: J. Appl. Cryst. **34**, 239–251 (2001)
52. E. Politsch, G. Cevc, A. Wurlitzer, M. Lösche: Macromolecules **34**, 1328–1333 (2001)
53. H.P. Schwefel: *Evolution and Optimum Seeking* (J. Wiley, New York, 1995)
54. K. Kjaer, J. Als-Nielsen, C.A. Helm, L.A. Laxhuber, H. Möhwald: Phys. Rev. Lett. **58**, 2224–2228 (1987)
55. D. Jacquemain, S.G. Wolf, F. Leveillier, M. Lahav, L. Leiserowitz, M. Deutsch, K. Kjaer, J. Als-Nielsen: J. de Physique **50** (Suppl. 10, Coll. C7), 29–37 (1989)
56. K. Kjaer, J. Als-Nielsen, C.A. Helm, P. Tippmann-Krayer, H. Möhwald: J. Phys. Chem. **93**, 3200–3206 (1989)
57. M. Lösche: "Surface-sensitive x-ray and neutron scattering characterization of planar lipid model membranes and lipid/peptide interactions", *Current Topics in Membranes* vol. 52, ed. by T.J. McIntosh and S.A. Simon (Academic Press, San Diego, 2002), pp. 117–161
58. G.H. Vineyard: Phys. Rev. B **26**(8), 4146–4159 (1982)
59. Y. Yoneda: Phys. Rev. **131**, 2010– (1963)
60. K. Kjaer: Physica B **198**, 100–109 (1994)
61. T.R. Jensen, K. Kjaer: "Structural properties and interactions of thin films at the air-liquid interface explored by synchrotron x-ray scattering", *Novel Methods to Study Interfacial Layers*, ed. by D. Möbius and R. Miller (Elsevier Science, Amsterdam, 2001), pp. 205–254
62. P. Krüger, M. Schalke, J. Linderholm, M. Lösche: Rev. Sci. Instrum. **72**, 184–192 (2001)
63. J. Majewski, R. Popovitz-Biro, W.G. Bouwman, K. Kjaer, J. Als-Nielsen, M. Lahav, L. Leiserowitz: Chem. Eur. J. **1**, 304–311 (1995)
64. D. Vaknin, P. Krüger, M. Lösche: Phys. Rev. Lett. **90**, 178102 (2003)
65. M. Lösche, E. Sackmann, H. Möhwald: Ber. Bunsenges. Phys. Chem. **87**, 848–852 (1983)
66. P. Krüger, M. Lösche: Phys. Rev. E **62**, 7031–7043 (2000)
67. S. Hénon, J. Meunier: Rev. Sci. Instrum **62**, 936–939 (1991)
68. D. Hönig, D. Möbius: J. Phys. Chem. **95**, 4590–4592 (1991)
69. P. Krüger, M. Schalke, Z. Wang, R.H. Notter, R.A. Dluhy, M. Lösche: Biophys. J. **77**, 903–914 (1999)
70. O. Albrecht, H. Gruler, E. Sackmann: J. Physique (France) **39**, 301–313 (1978)
71. R.S. Armen, O.D. Uitto, S.E. Feller: Biophys. J. **75**, 734–744 (1998)
72. H. Möhwald: Annu. Rev. Phys. Chem. **41**, 441–476 (1990)
73. C.A. Helm, P. Tippmann-Krayer, H. Möhwald, J. Als-Nielsen, K. Kjaer: Biophys. J. **60**, 1457–1476 (1991)
74. K. Hristova, S.H. White: Biophys. J. **74**, 2419–2433 (1998)
75. M. Schalke: (2000), "Konformation und Hydratation von Phospholipiden in Oberflächenmonoschichten: Röntgenreflexion und IR-Spektroskopie", Ph.D. thesis, Universität Leipzig
76. B. Alberts, A. Johnson, J. Lewis, M. Raff, K. Roberts, P. Walter: *Molecular Biology of the Cell* (Garland Science, London, 2002)
77. D.R. Patel, C.C. Jao, W.S. Mailliard, J.M. Isas, R. Langen, H. Haigler: Biochemistry **40**, 7054–7060 (2001)
78. G. Cevc: Biochim. Biophys. Acta **1031**, 311–382 (1990)
79. I. Weissbuch, R. Buller, K. Kjaer, J. Als-Nielsen, L. Leiserowitz, M. Lahav: Colloid Surfaces A: Physicochem Eng. Aspects **208**, 3–27 (2002)
80. J. Kmetko, C.J. Yu, G. Evmenenko, S. Kewelramani, P. Dutta: Phys. Rev. Lett. **89**, 186 102 (2002)

438   P. Krüger and M. Lösche

81. F. Leveiller, C. Böhm, D. Jacquemain, H. Möhwald, L. Leiserowitz, K. Kjaer, J. Als-Nielsen: Langmuir **10**, 819–829 (1994)
82. International Union of Pure and Applied Chemistry, and International Union of Biochemistry and Molecular Biology: *Biochemical Nomenclature and Related Documents* (Portland Press, London, 1992), pp. 156–157
83. M.J. Berridge, R.F. Irvine: Nature **341**, 197–205 (1989)
84. T. Takenawa, T. Itoh: Biochim. Biophys. Acta **1533**, 190–206 (2001)
85. J.P. Bradshaw, R.P. Bushby, C.C. Giles, M.R. Saunders, A. Saxena: Biochim. Biophys. Acta **1329**, 124–128 (1997)
86. J.P. Bradshaw, R.P. Bushby, C.C. Giles, M.R. Saunders: Biochemistry **38**, 8393–8401 (1999)
87. C. DeWolf, S. Leporatti, G. Kirsch, R. Klinger, G. Brezesinski: Chem. Phys. Lipids **97**, 129–138 (1999)
88. H. Mansour, D.S. Wang, C.S. Chen, G. Zografi: Langmuir **17**, 6622–6632 (2001)
89. M. Müller, O. Zschörnig, S. Ohki, K. Arnold: J. Membr. Biol. **192**(1), 33–43 (2003)
90. W. Caetano, M. Ferreira, M. Tabak, M.I. Mosquera-Sanchez, O.N. Oliveira Jr., P. Krüger, M. Schalke, M. Lösche: Biophys. Chem. **91**, 21–35 (2001)
91. K. Tatemoto, M. Carlquist, V. Mutt: Nature **296**, 659–660 (1982)
92. T. Adrian, J. Allen, S. Bloom, M. Ghatei, M. Rossor, G. Roberts, T. Crow, K. Tatemoto, J. Polak: Nature **306**, 584–586 (1983)
93. J. Allen, S. Bloom: Neurochem. Int. **8**, 1–8 (1983)
94. S. Monks, G. Karagianis, G. Howlett, R. Norton: J. Biomol. NMR **8**, 379–390 (1996)
95. R. Bader, A. Bettio, A. Beck-Sickinger, O. Zerbe: J. Mol. Biol. **305**, 307–329 (2001)
96. B. Rist, N. Ingenhoven, L. Scapozza, C. Peers, P.F. Vaughan, R.L. McDonald, H.A. Wieland, A.G. Beck-Sickinger: FEBS Lett. **394**, 169–173 (1996)
97. A. Bettio: "Biophysical investigations of neuropeptide Y interactions", Ph.D. thesis, University of Leipzig (2002)
98. P.J. Weber, J.E. Bader, G. Folkers, A.G. Beck-Sickinger: Bioorg. Med. Chem. Lett. **8**, 597–600 (1998)
99. N. Krasteva, D. Vollhardt, G. Brezesinski, H. Möhwald: Langmuir **17**, 1209–1214 (2001)
100. Z.W. Yu, P.J. Quinn: Biophys. J. **69**, 1456–1463 (1995)
101. Z.W. Yu, P.J. Quinn: Biophys. Biochim. Acta **1509**, 440–450 (2000)
102. D.D. Lasic, D. Martin: *Stealth Liposomes* (CRC Press, Boca Raton, 1995)
103. G. Blume, C. Cevc: Biochim. Biophys. Acta **1029**, 91–97 (1990)
104. S. Zalipsky, C.B. Hansen, J.M. Oaks, T.M. Allen: J. Pharm. Sci. **85**, 133–137 (1996)
105. P.G. de Gennes: J. Phys. (France) **37**, 1445–1452 (1976)
106. S. Alexander: J. Phys. (France) **38**, 983–987 (1977)
107. P.G. de Gennes: Adv. Colloid Interf. Sci. **27**, 189–209 (1987)
108. D.R. Woodle, D.D. Lasic: Biochim. Biophys. Acta **1113**, 171–199 (1992)
109. D.D. Lasic, D. Needham: Chem. Rev. **8**, 2601–2628 (1995)
110. S. Zalipsky: Adv. Drug Delivery Rev. **16**, 157–182 (1995)
111. S. Beugin, K. Edwards, G. Karlssona, M. Ollivon, S. Lesieur: Biophys. J. **74**, 3198–3210 (1998)
112. A. Wurlitzer, E. Politsch, S. Hübner, P. Krüger, M. Weygand, K. Kjaer, P. Hommes, O. Nuyken, G. Cevc, M. Lösche: Macromolecules **34**, 1334–1342 (2001)
113. C. Hiergeist, R. Lipowsky: J. Phys. II (France) **6**, 1465–1481 (1996)
114. T.R. Baekmark, T. Wiesenthal, P. Kuhn, T.M. Bayerl, O. Nuyken, R. Merkel: Langmuir **13**, 5521–5523 (1997)
115. T.R. Baekmark, T. Wiesenthal, P. Kuhn, A. Albersdörfer, O. Nuyken, R. Merkel: Langmuir **15**, 3616–3626 (1999)
116. T. Wiesenthal, T.R. Baekmark, R. Merkel: Langmuir **15**, 6837–6844 (1999)

Studying Lyotropic Crystalline Phases Using High-Resolution MAS NMR Spectroscopy

André Pampel[1] and Frank Volke[2]

[1] Universität Leipzig, Fakultät für Physik und Geowissenschaften,
anpa@physik.uni-leipzig.de
[2] Fraunhofer-Institut für Biomedizinische Technik IBMT,
frank.volke@ibmt.fraunhofer.de

Abstract. There is a large area, both in basic science and in applied research, in which lamellar liquid-crystalline phases play in important role. Dispersions of lipids and water, forming such phases, serve as models for biological membranes or because of there unique physico-chemical properties, they have been used as drug delivery systems. In this chapter the reader will be introduced in NMR methods for studying such systems. The chapter focuses on an experimental approach that is based on techniques that were developed for investigations of liquid-crystalline phases in combination with MAS. These techniques unsheathe properties that are related to the dynamic, liquid character of liquid-crystalline phases. Their general applicability and their limits are discussed. Their use is demonstrated with some examples covering biophysical studies as well as practical applications. The major focus is on problems that are related to molecules interacting with the lipid–water interfaces. The methods discussed reach from two-dimensional NOE spectroscopy for structure determination, up to the latest development, the combination of MAS with pulsed field gradient NMR spectroscopy to study diffusion properties.

1 Introduction

Lipids, especially phospholipids, a major component of cell membranes, have remarkable self-assembly properties which can be used to "design" biocompatible microstructures with different morphologies (phases) depending on, for example, the lipid itself, hydration, temperature, ion concentration and other parameters.

In this chapter, we concentrate exclusively on two types of phases, namely lamellar liquid-crystalline phases and bicontinuous cubic liquid-crystalline phases. The reader is referred to the chapter "Characterization of Floating Surface Layers of Lipids and Lipopolymers by Surface-Sensitive Scattering", where other important structures are introduced.

Lamellar liquid-crystalline phases are bilayer arrangements of amphiphilic molecules (such as phospholipids) which are organized in such a way that the polar molecular regions are oriented towards the polar liquid surrounding (water) and the hydrophobic regions (lipid hydrocarbon chains) form the inner region of the bilayer. The chemical structure of a typical phospholipid is shown in Fig. 4 below. Besides that kind of static "picture" there is a lot of molecular mobility involved: from segmental *trans–gauche* isomerizations, via molecular rotations around the long axis of the amphiphilic molecules, to lateral diffusion. A schematic structure

A. Pampel, F. Volke, Studying Lyotropic Crystalline Phases Using High-Resolution MAS NMR Spectroscopy, Lect. Notes Phys. **634**, 439–464 (2004)
http://www.springerlink.com/

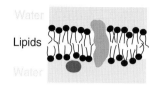

Fig. 1. Schematic view of a lamellar liquid-crystalline phase (left) and a bicontinuous cubic liquid-crystalline phase (right). The lamellar liquid-crystalline phase, for which we use the terms "membrane" or "bilayer", is composed of a double layer of amphiphilic lipids (black) and water (light gray), which separates the lipid double layers. The distance between the two water layers varies depending on the chemical and physical parameters in the range between 30 and 80 Å. The cubic phase structure shown here is described by the space group *Im3m*. In this structure, the lipid bilayer is arranged on a Schwartz surface. The sketch shows only the lipid bilayer, where the Schwartz surface has been approximated by the function $\cos(x) + \cos(y) + \cos(z) = 0$. The channel system is filled with water and water-soluble substances

is shown in Fig. 1. Lamellar phases are important in many fields. They are ideal models for biological membranes. To understand the action, structure, and function of membrane-associated biologically active molecules (drugs, polypeptides, sugars, and proteins), lamellar phases are used as a biological membrane mimetic. Furthermore, they may serve as templates for nano- and microstructure fabrication. At high water content, the lamellae form spherical structures, so-called vesicles or liposomes, which can be used to encapsulate several types of molecules and have found widespread application as drug delivery systems [1, 2].

Cubic liquid-crystalline phases (cubic phases) [3] show other fascinating physico-chemical properties. Here, we concentrate only on bicontinuous phases, as shown in Fig. 1 (right). Bicontinuous phases are cubic structures that are continuous with respect both to the polar (water) and the nonpolar (hydrocarbon) components. A bicontinuous phase is composed of a lipid bilayer that forms a periodic minimal surface, which is interlaced by water channels [3]. Such a porous system can be used as a host system for several drugs of different polarity [4, 5]. Recently, the use of cubic phases for protein crystallization has been described [6]. Besides, there are some reports about the physiological relevance of cubic phases, e.g. as intermediate states during cell fusion and during fat digestion [3].

There are many experimental methods to investigate the named systems, e.g. X-ray diffraction, neutron scattering, IR spectroscopy, fluorescence spectroscopy, calorimetry, and others [7]. All of them are well suited to enlightening us about several aspects of the physical properties of liquid-crystalline phases. Here, we concentrate on nuclear magnetic resonance spectroscopy (NMR). Again, the reader is referred to chapter "Characterization of Floating Surface Layers of Lipids and Lipopolymers by Surface-Sensitive Scattering" to get an impression about the abilities of other methods.

There are several NMR approaches that are highly suitable for studying the systems discussed. The traditional method is solid-state NMR spectroscopy. For instance, ^{31}P NMR spectroscopy is a very convenient tool for phase structure investigations [8]. Using ^{2}H solid-state NMR [9], it is possible to characterize the order parameter profile along lipid molecules and its segmental dynamics, utilizing the residual quadrupolar interaction. However, those methods do not allow the direct determination of the structure or of distances between certain molecular groups.

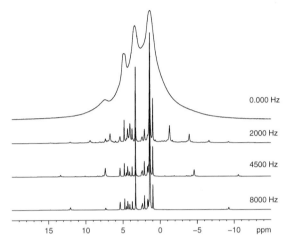

Fig. 2. ^{1}H NMR spectra of a lipid/water/peptide membrane dispersion measured at 750 MHz depending on the MAS rotation frequency. The static spectrum (top) is still dominated by the ^{1}H–^{1}H-dipolar couplings. At increasing rotation frequency, the influence of these couplings vanishes and almost all magnetically inequivalent signals become resolved. Further details are discussed in Sect. 2.2.

^{1}H NMR spectroscopy is a very sensitive technique even for low concentrations. However, residual magnetic dipolar interactions and influences of discontinuities of the magnetic susceptibility of liquid-crystalline samples broaden the ^{1}H NMR spectra and hamper the application of many experimental approaches that turned out to be quite useful in liquids.

The effects of the mechanism mentioned can be reduced by spinning the sample rapidly around an axis that is oriented at an angle of $54.7°$ to the direction of the magnetic field. By spinning at this so-called "Magic angle" at a rate larger than the anisotropic interactions, these interactions are averaged to their isotropic value, resulting in substantial line narrowing. Furthermore, magic-angle spinning (MAS) removes the magnetic susceptibility broadening (see chapter "Study of Conformation and Dynamics of Molecules Adsorbed in Zeolites by ^{1}H NMR"). We refer to other chapters of this book, where beneficial application of MAS methods to cartilage (chapter "NMR Studies of Cartilage - Dynamics, Diffusion, Degradation") and molecules adsorbed in zeolites (chapter "Study of Conformation and Dynamics of Molecules Adsorbed in Zeolites by ^{1}H NMR") are described.

The remaining proton–proton dipolar coupling for membrane molecules that are undergoing fast axial motion (i.e. correlation time smaller than 1.3×10^{-5}s) is about 10 kHz and can be easily averaged out by modest MAS rotation frequencies [10]. It has been demonstrated that many types of molecules embedded in membranes meet the condition of fast axial motion [11, 13, 16]. The ^{1}H MAS NMR spectra of these molecules in membranes provide enough resolution to identify the signals from different membrane components and to use an approach that uses high-resolution NMR techniques. Such methods have been termed high-resolution MAS NMR (HR-MAS). The increase in resolution of ^{1}H NMR spectra of a membrane when MAS is applied is shown in Fig. 2. Generally, a linewidth on the order of 10 Hz for some lipid resonances is possible. Glaubitz and Watts [14] have shown that the linewidth can be reduced further if the slow membrane motion is suppressed by using solid-supported membranes for MAS experiments.

The situation is even better for bicontinuous cubic phases. Here a much higher resolution in the range of 1 Hz for protons can be achieved, because of internal molecular motions and the geometry of the structure, which allows averaging of orientation-dependent interactions by diffusion along the strongly curved surfaces. In principle, it should be possible to acquire spectra with high resolution without MAS for cubic phases. However, the high stiffness of samples in cubic phases does not allow a proper preparation of samples for high-resolution NMR. The NMR tube will always contain air bubbles and cannot be filled homogeneously, which is essential for performing high-resolution ^{1}H NMR without MAS. Here, the application of MAS readily reduces the linewidth even at quite low spinning frequencies [15].

The resolution obtained for the ^{1}H spectra of lamellar and cubic phases allows the performance of many types of HR-MAS NMR experiments, which have found widespread application in the field of membrane research [13, 16]. By using a few examples from our own research, we show what kind of experiments can be performed and what kind of information about the systems under study can be gained.

2 Experiments for Elucidating Structural and Dynamical Properties – NOESY and ROESY

2.1 Theory of Cross-Relaxation

The high spectral resolution available by MAS in combination with high molecular mobility facilitates the application of experiments for determination of NMR relaxation parameters. Experiments that detect ^{1}H–^{1}H cross-relaxation effects are of particular interest, because they give access to structural as well as to dynamical information about liquid-crystalline systems and they can easily be performed in combination with MAS. The cross-relaxation phenomenon, often called the nuclear Overhauser effect (NOE), is the change of the net intensities of several resonances in one spectrum if one or more signal(s) of the spectrum is transferred into a nonequilibrium state [17]. The time evolution of z-magnetizations of a system containing N sites of different species (molecular subgroups) is described (neglecting the influence of J-couplings) by a system of differential equations [18]

$$\frac{\mathrm{d}}{\mathrm{dt}}\Delta\mathbf{M}_z(t) = -\mathbf{R}\Delta\mathbf{M}_z(t), \tag{1}$$

where the deviation from thermal equilibrium is $\Delta\mathbf{M}_z(t) = \mathbf{M}_z(t) - \mathbf{M}_0$. The equilibrium z-magnetizations of the N sites are collected in the vector $\mathbf{M}_0$, with elements $M_{l0} = n_l M_0$ that are proportional to the number n_l of equivalent nuclei in the site l and to the equilibrium magnetization M_0 [18]:

$$\mathbf{M} = \begin{pmatrix} M_1 \\ M_2 \\ \vdots \\ M_n \end{pmatrix}, \quad \mathbf{M_0} = \begin{pmatrix} M_{10} \\ M_{20} \\ \vdots \\ M_{n0} \end{pmatrix}, \quad \mathbf{R} = \begin{pmatrix} \rho_{11} & \sigma_{12} & \cdots & \sigma_{1n} \\ \sigma_{21} & \rho_{22} & \cdots & \sigma_{2n} \\ \vdots & \vdots & \ddots & \vdots \\ \sigma_{n1} & \sigma_{n2} & \cdots & \rho_{nn} \end{pmatrix}. \tag{2}$$

For a homonuclear two-spin system, σ_{ij} is defined by

$$\sigma_{ij} = \left(\frac{\mu_0}{4\pi}\right)^2 \left(\frac{2\pi}{5}\right) \gamma^4 \hbar^2 [3 J_{ij}(2\omega_0) - \frac{1}{2} J_{ij}(0)], \tag{3}$$

where the Larmor frequency is $\omega_0 = -\gamma \mathbf{B}_0$ and the gyromagnetic ratio is γ. The following symmetry relation holds:

$$n_j \sigma_{ij} = n_i \sigma_{ji}. \tag{4}$$

The spectral density function $J_{ij}(\omega)$ is defined by

$$J_{ij}(\omega) = \mathrm{Re} \left\{ \int_{-\infty}^{\infty} C_{ij}(\tau) \exp(-i\omega\tau)\,\mathrm{d}\tau \right\}, \tag{5}$$

where the autocorrelation function of the homonuclear dipole-dipole interaction, expressed as

$$C_{ij}(\tau) = \frac{4}{5} \sum_i \sum_j \left\langle \frac{Y_{20}(\mathbf{r}(0))}{r^3(0)} \frac{Y_{20}(\mathbf{r}(\tau))}{r^3(\tau)} \right\rangle, \tag{6}$$

characterizes the motion of the vector $\mathbf{r}$ connecting the dipoles. The function Y_{20} is defined by

$$Y_{20}(\mathbf{r}) = \sqrt{\frac{5}{16\pi}} (3\cos^2(\theta) - 1), \tag{7}$$

where θ is the angle between the magnetic field and $\mathbf{r}$. The summations over i and j include all magnetically equivalent protons of each resonance.

In the case of lamellar membranes or cubic phases, the interpretation of σ_{ij} is different from the interpretation that is commonly used for macromolecules, e.g. proteins. In proteins or other large, complex molecules, the length of $\mathbf{r}$ is not changing.

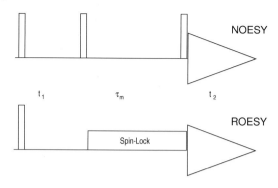

Fig. 3. Schematic drawing of the NOESY and ROESY pulse sequences. All pulses are 90° pulses. The phases of the pulses and of the detector are cycled to select the desired coherences and to suppress unwanted ones

Therefore only the change of the orientation has to be estimated (for instance by the use of a constant overall tumbling time), which allows the direct determination of distances. This approach has found widespread application for structure determination of biomolecules. However, it is unwise and impossible to describe a dynamical system such as a lipid bilayer by a single structure. Therefore, Gawrisch and Huster [13, 19, 20] have suggested relating the cross-relaxation rates not to a single, constant distance but to the probability of close contact or approach between several molecular subgroups. In this sense, the factor $C_{ij}(0)$, which is extremely sensitive to the number of very close distances because of its $1/r_{ij}^6$ dependence, describes the probability of close contact.

The easiest way to measure $\mathbf{R}$ is the two-dimensional NOESY (nuclear Over-hauser enhancement spectroscopy) experiment. The pulse sequence is shown in Fig. 3 (top). It starts with a longitudinal magnetization, which is converted by a 90° pulse into a transverse magnetization. The magnetization evolves freely under the influence of the chemical shift and J-coupling. In NOESY, the magnetization is flipped back into the z-direction and evolves under the influence of longitudinal relaxation as described in (1). The final pulse flips the magnetization again into the xy plane, where it is detected.

By performing Fourier transformation along both time axes t_2 and t_1, one obtains a two-dimensional spectrum, the peak amplitudes $\mathbf{A}(\tau_m)$ can be calculated from

$$\mathbf{A}(\tau_m) = \exp(-\mathbf{R}\tau_m)\mathbf{A}(0), \tag{8}$$

where $\mathbf{A}(0)$ are the peak volumes at mixing time zero. By inverting (8), the relaxation rate matrix $\mathbf{R}$ can be calculated from the spectra measured at several mixing times. A more reliable approach is to record spectra at many different mixing times and fit $\mathbf{R}$ to the data. At short mixing times, the so-called initial-rate approximation, the peak amplitudes are directly proportional to the relaxation rates, i.e. the spectrum can be understood as a pictorial representation of the rate matrix $\mathbf{R}$. This is not valid at longer mixing times, where so-called spin diffusion processes, which are caused by multiple, consecutive steps of magnetization transfer, can occur and influence the peak amplitudes significantly [18]. As a consequence, it might happen that cross-peaks between protons are observed although their respective cross-relaxation rate is zero. When NOESY experiments on membranes were described for the first time by many authors, there was strong concern that the cross-relaxation rates obtained did not represent the membrane structure but are dominated by such spin diffusion processes. Here Huster, Gawrisch [13, 19, 20] and coworkers, using an ingenious concept of diluting ^{1}H spin systems and carefully performing and analyzing experiments, could prove the assumptions about the relevance of NOESY experiments. In particular they showed that the spin diffusion in most membranes is insignificant at mixing times up to 300 ms. An example of NOESY experiments on membranes is discussed in Sect. 2.2.

There is an equivalent experiment to NOESY, the ROESY experiment (rotating frame nuclear Overhauser enhancement spectroscopy) [21]. While in NOESY the longitudinal, i.e. z-magnetization is investigated, ROESY detects the cross-relaxation in the rotating frame. The experiment, shown in Fig. 3, starts also with a single 90° pulse, but after t_1, the magnetization remains in the xy plane where an applied r.f. field during τ_m prevents evolution by the chemical shift, enabling effective cross-relaxation, which would be suppressed otherwise. The r.f. field, the so-called "spin-lock" can be applied using repetitive application of pulses or in continuous manner. Immediately after τ_m, the detection starts. In both sequences, appropriate phase cycling allows the selection of desired coherence pathways while suppressing unwanted ones. For details of the phase cycling and variants of the NOESY and ROESY sequences, the reader is referred to the literature [21]. There is an equivalent equation to (1) that describes the dynamics of the spin-locked magnetization neglecting influences of J-coupling:

$$\frac{d}{dt}\mathbf{M}^T(t) = -\mathbf{R}^T\mathbf{M}^T(t). \tag{9}$$

The cross-relaxation rate in the rotating frame is given by

$$\sigma^{ROE} = \left(\frac{\mu_0}{4\pi}\right)^2 \left(\frac{2\pi}{5}\right) \gamma^4 \hbar^2 [J(0) + \frac{3}{2}J(\omega_0)]. \tag{10}$$

The NOESY experiment is a quite robust experiment that can be applied to study membranes and cubic phases. The ROESY experiments, however, cannot be performed on membranes because during the long mixing time the spin-locked magnetization decays too fast to give a reliable signal.

A remarkable difference between ROESY and NOESY is the magnitude and sign of respective cross-relaxation rates and cross-peaks (compare (3) and (10)). Generally, cross-peaks in NOESY may have a positive sign, reflecting a large negative cross-relaxation rate, or a negative sign, reflecting a smaller, positive cross-relaxation rate. Especially for molecular systems with fast motions, the longitudinal cross-relaxation rate can become very small or even vanish, thus resulting in undetectable cross-peaks. The cross-relaxation rate in the rotating frame σ^{ROE} is always

positive and may not become zero. Indeed, both experiments are rather complementary, because systems with fast molecular motions that are not accessible with
NOESY are well suited to being investigated with ROESY. As an added advantage,
ROESY spectra are hardly influenced by spin-diffusion, which allows a straightforward interpretation of observed cross-peaks. An example of ROESY experiments on
cubic phases is discussed in Sect. 2.3.

2.2 Location of Molecules in the Membrane

Here, we show, using an example of the peptide AFAB (Ala-Phe-Ala-*O-tert*-butyl)
bound to a membrane, how the embedding of molecules can be measured using
NOESY experiments. The system lipid/AFAB/HDO is a model system to investigate
lipid–peptide interactions. It has already been investigated using neutron scattering
and high-resolution MAS NMR [22, 23]. The structures of AFAB and POPC (1-
palmitoyl-2-oleoyl-*sn*-glycero-3-phosphatidylcholin), a common lipid that was used
in this study, are shown in Fig. 4.

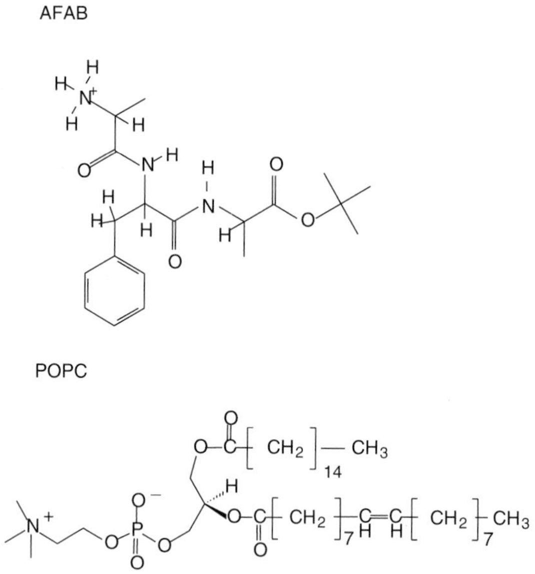

Fig. 4. Schematic structure of AFAB (top) and POPC (bottom)

The NMR spectra shown here were measured using a DMX 750 spectrometer
(Bruker, Germany) operating at 750.13 MHz proton frequency. The probe was a
narrow-bore 4 mm CP MAS probe (Bruker). The X-channel was tuned to ^{2}H and
used as a lock channel for field stabilization. Samples were prepared by mixing the
appropriate amounts of POPC and AFAB (molar ratio 6:1) in methanol. Methanol
was evaporated under high vacuum overnight. Water (HDO) was added to the dry

powder. Sample homogenization was achieved by applying several freeze–thawing cycles.

The ^{1}H spectrum of the sample at 310 K observed using 8 kHz sample spinning is shown in Fig. 5. The unambiguous assignment was obtained from two-dimensional COSY spectra (not shown here) performed under the same conditions.

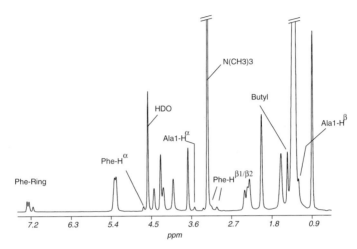

Fig. 5. ^{1}H MAS NMR spectrum of AFAB attached to a lamellar liquid-crystalline phase (membrane) composed of POPC and water (HDO). The high resolution obtained allows the assignment of ^{1}H signals, except those of the NH groups. NH groups give only one single, broad signal, which cannot be observed if HDO is used

It should be noted that the high resolution shown here is only possible if the NMR sample is carefully prepared. For this purpose, we use so-called HR-MAS rotors, which allow the location of the sample material within a sphere (radius 3 mm) in the center of the 4 mm ZrO rotor. This ensures that the sample is almost free of air bubbles and is kept within the rotor during spinning; it allows good field homogenization (shimming) and also minimizes the temperature gradient.

As already mentioned above, the question about the absolute structure of the membrane makes no sense, because a membrane and all its components form a dynamic structure. An interesting question is whether AFAB, or any other molecule under different conditions, is positioned within the POPC bilayer, and if that is the case, how this position can be characterized. Such a knowledge is also very important for characterizing drug delivery systems. The cross-relaxation NMR experiments introduced, for membranes the NOESY experiment, are well suited to addressing such questions.

The NOESY spectrum of the POPC/AFAB/HDO sample measured at a mixing time of 200 ms is shown in Fig. 6. A close inspection of this spectrum shows that there are cross-peaks between all groups of the lipid molecule, even between those of the terminal CH_3 group and the $N(CH_3)_3$ group. These cross-peaks reflect the disorder

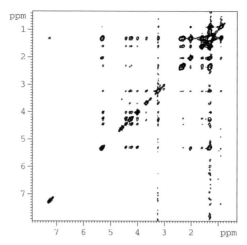

Fig. 6. ^{1}H NOESY MAS NMR spectrum of a membrane composed of POPC, AFAB, and HDO measured at 310 K, applying MAS using 8 kHz rotation frequency

in the lipid matrix, as it has been discussed above. If we discard spin-diffusion influences, which is reasonable at this mixing time, the spectrum provides evidence that there is a probability of close contact between the terminal CH_3 group and the $N(CH_3)_3$ group of POPC. This observation was surprising for researchers in the field of biomembranes, but it has been confirmed for many other systems of the same kind [11,13]. It is in discrepancy with the more static membrane picture provided in Fig. 1. Quantification of the cross-relaxation rates shows, that this contact is of course not the most probable one, but that it cannot be neglected. Molecular dynamics simulations have shown that this contact occurs between neighboring molecules [11,20].

In this sense, the question about the most probable location of AFAB within the POPC bilayer is a question about the largest cross-relaxation rate between peptide and lipid signals. Generally one would calculate all rates by measuring cross-peak amplitudes at several mixing times and inverting (8). This approach is somewhat hampered for our systems, because all peptide signals are at least partially overlapped by lipid signals (compare Fig. 5), which complicates the exact determination of the amplitudes. Furthermore, the amplitude of the peptide signal is rather small. Therefore, we concentrate exclusively on cross-peaks that involve the signals of the aromatic ring of AFAB. The only cross-peak visible in the spectra is a cross-peak with the CH_2-groups of POPC. At a lower threshold, there are also cross-peaks with signals of the glycerol backbone of POPC visible. These results lead to the conclusion that the most probable location of the aromatic ring of AFAB is the upper region of the CH_2 chains of the lipid bilayer. This observation is in agreement with previous observations [22].

In general, the interaction of molecules with lipid bilayers (membranes) and their most probable position within them can be nicely studied by NOESY in combination with MAS. The results obtained are more stringent, as the quality of the spectra increases and thus the measurement of the cross-relaxation rates can be performed. The necessary conditions cannot always be fulfilled, because the maximum resolution available in the ^{1}H spectra of membranes is limited. This is caused not only by the linewidth but also by the fact that some groups, e.g. the CH_2 chain of lipids result in mainly one signal. Here, isotopic labeling could improve the situation, which however could result in enormous costs. Nevertheless, often the observation of a single cross-peak can already give valuable information about the system under study, for instance if a certain molecules interacts with parts of the bilayer or even if it does not. Such information must not be underestimated in lipid bilayer research, especially in studies of drug delivery systems. A further example of such an "minimal" approach is given in the next section.

2.3 Lipid–Drug Interaction in Drug Delivery Systems in Cubic Liquid-Crystalline Phases

The porous system of a cubic phase can be used as a host system for drugs because it forms a stiff structure that exists even in an excess of water. For medical applications the cubic phase is prepared using water containing drugs, which fills the channel system. This dispersion can be spread as an ointment or implanted in the body, where the porous structure allows a slow, continuous release of the compounds included. Besides, one could use the lipid matrix as a host for water-insoluble drugs (e.g. Paclitaxel) to form a double release system, which would have many advantages, especially in anticancer therapy [4, 5].

To develop such a release system, it is essential to determine its biophysical and chemical properties, before clinical tests can be performed. It is important to have information about the structure of the incorporated drugs, about their interaction with the lipid matrix, and about their mobility (diffusibility). The latter is of direct importance for the pharmacokinetical properties and will be discussed in Sect. 3.1. Because the lipid matrix of a cubic phase is assumed to be stable in excess water for a longer time, the diffusivity of the water-soluble drugs within the water channels is a determining factor of the release kinetics over the first hours/days [4, 5].

Here we show how HR-MAS experiments, especially ROESY, can be applied to study such systems.

The system that is discussed here is a cubic phase composed of monoolein (1-monooleoyl-*sn*-glycerol, MO), MPEG-DSPE (1,2-distearoyl-*sn*-glycero-3-phospho-ethanolamine-*N*-[methoxy (polyethylene glycol)-2000]), D_2O and the anticancer drug carboplatin in the molar ratio 40.3 : 1.45 : 1 : 471 [24]. It was found that it forms a cubic phase, which is described by the space group *Im3m*. The chemical structures of MPEG-DSPE and carboplatin are shown in Fig. 7.

For the following, all spectra were recorded on samples of this composition at 310 K on a Bruker DMX 600 spectrometer operating at 600.13 MHz for ^{1}H. The probe was a Bruker 4 mm HR-MAS probe, equipped with ^{1}H, ^{13}C, and ^{2}H lock channel and pulsed-field-gradient capabilities. The typical 90° pulse length was about 10.5 µs.

A proton NMR spectrum measured under MAS using a 3500 Hz rotation frequency of the system is shown in Fig. 8. The linewidth obtained for the terminal

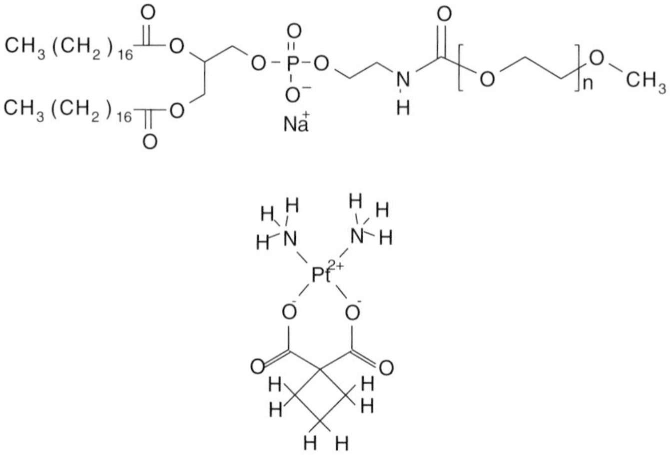

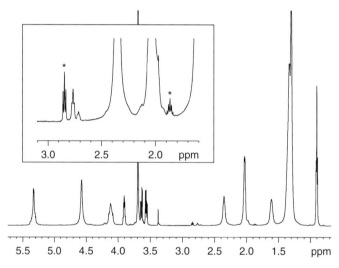

Fig. 7. Structures of MPEG-DSPE (top) and carboplatin (bottom).

Fig. 8. : ^{1}H MAS NMR spectrum (600.13 MHz) of a monoolein, MPEG-DSPE, carboplatin and water (D_2O) dispersion at 310 K and 3500 Hz MAS rotation frequency. The spectrum can be divided roughly into two main spectral groups, aliphatic fatty acid chains (0.9–2.5 ppm) and the glycerol backbone and MPEG chain (3–4.5 ppm). Beside these, there are signals of water (HDO, 4.57 ppm) and the CH=CH group (5.33 ppm). The inset displays the extended spectra of carboplatin (marked by asterisks), impurities in the lipid material (at 2.76 and 2.71 ppm), and some of the strong lipid signals. The signal of the $(-CH_2CH_2O-)$ moiety of MPEG-DSPE occurs at 3.694 ppm. Sample preparation: the lipids were codissolved in ethanol. After vortexing, the solvent was removed using high vacuum over several hours. The D_2O/carboplatin solution was added to the dry lipid mixture. Finally, the dispersion was homogenized by freeze–thawing cycles. For the NMR experiments, the material was transferred into Bruker HR-MAS rotors and measured immediately

methyl group of MPEG was 1.5 Hz, which provides excellent conditions for performing HR-MAS experiments. The complete assignment is given in the literature [15]. Only a few important signals are indicated.

It has been observed that carboplatin has different pharmaceutical and pharmacokinetical properties in the presence of polyethylene glycol (PEG) chains. This leads to the question of whether there is a special interaction between Carboplatin and PEG, which triggers the change of the properties of the drug. Such interactions can be visualized by cross-relaxation experiments. In the highly mobile cubic phase, it is of advantage to perform ROESY experiments. It is rather complicated to quantify cross-relaxation spectra of liquid-crystalline phase systems, owing to the lack of exact knowledge about the correlation function (see Sect. 2.1). However, a negative cross-peak in a ROESY spectrum between uncoupled protons can only be caused by a spatial proximity. Indeed, such proximity could be observed in two-dimensional ^{1}H ROESY spectra of the sample studied. The spectrum is shown in Fig. 9.

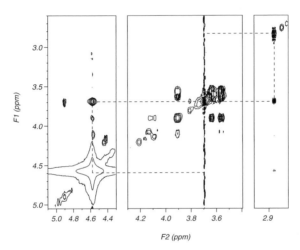

Fig. 9. Parts of the ^{1}H ROESY MAS-NMR spectrum of a monoolein, MPEG-DSPE, carboplatin, and water (D_2O) dispersion at 310 K and 3500 Hz MAS rotation frequency. Spins were locked by a continuous 2.7 kHz pulse of 200 ms duration. 32 scans with 2.5 s repetition time were detected for each increment. The 8192×320 complex data points were multiplied by a $\cos^2$ window function along t_1 and exponential line broadening (2 Hz) along the t_2 dimension before Fourier transformation. To avoid interference with rotational sidebands, the spinning speed and the spectral width in t_1 were set to 5 kHz [25]. The dotted lines mark the cross-peaks between PEG (3.694 ppm)/carboplatin and PEG/HDO. Reprinted from [24] with permission of Elsevier

Negative cross-peaks between the PEG chain of MPEG-DSPE and carboplatin, as well as between PEG and HDO, are visible. Both cross-peaks cannot be detected in the corresponding NOESY spectra (not shown), indicating a rather small NOE effect caused by fast molecular motions.

Thus, an interaction between PEG and carboplatin could be established. Whether this interaction readily influences the mobility of the drug, i.e. whether it is suited to hindering and slowing down its motion, cannot be concluded from cross-relaxation experiments. In Sect. 3.4 it will be shown how measurements of the diffusion coefficient can contribute to answering this question.

3 Pulsed Field Gradient NMR in Combination with MAS

3.1 Diffusion Studies in Liquid-Crystalline Phases

The diffusion of molecules incorporated in or attached to lipid liquid-crystalline systems is an important characteristic for understanding the structure, dynamics, and function of such systems. Knowledge of the lateral diffusion coefficients of all membrane components can provide important additional information about the biophysical properties and it is suited to gaining a more detailed understanding of membrane function. For instance, lateral segregation and aggregation of membrane molecules is commonly assumed to give rise to biologically differentiated regions. Segregation processes would lead to the formation of domains, and the length scales of these domains may be as small as 10–100 nm, which is difficult to detect with other experimental techniques. The lateral diffusion coefficient could provide more information if molecules were embedded in such domains, as would be commonly observed using quadrupolar splittings or relaxation parameters, which reflect mainly local properties such as the orientation of molecular moieties or the correlation time of their anisotropic reorientation. The lateral diffusion, however, reflects the long-range characteristics of the membrane components and the long-term behavior of molecules bound to the membrane.

The latter is of crucial importance for the pharmacokinetical properties of drug release systems. Because the lipid matrix of a cubic phase is assumed to be stable in excess water for a long time, the diffusivity of the water-soluble drugs within the water channels is a release-kinetics-determining factor over the first hours/days [5]. Furthermore, the diffusional behavior also reflects structural properties of molecules, which cannot be easily detected by NOESY or ROESY spectroscopy.

As already discussed in chapter "Structure-Mobility Relations of Molecular Diffusion in Interface Systems", diffusion processes can be studied by using pulsed field gradient (PFG) NMR experiments. It is strongly recommended to read the excellent reviews by Johnson [26], Kärger [27], and Price [28, 29] to get a detailed introduction to the method. As a prerequisite to investigating mixtures of different molecules, the signals arising from the different molecules should be clearly separated by the chemical shift before the analysis of the diffusion is performed [30]. If there is no or just limited spectral resolution available, a multicomponent analysis of the signal decay curves has to be performed, which requires many data points and good signal-to-noise-ratios. However, in the systems discussed here, the signals of the lipids tend to cover the complete spectral range from 0.8 to 5 ppm in proton spectra with very strong signals. In this case the performance of a multicomponent analysis will be very limited, because the data will be dominated by the lipid signals. Hence, it will be difficult to find the small contributions from the incorporated molecules to the overall decay curve, and even if it were possible to get reasonable multicomponent fits, there is still an assignment problem, for example owing to the presence of signals from impurities. It is essential to assign the components to the diffusion coefficient obtained to get reliable data. Without spectral resolution, an assignment is only possible if special isotopic labels are used, but such studies have the disadvantage that the diffusion of all components cannot be obtained from one sample. Furthermore, it is not unproblematic to introduce such labels, because they may change the physical character of molecules.

3.2 PFG NMR in Combination with Magic-Angle Spinning

The standard application of PFG echo sequences (see chapter "Structure-Mobility Relations of Molecular Diffusion in Interface Systems") is not practical for molecules embedded in biological or model membranes. As discussed in Sect 1, the NMR properties of the systems considered result not only in a fast decay of the transverse magnetization during the time course of the experiment, but also in short echo signals and thus broadened spectra. Therefore, for studies of membrane systems, the PFG methods have to be combined with techniques for suppressing dipolar couplings in particular. This can be accomplished by performing experiments using solid-supported membranes or using decoupling techniques [31, 33], but these two methods are often not sufficient to a provide spectral resolution that is suitable for the spectroscopic separation of signals from many components. A more efficient way to obtain both high spectral resolution and measuring of diffusion would be the combination of MAS and PFG.

The application of PFG techniques requires probes and electronic devices that facilitate the application of pulsed field gradient pulses in combination with sophisticated r.f. pulse sequences. In order to enable the combination of pulsed field gradients with MAS, the field gradient has to be oriented parallel to the spin axes of the rotor. The MAS probes nowadays commercially available allow the application of field gradients in the range of 0.5 T/m, which is small compared with equipment optimized for measuring diffusion exclusively. The small gradient strength available limits diffusion processes that can be studied. Regarding the minimum displacements that can be monitored by PFG with these gradients, it is clear that only diffusion processes described by diffusion coefficients larger than $10^{-12} \, \text{m}^2/\text{s}$ can be investigated. Slower diffusion will not result in measurable spin echo attenuation.

For obtaining accurate data, a large spin echo attenuation is required; however, not only does the reliability of the diffusion coefficients depend on the strong decay of the spin echo amplitude, but also the quality of the spectra is of crucial importance. A stimulated echo sequence using bipolar sine-shaped gradients and longitudinal eddy current delay meets both conditions. The pulse sequence is shown in Fig. 10. Here, applying two pairs of pulsed gradients with different polarity separated by a 180° pulse creates the stimulated echo. The sine shape of the gradient reduces the induction of eddy currents, while their different polarity minimizes the impact on the deuterium

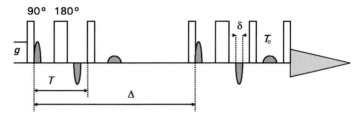

Fig. 10. Stimulated echo sequence using bipolar sine-shaped gradients and longitudinal eddy current delay

lock signal. As an added advantage, the 180° pulse between the gradient pair also minimizes the influence of cross-relaxation effects on the measurements [32]. In order to reduce transverse components, spoil gradients are applied during the observation time and during the eddy current delay. An eddy current delay T_e of 5 ms, during which the magnetization is stored in the z-direction, is quite important for obtaining a sufficient quality of spectra.

It is recommended that one restricts the power of the applied pulse in order to prevent destruction of the coil and the MAS stator. It was found that a duration of 3 ms for a single gradient pulse in the pulse sequence shown in Fig. 10 is feasible for most of the applications described here if the maximum applied strength is restricted to 80% of the maximum available current (10 A) in the amplifier unit. Higher power or using multiple bipolar pulses might be possible, but it is strictly recommend that one discusses the details with the probe manufacturer. Generally, the gradient strength and duration should be kept as small as possible.

The gradient strength, which is a crucial parameter for all PFG experiments, can be calibrated by recording the ^1H spectrum with the gradient on using a water-filled HR-MAS rotor (Bruker). The spherical form of the sample allows the calculation of the gradient strength from the line shape of the spectrum [29]. The gradient strength obtained can be checked using liquid samples with a known diffusion coefficient, but great care has to be taken because of temperature control and sample instabilities due to poor rotor filling.

3.3 Pulsed Field Gradient NMR in Liquid Crystalline Phases

The theoretical background of pulsed field gradient NMR self-diffusion chapter "Structure-Mobility Relations of Molecular Diffusion in Interface Systems". Here we focus on the description that is relevant for lamellar and cubic liquid-crystalline phases. Pulsed field gradient NMR self-diffusion measurements are based on NMR pulse sequences that generate spin echoes of the magnetization of the resonant nuclei. By appropriate addition of pulsed magnetic field gradients of duration δ, intensity g, and separation Δ (the observation time) in the defocusing and refocusing period of the NMR pulse sequence, the observed spin echo becomes sensitive to the translational motion of the molecules which carry the nuclear spin under investigation. In the case of self-diffusion, the spin echo intensity $M(\delta g, \Delta)$ is attenuated. Assuming that a constant pulsed field gradient is applied along the z-axis of the laboratory frame of reference, the attenuation factor is given by (11), where $\langle z^2 \rangle$ denotes the mean square displacement of the diffusing molecules in the direction of the pulsed field gradient during the observation time and γ is the gyromagnetic ratio of the type of nuclei observed.

$$\Psi(\delta g, \Delta) = \frac{M(\delta g, \Delta)}{M(\delta g = 0, \Delta)} = \exp\left[-(\gamma\delta g)^2 \frac{1}{2} \langle z^2(\Delta) \rangle\right]. \tag{11}$$

For any diffusion process, one may define an apparent time-dependent self-diffusion coefficient $D_z(\Delta)$ along the z-direction, whereby for normal (unrestricted) self-diffusion D_z is independent of the observation time:

$$\langle z^2(\Delta) \rangle = 2D_z(\Delta)\Delta. \tag{12}$$

With (12), the time-dependent self-diffusion coefficient or the mean square displacement can be used to describe the self-diffusion process. According to (11) and (12), both values may be measured by PFG NMR from the slope of the semi-logarithmic plot of the spin echo attenuation as a function of the square of the applied pulsed-field-gradient strength, $\ln(\Psi)$ vs. g^2. The displacement of molecules is monitored in one spatial dimension, which is defined by the direction of the pulsed field gradient. For anisotropic motion of molecules embedded in lipid membranes, the diffusion coefficient defined by (12) is not the diffusion coefficient that describes the motion of the molecules within the bilayer. Molecular motion in a membrane system x', y', z' (see Fig. 11) oriented at an angle Θ with respect to the gradient direction results in z-displacement along the gradient:

$$\langle z^2(\Delta) \rangle = z'^2 \cos^2(\Theta) + x'^2 \sin^2(\Theta). \tag{13}$$

If the motion is related to a diffusion process, one obtains

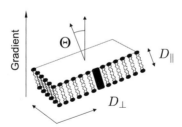

Fig. 11. Sketch of diffusion processes in a lipid bilayer. The bilayer normal, which defines the z'-direction, is oriented at an angle Θ with respect to the gradient direction. The gradient direction defines the z-direction of our laboratory frame

$$\langle z^2(\Delta) \rangle = 2D_\parallel(\Delta)\Delta \cos^2(\Theta) + 2D_\perp(\Delta)\Delta \sin^2(\Theta). \tag{14}$$

$D_\parallel$ and $D_\perp$ are the self-diffusion coefficients describing the diffusion parallel and perpendicular to the membrane bilayer, respectively. Compared with (11) and (12), the damping factor of the spin echo decay now depends on the orientation of the bilayer normal to the field gradient. In a nonoriented powder sample, the bilayer normal is randomly oriented with respect to the gradient direction and the spin echo decay has to be calculated by integration over all possible orientations. The spin echo attenuation is expressed as

$$\Psi(k) = \frac{1}{2} \int_0^\pi \exp\left\{-k \left(D_\parallel \cos^2(\Theta) + D_\perp \sin^2(\Theta)\right)\right\} \sin\Theta \, d\Theta \qquad (15)$$

The factor k depends on the experimental setup chosen, and in the experiment used here (see Fig. 10), it is given by

$$k = \left(\frac{4\delta\gamma g}{\pi}\right)^2 \left(\Delta - \frac{T}{6} - \frac{\delta}{8}\right) \quad \text{and defining} \qquad (16)$$

$$T = p^\pi + 2\delta + 2\delta_{\text{rec}}, \qquad (17)$$

where δ denotes the duration of the gradient pulse and δ_{rec} the recovery time between a gradient pulse and the subsequently applied r.f. pulse. It takes into account that sine-shaped gradients were applied in a bipolar manner with duration δ, with inverted sign and separated by a 180° pulse of duration p^π [34].

The formal solution of (15) is given by

$$\Psi(k) = \frac{\exp(-kD_\perp)\sqrt{\pi}\,\text{erf}\left\{\sqrt{k(D_\parallel - D_\perp)}\right\}}{2\sqrt{k(D_\parallel - D_\perp)}}, \qquad (18)$$

with the error function expressed by

$$\text{erf}(z) = \frac{2}{\sqrt{\pi}} \int_0^z \exp(-t^2)\, dt. \qquad (19)$$

For data evaluation purposes the integrals contained in (15) or (18) have to be solved numerically or expressed as sums.

For molecules confined in the membrane system, displacements parallel to the bilayer normal are considered as not contributing to spin echo attenuation, which is a reasonable assumption keeping in mind the membrane thickness and the space which is available for the water molecules between the bilayers. In this sense, the diffusion in a membrane system is described as a two-dimensional process, i.e. $D_\parallel$ is neglected, thus resulting in [35]

$$\Psi(k) = \frac{\exp(-kD_\perp)\sqrt{\pi}\,\text{erf}\left\{\sqrt{-kD_\perp}\right\}}{2\sqrt{-kD_\perp}}. \qquad (20)$$

Equation (20) can be expanded in a series, thus giving, in very good approximation for small gradient values,

$$\ln[\Psi(k)] = -\frac{2}{3}(kD_\perp) + \frac{2}{45}(kD_\perp)^2 + O(kD_\perp)^3. \qquad (21)$$

For a larger gradient strength/diffusion coefficient or a longer observation time, the deviation from the linear behavior increases.

The description of the PFG experiment by (14), (15), (18), and (20) is only valid if the bilayer orientation for each diffusing molecule remains constant. However, under the real conditions occurring in lipid/water systems, the lipid bilayers form structures with bends and curvatures. If one assumes that the rms displacement of a diffusing molecule within the bilayer during the observation time is small compared with a certain domain size, it is reasonable to describe these structures by planar domains in which the molecules are confined [35]. At longer observation times a molecule diffusing along such a structure will sample different director orientations. The amount of sampled orientations N depends on the diffusitivity and the domain sizes. Callaghan and Söderman [35] distinguished and discussed two cases: sampling a large amount of domains, i.e. the case $N \to \infty$, and the case where only a few domains are sampled. For $N \to \infty$ they found that the observed spin echo decay would be Gaussian, i.e. the semilogarithmic plot $\ln[\Psi(k)]$ vs. g^2 would become linear. This case is not relevant here, as will be shown below. If N is small but greater than one, they argued that it is reasonable to assume that the observed diffusion coefficient is inversely proportional to the number of domains sampled by each spin, thus obtaining

$$\Psi(k) = \frac{\exp(-kD_\perp/N)\sqrt{\pi}\,\mathrm{erf}\left\{\sqrt{-kD_\perp/N}\right\}}{2\sqrt{-kD_\perp/N}} \qquad (22)$$

and, equivalently to (21)

$$\ln[\Psi(k)] = -\frac{2}{3}(kD_\perp/N) + \frac{2}{45}(kD_\perp/N)^2 + O(kD_\perp/N)^3. \qquad (23)$$

As a consequence of the sampling of larger amounts of domains at longer observation times, the diffusion coefficient decreases with increasing Δ. If the data are fitted using (23) an apparent diffusion $D_{\mathrm{app}}(\Delta) = D_\perp/N$ will be obtained. To determine $D_\perp$, the factor N has to be estimated. This is a critical point for diffusion measurements that will be discussed below.

Also, in bicontinuous cubic phases, the lipid diffusion can be modeled as two-dimensional diffusion of a particle confined to the bilayer surface, with a constant diffusion coefficient D_0. Anderson and Wennerström [36] showed, by analytically solving the surface diffusion equation that the observed diffusion coefficient D_{eff} of a particle diffusing over any minimal surface of cubic symmetry is exactly $\frac{2}{3}D_0$. The translational dynamics of water and water-soluble substances within the labyrinthine subvolumes formed by the lipid bilayer can be described as a quasi-free three-dimensional dynamics. For long observation times the trajectory of any diffusing particle averages over the local topology, thus leading to an effective diffusion coefficient

$$D_{\text{eff}} = \beta D_0. \tag{24}$$

The obstruction factor β depends on details of the topology of the phase and is a measure of the reduction of the diffusion coefficient. The spin echo decay for molecules diffusing in cubic phases is thus described by

$$\Psi(k) = \exp(-kD_{\text{eff}}). \tag{25}$$

3.4 Determining Diffusivities of Molecules in a Cubic Phase Drug Delivery System

As an example of PFG measurements of cubic phases, we discuss experiments on the cubic-phase drug delivery system already described. In general, PFG diffusion experiments were performed using the modified stimulated-echo pulse sequence using bipolar sine-shaped gradient pulses including longitudinal eddy current delay and two additional spoil gradients, which is shown in Fig. 10. Progressive gradient strengths of 2 to 95% maximal (maximum = 0.48 T/m) were used to acquire a series of one-dimensional spectra. Gradient pulse lengths (δ) of 1 or 3 ms were used with several observation times (Δ) between 50 and 1500 ms. Equation (25) was fitted using the data.

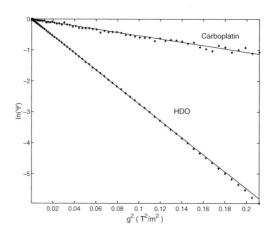

Fig. 12. Semilogarithmic plot of the spin echo attenuation of the water and carboplatin signals vs. g^2. $\Delta = 150$ms, $\delta = 2$ms

In Fig. 12, the semilogarithmic plot of the spin echo decay vs. g^2 for the integrated intensity of the water signal and the carboplatin signal is shown. The self-diffusion coefficient of $6.1 \times 10^{-10}\,\mathrm{m}^2/\mathrm{s}$ obtained is in close agreement with the data given in [12], which supports the assumption that the diffusion measurements on cubic phases can be performed in combination with sample spinning. Compared with the signals of water and the lipids, the carboplatin signals are rather small; therefore the influence of noise becomes notable. However, this data also fits reasonably to (25). It is important to note that one does not observe any deviations from the expected

Table 1. Diffusion coefficients for water-soluble molecules in the cubic phase and in solution

	D_{sol} $10^{-10}\,\mathrm{m}^2/\mathrm{s}$	D_{cub} $10^{-10}\,\mathrm{m}^2/\mathrm{s}$	$D_{\mathrm{cub}}/D_{\mathrm{sol}}$
HDO	24	6.1	0.25
Carboplatin	7.8	1.2	0.15

behavior. Diffusion coefficients of water-soluble molecules in the cubic phase are listed in Table 1.

Compared with the solution, one finds a reduction of the diffusion coefficients of water and carboplatin in the cubic phase, where the reduction is stronger for carboplatin, which indicates an interaction with lipids that influences its mobility. Theoretical and experimental analyses of diffusion phenomena of water in cubic phases suggest a simple two-site model [12], which describes the measured water diffusion coefficient in cubic phases:

$$D_{\mathrm{cub}}^{\mathrm{water}} = p_{\mathrm{B}} D_{\mathrm{cub}}^{\mathrm{lipid}} + (1 - p_{\mathrm{B}}) \beta D_0^{\mathrm{water}}. \tag{26}$$

It is assumed that water that is bound to lipid molecules diffuses with the lipid diffusion coefficient. The parameter p_{B} is the fraction of water that is bound to the lipids, but it should be interpreted as the extent to which the lipid surface influences the dynamics of the water, and not as a prediction of a specific number of bound water molecules per lipid headgroup. The factor β is the obstruction factor for the diffusion of unbound water in the cubic phase, and D_0^{water} is the diffusion coefficient of bulk water. This model can be easily extended to a three-site model, which describes the diffusion of water-soluble substances in a cubic phase composed of two lipid components, namely MO and MPEG-DSPE, with different diffusion coefficient. The diffusion coefficient of water (HDO) and carboplatin in the cubic phase is then given by

$$D_{\mathrm{cub}}^{\mathrm{HDO,CP}} = p_{\mathrm{MO}} D_{\mathrm{cub}}^{\mathrm{MO}} + p_{\mathrm{PEG}} D_{\mathrm{cub}}^{\mathrm{PEG}} + (1 - p_{\mathrm{MO}} - p_{\mathrm{PEG}}) \beta D_0^{\mathrm{HDO,CP}}, \tag{27}$$

where p_{X} is the fraction of HDO/carboplatin bound to MO or MPEG-DSPE, $D_{\mathrm{cub}}^{\mathrm{HDO,CP}}$ are the measured lipid diffusion coefficients in the cubic phase, β is the obstruction factor for the diffusion of unbound water/carboplatin, and D_0^{X} is the diffusion coefficient in bulk water. However, the measurement of the diffusion coefficient in the cubic phase does not provide enough information to determine all parameters in (27), which would allow an estimation of possible effects of interaction between water/carboplatin and the lipids. Nevertheless, if one assumes the same β for water and carboplatin, one can conclude that there are differences in the extent to which the lipids influence the dynamics of both molecules. For this influence, interactions between water/carboplatin and the lipid headgroups come into question, which result in temporary binding of molecules and blocking of the diffusive motion. The observations can be tentatively explained as follows. Water and carboplatin diffuse in the cubic phase channel system. The channel system is obstructed by PEG tails, which results in collisions and blocking of the diffusive motion of both molecules. The interaction is not strong enough to form a complex that exists for a longer time. The motion of carboplatin is somewhat more hindered, resulting in a stronger reduction of its diffusion coefficient compared with water. This result complies with the observations discussed in Sect. 2.3. Now, it becomes clearer to what amount the PEG chains are suited to binding the drug and to reducing its release kinetics.

In Fig. 13 the semilogarithmic plot of the spin echo decay of the integrated intensity of the monoolein lipid signal is shown.

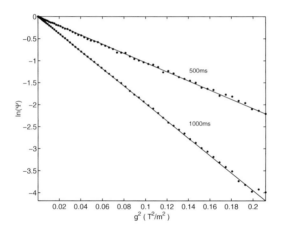

Fig. 13. Semilogarithmic plot of the spin echo attenuation of one signal of monoolein in cubic phase vs. g^2. $\delta = 2$ms

The diffusion coefficient of $25 \times 10^{-12}\,\mathrm{m^2/s}$ obtained is in agreement with previously published data [37]. Note that one clearly sees the linear behavior of the data at both observation times. This behavior is an important indicator that a molecule is indeed incorporated into the lipid bilayer of the cubic phase, because such a translational motion along the direction of the gradient could be hardly performed if the molecule was for instance, located in the lamellar phase or entrapped in microdomains. Such information is important during the development of new encapsulation systems based on cubic phases.

3.5 Diffusion Measurements in Membranes

As an example of measurements of diffusion coefficients of membrane components we discuss experimental results that were obtained on a simple model system composed of POPC and water (HDO) in the molar ratio $1:4$. All data were measured at 300 K on a Bruker DMX 600 spectrometer operating at 600.13 MHz proton Larmor frequency. The probe was a Bruker 4 mm HR-MAS probe, equipped with ^{1}H, ^{13}C, and ^{2}H lock channel and pulsed field gradient capabilities. The typical 90° pulse length was about 10.5 µs.

The semilogarithmic plot of the spin echo attenuation of the water signal measured at different observation times is shown in Fig. 14. For fitting (15) to the data,

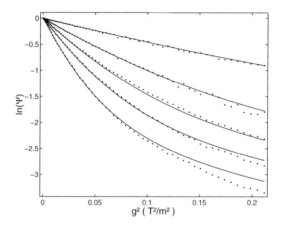

Fig. 14. Semilogarithmic plot of the spin echo attenuation of the water signal vs. g^2 at observation times Δ of 20 ms (top), 50 ms, 80 ms, 120 ms, and 180 ms (bottom). $\delta = 3$ ms was used. The solid lines are the result of the fit of (15) to the experimental data

the integral was approximated by a sum of 100 terms. The data can be reasonably described by (15), which shows that we really can observe the two-dimensional diffusion of interbilayer water. However, with increasing Δ, increasing deviations from the line fitted to the data occur. The apparent lateral diffusion coefficients that were calculated from the fit are shown in Fig. 15.

As was predicted by Callaghan and Söderman [35] and expressed in (22), the apparent lateral diffusion coefficient decreases with increasing observation time. However, in contrast to them, we cannot observe a discrete change in D_{app}, but observe a continuous one.

The disagreement may have several reasons. The first is rather obvious, the deviation from the behavior of $\ln[\Psi(k)]$ vs. g^2 at increasing Δ as would be calculated by assuming a pure two-dimensional process. The second reason is the simplicity of the model that describes the curved and bent membrane structures by planes us-

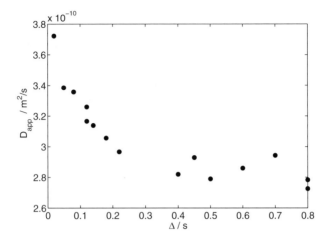

Fig. 15. Apparent diffusion coefficients in dependence on the observation time Δ

ing a fixed domain size. It is not unreasonable to assume that such a model cannot give a satisfactory description of our system as was the case for the system studied in [35]. Taking the discrepancy between theory and observation into account, it is not yet possible to obtain the true lateral diffusion coefficient $D_\perp$ from the observed $D_{app}(\Delta)$.

It is assumed that it will be necessary to take the real structures, as known from electron microscopy pictures of lipid bilayers, into account and to perform random-walk simulations or exact calculations (if possible) in order to obtain a reliable description of the MAS PFG NMR experiment. However, in the sense of "minimal approach" the experiment can already be used, for instance to estimate what amount of different types of water soluble molecules are partially attached or dunked into the bilayer, because this would change their diffusion properties remarkably. Such remarkable changes would be easy to observe and would not require sophisticated models to describe the data.

The application of MAS PFG NMR for investigations of transversal diffusion process in membranes is currently being investigated. To give actual information into the reader's hand, this first results are also included in this book. For applications of this new methods, the reader is referred to forthcoming publications by the authors [38].

4 Conclusions

The examples presented of HR-MAS NMR studies of complex membrane systems with great potential for technological and biomedical use demonstrate the power of sophisticated HR-MAS NMR techniques in combination with multidimensional NMR methods and pulsed-field-gradient support. The HR-MAS NMR complements other techniques, with the advantage of yielding structural, diffusion and dynamic details on a molecular level within one experimental setup. Especially, the observation of the physico-chemical behavior of membrane-associated, biologically active molecules opens up a wide field for a much more detailed study of pharmaceutical products interacting with interfaces. We have applied these methods to study other "soft matter" systems [39, 40] ranging from alginate microcapsules to exotic plants and cell aggregations that allowed, for example the differentiation of gene-manipulated tiny seeds. All new applications require careful preparation of the objects as well as smart new ideas to set up the NMR experiment. Both will lead to an extension of such studies to many fields in the materials and life sciences, as well as to better HR-MAS technical equipment. It is obvious that with the introduction of these highly sophisticated techniques in many life- and materials-science-oriented laboratories, new pulse sequences and exciting results are expected in the near future.

Acknowledgements

This work was supported by the Deutsche Forschungsgemeinschaft. We would like to thank our partners Dr. R. Reszka (Max-Delbrück-Centrum für Molekulare Medizin Berlin), Prof. Dr. G. Lindblom (University of Umeå), Dr. E. Strandberg (University of Umeå) and their respective teams. We also thank Dr. P. Galvosas, Prof. Dr. J. Kärger, Prof. Dr. D. Michel, and Dr. F. Stallmach, for fruitful discussions and support. Special thanks to Bertram Manz for critical reading and constructive suggestions.

References

1. D. Lasic: *Liposomes: from Physics to Applications*, 1st edn (Elsevier, Amsterdam, 1993)
2. D. Lasic: *Liposomes in Gene Delivery*, 1st edn (CRC, Boca Raton, 1997)
3. G. Lindblom, L. Rilfors: Biochim. Biophys. Acta **988**, 221 (1988)
4. J.C. Shah, Y. Sadhale, D.M. Chilukuri: Adv. Drug. Deliv. Rev. **47**, 229 (2001)
5. B. Ericsson, P.O. Eriksson, J.E. Löfroth, S. Engström: ACS Symp. Ser. **469**, 251 (1991)
6. G. Rummel, A. Hardmeyer, C. Widmer, L.M. Chiu, P. Nollert, K.P. Locher, I. Redruzzi, E.M. Landau, J.P. Rosenbusch: J. Struct. Biol. **121**, 82 (1998)
7. G. Cevc, D. Marsh: *Phospholipid Bilayers*, (Wiley, New York, 1987)
8. K. Arnold, K. Gawrisch, F. Volke: Stud. Biophys. **76**, 85 (1979)
9. J.H. Davis: "Membranes: Deuterium NMR", in *Encyclopedia of Nuclear Magnetic Resonance*, Vol. 5, ed. by R.K. Harris, M. Grant (Wiley, Chicester, 1996), pp. 3008–3015
10. J.H. Davis, M. Auger, R.S. Hodges: Biophys. J. **69**, 1917 (1995)
11. F. Volke, A. Pampel: Biophys. J. **68**, 214 (1995)
12. P.O. Eriksson, G. Lindblom: Biophys. J. **64**, 129 (1993)
13. K. Gawrisch, N.V. Eldho, I.V. Polozov: Chem. Phys. Lipids **116**, 135 (2002)
14. C. Glaubitz, A. Watts: J. Magn. Reson. **130**, 305 (1998)
15. A. Pampel, E. Strandberg, G. Lindblom, F. Volke: Chem. Phys. Lett. **287**, 468 (1998)
16. A. Pampel, F. Volke, F. Engelke: Bruker Rep. **145**, 23 (1998)
17. D. Neuhaus, M. Williamson: *The nuclear overhauser effect in structural and conformational analysis* (VCH, New York, 1989)

18. R.R. Ernst, G. Bodenhausen, A. Wokaun: *Principles of Nuclear Magnetic Resonance in One and Two Dimensions* (Clarendon, Oxford, 1989)
19. D. Huster, K. Gawrisch: J. Am. Chem. Soc. **121**, 1992 (1999)
20. S. Feller, D. Huster, K. Gawrisch: J. Am. Chem. Soc. **121**, 8963 (1999)
21. A. Bax, S. Grzesiek: "ROESY", in *Encyclopedia of Nuclear Magnetic Resonance*, Vol. 7, ed. by R.K. Harris, M. Grant (Wiley, Chicester, 1996) pp. 4157–4167
22. R.E. Jacobs, S.H. White: Biochemistry **28**, 3421 (1989)
23. J.W. Brown, W.H. Huestis: J. Phys. Chem. **97**, 2967 (1993)
24. A. Pampel, R. Reszka, D. Michel: Chem. Phys. Lett. **357**, 131 (2002)
25. A. Pampel, F. Volke: J. Magn. Reson. Anal. **3**, 193 (1997)
26. C.S. Johnson: "Diffusion measurements by magnetic field gradient methods", in *Encyclopedia of Nuclear Magnetic Resonance*, Vol. 3, ed. by R.K. Harris, M. Grant (Wiley, Chicester, 1996) pp. 1626–1644
27. J. Kärger: "Diffusion in porous media", in *Encyclopedia of Nuclear Magnetic Resonance*, Vol. 3, ed. by R.K. Harris, M. Grant (Wiley, Chicester, 1996) pp. 1656–1663
28. W.S. Price: Concepts Magn. Reson. **9**, 299 (1997)
29. W.S. Price: Concepts Magn. Reson. **10**, 197 (1998)
30. C.S. Johnson: Prog. NMR Spectrosc. **34**, 203 (1999)
31. I. Furo, S.V. Dvinskikh: Magn. Reson. Chem. **40**, 3 (2002)
32. S.V. Dvinskikh, I. Furo: J. Magn. Reson. **146**, 283 (2000)
33. G. Orädd, G. Lindblom, P.W. Westerman: Biophys. J. **83**, 2702 (2002)
34. D. Wu, A. Chen, C.S. Johnson Jr.: J. Magn. Reson. A **115**, 260 (1995)
35. P.T. Callaghan, O. Söderman: J. Phys. Chem. **87**, 1737 (1983)
36. D.M. Anderson, H. Wennerström: J. Phys. Chem. **94**, 8683 (1990)
37. G. Orädd, G. Lindblom, K. Fontell, H. Ljusberg-Wahren: Biophys. J. **68**, 1856 (1995)
38. A. Pampel, J. Kärger, D. Michel: Chem. Phys. Lett. **379**, 379 (2003)
39. H. Schneider, B. Manz, M. Westhoff, S. Mimietz, M. Szimtenings, T. Neuberger, C. Faber, G. Krohne, A. Haase, F. Volke, U. Zimmermann: New Phytol. **159**, 487 (2003)
40. U. Zimmermann, U. Leinfelder, M. Hillgärtner, B. Manz, H. Zimmermann, F. Brunnenmeier, M. Weber, J.A. Vásquez, F. Volke, C. Hendrich: "Homogeneously cross-linked scaffolds based on clinical-grade alginate for transplantation and tissue engineering" in *Cartilage Surgery and Future Perspectives*, Vol. 9, ed. by C. Hendrich, U. Nöth, J. Eulert (Springer, Berlin,Heidelberg, 2003) pp. 77–86

NMR Studies of Cartilage –
Dynamics, Diffusion, Degradation

Daniel Huster[1], Jürgen Schiller[2], Lama Naji[3], Jörn Kaufmann[4], and Klaus Arnold[5]

[1] Universität Leipzig, Biotechnologisch-Biomedizinisches Zentrum, Nachwuchsgruppe "Strukturaufklärung membranassoziierter Proteine mittels Festkörper-NMR", husd@medizin.uni-leipzig.de
[2] Universität Leipzig, Institut für Medizinische Physik und Biophysik, schij@medizin.uni-leipzig.de
[3] Universität Leipzig, Institut für Medizinische Physik und Biophysik, najil@medizin.uni-leipzig.de
[4] Universität Magdeburg, Klinik für Neurologie II, jkauf@neuro2.med.uni-magdeburg.de
[5] Universität Leipzig, Institut für Medizinische Physik und Biophysik, arnold@medizin.uni-leipzig.de

Abstract. An increasing number of people is suffering from rheumatic diseases, and, therefore, methods of early diagnosis of joint degeneration are urgently required. For their establishment, however, an improved knowledge about the molecular organisation of cartilage would be helpful. Cartilage consists of three main components: Water, collagen and chondroitin sulfate (CS) that is (together with further polysaccharides and proteins) a major constituent of the proteoglycans of cartilage.

^{1}H and ^{13}C MAS (magic-angle spinning) NMR (nuclear magnetic resonance) opened new perspectives for the study of the macromolecular components in cartilage. We have primarily studied the mobilities of CS and collagen in bovine nasal and pig articular cartilage (that differ significantly in their collagen/polysaccharide content) by measuring ^{13}C NMR relaxation times as well as the corresponding ^{13}C CP (cross polarisation) MAS NMR spectra. These data clearly indicate that the mobility of cartilage macromolecules is broadly distributed from almost completely rigid (collagen) to highly mobile (polysaccharides), which lends cartilage its mechanical strength and shock-absorbing properties.

Additionally, the diffusion behaviour of ions and polymers in cartilage was also studied by PFG (pulsed-field gradient) NMR to clarify the influence of charges and a varying molecular weight of the diffusing species: Only at longer observation times, however, structural parameters of cartilage play a significant role, whereas at shorter observation times, the water content has the most tremendous impact on diffusion, equally what kind of molecule is observed.

Finally, it will also be shown that both - NMR spectroscopy in solution as well as in the solid state - are very promising tools to investigate the molecular pathways of cartilage degradation in rheumatic diseases.

1 Introduction

Cartilage is a complex biological tissue that covers the ends of the bones but also occurs in the noses and ears of most mammalian species. The most obvious functions of cartilage are shape retention and the shock-absorbing capacity of the tissue. Thus,

D. Huster, J. Schiller, L. Naji, J. Kaufmann, K. Arnold, NMR Studies of Cartilage – Dynamics, Diffusion, Degradation,
Lect. Notes Phys. **634**, 465–503 (2004)
http://www.springerlink.com/

the mechanical energy of motions is distributed homogeneously over the bone in a time-retarded fashion, which prevents the destruction of the bone ends as a reaction to motions and mechanical strain. Several cartilage diseases are known that destroy the cartilage layer, and painful arthritis is the result. It is assumed that about 15% of the population of the industrialized countries are suffering from some kind of joint disease [1], and it is a common pattern of all rheumatic diseases that the thickness of the cartilage layers of the joints is progressively reduced. Therefore, one very active field of research is tissue engineering for replacement of diseased cartilage by artificial cartilage tissue [2].

Articular cartilage is a water-rich tissue that contains only a small number of cells, the chondrocytes, that synthesize the extracellular matrix. From the physical point of view, cartilage is a gel formed by a heterogeneous assembly of partially crosslinked macromolecules. Some of these polymers are polyelectrolytes that are responsible for the swelling and water uptake of the tissue. A schematic sketch of the complex cartilage architecture and a simplified image of the cartilage proteoglycans are given in Fig. 1. The main macromolecular constituents of cartilage are the collagen and the proteoglycans (PGs), which have to be discussed in more detail.

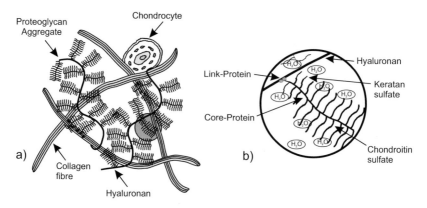

Fig. 1. Sketch of the macromolecular organization of articular cartilage (a), comprising collagen fibres, hyaluronan, chondrocytes, proteoglycan, and water. In (b), the molecular organization of a typical proteoglycan aggregate is shown

Different types of collagen occur in articular cartilage, with collagen type II being the most abundant species [3]. This collagen type has the capability to bind further matrix molecules and is, therefore, essential for the stability of cartilage [4]. Collagens represent the most frequently occurring protein on earth and exhibit a highly specialized and complex organization. Collagen tripel helices are formed from three individual polypeptide chains (see Fig. 2). These collagen triple helices, with molecular dimensions of about 1.5 nm × 300 nm, are again aligned in the form of the so-called collagen fibrils that comprise ~20 wt% of articular cartilage.

(a)

(b)

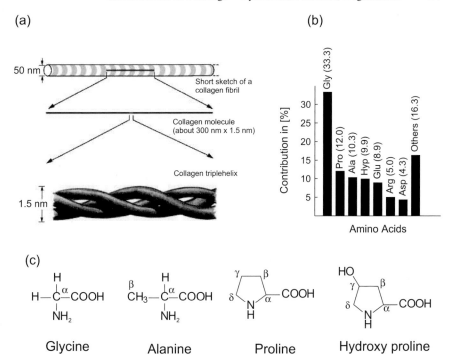

(c)

Glycine   Alanine   Proline   Hydroxy proline

Fig. 2. Overview of the structure of fibril-forming collagen. Collagen consists of triple helices of polypeptide chains that are aligned into fibrils (a). The molecular dimensions of the individual subunits are given, as well as the amino acid composition of collagen type II (b). Bars are labeled according to the three-letter amino acid code. Hyp represents hydroxyproline. The chemical structures of the most abundant amino acids of collagen and the nomenclature of their carbon sites are shown in (c)

The highly packed and curved structure of the polypeptide chains in collagen cannot be implemented by any random amino acid composition but, rather, by only considerable amounts of small amino acid side chains. Therefore, glycine (the smallest amino acid) constitutes about one third of the amino acid residues in all collagen types [5]. Only four different amino acids make up about two thirds of the amino acid residues in the collagen molecule. Additionally, collagen contains very special amino acids such as hydroxyproline (Hyp) and hydroxylysine. Whereas the hydroxylysine residues are covalently attached to the carbohydrates of the PG (see below) [6] and mediate the supramolecular structure of cartilage, the Hyp residues are essential for the stability of the collagen as such.

The collagen moiety is crucial for the stability of cartilage and for the limitation of its swelling behavior, i.e. the water uptake of cartilage: it is well known that the degradation of cartilage collagen by collagenase results in a strongly increased water content [7]. Native triple helical collagen is not digestible by the majority of proteolytic enzymes, with the exception of collagenase, whereas denatured collagen (commonly termed "gelatine", produced upon heating of collagen above 65 °C) is highly susceptible to most proteases.

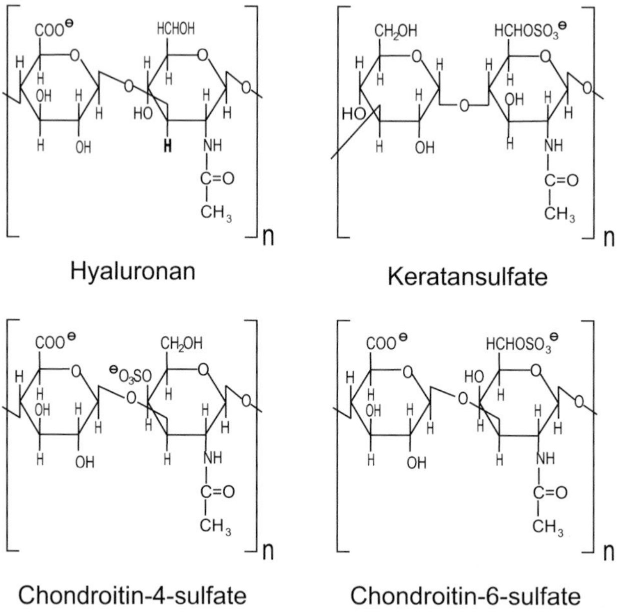

Fig. 3. Chemical structures of the subunits of polysaccharides of proteoglycan in articular cartilage. Both isomers of CS, the 4- and 6-isomers, are shown

Hyaluronan strands are noncovalently attached to the collagen network and, therefore, hyaluronan is the mediator of the interaction between the proteoglycans and the collagen (see Fig. 1). Hyaluronan is a polysaccharide that consists of dimeric subunits composed of alternating β-D-N-acetylglucosamine and β-D-glucuronic acid residues [8]. These subunits are connected by β-1,3- and β-1,4-glycosidic linkages (Fig. 3). The molecular weight of hyaluronan within articular cartilage is $\sim 10^7$ dalton (Da). At physiological pH, the majority of carboxyl groups are deprotonated and, therefore, hyaluronan is negatively charged. Proteoglycans are large aggregates of proteins and carbohydrates [9] that are primarily responsible for the high water-binding ability of cartilage. Proteoglycans consist of a protein backbone, the core protein, to which unbranched glycosaminoglycans are covalently attached. The prime glycosaminoglycans of human articular cartilage are keratan sulfate (KS) and chondroitin sulfate (CS). KS consists of alternating disaccharides of β-D-galactose and sulfated β-D-N-acetylglucosamine (see Fig. 3). The molecular weight of KS is rather low, on the order of 5 kDa. KS chains are located primarily near the hyaluronan (HA) strand. CS, on the other hand, is more distant from the HA strand and consists of alternating disaccharides of β-D-glucuronic acid and sulfated β-D-N-acetylgalactosamine (Fig. 3). The molecular weight of CS (~ 50 kDa) is significantly higher in comparison with the KS. There are two frequently occurring isomers of CS, the 4- and the 6-sulfate. It is not known why only these two isomers are synthesized under *in vivo* conditions, although it is well known that depending on the species from which the cartilage is sampled, both isomers may occur in equal proportions (e.g. pig articular cartilage) or the 4-sulfate isomer may be in vast excess over the 6-sulfate isomer (e.g. bovine nasal cartilage).

To understand the complex architecture and function of cartilage, to comprehend the ubiquitous cartilage degradation pathways, and to set the basis for tissue engineering of artificial cartilage, structural and dynamical studies of native and degraded cartilage tissue with atomic resolution are indispensable. NMR spectroscopy represents a versatile and powerful tool to study macromolecular assemblies and tissues quantitatively.

Several NMR approaches are suitable for studying cartilage tissue. The simplest technique is high-resolution (HR) NMR. Since low-power decoupling is applied in HR NMR, only signals of those molecules that reorient fast on the NMR timescale are detected. Therefore, only the mobile molecules in cartilage can be detected. However, considering the high water content of cartilage, several molecules should have sufficient motilities. Recently, high-resolution magic-angle spinning (HR MAS) NMR has been introduced to improve the resolution of HR NMR spectra of biological tissues [10] (see also the chapters by Michel et al. and Pampel and Volke in this volume). In this approach, the sample is spun at an angle of $54.7°$ with respect to the external magnetic field (the "magic angle"). Thus, line-broadening effects of isotropic susceptibility differences are averaged out, resulting in better-resolved NMR spectra of biological tissues.

Another very useful NMR tool for cartilage research is pulsed field gradient (PFG) NMR, which allows one to study the diffusion of ions and molecules of varying molecular weight in the aqueous phase of cartilage tissue [11, 12] (see also the chapter by Kärger et al. in this volume).

For the investigation of rigid molecules, solid-state NMR approaches based on cross-polarization (CP), high-power dipolar decoupling, and MAS are available [13] (see also the chapter by Freude and Loeser in this volume). Here, completely rigid molecules can be investigated, however, the typical resolution is inferior to that in HR (MAS) spectra.

Finally, magnetic-resonance imaging (MRI) approaches have been widely applied in the characterization of cartilage, as well as in clinical studies [14–16]. The obvious advantage of MRI is that *in vivo* studies on human joints can be performed.

In this chapter, we describe HR, HR MAS, PFG, and solid-state NMR experiments to study structural and dynamical aspects of native bovine nasal and pig articular cartilage. The aim of this research project in the *Sonderforschungsbereich* was the comprehensive description of the dynamics of the molecules and ions that form cartilage. Further, the enzymatic degradation of cartilage typically occurring as a consequence of disease has been investigated, and degradation products have been analyzed by HR (MAS) NMR.

2 Characterization of the Dynamics of the Macromolecules in Cartilage

As stated before, cartilage tissue consists of two classes of macromolecules, the proteoglycan and the collagen moiety. While the polyelectrolytes of the proteoglycan are responsible for the swelling and the high water uptake of the tissue, the collagen represents the molecular scaffold of the tissue necessary for shape retention and mechanical stability. It is obvious that these very different functions of the macromolecules can only be manifested by very unique physical properties. Owing to their very different molecular structure, collagen and solutions of polyelectrolytes possess completely different dynamical properties. While collagen forms relatively rigid insoluble fibrils, the proteoglycans of cartilage are highly charged, which allows them to take up water and to swell. As a consequence, these molecules exhibit a high molecular mobility.

In accordance with the large motional diversity of these macromolecules, two different NMR techniques are applied that allow one to separate signals from the collagen moiety and from the proteoglycans spectroscopically. While HR NMR and HR MAS NMR methods are suitable for studying the mobile cartilage polysaccharides [17,18], solid-state NMR equipment and the application of CP MAS techniques are required to resolve the signals of the cartilage collagen [19]. Thus, the dynamics of each macromolecular cartilage component can be studied individually by spectroscopic separation of their signals.

2.1 Mobility of Cartilage Polysaccharides by High-Resolution NMR

Owing to the high water content of cartilage, the polysaccharides of the proteoglycans can be detected by simple HR NMR methods [17,18,20,21]. In solution, molecules yield HR NMR signals as long as they reorient quickly on the NMR timescale (typically on the order of several nanoseconds). In larger molecules, the dipolar couplings and chemical-shift anisotropy (CSA) induce line broadening through relaxation mechanisms that limit the application of HR NMR to soluble proteins of $\sim$40 kDa molecular weight. Polymers are linear molecules with high segmental motions. Therefore, their ^{13}C HR NMR spectra can be acquired relatively easily [22].

In Fig. 4, typical ^{13}C NMR spectra of pig articular and bovine nasal cartilage are shown. In particular, bovine nasal cartilage spectra are well resolved, owing to their higher water content, which allows faster molecular reorientation of the macromolecular segments. Pig articular cartilage is richer in collagen and has a lower water content. Accordingly, a lower resolution is obtained for this tissue [18].

Recently, a new technique has emerged that is very well suited for the investigation of biological tissues. In this approach, the spectrum is recorded under high-resolution conditions (i.e. using low-power scalar decoupling), but magic-angle sample spinning is also applied at spinning frequencies on the order of 5–10 kHz. While these rotational frequencies are typically not sufficient to average out dipolar couplings and CSAs of the rigid molecules, they allow one to abolish the line-broadening effects of isotropic susceptibility differences in the sample. This technique, called

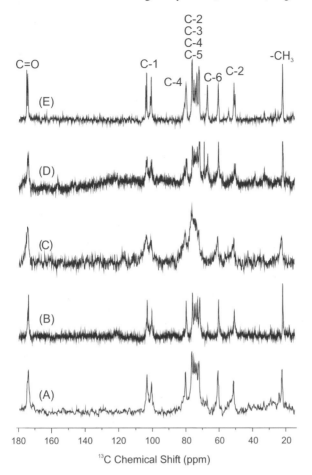

Fig. 4. Proton-decoupled 150.9 MHz ^{13}C NMR spectra of bovine nasal cartilage (A, B), pig articular cartilage (C, D), and an 8 wt% chondroitin sulfate solution in D_2O (E). Spectra (B, D) were recorded under HR MAS conditions at a rotational frequency of 5 kHz. All spectra were recorded at a temperature of 37 °C

high-resolution magic-angle spinning (HR MAS), typically results in ^{1}H and ^{13}C NMR spectra that are much better resolved than conventional high-resolution NMR spectra [10]. The application of HR MAS to cartilage tissue is also demonstrated in Fig. 4. Significant improvements in spectral resolution are achieved, especially for pig articular cartilage.

Next, the question arises of which signals, i.e. which molecular species, can be detected by applying HR (MAS) NMR techniques. For comparison, a ^{13}C NMR of chondroitin sulfate is shown in Fig. 4(E). From the good agreement between the cartilage spectrum and the spectrum of chondroitin sulfate, it can be concluded that HR (MAS) NMR spectra of nasal or articular cartilage are dominated by the signals from chondroitin sulfate, which consequently must represent the most mobile macromolecule in intact cartilage tissue. No signals of the more rigid polysaccharides (hyaluronan) or proteins (collagen, link or core protein) can be detected by this approach because the decoupling fields are not sufficient to average out the strong heteronuclear dipolar couplings present in these rigid molecules. The assignment of the cartilage spectra is, therefore, given by the assignment of the chondroitin sulfate spectrum (see Fig. 4) in accordance with the literature [20]. It has been estimated that under these conditions ~80–100 % of the cartilage condroitin sulfate can be detected by HR NMR methods. The residual portion of the chondroitin sulfate molecules is not amenable to HR NMR because its molecular reorientation is slow on the NMR timescale, which leads to signals that are broadened beyond the limit of detection.

After clarifying the origin of the HR NMR spectra of cartilage, the next aim has been the comprehensive analysis of the motions of these molecules. In HR NMR, the standard approach to obtain information about the dynamics of macromolecules is the quantitative analysis of relaxation time measurements [23–25]. Nuclear spins relax owing to fluctuations of NMR interaction fields caused by the dipolar interaction or the CSA. The fluctuation of these fields is induced by the molecular motions. Therefore, relaxation parameters contain information about the timescales and amplitudes of the molecular motions. If the motions are fast, the relaxation times (which are on the order of ms to s) are inversely proportional to the correlation times of motions (typically on the order of ps to ns for organic molecules); for slower motions, a more complicated dependence on the correlation time applies. The theoretical basis for nuclear relaxation has been worked out in the Redfield relaxation theory [26]. For a ^{13}C nucleus to which one or more ^{1}H nuclei are bound, the expressions for the relaxation rates are given as [23]

$$\frac{1}{T_1} = \frac{\mu_0^2 \hbar^2 \gamma_H^2 \gamma_C^2}{64\pi^2 r^6} \left[J(\omega_H - \omega_C) + 3J(\omega_C) + 6J(\omega_H + \omega_C) \right] + \frac{\omega_C^2 \Delta\sigma^2}{3} J(\omega_C), \quad (1)$$

$$\frac{1}{T_2} = \frac{\mu_0^2 \hbar^2 \gamma_H^2 \gamma_C^2}{128\pi^2 r^6} \left[4J(0) + J(\omega_H - \omega_C) + 3J(\omega_C) + 6J(\omega_H) + 6J(\omega_H + \omega_C) \right]$$
$$+ \frac{\omega_C^2 \Delta\sigma^2}{18} \left[4J(0) + 3J(\omega_C) \right], \quad (2)$$

and the nuclear Overhauser effect (NOE) at high magnetic field follows from [23]:

$$\text{NOE} = \frac{\gamma_H}{\gamma_C} \frac{6J(\omega_H + \omega_C) - J(\omega_H - \omega_C)}{J(\omega_H - \omega_C) + 3J(\omega_C) + 6J(\omega_H + \omega_C)}. \quad (3)$$

In (1)–(3), μ_0 is the permeability of free space, $\hbar$ is Planck's constant divided by 2π, γ_H and γ_C are the gyromagnetic ratios of ^{1}H and ^{13}C, respectively, r is the CH bond length, ω_H and ω_C are the Larmor frequencies of ^{1}H and ^{13}C, respectively, and $\Delta\sigma$ is the span of the CSA tensor of the ^{13}C spin. The spin–lattice relaxation time T_1 describes the relaxation of longitudinal magnetization towards the lattice and can be measured by the inversion recovery technique [27]. The spin-spin relaxation time T_2 describes the relaxation of transverse magnetization and is measured by the Carr–Purcell–Meiboom–Gill pulse sequence [28, 29]. Finally, the NOE is a cross-relaxation phenomenon describing magnetization transfer between ^{1}H and ^{13}C, which is most easily measured by the steady-state NOE technique [30].

To analyze relaxation data at high Larmor frequency, information about the CSA of the respective ^{13}C nucleus is necessary. We have measured the CSAs of the ring carbons of chondroitin sulfate by a recoupling technique as described in [31, 32]. Typically, the $\Delta\sigma$ values were between 26 and 32 ppm.

The spectral density functions $J(\omega)$ in (1)–(3) is given by the Fourier transform of the correlation function $C(\tau)$ describing the molecular motion that leads to fluctuation of the dipolar and CSA fields, according to

$$J(\omega) = \frac{2}{5} \int_{-\infty}^{\infty} C(t) \cos(\omega t)\, \mathrm{d}t. \tag{4}$$

Relaxation parameters were measured at varying temperatures for bovine nasal cartilage and a chondroitin sulfate solution of 8 wt%. The latter has been used as a simple model substance for the motional behavior of chondroitin sulfate in cartilage for comparison. Typical relaxation time and NOE values are given in Table 1. While the T_1 and NOE values are comparable for chondroitin sulfate in cartilage and in aqueous solution, longer T_2 values are observed for pure chondroitin sulfate solutions.

On a physical basis, molecular motions are described by correlation functions that provide the basis for a quantitative analysis of relaxation data according to (1)–(3). This provides information about the correlation times and amplitudes of molecular motions. For quantitative data analysis, two different motional models have been applied to describe the molecular motions of chondroitin sulfate in cartilage and in solution. The first model is the Lipari–Szabo model-free approach [33, 34]. In this model, originally developed for soluble proteins, the molecular motion is described by two correlation times (one for molecular reorientation of the entire molecule, and one for internal motions of the CH bond vectors) and an order parameter for the amplitude of the internal motions of the CH bond vectors, according to

$$J(\omega) = \frac{2}{5} \left[S^2 \frac{\tau_c}{1 + \omega^2 \tau_c^2} + (1 - S^2) \frac{\tau}{1 + \omega^2 \tau^2} \right], \tag{5}$$

where τ_c is the overall correlation time and τ_{int} the internal correlation time according to

$$\tau^{-1} = \tau_c^{-1} + \tau_{int}^{-1}, \tag{6}$$

and S is the molecular order parameter of the CH bond vector.

Table 1. Relaxation times and NOEs for the C-2 carbon in chondroitin sulfate of bovine nasal cartilage and a chondroitin sulfate solution (8 wt% in D_2O) at varying temperatures, measured at a ^{13}C Larmor frequency of 150.9 MHz

Temperature (K)	Bovine nasal cartilage			Chondroitin sulfate (8 wt%)		
	T_1 (sec)	T_2 (sec)	NOE	T_1 (sec)	T_2 (sec)	NOE
278	0.66	0.012	1.3	0.63	0.017	1.6
288				0.52	0.026	1.8
295	0.41	0.013	1.2	0.51	0.025	1.4
310	0.45	0.015	1.5	0.54	0.035	1.6
325	0.72	0.019	1.4	0.43	0.045	1.8
335				0.45	0.041	1.7
348				0.55	0.060	1.4

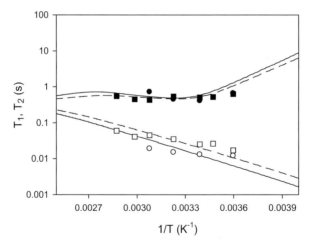

Fig. 5. T_1 (filled symbols) and T_2 (open symbols) relaxation times of chondroitin sulfate in bovine nasal cartilage (circles) and in an 8 wt% D_2O solution (squares). Lines represent the best-fit simulations of the temperature dependence for chondroitin sulfate in cartilage (solid line) and in solution (dashed lines) according to the model-free approach of Lipari and Szabo ((1), (2), and (5))

The three model-free parameters were determined from a nonlinear minimization procedure [35] and fit the temperature dependence of the experimental relaxation time data reasonably well (see Fig. 5). From this figure, it is clear why T_1 relaxation times do not show a temperature dependence because they scatter around a minimum of the T_1 curve, while T_2 relaxation times increase with increasing temperature for both samples.

The correlation times and order parameters calculated from the model-free analysis are given in Table 2. For both cartilage and chondroitin sulfate solutions, the correlation times decrease with increasing temperature as expected showing a reasonable Arrhenius behavior (see Fig. 6). With increasing temperature, the order parameters increase indicating that the amplitude of the motions of the CH bond vectors decreases with increasing temperature. The internal correlation times of the motions of the CH bond vectors are approximately equal for pure chondroitin sulfate and chondroitin sulfate in bovine nasal cartilage. However, the overall correlation times in cartilage are a factor of ~2–3 larger compared with the chondroitin sulfate solution. This strongly indicates slower motions for chondroitin sulfate molecules in cartilage compared with chondroitin sulfate in solution.

Table 2. Correlation times and order parameters for the C-2 carbon in chondroitin sulfate of bovine nasal cartilage and a chondroitin sulfate solution (8 wt% in D_2O) at varying temperatures, according to the Lipari–Szabo model-free analysis of the data shown in Table 1

Temperature (K)	Bovine nasal cartilage			Chondroitin sulfate (8 wt%)		
	τ_c (ns)	τ_{int} (ns)	S^2	τ_c (ns)	τ_{int} (ns)	S^2
278	52.2	3.38	0.34	32.2	3.23	0.36
288				31.2	3.26	0.30
295	49.9	1.65	0.33	29.9	3.26	0.13
310	33.1	0.60	0.46	9.7	0.46	0.69
325	16.6	0.62	0.72	8.1	0.39	0.53
335				6.5	0.41	0.61
348				4.1	0.12	0.87

Two possible reasons would explain these differences in the correlation times. First, owing to the rigid collagen moiety in cartilage, chondroitin sulfate motions are restricted, which yields longer molecular reorientation times. Second, the viscosity of the 8 wt% chondroitin sulfate solution may vary from that of cartilage, which influences the reorientation times of the macromolecules. It is obvious that solutions of single cartilage components do not satisfactorily represent the behavior of the complex cartilage tissue.

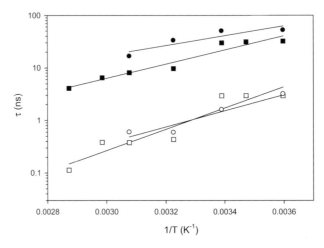

Fig. 6. Overall correlation times (filled symbols) and internal correlation times (open symbols) for chondroitin sulfate in bovine nasal cartilage (circles) and in 8 wt% D_2O solution (squares) for the C-2 carbon signal, obtained from the Lipari–Szabo model-free analysis of relaxation data. The straight lines are linear-regression fits

The interpretation of the model-free parameters for soluble polymers requires slight modifications from the application to soluble proteins. Owing to the high concentration of these macromolecules with a typical molecular weight of ~50 kDa, a fast isotropic reorientation of the molecules on the NMR timescale does not occur. Rather, it can be assumed that the overall correlation time represents a measure of the reorientation of larger molecular segments of the linear polyelectrolyte, which occurs on a typical timescale of a few to a few tens of nanoseconds. Internal motions of the bond vectors occur in a time window of several hundreds of picoseconds. These internal motions have also been observed in molecular dynamics simulations of glycosaminoglycans in water [36, 37]. The internal motions can be visualized by root mean square amplitude fluctuations of the C–H bond vector in a cone with an opening angle on the order of 20–60° (see below).

While the Lipari–Szabo model provides a reasonable fit for the relaxation data of chondroitin sulfate, revealing the mobility of the molecule, both in cartilage and in D_2O solution (see Fig. 5), it remains questionable whether just three motional parameter describe sufficiently the complicated dynamics of an elongated macromolecule with a molecular weight on the order of ~50 kDa. According to the proteoglycan model (see Fig. 1), the chondroitin sulfate strands are attached on one side to the core protein of the aggregate. Since the core protein and the hyaluronan represent rather stiff molecules, the mobility of the chondroitin sulfate segments should increase with increasing distance from the point of attachment. This would result in a variation of the reorientation times from very long ones at the attachment points to very short ones at the end of the molecule. This property is not modeled by the Lipari–Szabo approach, which only considers two fixed correlation times of motion. For the interpretation of these correlation times, they would have to be considered as average values. To model such a variation of correlation times within the chondroitin sulfate molecules, the relaxation data have been analyzed by means of a second model. This model now considers any given distribution function $G(\tau)$ of the correlation times, which yields spectral density functions of the form

$$J(\omega) = \int\limits_0^\infty \frac{G(\tau)\tau}{1 + \omega^2\tau^2}\mathrm{d}\tau. \tag{7}$$

The special form of the distribution function $G(\tau)$ depends on the given molecular system. While it is frequently assumed that physical quantities are normally distributed, for the condroitin sulfate chains attached to a rigid molecule on one side, an asymmetric distribution function would be more appropriate. That is because most segments are far away from the point of attachment and would have a rather short correlation time. For the segments closer to the core protein, correlation times would gradually increase, resulting in a long tail of the distribution towards longer correlation times. These features are well represented in the χ^2 distribution, which has the form [38, 39]

$$G(\tau)\,\mathrm{d}\tau = \frac{1}{\Gamma(p)}(p\tau)^{p-1}e^{-p\tau}p\,\mathrm{d}\tau. \tag{8}$$

The distribution is defined by its width and its mean. The width is characterized by the parameter p. The larger p becomes, the narrower is the distribution, eventually approaching a δ-function for $p \to \infty$, yielding spectral density functions for isotropic reorientation with one correlation time τ (see Fig. 7).

The chondroitin sulfate relaxation data can be fitted using the same χ^2 distribution model for cartilage and the solution (see Fig. 8). A p value of 11 results in a distribution with a width of approximately 10 ns. While a rather steep slope is observed for the short correlation times, the distribution approaches zero very gradually towards longer correlation times, mimicking the segments that are closer to the core protein in proteoglycan aggregates (note that the time axis in Fig. 8 is scaled logarithmically). Most segments of the chondroitin sulfate have a mean correlation time on the order of a few hundreds of picoseconds up to about one nanosecond depending on temperature.

Both the Lipari–Szabo and the χ^2 distribution model yield reasonable fits of the experimentally obtained relaxation and NOE data. However, the latter model appears to describe the complicated dynamics of chondroitin sulfate in cartilage more realistically because the correlation times of chondroitin monomers are broadly distributed and are expected to decrease from the point of attachment at the link

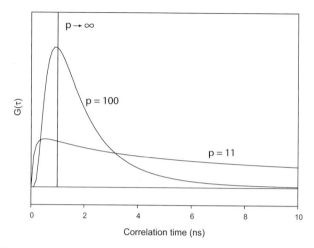

Fig. 7. χ^2 distributions of correlation times for three different values of the parameter p according to (8)

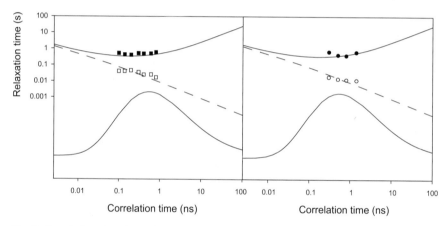

Fig. 8. Correlation time dependence of T_1 (filled symbols) and T_2 (open symbols) for chondroitin sulfate solutions (left panel, squares) and chondroitin sulfate in bovine nasal cartilage (right panel, circles). Data were fitted using a χ^2 correlation time distribution model with a width of ~ 10 ns. The correlation time distribution function is also shown in each graph.

protein towards the end of the molecular chain. Most of the chondroitin sulfate segments are expected to move almost isotropically.

HR (MAS) NMR techniques have proven to be very useful for detecting the signals of the mobile macromolecules in cartilage. These are, in particular, the chondroitin sulfate segments. Relaxation analysis has shown that segmental reorientations occur on a fast timescale of a few to a few tens of nanoseconds. The mobility of the chondroitin sulfate varies and is best described by a broad distribution function with a long tail towards longer correlation times, as for instance in a χ^2 distribution. In the next section, the mobility of the collagen moiety of cartilage will be investigated.

2.2 Mobility of Cartilage Collagen by Solid-State NMR

As outlined before, signals from the collagen moiety cannot be detected by means of HR (MAS) NMR approaches, and solid-state NMR equipment is required to detect these signals. The reason for the failure of the HR techniques is that collagen forms a rather rigid lattice, in which dipolar couplings and the CSA are not averaged out by any motions. Therefore, the NMR frequencies depend on the orientation of the molecules with regard to the laboratory frame, yielding broad NMR signals with a width on the order of 10–50 kHz. Consequently, a superposition of very broad anisotropic powder patterns results. Nevertheless, NMR powder patterns contain both structural and dynamical information because any molecular motion scales down these anisotropic interactions. To separate these interactions, the strong dipolar couplings can be averaged out by dipolar decoupling, and pure CSA line shapes result. To avoid signal superposition in cartilage spectra, single-site isotopic labeling has been achieved by feeding animals amino acids labeled singly with ^{13}C, ^{15}N, or ^{19}F, which would grow isotopically labeled cartilage [40–42]. Also, motionally averaged ^2H quadrupolar splittings have been analyzed [42]. While this approach is tedious and very expensive, detailed dynamic information has been acquired for the collagen sites investigated. It was found that all sites selected by isotopic labeling undergo small-amplitude root mean square angle fluctuations with correlation times on the order of ns [40–42].

The next step was the application of cross-polarization MAS NMR (CP MAS NMR) to study isolated collagen [45,46]. In the CP MAS experiment, the anisotropy of the chemical shift was averaged out by fast MAS and the dipolar interactions were decoupled by dipolar decoupling [13]. Thus, well-resolved spectra have been obtained. The ^{13}C signals are observed with higher sensitivity because polarization from the protons is transferred to ^{13}C. Nevertheless, with the equipment available more than 10 years ago, no dynamical characterization of either pure collagen or cartilage collagen was feasible.

In our approach, we have investigated natural cartilage by high-field CP MAS techniques. The application of highest-magnetic-field solid-state NMR (^1H Larmor frequency of 750 MHz) leads to a significant sensitivity enhancement that not only allows one to study collagen in natural cartilage but also allows the application of two-dimensional NMR techniques that allow a comprehensive investigation of collagen dynamics in cartilage in a single experiment without isotopic labeling (see below).

Figure 9 shows a ^{13}C CP MAS spectrum of natural pig articular cartilage (A) and isolated collagen fibrils from bovine Achilles tendon (B). The cartilage spectrum is dominated by the collagen signals, as seen by comparison with spectrum (B). The peak assignment was carried out according to [45]. The most intense peak is assigned to glycine (Gly) Cα at 43.0 ppm. Gly is the most abundant amino acid in collagen and represents about one third of all amino acid residues (see Fig. 2). Also, resonances of proline (Pro), alanine (Ala), and hydroxyproline (Hyp) can be assigned [45]. These

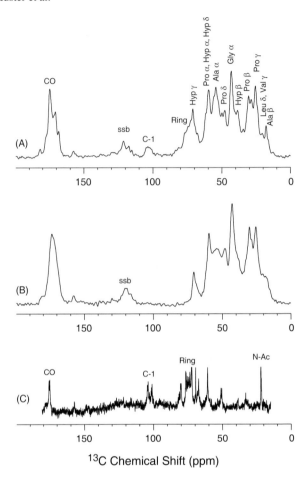

Fig. 9. Proton-decoupled ^{13}C CP MAS spectra of pig articular cartilage (A) and collagen from bovine Achilles tendon (B) at a resonance frequency of 188.6 MHz and a spinning speed of 10 kHz at 20 °C. The ^{1}H decoupling field was $\gamma B_1/2\pi \sim 60$ kHz. The most prominent peaks are assigned according to [45] (see also Table 3). For comparison, a 150.9 MHz HR MAS ^{13}C NMR spectrum of pig articular cartilage with single-pulse excitation and low-power decoupling ($\gamma B_1/2\pi \sim 3.5$ kHz) is also shown (C); ssb, spinning sideband

four residues account for about 65% of all amino acids in collagen. A somewhat better resolution is observed for the cartilage sample compared with pure collagen, indicating the presence of fast molecular motions in cartilage collagen.

Further, in CP MAS spectra of articular cartilage, signals of polysaccharides are detected at about 103 ppm and between 70 and 80 ppm, superimposed on the Hyp Cγ resonance at 71.1 ppm. These resonances must be assigned to rigid molecules of the other major component of cartilage, the proteoglycans. Indeed, if we compare the cartilage CP MAS spectrum with the single-pulse excitation HR MAS ^{13}C

NMR spectrum of pig articular cartilage for low-power decoupling (C), good correspondence is obtained for the polysaccharide signals. Only the almost isotropically mobile polysaccharides of the proteoglycans can be detected by HR MAS NMR, because dipolar couplings are averaged out by the fast motions. By CP MAS, only rigid molecules with strong dipolar couplings are detected, because motions average out those couplings necessary for the transfer of polarization from ^{1}H to ^{13}C. In addition, CP enhances signals because it reduces the T_1 bottleneck that often occurs with ^{13}C and other rare spins in solids. Consequently, no signals from the mobile chondroitin sulfates are detected. The small carbohydrate signals observed in the CP MAS spectra of cartilage must be caused by the rather rigid hyaluronan that forms the backbone of proteoglycans, and perhaps the less mobile keratan sulfate that is located closer to the proteoglycan backbone and is consequently motionally more restricted. Therefore, the mobile and the rigid components of cartilage can be separated spectroscopically, allowing one to study the motions of each component individually.

Several NMR methods allow the studying of the fast motions in solid materials [23, 43, 44]. The common effect that these methods exploit is that molecular motions scale down anisotropic interactions, for instance the dipolar couplings between nuclear spins. The higher the mobility of the respective site, the smaller is the residual dipolar interaction, providing a measure of the motional amplitude, typically expressed by an order parameter. It is desired to measure a dipolar coupling for each molecular site, providing motional information for the entire molecule. However, the various anisotropic line shapes of rigid molecules represent a superposition of many dipolar spectra and cannot be analyzed quantitatively. A possible solution to that problem would be site-specific isotopic labeling, which is tedious, expensive, and particularly difficult for biological tissue samples. Alternatively, the idea of separating anisotropic interactions in a two-dimensional (2D) experiment evolved in the 1980s on the basis of the ideas of Jeener, Ernst, Waugh, and their coworkers [47]. To date, a number of 2D pulse sequences are available, in which the MAS spectrum is detected in the direct dimension and an anisotropic spectrum is measured in the indirect dimension for each resolved site. Thus, many motional parameters can be obtained site-specifically in a single experiment, ideally without isotopic labeling.

For cartilage, we have measured ^{1}H–^{1}H and ^{1}H–^{13}C dipolar couplings [19, 48]. While ^{1}H–^{1}H dipolar couplings provide qualitative mobility information, ^{1}H–^{13}C dipolar couplings are truly quantitative. The known geometry of the heteronuclear dipolar-interaction tensor along the C–H bond vector allows the precise determination of the molecular order parameter of that bond from the motionally averaged coupling strength, since the dipolar coupling measured for a ^{13}C nucleus is determined only by the nearest covalently attached proton.

Recently, a new method has been introduced to measure dipolar couplings at a high MAS rotational frequency [49]. By locking the ^{1}H magnetization along an effective field at $54.7°$ with respect to the B_0 field in a Lee Goldburg (LG) CP experiment, ^{1}H–^{1}H spin diffusion is suppressed and ^{13}C magnetization buildup is observed in a time-oscillatory manner. The Fourier transformation of that magnetization buildup curve results in a powder pattern with a splitting that corresponds to the dipolar coupling. Since ^{13}C magnetization is detected under MAS conditions in the direct dimension, these powder patterns can be measured with site resolution to yield the ^{1}H–^{13}C dipolar coupling for each resolved ^{13}C peak. Therefore, the experiment correlates dipolar coupling and isotropic chemical-shift information.

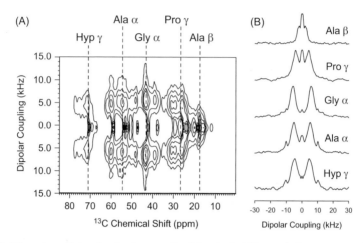

Fig. 10. 2D contour plots of a dipolar-coupling/chemical-shift correlation spectrum (2D LG CP spectrum) of pig articular cartilage (A). The spectrum was acquired at a spinning speed of 10 kHz and a temperature of 20 °C. A total of 80 t_1 points were acquired, with 800 transients per increment and a recycle delay of 2.7 s, yielding total acquisition times of ~48 h. Vertical lines indicate where 1D dipolar spectra have been extracted. These dipolar spectra for representative ^{13}C sites are shown on the right-hand side (B)

In Fig. 10, the 2D dipolar correlation spectrum of pig articular cartilage is shown. In the F2 domain, the resolved isotropic ^{13}C MAS spectrum is measured, while dipolar splittings can be read off in the F1 domain for each resolved ^{13}C signal. Therefore, the dipolar couplings of individual peaks can be measured site-specifically by analyzing slices of the 2D spectra extracted for each resolved peak in the ^{13}C dimension.

In Fig. 11, a typical slice from a 2D LG CP spectrum is extracted. The distance between the characteristic "horns" in the dipolar spectra defines the dipolar coupling scaled by $\sin(54.7°)$. Therefore, the strength of the dipolar coupling is directly available from the experimental spectrum. However, the line shape can also be simulated numerically, following the treatment in [49, 50]. In the simplest case, the oscillation frequency of the CH dipolar interaction is given by

$$\omega_{CH,LG} = \frac{1}{2}\bar{\delta}_{CH}\sin\theta_m$$

$$\times \sqrt{\frac{1}{2}\sin^2(2\beta)\left(1 + \frac{\bar{\eta}\cos(2\alpha)}{3}\right)^2 + \frac{2}{9}\bar{\eta}^2\sin^2(\beta)\sin^2(2\alpha)}, \quad (9)$$

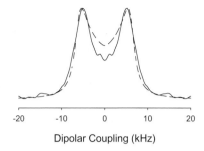

Dipolar Coupling (kHz)

Fig. 11. Typical experimental line shape from a 2D LG CP experiment for Pro Cγ in pure collagen. The dashed line represents a numerical simulation according to (9) using a motionally averaged dipolar coupling of $\bar{\delta}_{CH} = 21.3$ kHz and an asymmetry parameter of $\bar{\eta} = 0$

Table 3. Motionally averaged ^{13}C–^{1}H dipolar couplings ($\bar{\delta}_{CH}$) and order parameters (S_{CH}) at 20 °C for resolved signals of bovine Achilles tendon collagen and pig articular cartilage, measured from the 2D LG CP experiment

Chemical shift	Assignment	$\bar{\delta}_{CH}$ (cartilage)	$\bar{\delta}_{CH}$ (collagen)	S_{CH} (cartilage)	S_{CH} (collagen)
14.4 ppm	Ala β	8.0 kHz	7.5 kHz	–	–
25.5 ppm	Pro γ	18.0 kHz	21.3 kHz	0.77	0.91
30.4 ppm	Pro β	19.5 kHz	22.0 kHz	0.84	0.94
43.0 ppm	Gly α	22.4 kHz	23.3 kHz	0.97	1.00
47.7 ppm	Pro δ	22.0 kHz	22.7 kHz	0.96	1.00
43.0 ppm	Ala α	21.0 kHz	22.4 kHz	0.92	0.98
59.5 ppm	Pro α, Hyp α, Hyp δ	21.0 kHz	22.4 kHz	0.92	0.98
71.1 ppm	Hyp γ	20.5 kHz	20.8 kHz	0.90	0.91
71.4 ppm	Carbohydrate (ring)	21.0 kHz	–	0.92	–
103.0 ppm	Carbohydrate (C-1)	19.0 kHz	–	0.83	–
103.0 ppm	Carbohydrate (C-1)	11.0 kHz	–	0.48	–

where $\bar{\delta}_{CH}$ is the motionally averaged dipolar coupling and θ_m is the magic angle (54.7°). If the motionally averaged anisotropy parameter $\bar{\eta} = 0$, (9) is simplified and only the Euler angle β has to be considered for powder averaging. An example of the numerical line shape simulation for the experimental spectrum is also given in Fig. 11.

From the simulated spectra, the values of the dipolar coupling can be extracted for each resolved signal from a single LG CP spectrum. Values for the motionally averaged dipolar couplings for pig articular cartilage and collagen are given in Table 3.

From the measurement of motionally averaged dipolar couplings in the LG CP experiment, the molecular order parameters (S_{CH}) along the CH bond can be determined according to the following equation:

$$S_{CH} = \bar{\delta}/\delta. \tag{10}$$

The experimental order parameters for bovine Achilles tendon collagen and pig articular cartilage are also given in Table 3. For the CH_3 groups of Ala, Val, and Leu, dipolar couplings of approximately 7.5 kHz are measured. These groups perform fast threefold hopping motions and symmetric rotations about the $C\alpha$–$C\beta$ bond axis, which scales the dipolar coupling by a factor of $1/3$ [51]. This scaling of the dipolar coupling due to axially symmetric motions should not be described in terms of an order parameter. Typical order parameters for CH and CH_2 side-chain signals in pure collagen fibrils are between 0.91 and 0.98. Significantly smaller values, between 0.77 and 0.90, are measured for collagen in the native cartilage tissue. The backbone CH bond vectors represent the most rigid part of the collagen molecules, yielding order parameters between 0.98 and 1.00 for pure collagen and between 0.92 and 0.97 for collagen in cartilage. For the CO groups, a dipolar coupling of 2.0 kHz is measured for both samples. However, this small value is due to the rather large distance between ^{13}C within CO and ^1H, which results in a much smaller dipolar coupling, and no molecular order parameters can be extracted from these dipolar couplings.

In the CP MAS spectra of pig articular cartilage, there are also weak signals of carbohydrates that are due to immobile glycosaminoglycans (mostly hyaluronan and keratan sulfate) (see Fig. 9). Since the rigid glycosaminoglycan content of cartilage is much lower than the rigid moiety of collagen, these spectra are more noisy. It has been estimated that about 80% of the glycosaminoglycan molecules are highly mobile [20] and, therefore, only a small fraction of the glycosaminoglycans can be detected. For these polysaccharide signals, ^{13}C–^1H couplings between 11 and 21 kHz are measured, which translates into order parameters between 0.48 and 0.92 (Table 3). This indicates that the cartilage polysaccharides that can be detected by CP MAS represent only the rigid molecules. On average, these molecules are more mobile than the collagen but, by far, not as mobile as the chondroitin sulfate molecules that were detected by HR MAS methods.

Using a simple model, these order parameters can be converted into root mean square (rms) amplitude fluctuations of the C–H bond vector according to [23]

$$S^2 = 1 - \frac{3}{2} \langle \theta^2 \rangle. \tag{11}$$

Root mean square amplitude fluctuations are reported in Table 4 for all resolved signals of collagen and cartilage. For the collagen backbone, angle fluctuations of

Table 4. Root mean square amplitude fluctuations of the C–H bond vector calculated from order parameters for pig articular cartilage and bovine tendon collagen at 20 °C

Chemical shift	Assignment	$\sqrt{\langle\theta^2\rangle}$ (cartilage)	$\sqrt{\langle\theta^2\rangle}$ (collagen)
25.5 ppm	Pro γ	$29° \pm 2°$	$17° \pm 5°$
30.4 ppm	Pro β	$24° \pm 2°$	$13° \pm 5°$
43.0 ppm	Gly α	$11° \pm 3°$	$0° \pm 11°$
47.7 ppm	Pro δ	$13° \pm 4°$	$0° \pm 13°$
54.3 ppm	Ala α	$18° \pm 3°$	$9° \pm 5°$
59.5 ppm	Pro α, Hyp α, Hyp δ	$18° \pm 3°$	$9° \pm 5°$
71.1 ppm	Hyp γ	$20° \pm 4°$	$21° \pm 3°$
71.4 ppm	Carbohydrate (ring)	$19° \pm 5°$	–
103.0 ppm	Carbohydrate (C-1)	$28° \pm 5°$	–
103.0 ppm	Carbohydrate (C-1)	$41° \pm 7°$	–

up to 9° occur. The collagen side chains are somewhat more mobile, resulting in rms fluctuations between 13° and 21°. The motion of the cyclic Pro and Hyp residues in the collagen triple helix can be visualized as rapid puckering motions [42, 45, 52].

Cartilage collagen is more mobile than isolated collagen fibrils. Both the backbone and the side-chain order parameters of cartilage collagen are somewhat lower than in bovine Achilles tendon collagen. The rms amplitude fluctuations typically double for cartilage collagen, with values between 11° and 18° for the collagen backbone and 20° and 29° for the side chains. This represents bond fluctuations over a solid angle that is about 3–4 times larger than for isolated collagen fibrils. The CH_3 groups of Ala, Leu, and Val show the same scaling of the dipolar coupling of $1/3$ for both cartilage collagen and bovine Achilles tendon collagen, indicating fast threefold hopping motions and rotations of the methyl group about the $C\alpha$–$C\beta$ bond axis.

The carbohydrate rings the of rigid glycosaminoglycans of cartilage show order parameters that are comparable to or smaller than those for collagen. These signals, detected by CP MAS, must be due to largely rigid hyaluronan and other less mobile glycosaminoglycans exhibiting order parameters between 0.92 and 0.48. These order parameters translate into rms amplitudes between 18° and 41°.

What is the reason for the increased mobility of cartilage collagen? First of all, cartilage has a much higher water content compared with the dry collagen fibers that were investigated for comparison. It has been shown in the literature that mineralized and water-depleted collagen, as it occurs in bone or teeth, is even more rigid than the isolated collagen investigated here [41, 53]. Indeed, an improvement in spectral resolution has been found for hydrated collagen fibrils, indicating that hydration plays an important role in collagen dynamics [45]. A similar increase in mobility with increasing hydration is also known for elastin [54]. Another contribution to the increased collagen mobility in cartilage may arise from collisions with the highly mobile glycosaminoglycan molecules in the cartilage matrix. These carbohydrates show isotropic mobility and are located in the close vicinity of the collagen (see Fig. 1). Considering the high water content of cartilage and the almost isotropic mobility of the glycosaminoglycan molecules, it is remarkable how little this affects collagen mobility! In agreement with MRI studies, this investigation has shown that collagen remains substantially ordered and rather rigid in soft cartilage tissue [16].

3 Diffusion Processes in Cartilage

Since cartilage tissue contains neither blood nor lymph vessels, diffusion processes are of paramount importance for cartilage nutrition, the removal of metabolic waste products and the supply of the cartilage with molecules with regulatory functions, e.g. cytokines [55]. Because of the extreme relevance of diffusion processes for cartilage physiology, it is even assumed that degenerative joint diseases such as rheumatoid arthritis or osteoarthritis are – among further factors – also caused by changes in the metabolic transport systems of cartilage [56]. However, details of the mechanisms of these diseases are so far not completely understood [57], and further attempts are clearly necessary to clarify these aspects in more detail.

A large number of diffusion studies on cartilage (summarized in [58]) have already been performed in the past and the majority of these studies used previously frozen cartilage as a membrane and determined the flux of (normally) radioactively labeled molecules across this membrane. Under these conditions, Fick's first law holds and the corresponding diffusion coefficient can be easily calculated if the applied concentration gradient is known [59]. Unfortunately, this technique has considerable disadvantages: for instance, self-diffusion coefficients (SDCs) cannot be measured, and the application of radioactive isotopes is required.

Both drawbacks can be overcome by applying pulsed-field-gradient (PFG) NMR [12]. Besides being nondestructive, one further very important feature of PFG NMR is that time-dependent diffusion measurements can be performed, allowing the estimation of barrier distances. A rough illustration of the influence of solid barriers on the diffusion behavior is given in Fig. 12.

Normally, the mean square displacement $\langle r^2 \rangle$ of a diffusing molecule is given by the equation $\langle r^2 \rangle = 6 Dt$ (D is the diffusion coefficient and t the time) and increases with increasing observation time (see Fig. 12a). For a given molecule in a certain environment one would, therefore, expect a linear relationship between the pathway and the observation time [60], i.e. the longer a molecule is observed, the larger is the distance it will diffuse [57]. If the linear correlation breaks down, there must be barriers within the system, and the SDC becomes dependent on the observation time ("restricted diffusion"). Such SDCs are normally termed "apparent SDCs". It has been shown that the SDC of water in the presence of certain barriers (D) (see

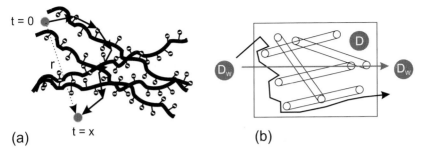

Fig. 12. "Free" self-diffusion of water: (a) independent of diffusion time in a given polymer matrix, and (b) restricted diffusion of water in an environment with barriers, e.g. the collagen fibrils of cartilage

Fig. 12b) is smaller than the SDC (D_W) of pure water and can be described with the following equation [57]:

$$\frac{D}{D_W} = \left[\frac{1 - V_P}{1 + V_P}\right]^2,\tag{12}$$

where V_P is the contribution of the volume of the barriers in comparison with the total volume. Normal values of this ratio are between 0.4 and 0.5 [58]. Therefore, the observation of restricted diffusion is a quite useful technique to obtain information the distance between two collagen fibrils within the cartilage [61].

We shall show in the next sections that the long-time self-diffusion behavior of organic ions, water, and uncharged polymers in cartilage is quite similar, indicating that the chemical structure of the diffusing species plays only a minor role. In contrast, the short-time SDC is primarily influenced in all cases by the water content of the sample and is, therefore, useful for determining the water content of cartilage samples [57].

3.1 Diffusion of Ions by PFG NMR

Ions, and especially cations, are of considerable interest in the field of cartilage research since the polysaccharides of cartilage possess a strong negative fixed charge density [62–64], which strongly binds cations. For instance, the calcification of cartilage that is known to occur upon aging represents an initial process in the development of arthritic joint diseases. Typical data about the ion contents of articular cartilage can be found in [65, 66].

In our studies on the diffusion behavior of ions in cartilage, we have applied PFG NMR measurements [58]. Unfortunately, the ions that are of the highest relevance for cartilage research, sodium and potassium ions, possess rather unfavorable NMR properties, i.e. a large quadrupole moment and, consequently, very short relaxation times [67]. This limits their applicability considerably, although in the past a few diffusion NMR studies were performed by ^{23}Na NMR spectroscopy. ^{19}F NMR has also been used and trifluoroacetic acid was in this case the diffusing species of interest [68].

In our studies, tetramethylammonium (TMA) chloride ($N(CH_3)_4Cl$) and tetra-ethylammonium (TEA) chloride ($N(C_2H_5)_4Cl$) were used. Both compounds are readily soluble in water [69]. They represent electrolytes and dissociate in water to give the TMA cation $[N(CH_3)_4]^+$ and the TEA cation $[N(C_2H_5)_4]^+$, respectively, and the Cl^- anion. These organic cations were chosen because they can be detected by ^{1}H NMR spectroscopy. Additionally, their protons do not show exchange with the water and, therefore, when all the solutions are prepared in deuterated water (D_2O), the concentration of HDO within the samples can be minimized [69]. Since the water content (see below) has a very important impact on the measured diffusivities in cartilage, all samples were prepared by the application of the osmotic stress technique [70,71]. For that purpose, a certain amount of polyethylene glycol (PEG) was dissolved in water (or D_2O) with the appropriate concentration of the corresponding salts. Cartilage specimens were incubated in these PEG solutions to ensure a fixed osmotic pressure, which is primarily dependent on the PEG concentration [72] and to a lesser extent on the ionic strength.

Under these conditions, the cartilage specimens will absorb or lose water until their osmotic pressure is in equilibrium with that of the surrounding solution [70]. To prevent the PEG from penetrating into the cartilage, the cartilage is separated from the PEG solution by a dialysis tube with a sufficiently low molecular-weight cutoff [57]. The reader should note that various cartilage species behave differently when they are exposed to the same osmotic pressure. We have again used the two different cartilage types of pig articular and bovine nasal cartilage. The two cartilage types differ significantly in their collagen and glycosaminoglycan content [18,20,73].

In Fig. 13, typical data for the water uptake of cartilage specimens as a function of the applied osmotic pressure are given.

It is evident that higher osmotic pressures are required to ensure the same water content of bovine nasal cartilage in comparison with the pig articular cartilage, since bovine nasal cartilage has a higher glycosaminoglycan content and, accordingly, a higher tendency to bind water [57]. There are also pronounced differences with respect to the age of the animals from which the cartilage specimens were taken (Fig. 13). Juvenile cartilage has a higher water content at the same osmotic pressure than adult cartilage. This is most probably caused by increased amounts of calcium and other ions in the adult cartilage and, accordingly, a shielding of the charges that leads to a weaker water-binding ability [58].

The raw data for the cation diffusion of TMA and TEA in cartilage, i.e. the attenuation of the spin echo amplitude at increasing gradient strengths (data not shown) [11], can be well described by a biexponential fit, obviously resulting from two different species of diffusing particles. The more intense contribution, with a D of $4.8 \times 10^{-10}\,m^2\,s^{-1}$ in the case of TMA (or $3.0 \times 10^{-10}\,m^2\,s^{-1}$ in the case of TEA), can be assigned to the corresponding cations, while the less intense component, with a D of $3.7 \times 10^{-11}\,m^2\,s^{-1}$, is most probably caused by residual water protons.

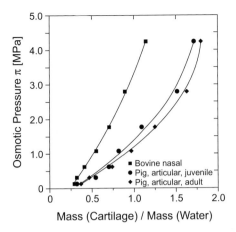

Fig. 13. Cartilage/water weight ratios of different kinds of cartilage, plotted against the osmotic pressure that is necessary to establish the corresponding water contents. Different osmotic pressures are required for different cartilage types to ensure the same water contents. For details see the text

These data correspond well to the data found for cartilage collagen [61]. As outlined below, the water content has the highest impact on the short-time diffusion behavior in cartilage, and this is also obvious from Fig. 14.

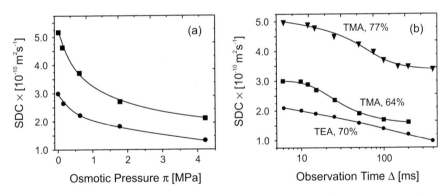

Fig. 14. Self-diffusion coefficients of TMA (squares) and TEA (circles) in bovine nasal cartilage dependening on the applied osmotic pressure (a). In (b), the apparent self-diffusion coefficients of 0.15 M TMA in cartilage with a water content of 77 wt% (triangles) and 64 wt% (squares) as a function of the observation time Δ are shown. For comparison, the diffusion coefficient of 0.15 M TEA (circles) in cartilage with a water content of 70 wt% as a function of the observation time is also shown. The lines represent spline curves

Here (Fig. 14a), the diffusion coefficients of TMA and TEA in dependence on the applied osmotic pressure in bovine nasal cartilage are shown [69]. The water content of the cartilage was adjusted by the osmotic stress technique, while the ion content was fixed in all cases [69]. In the same way as the water content of cartilage decreases when the applied osmotic pressure increases (data not shown), the diffusion coefficients also decrease. This holds for TMA as well as for TEA; however, according to its higher molecular weight, TEA exhibits a lower diffusion coefficient in comparison with TMA.

As expected, the measured apparent SDCs of both cations in the cartilage were strongly dependent on the time Δ over which the diffusion was observed (Fig. 14b). When the diffusion time Δ is increased, the apparent SDC of, for instance, the TMA cation also decreases significantly. As expected, the TEA exhibits a lower SDC in comparison with TMA when a comparable water content is used. Additionally, the effect of restricted diffusion is more pronounced for lower water contents of the cartilage (Fig. 14b) [69]. Such an effect has already been observed by Burstein et al. about ten years ago [68] by using sodium and fluorine diffusion NMR techniques. Therefore, the use of TMA and TEA yields comparable results but in a much more convenient way.

From the data given in Fig. 14b, we have also been able to calculate the mean square displacements of the diffusing species. This calculation yields that at the lower water contents (64 wt%), cations are free for a diffusion pathway of just about 2.5 μm, while at higher water contents (77 wt%) corresponding to the native water content of bovine nasal cartilage, free diffusion over a distance of about 9 μm has been observed. This corresponds to the smaller distance between two collagen fibrils at higher osmotic pressure, since the collagen fibrils of cartilage are the main reason for the occurrence of restricted diffusion [57, 61].

Surprisingly, the charges of the ions applied have only a minor influence on the SDC and, accordingly, nearly the same information can be drawn from the observation of the water diffusion, as discussed in the next section.

3.2 Diffusion of Small Uncharged Molecules by PFG NMR

Healthy articular cartilage has a water content of about 70–80 wt%, but this water content can be reduced when a higher osmotic pressure is applied from outside [70]. Vice versa, incubating cartilage with deionized water results in an increased water content. The investigation of this behavior is of high relevance since cartilage is steadily exposed to external pressures and, for instance, compressive stresses as high as $\sim$20 MPa in the hip have been reported [74].

Experiments on the compression behavior of cartilage have often been performed by mechanical [75] and by osmotic [70] compression. However, it seems that both types of compression are rather difficult to compare and a comparison is only useful when the tissue is mechanically loaded in an isotropic and uniform manner, with the applied load is equal to that of the osmotic pressure [76].

Although the determination of the water content sounds trivial, it is crucial for performing reliable NMR relaxation [70] and diffusion measurements. Simultaneously, the water content of cartilage depends on the ion content. Since water binds primarily to the negatively charged polysaccharides of cartilage, the screening of these charges by ions, and especially by divalent ions such as Ca^{2+}, reduces the water content of cartilage considerably. This is the reason why calcification processes of cartilage are always accompanied by a decreased water content [70].

There are different approaches to how the water content of a given cartilage sample can be determined. The most simple way is to determine the wet weight of cartilage, to remove the water under vacuum and to reweigh the cartilage. The weight difference corresponds to the amount of water in the cartilage and is often related to the dry weight of the tissue [70]. It is, however, possible that under these conditions not all the water in the cartilage is determined, since some water may be bound very tightly to the cartilage polymers and can, therefore, not be removed by applying vacuum and/or heating. An alternative possibility is to treat the cartilage with organic solvents (e.g. methanol or DMSO). The organic solvent removes the water from the cartilage, and the water content within the organic solvent can be easily determined by, for example, Karl Fischer titration [77] . As will be outlined below, NMR is also suitable for determining the water content of a given cartilage sample.

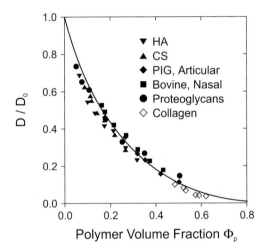

Fig. 15. Relative self-diffusion coefficient D/D_0 of water in cartilage, collagen, and cartilage-relevant polymers as a function of the volume fraction (Φ_p) of the solid component. The observation time Δ was 13 ms in all cases. D_0 is the diffusion coefficient of pure water. The solid line represents the curve calculated according to the model of Mackie and Meares [78]

Figure 15 shows the dependence of the ratio of the self-diffusion coefficients D/D_0 of water in pig articular and bovine nasal cartilage, as well as in different cartilage components (hyaluronan, chondroitin sulfate, proteoglycans, and collagen), on the polymer volume fraction. D_0 represents the self-diffusion coefficient of pure water (2.3×10^{-9} m^2 s^{-1} at 293 K) and D is the actual water SDC measured in the given system [57].

Although the polymeric systems shown in Fig. 15 differ considerably in their chemical structures, all the diffusion data obtained can be easily described by nearly the same dependence on the polymer volume fraction which relates closely to the water content of the samples. The solid line given in Fig. 15 shows the theoretical curve derived by the use of the model of Mackie and Meares [78]:

$$\frac{D}{D_0} = \left[\frac{1 - \Phi_\text{p}}{1 + \Phi_\text{p}}\right]^2. \tag{13}$$

This model predicts that the ratio D/D_0 depends simply on the polymer volume fraction Φ_p that is described by $\Phi_\text{p} = V_\text{p}/(V_\text{p} + V_\text{water})$. V_p and V_water are the polymer volume and the water volume, respectively. The water volume fraction Φ_w is correlated with the polymer volume fraction Φ_p by the equation $\Phi_\text{p} + \Phi_\text{w} = 1$. The volume fractions can, therefore, be calculated when the mass fraction and the densities of the polymers are known [68].

It is surprising that such a simple equation reliably describes the diffusion behavior in relatively complex systems. However, this holds exclusively for short observation times, while the situation is more complex at longer diffusion times.

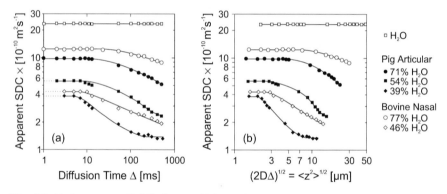

Fig. 16. (a) Apparent self-diffusion coefficient of water in pig articular and bovine nasal cartilage as a function of the diffusion time Δ at various water contents. The upper curve shows the behavior of free water, which is, of course, not affected by the observation time. (b) The apparent self-diffusion coefficients of water in pig articular and bovine nasal cartilage with various water contents, plotted against the mean path length of the diffusing water molecules

Figure 16 shows the dependence of the measured SDC on the diffusion time Δ in cartilage with different water contents (Fig. 16a) and the dependence of the SDC on the mean free path length (Fig. 16b). It is obvious that the measured SDC depends strongly on the observation time Δ and should, therefore, be termed the apparent self-diffusion coefficient D_app. The possibility of applying very short field gradient pulses of large magnitude enables the measurement of very short diffusion times [11]. Using a home-built PFG NMR spectrometer (FEGRIS 400) allows the measurement of diffusion coefficients in native cartilage ($\sim$70–80 wt% water) down to diffusion times as low as 1 ms [11].

Additionally, information about the characteristic length scales of restricted diffusion can also be provided by the representation of the data shown in Fig. 16b. From the D_{app} versus $\langle z^2 \rangle^{1/2}$ curves, it can clearly be seen that the diffusion is not restricted over approximately 6 μm in the native pig articular cartilage (water content about 71 wt%), while in the bovine nasal cartilage (water content about 77 wt%) the influence of the obstacles on the pathway of the diffusing water molecules becomes obvious only after about 10 μm. Since bovine nasal cartilage possesses a lower collagen content than does pig articular cartilage, this behavior was expected because the collagen fibrils are the most relevant barriers within the cartilage matrix [61].

Therefore, one must conclude that the long-time SDCs provide much more structural information than do the short-time diffusion coefficients, which, in contrast, allow the determination of the water content of a given sample. However, the long-time diffusion behavior is much more influenced by relaxation effects, and PFG NMR studies should always be combined with measurements of the water relaxation times in the corresponding samples [70].

Because of the observed strong dependence of the SDC on the water content, one should analyze diffusion data obtained with enzymatically digested cartilage with great caution [79]. When, for instance, the collagenous network of cartilage is digested by an enzyme, the cartilage sample will absorb much more water than under native conditions. Therefore, if the diffusion measurement is carried out without a careful control of the water content, one will determine a higher SDC. On the other hand, however, when the water content is carefully adjusted to the previous content (i.e. by the application of a higher osmotic pressure), then one will not notice any differences, regardless of whether cartilage was digested enzymatically or not [58].

3.3 Diffusion of Polymers

The considerable dependence of the short-time SDC on the water content is also observed when the diffusion of polymers in the cartilage is investigated [80]. For that purpose, we have used polyethylene glycol as a model compound, which is a relatively flexible polymer [81].

Owing to the extremely strong field gradients that can be applied by our diffusion NMR equipment [11], we have been unable to perform spectroscopically resolved measurements. Therefore, measurements are complicated if there is more than one NMR-detectable species, for instance the polymers of interest and residual water. However, this problem can be overcome when deuterated water instead of H_2O is used or when the fast-decaying component of the spin echo attenuation plots can be clearly assigned to water with a defined SDC.

It is obvious from Fig. 17a that a strong dependence of the PEG self-diffusion on the molecular weight (M) and on the water content of the cartilage sample exists. The diffusion data obtained (at the same water contents) can be fitted by the function $D \propto M^{-0.9}$, i.e. there is a nearly linear dependence of the observed SDC on the molecular weight of the PEG applied. However, it is surprising that the diffusion behavior of the PEG is nearly the same in a polymer solution and within the cartilage

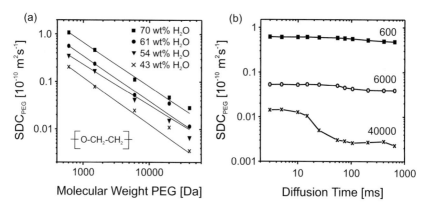

Fig. 17. Polymer self-diffusion of polyethylene glycol in bovine nasal cartilage depending on the water content and the molecular weight (diffusion time $\Delta = 13$ ms) (a), and (b) depending on the diffusion time, for the example of PEG 600, 6000, and 40000 at a fixed water content of 60 wt%

(data not shown). From this result, one must conclude that the mobility of the PEG is reduced only to a very small extent by the inner structure of the cartilage [80].

In Fig. 17b, the effect of the diffusion time Δ on the diffusion behavior of PEG with different molecular weights in cartilage at a fixed water content of 60% is shown. It is obvious that the reduction of the SDC at longer observation times is the more distinct the higher the molecular weight (given in Fig. 17b) of the diffusing species. One can conclude that the influence of the diffusion time Δ is most pronounced between observation times of about 10 and 20 ms. Therefore, these data correlate favorably with the data obtained by observing the water [57] and the cation [69] diffusion behavior.

4 Degradation of Cartilage by Enzymes

There are two different opinions about in what way the cartilage layers of the joints may be destroyed in rheumatic diseases. The first is that the cartilage cells, i.e. the chondrocytes that synthesize the cartilage extracellular matrix, are damaged by inflammatory products leading to an imbalance between catabolic and anabolic processes [82]. Another opinion is that products (enzymes and reactive oxygen species (ROS)) of inflammation cells (macrophages, T-cells, neutrophilic granulocytes) that invade the joint space in acute inflammation are the main and most direct effectors of cartilage degradation [83]. Especially, the contributions of ROS have been emphasized in the past by a number of publications of our own group [84–86].

Inflammation cells also release proteolytic enzymes such as elastase or matrix metalloproteinases such as collagenase. There are many indications that elastase is primarily responsible for cartilage damage [87], while the potential interaction between ROS and released enzymes is so far controversial [86].

Many different agents that are potentially responsible for *in vivo* cartilage degeneration have been applied, including hypochlorous acid [84], hydroxyl radicals [85], proteolytic enzymes [18, 87] and stimulated neutrophilic granulocytes [86, 87]. The effect of all these different agents could be easily differentiated by means of the corresponding NMR spectra of the supernatants and, therefore, NMR spectroscopy seems to represent a convenient method to assess cartilage degradation processes [83].

Additionally, these investigations are also of special interest since the composition of the supernatants of cartilage may be regarded as a suitable model of the composition of synovial fluids from patients suffering from rheumatic diseases [86].

In the present paper we shall focus exclusively on the effects of selected artificial proteolytic enzymes on pig articular cartilage, and the induced enzymatic effects will be investigated using the cartilage supernatants as well as the cartilage as such, i.e. by standard HR and HR MAS NMR, respectively.

The enzymes applied (papain and collagenase) were chosen because they cause completely different degrading effects on the individual cartilage components [18]. While collagenase is one of the few enzymes that degrades the native collagen of cartilage, papain affects exclusively the core and the link proteins of the proteoglycans of cartilage, while the collagen is not affected at all [88].

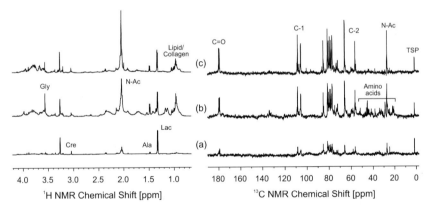

Fig. 18. ^{1}H (left) and ^{13}C (right) HR NMR spectra of cartilage supernatants depending on pretreatment conditions: (a) represents a cartilage sample that was treated exclusively with buffer, while (b) was digested with the enzyme collagenase and (c) with the enyzme papain. Abbreviations used in peak assignment: Ala, alanine; Cre, creatine; Gly: glycine; Lac, lactate; N-Ac, *N*-acetyl groups of cartilage polysaccharides; TSP, trimethylsilyl propionic acid. In the ^{13}C NMR spectra, individual resonances of chondroitin sulfate (CS) are marked

In Fig. 18, typical NMR spectra of the supernatants of pig articular cartilage incubated with pure buffer (a) and subsequent to enzymatic digestion with collagenase (b) or papain (c) are shown. On the left-hand side, the ^{1}H NMR spectra are provided, while on the right-hand side the corresponding ^{13}C NMR spectra of the cartilage supernatants are shown. It is obvious that the individual spectra differ considerably, depending on the enzyme applied. In the supernatant obtained upon cartilage incubation with pure buffer, only very few typical metabolites are detectable. Compounds such as lactate, alanine and creatine are characteristic but less specific metabolites and can be detected in nearly all kinds of body fluids or tissue extracts [87].

One should notice that the signal of the N-acetyl groups (at about 2.0 ppm in ^{1}H NMR spectra and 23.7 ppm in ^{13}C NMR spectra) of cartilage polysaccharides is very small. This indicates that there are either no polysaccharides of cartilage in the supernatant or the molecular weight of these polysaccharides is beyond the detection limit of high-resolution NMR [87].

It is obvious that there are far more resonances when the supernatant of collagenase-digested cartilage is investigated (Fig. 18b): a large number of amino acids can be detected under these conditions, where especially the generation of the glycine resonance (at 3.55 ppm) is indicative, since glycine is a very important constituent of cartilage and represents about every third amino acid residue of collagen [55]. A similar situation can also be found when cartilage is digested with papain (Fig. 18c), although the glycine resonance here is weaker, while the N-acetyl resonance of the cartilage polysaccharides is more intense. This may be explained by the different specificities of the two enzymes. Papain is capable only of cleaving the core and the link proteins of cartilage, which simultaneously leads to a release of chondroitin sulfate and keratan sulfate [88]. One peak that has to be discussed in more detail is the broad resonance at about 0.9 ppm. As shown by two-dimensional NMR techniques [18] (data not shown) this resonance represents a superposition of amino acids such as valine, leucine, and isoleucine. One further unexpected result was that this resonance represents also to some extent the aliphatic protons of the fatty acid residues of lipids. Although lipids and phospholipids are not highly abundant in the cartilage, they nevertheless seem to be detectable by NMR [18].

Similar data can also be derived by ^{13}C NMR spectroscopy, although under these conditions primarily the glycosaminoglycans of cartilage are detected. It is obvious on the right side of Fig. 18 that all spectra resemble each other closely, while the intensity of the individual resonances in comparison with the TSP (trimethylsilyl propionic acid) standard is different. In (a), the spectrum of the supernatant after incubation of cartilage with pure buffer is shown. This spectrum exhibits the resonances of the lowest intensities, which can all be explained by the presence of chondroitin sulfate, which is a major constituent of cartilage [18]. The low intensity of these resonances is not surprising, since in buffer only a small amount of the CS of cartilage is dissolved (cf. the corresponding ^{1}H NMR spectra).

In contrast, the digestion of cartilage with papain provides the most pronounced effect, i.e. the highest release of chondroitin sulfate. This is in very good agreement with the proton NMR spectra, which also provided evidence of an enhanced glycosaminoglycan release. The collagenase treatment of cartilage also leads to an enhanced glycosaminoglycan release, while under these conditions the resonances of different amino acids of collagen are also detectable (Fig. 18b). In contrast to ^{1}H NMR spectroscopy, the detection of abundant amino acids of collagen such as hydroxyproline is possible by means of ^{13}C NMR.

Although it is a very commonly used approach to investigate the supernatants of cartilage and to draw conclusions about the processes of cartilage degradation, these experiments are limited in the sense that the results obtained are strongly influenced by the extraction behavior of the cartilage. Therefore, the investigation of the solid cartilage specimens would provide a much more direct approach. Unfortunately, this cannot be performed by standard high-resolution NMR spectroscopy [18]. It could be shown that under HR MAS conditions primarily the quality of the spectra of pig articular cartilage was enhanced, whereas the improvement was less pronounced for bovine nasal cartilage [18]. Somewhat later, HR MAS was also applied to monitor the changes of cartilage after enzymatic digestion, and this was the very first time when resolved proton NMR spectra of cartilage could be obtained [18].

Although HR MAS NMR spectra are not useful for detecting the native collagen of cartilage, this technique is suitable for detecting fragmentation products of cartilage subsequent to enzymatic digestion.

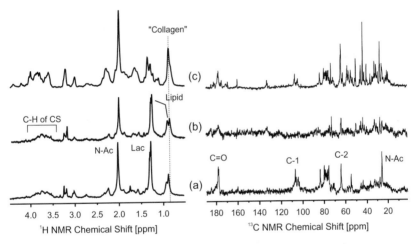

Fig. 19. ^{1}H (left) and ^{13}C (right) HR MAS NMR spectra of cartilage depending on pretreatment conditions: (a) represents a cartilage sample that was treated exclusively with buffer, while (b) was digested with the enzyme papain and (c) with the enzyme collagenase. Abbreviations used in peak assignment: Lac, lactate; N-Ac, N-acetyl groups of cartilage polysaccharides. In the ^{13}C NMR spectra, individual resonances of chondroitin sulfate (CS) are marked

In Fig. 19, the ^{1}H (left) and ^{13}C (right) HR MAS NMR spectra of cartilage are shown. Spectrum (a) represents the control sample that was treated exclusively with buffer, while in (b) cartilage was digested with papain and in (c) with collagenase. Both enzymes were used, since they yield remarkable differences. Papain leads only to small differences in the ^{1}H NMR spectra, while the ^{13}C NMR spectrum shows a considerable loss of the signals of the chondroitin sulfate. This is caused by the cleavage of the core protein and the release of the corresponding degradation products into the supernatant. In contrast, collagenase digestion is accompanied by marked changes in the proton and the ^{13}C NMR spectrum, since a number of smaller peptides are generated. The ^{13}C NMR spectra also provide the resonances of the most abundant amino acids of collagen. In this way, it is possible to differentiate the effects of individual enzymes and their effects on cartilage by means of HR MAS NMR [18].

Finally, the reader should note that there is one important difference between NMR spectroscopy and biochemical approaches such as electrophoresis. Electrophoresis is, owing to the properties of the gels applied, more suitable for detecting compounds with higher molecular weights, while NMR detects fragmentation products of cartilage more sensitively the lower their molecular weight is. Therefore, NMR is the detection technique of choice when small fragmentation products of cartilage are expected [86].

5 Conclusions

Cartilage is a complex connective tissue with great functional importance for all mammalian species. The most obvious function of cartilage is shape retention of the tissue in combination with its shock-absorbing capacity. Especially, articular cartilage is crucial for low-friction movement of the bones in the joints. Several serious cartilage diseases are known. A common pattern of all these rheumatic diseases is that the thickness of the cartilage layers of the joints is progressively reduced. Therefore, the design of artificially engineered cartilage for replacements is a field of active research, especially towards joint diseases.

In our project in the *Sonderforschungsbereich*, we have described several different NMR techniques to study the molecular motions of the cartilage macromolecules, the diffusion of water, ions, and polymers, and the enzymatic degradation of the tissue *in vitro*. Like no other physical method, NMR spectroscopy provides deep quantitative insight into all these different processes on a molecular level with angstrom resolution. Both rigid and highly mobile molecules that are part of the complicated architecture of the tissue have been investigated in the project.

The main components of cartilage can be easily differentiated by HR and solid-state NMR techniques: the polysaccharide moiety of cartilage is amenable to single-pulse excitation ^{13}C HR MAS NMR with scalar decoupling, but detection of the rigid collagen fibers requires solid-state NMR methods. Both approaches allow spectroscopic separation and individual characterization of the dynamics of the major macromolecular cartilage components.

Partially resolved NMR spectra of the cartilage carbohydrates can be obtained by high-resolution ^{13}C NMR, indicating that they are highly mobile. The resonances obtained can be assigned to chondroitin sulfate, the most mobile macromolecular component of cartilage. To characterize the timescales and amplitudes of molecular motions, we have also measured T_1 and T_2 relaxation times and nuclear Overhauser effects as a function of temperature and analyzed these by two different motional models. Typical correlation times for the segmental motions of chondroitin sulfate are on the order of nanoseconds, while segmental motions show typical correlation times on the order of 100 ps. Rather large amplitudes of the internal motions have been found. A broad distribution of correlation times has to be assumed to describe the motions of chondroitin sulfate in cartilage adequately.

For the dynamical characterization of the cartilage collagen, CP MAS and high-power decoupling are indispensable. CP MAS spectra of cartilage are dominated by the rigid collagen, while only low-intensity signals from the cartilage polysaccharides are observed. The spectral resolution of collagen fibrils in native cartilage is somewhat higher than for isolated collagen investigated for comparison, indicating the presence of fast motions in cartilage collagen. Typical order parameters for cartilage collagen are 0.77–0.90 for the side chain and 0.92–0.97 for the backbone. The only polysaccharide signals that could be detected by CP MAS showed order parameters of 0.48–0.92 and were assigned to rigid hyaluronan.

Considering the high water content of cartilage and the almost isotropic mobility of the chondroitin sulfate molecules it is remarkable how little this affects the collagen mobility. The observed fast low-amplitude motions of collagen in cartilage are presumably due to the high water content of cartilage and collisions with the isotropically mobile polysaccharides. These motions allow cartilage to absorb mechanical energy generated from shocks and tension. It appears that the mobility of the macromolecules in cartilage is heterogeneous and broadly distributed, from the rather rigid collagen, through the more mobile hyaluronan and other glycosaminoglycans, to the nearly isotropically mobile chondroitin sulfate. In this manner, the viscoelastic properties of cartilage tissue come about, which are required to sustain the various mechanical stresses acting on that tissue.

The second aim of this project was the investigation of diffusion processes in cartilage. Since cartilage contains neither blood nor lymph vessels, diffusion is the only transport mechanism for the supply of the cartilage cells with nutrients and the removal of metabolic waste products. Although a large variety of methods are available for studying diffusion processes, PFG NMR offers a number of advantages. For instance, no radioactive labeling of the compound of interest is required, real self-diffusion coefficients (that are not based on concentration gradients) can be determined, and the observation time can be easily varied to monitor internal distances. Therefore, this diffusion study has been exclusively focused on PFG NMR.

Besides the "classical" water diffusion (which is of highest interest for a potential *in vivo* application), the diffusion behavior of ions (tetramethylammonium and tetraethylammonium) and one selected polymer (polyethylene glycol) in cartilage was also measured to clarify the influence of charges and a varying molecular weight of the diffusing species.

Surprisingly, for short observation times of a few milliseconds, diffusion is primarily determined by the water content of the cartilage sample regardless of whether water, ion, or polymer diffusion is observed. The self-diffusion coefficients of all individual diffusing species were found to decrease steadily with decreasing water content. However, at longer observation times, the diffusion coefficients reflect structural properties of the cartilage, and restricted diffusion occurred already after about 6 μm. This number provides an estimate of the mesh size of the cartilage network. It is most likely that the collagen network of cartilage is responsible for the observed restricted diffusion, since enzymatic digestion leads to large free pathways of unrestricted diffusion. The extent of restriction is, however, also influenced by the water content of the cartilage.

Finally, it could be shown that modern NMR techniques are also useful for improving the further understanding of cartilage degradation. In order to induce well-defined cartilage damage, two enzymes have been applied that differ in their selectivities against the cartilage polysaccharides (papain) and collagen (collagenase). These enzymatic effects were compared by the investigation of the cartilage supernatants as well as the solid cartilage tissue. The effect of proteoglycan degradation could be monitored by both techniques, while for the evaluation of the fragmentation of the cartilage collagen, the NMR spectra of the solid cartilage tissue were much more indicative, since these spectra are not influenced by the extraction behavior of the cartilage at all. Therefore, a comparison of both approaches seems to provide the most complete information for the further elucidation of cartilage degeneration in rheumatic diseases. These NMR approaches were found to be much more suitable for the detection of smaller fragmentation products than were biochemical essays.

It is our aim for the future to apply our know-how established so far for the investigation of artificially engineered cartilage and to compare in this way its properties with native human cartilage. We hope that this approach will stimulate further improvements in cartilage replacement and, therefore, the clinical treatment of various rheumatic diseases.

Acknowledgments

This work was supported by the Sächsisches Ministerium für Wissenschaft und Kunst, the Deutsche Forschungsgemeinschaft (SFB 294/G5), the Bundesministerium für Bildung und Forschung (BMBF), and the Interdisciplinary Center for Clinical Research (IZKF) at the University of Leipzig (01KS9504/1, Project A17). We would like to thank Prof. Dr. Berger, Dr. Findeisen, Prof. Dr. Häntzschel, Prof. Dr. Kärger, Prof. Dr. Michel, Dr. Pampel, Dr. Stallmach, and Dr. Wagner, for fruitful discussions and technical support. The helpful advice of Mr. Müller in all aspects of formatting LaTeX files is gratefully acknowledged.

Abbreviations and Acronyms

B_0, magnetic field strength; **CP**, cross-polarization; **CSA**, chemical-shift anisotropy; **CS**, chondroitin sulfate; **D**, diffusion coefficient; **Da**, dalton; **HR MAS**, high-resolution magic-angle spinning; **Hyp**, hydroxyproline; **KS**, keratan sulfate; **LG CP**, Lee Goldburg cross-polarization; **MRI**, magnetic-resonance imaging; **NMR**, nuclear magnetic resonance; **NOE**, nuclear Overhauser effect; **PEG**, polyethylene glycol; **PFG**, pulsed field gradient; **PG**, proteoglycan; **ROS**, reactive oxygen species; **rms**, root mean square; **SDC**, self-diffusion coefficient; T_1, spin–lattice relaxation time; T_2, spin–spin relaxation time; **TEA**, tetraethylammonium; **TMA**, tetramethylammonium.

References

1. M.P. Pagano: Clinician Rev. **6**, 65 (1996)
2. H. Kuhn: Bioworld **5**, 16 (2001)
3. D. Eyre: Arthritis Res. **4**, 30 (2002)
4. J.E. Scott: Pathol. Biol. (Paris) **49**, 284 (2001)
5. G.Ebert: *Biopolymere* (Teubner, Stuttgart, 1993)
6. J.E. Scott:, J. Anat. **187**, 259 (1995)
7. T. Linsenmayer: "Collagen", in *Cell Biology of the Extracellular Matrix*, ed. by E.D. Hay (Plenum Press, New York, 1991), pp. 7–25
8. B. Chakrabarti, J.W. Park: Crit. Rev. Biochem. **8**, 225 (1980)
9. A.R. Poole: Biochem. J. **236**, 1 (1986)
10. A.M. Tomlins, P.J.D. Foxall, J.C. Lindon, M.J. Lynch, M. Spraul, J.R. Everett, J.K. Nicholson: Anal. Commun. **35**, 113 (1998)
11. J. Kärger, H. Pfeiffer, W. Heink: "Principles and applications of self-diffusion measurements by nuclear magnetic resonance", in *Advances in Magnetic Resonance* , ed. by J.S. Waugh (Academic Press, London, 1988), pp. 1–99
12. J. Kärger, P. Heitjans, R. Haberlandt: *Diffusion in Condensed Matter* (Vieweg, Braunschweig, 1988)
13. E.O. Stejskal, J. Schaefer: J. Am. Chem. Soc. **98**, 1031 (1976)
14. D. Burstein, A. Bashir, M.L. Gray: Invest. Radiol. **35**, 622 (2000)
15. J.G. Waldschmidt, E.M. Braunstein, K.A. Buckwalter: Rheum. Dis. Clin. North Am. **25**, 451 (1999)
16. W. Gründer, M. Wagner, A. Werner: Magn. Reson. Med. **39**, 376 (1998)
17. L. Naji, J. Kaufmann, D. Huster, J. Schiller, K. Arnold: Carbohydr. Res. **327**, 439 (2000)
18. J. Schiller, L. Naji, D. Huster, J. Kaufmann, K. Arnold: MAGMA **13**, 19 (2001)
19. D. Huster, J. Schiller, K. Arnold: Magn. Res. Med. **48**, 624 (2002)
20. D.A. Torchia, M.A. Hasson, V.C. Hascall: J. Biol. Chem. **252**, 3617 (1977)
21. C.F. Brewer, H. Keiser: Proc. Natl. Acad. Sci. U.S.A. **72**, 3421 (1975)
22. M. Tylianakis, A. Spyros, P. Dais, F.R. Taravel, A. Perico: Carbohydr. Res. **315**, 16 (1999)
23. A.G. Palmer III, J. Williams, A. McDermott: J. Phys. Chem. **100**, 13293 (1996)
24. A.G. Palmer III: Annu. Rev. Biophys. Biomol. Struct. **30**, 129 (2001)
25. R. Tycko: *Nuclear Magnetic Resonance Probes Molecular Dynamics* (Kluwer Academic, Dordrecht, 1994)
26. A.G. Redfield: Adv. Magn. Reson. **1**, 1 (1965)
27. R.L. Vold, J.S. Waugh, M.P. Klein, D.E. Phelps: J. Chem. Phys. **48**, 3831 (1968)
28. H.Y. Carr, E.M. Purcell: Phys. Rev. **94**, 630 (1954)
29. S. Meiboom, D. Gill: Rev. Sci. Instrum. **29**, 688 (1958)
30. J.H. Noggle, R.E. Shirmer: *The Nuclear Overhauser Effect: Chemical Applications* (Academic Press, New York, 1971)
31. M. Hong: J. Am. Chem. Soc. **122**, 3762 (2000)
32. D. Huster, X. Yao, K. Jakes, M. Hong: Biochim. Biophys. Acta **1561**, 159 (2002)
33. G. Lipari, A. Szabo: J. Am. Chem. Soc. **104**, 4546 (1982)
34. G. Lipari, A. Szabo: J. Am. Chem. Soc. **104**, 4559 (1982)

35. A.G. Palmer, P.E. Wright, M. Rance: J. Am. Chem. Soc. **113**, 4371 (2000)
36. J. Kaufmann, K. Möhle, H.J. Hofmann, K. Arnold: J. Mol. Struct. **422**, 109 (1998)
37. J. Kaufmann, K. Möhle, H.J. Hofmann, K. Arnold: Carbohyd. Res. **318**, 1 (1999)
38. J. Schaefer: Macromolecules **6**, 882 (1973)
39. J.R. Lyerla, D.A. Torchia: Biochemistry **14**, 5175 (1975)
40. S.K. Sarkar, C.E. Sullivan, D.A. Torchia: Biochemistry **24**, 2348 (1985)
41. S.K. Sarkar, C.E. Sullivan, D.A. Torchia: J. Biol. Chem. **258**, 9762 (1983)
42. S.K. Sarkar, Y. Hiyama, C.H. Niu, P.E. Young, J.T. Gerig, D.A. Torchia: Biochemistry **26**, 6793 (1987)
43. S.J. Opella: Method. Enzymol. **131**, 327 (1986)
44. D.A. Torchia: Ann. Rev. Biophys. Bioeng. **13**, 125 (1984)
45. H. Saitô, M. Yokoi: J. Biochem. **111**, 376 (1992)
46. H. Saitô, R. Tabeta, A. Shoji, T. Ozaki, I. Ando, T. Miyata: Biopolymers **23**, 2279 (1984)
47. K. Schmidt-Rohr, H.-W. Spiess: *Multidimensional Solid-State NMR and Polymers* (Academic Press, San Diego, 1994)
48. D. Huster, J. Schiller, K. Arnold: "Solid-state NMR to study the dynamics of cartilage polymers", in *Osteoarthritis: Methods and Protocols*, ed. by F. De Ceuninck, P. Pastoureau, M. Sabatini (Humana, Totowa, in press)
49. B.J. van Rossum, C.P. de Groot, V. Ladizhansky, S. Vega, H.J.M. de Groot: J. Am. Chem. Soc. **122**, 3465 (2000)
50. M. Hong, X. Yao, K. Jakes, D. Huster: J. Phys. Chem. B **106**, 7355 (2002)
51. K. Beshah, E.T. Olejniczak, R.G. Griffin: J. Chem. Phys. **86**, 4730 (1987)
52. S.K. Sarkar, P.E. Young, D.A. Torchia: J. Am. Chem. Soc. **108**, 6459 (1986)
53. R. Fujisawa, Y. Kuboki: Biochem. Biophys. Res. Commun. **167**, 761 (1983)
54. A. Perry, M.P. Stypa, B.K. Tenn, K.K. Kumashiro: Biophys. J. **82**, 1086 (1998)
55. A. Maroudas: "Physicochemical properties of articular cartilage", in *Adult Articular Cartilage*, ed. by M.A.R. Freeman (Pitman Medical, Turnbridge, 1979), pp. 131–170
56. L.A. Flugge, L.A. Miller-Deist, P.A. Petillo: Chem. Biol. **6**, R157 (1999)
57. R. Knauss, J. Schiller, G. Fleischer, J. Kärger, K. Arnold: Magn. Reson. Med. **41**, 285 (1999)
58. L. Naji, R. Trampel, W. Ngwa, R. Knauss, J. Schiller, K. Arnold: Z. Med. Phys. **11**, 179 (2001)
59. A. Maroudas, P.D. Weinberg, K.H. Parker, C.P. Winlove: Biophys. Chem. **32**, 257 (1988)
60. L. Naji, J. Schiller, J. Kaufmann, F. Stallmach, J. Kärger, K. Arnold: Biophys. Chem. **104**, 131 (2003)
61. R. Knauss, G. Fleischer, W. Gründer, J. Kärger, A. Werner: Magn. Reson. Med. **36**, 241 (1996)
62. M.D. Buschmann, A.J. Grodzinsky: J. Biomech. Eng. **117**, 179 (1995)
63. K. Dähnert, D. Huster: J. Colloid Interface Sci. **215**, 131 (1999)
64. K. Dähnert, D. Huster: J. Colloid Interface Sci. **228**, 226 (2000)
65. K.P.H. Pritzker, J.M.D. Chateauvert, M.D. Grynpas: J. Rheumatol. **14**, 806 (1987)
66. J.M.D. Chateauvert, K.P.H. Pritzker, M.J. Kessler, M.D. Grynpas: J. Rheumatol. **16**, 1098 (1989)
67. H.-O. Kalinowski, S. Berger, S. Braun: *150 and More Basic NMR Experiments* (Wiley-VCH, Stuttgart, 1998)
68. D. Burstein, M.L. Gray, A.L. Hartman, R. Gipe, B.D. Foy: J. Orthop. Res. **11**, 465 (1993)
69. W. Ngwa, O. Geier, F. Stallmach, L. Naji, J. Schiller, K. Arnold: Eur. Biophys. J. **31**, 73 (2002)
70. S. Lüsse, R. Knauss, A. Werner, W. Gründer, K. Arnold: Magn. Reson. Med. **33**, 483 (1995)

71. V.A. Parsegian, R.P. Rand, N.L. Fuller, D.C. Rau: Method. Enzymol. **127**, 400 (1986)
72. K. Arnold, L. Pratsch, K. Gawrisch: Biochim. Biophys. Acta **728**, 121 (1983)
73. F.A. Meyer: "The use of enzyme-modified tissues to study selected aspects of tissue structure and function", in *Methods in Cartilage Research*, ed. by A. Maroudas, K. Kuettner (Academic Press, London, 1990), pp. 222–227
74. W.A. Hodge, R.S. Fijan, K.L. Carlson, R.G. Burgess, W.H. Harris, R.W. Mann: Proc. Natl. Acad. Sci. U.S.A. **83**, 2879 (1986)
75. W. Gründer, M. Kanowski, M. Wagner, A. Werner: Magn. Reson. Med. **43**, 884 (2000)
76. W.M. Lai, W.Y. Gu, V.C. Mow: J. Biomech. **31**, 1181 (1998)
77. D. Citterio, T.K. Minamihashi, Y. Kuniyoshi, H. Hisamoto, S. Sasaki, K. Suzuki: Anal. Chem. **73**, 5399 (2001)
78. J.S. Mackie, P. Meares: Proc. R. Soc. London A **232**, 498 (1955)
79. Y. Xia, T. Farquharn, N. Burton-Wurster, M. Vernier-Singer, G. Lust, L.W. Jelinski: Arch. Biochem. Biophys. **323**, 323 (1995)
80. R. Trampel, J. Schiller, L. Naji, F. Stallmach, J. Kärger, K. Arnold: Biophys. Chem. **97**, 251 (2002)
81. J.P.G. Urban, "Solute transport between tissue and environment", in *Methods in Cartilage Research*, ed. by A. Maroudas, K. Kuettner (Academic Press, London, 1990), pp. 241–248
82. C.J. Malemud, V.M. Goldberg: Front. Biosci. **4**, D762 (1999)
83. J. Schiller, B. Fuchs, J. Arnhold, K. Arnold: Curr. Medic. Chem. **10**, 2123 (2003)
84. J. Schiller, J. Arnhold, K. Arnold: Eur. J. Biochem. **233**, 672 (1995)
85. J. Schiller, J. Arnhold, J. Schwinn, O. Brede, H. Sprinz, K. Arnold: Free Radical Res. **28**, 215 (1998)
86. J. Schiller, S. Benard, S. Reichl, J. arnhold, K. Arnold: Chem. Biol. **7**, 557 (2000)
87. N. Hilbert, J. Schiller, J. Arnhold, K. Arnold: Bioorg. Chem. **30**, 119 (2002)
88. J. Schiller, J. Arnhold, K. Arnold: Z. Naturforsch. **53c**, 1072 (1998)

Index

Lecture Notes in Physics

For information about Vols. 1–590
please contact your bookseller or Springer-Verlag
LNP Online archive: springerlink.com